WITHDRAWN
UTSA LIBRARIES

RENEWALS 458-4574

DATE DUE

GAYLORD			PRINTED IN U.S.A.

New Technologies, Mobility and Security

New Technologies, Mobility and Security

Edited by

Houda Labiod
ENST, Paris, France

Mohamad Badra
CNRS, LIMOS Laboratory-UMR 6158, Aubiere, France

A C.I.P. Catalogue record for this book is available from the Library of Congress.

ISBN 978-1-4020-6269-8 (HB)
ISBN 978-1-4020-6270-4 (e-book)

Published by Springer,
P.O. Box 17, 3300 AA Dordrecht, The Netherlands.

www.springer.com

Printed on acid-free paper

Library
University of Texas
at San Antonio

All Rights Reserved
© 2007 Springer
No part of this work may be reproduced, stored in a retrieval system, or transmitted
in any form or by any means, electronic, mechanical, photocopying, microfilming, recording
or otherwise, without written permission from the Publisher, with the exception
of any material supplied specifically for the purpose of being entered
and executed on a computer system, for exclusive use by the purchaser of the work.

First IFIP International Conference on New Technologies, Mobility and Security, 2nd - 4th May 2007, Telecom- Paris (ENST), Paris, France

Conference Organisers

NTMS'2007 PARTNERS

SUN Microsystems
NTMS'2007 Platinum sponsor

IEEE France
Section 8

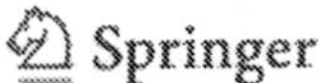

Springer Edition

International Federation for Information Processing

NTMS'2007 ORGANIZERS

Ecole Nationale Supérieure des Télécommuncations

Engineering and Scientific Research Groups

Lebanese University

École Nationale Supérieure des Télécommunications de Bretagne

NTMS'2007 IS SUPPORTED BY

Groupe des écoles des télécommunications

Arabic Computer Society (ACS)

PREFACE

This volume contains the proceedings of NTMS'2007, the first IFIP International Conference on New Technologies, Mobility and Security. The conference took place at TELECOM PARIS in Paris, France (May 2–4, 2007). It was technically co-sponsored by IFIP and IEEE France and supported by SUN Microsystems and GET (Groupe des Ecoles des Telecommunications).

NTMS'2007 aimed at fostering advances in the areas of *New Technologies, Wireless Networks, Mobile Computing, Ad hoc and Ambient Networks, QoS, Network Security and E-commerce*, to mention only a few, and provided a dynamic forum for researchers, students and professionals to present their state-of-the-art research and development in these interesting areas. The event was combined with tutorial sessions and a workshop. The tutorials preceded the main conference, aiming at the dissemination of mature knowledge and technology advances in the field. The workshop was held after the conference offering the opportunity for a more focused exchange of ideas and presentation of ongoing research relevant to multi-agent systems challenges for ubiquitous and pervasice computing. During NTMS'2007, poster sessions were also arranged to give opportunity to R&D labs and consortia to present their ongoing work.

The conference was organized in single or multiple-track sessions with presentation of invited and regular papers from worldwide institutions. NTMS'2007 concentrated on papers addressing future proposals and paradigms related to new technologies, mobility and security. 102 manuscripts had been submitted from authors in many countries from all over the world, including Europe, United States, Asia and Africa. 51 papers were finally accepted by the Technical Program committee for publication in this volume. In addition to that, abstracts of 9 posters presentations have been included in this volume.

Three half-day tutorials were held:

Peer-to-Peer Networking: State of the art and research challenges.
Prof. Raouf Boutaba, University of Waterloo – Ontario, Canada

Cognitive Radio Networks.
Prof. Ian F. Akyildiz, Broadband and Wireless Networking. School of Electrical and Computer Engineering, Georgia Institute of Technology – Atlanta, USA

Mobile Terminal Device Architecture: Present and Future.
S. Vijay Anand, Quasar Innovations Pvt. Limited – Bangalore, India

We would like to thank all members of the Technical Program Committee and the additional reviewers for their support and effort. We truly believe that thanks to all these efforts the program reflects a high quality of contributinons covering a broad spectrum of research and industrial key issues.

Finally, we would like to thank our sponsoring institutions. Special thanks go to ENST/INFRES department, the Steering Board and the Organizing Committee, in particular to Dr. Ibrahim Hajjeh for his very valuable contribution.

The first IFIP NTMS conference was a successful international event with fruitful discussions between academia and industry. It has provided an excellent opportunity for future cooperation.

Dr. Houda Labiod, Associate Professor, General Chair

Prof. Otto Spaniol, RWTH Aachen University, IFIP TC6 (Communication Systems)

Conference Program Chairs

Houda Labiod *ENST Paris* *labiod@enst.fr* **General Chair**	**Bassel Souleiman** *ENST Bretagne* *Basel.Solaiman@enst-bretagne.fr* **Vice-chair**	**Bilal Chebaro** *Lebanese University* *bchebaro@ul.edu.lb* **Vice-chair**

Tutorials and Workshops chair

Michel Riguidel - (ENST Paris), **France** - *riguidel@enst.fr*

Mohamad Badra - (CNRS-LIMOS), **France** - *badra@isima.fr*

Steering Board

Prof. Houda Labiod	(ENST Paris), **France**
Prof. Algirdas Pakstas	(London Metropolitan University), **England**
Prof. Stamatios Kartalopoulos	(University of Oklahoma), **USA**
Prof. Bassel Souleiman	(ENST Bretagne), **France**
Dr. Mohamad Badra	(CNRS-LIMOS), **France**
Dr. Jacques Demerjian	(Altran Telecom & Media), **France**
Mr. Samer El Sawda	(ENST Paris), **France**
Dr. Ibrahim Hajjeh	(ESRGroups), **France**
Dr. Ouahiba Fouial	(ESRGroups), **France**

Organisation committee

Nadine Akkari	(ENST Paris), **France**
Vincent Toubiana	(ENST Paris), **France**
Hai Lin	(ENST Paris), **France**
Ahmad Fadlallah	(ENST Paris), **France**
Miguel Garcia	(ENST Paris), **France**
Bennet Fischer	(ENST Paris), **France**
Ktari Salma	(ENST Paris), **France**
Tchepnda Christian	(ENST Paris), **France**
Irfan Hamid	(ENST Paris), **France**

Technical Program Co-Chairs

Houda Labiod	(ENST Paris), **France**
Stamatios Kartalopoulos	(University of Oklahoma), **USA**
Pascal Lorenz	(Université de Haute Alsace), **France**

Technical Program Committee

Akmal Abdelfatah (American University of Sharjah), **UAE**
Khalil A. Abuosba (Philadelphia University), **Jordan**
Mohamed Achemlal (France Telecom R&D Caen), **France**
Khaldoun Al Agha (University Paris-Sud), **France**
Kablan Barbar (Lebanese University), **Lebanon**
Carlo Blundo (Di Salerno University), **Italy**
Raouf Boutaba (University of Waterloo), **Canada**
Azzedine Boukerche (University of ottawa), **Canada**
Rajkumar Buyya (University of Melbourne), **Australia**
Haidar Chamas (Verizon Communications), **USA**
Jacques Demerjian (Altran Telecom & Media), **France**
Madiagne Diallo (IND PUC), **Brazil**
Mahmoud Doughan (Lebanese University), **Lebanon**
Bertrand du Castel (Schlumberger Fellow, Schlumberger), **USA**
Bachar El Hassan (Lebanese University), **Lebanon**
Khaled Fouad Elsayed (Cairo University), **Egypt**
Ahmad Fadlalla (ENST, Paris), **France**
Stephan Flake (ORGA Systems), **Germany**
Steve Furnell (University of Plymouth), **United Kingdom**
Giulio Galante (Istituto Superiore Mario Boella), **Italy**
Wassim Haddad (Ericsson Research), **Canada**
James Hughes (Sun Microsystems), **USA**
Robert S. H. Istepanian (MINT - Kingston University), **United Kingdom**
Bilel Jamoussi (Nortel - Northern Telecom), **Canada**
Joe Khalife (Lebanese American University), **Lebanon**
Yvon Kermarrec (ENST Bretagne), **France**
Bo-Kyung Lee (Korea Polytechnic University), **Korea**
Jean Leneutre (ENST Paris), **France**
Seng Loke (La Trobe University), **Australia**
Maryline Maknavicius (INT Evry),**France**
Muneer Masadah (University of Glasgow), **United Kingdom**
Mohamed Salim BOUHLEL (Sfax University),**Tunisie**
Imad Mougharbel (Lebanese University), **Lebanon**
Chafik Moukbel (Balamand University), **Lebanon**
Hassnaa Moustafa (France Telecom R&D Paris), **France**
Elie Najm (ENST Paris), **France**

Ahmad Nasri (American University of Beirut), **Lebanon**
Jose Marcos Nogueira (Federal Univ. of Minas Gerais) , **Brasil**
Alessandro Nordio (Polytechnico di Torino), **Italy**
Pierre Paradinas (CNAM, INRIA), **France**
Guy Pujolle (Paris 6 Univeristy),**France**
Alain Quillot (CNRS-LIMOS),**France**
Francis Rousseaux (Université de Reims), **France**
Pedro Ruiz (University of Murcia), **Espagne**
Kassem Saleh (American University of Sharjah), **UAE**
Yahya Sanadidi (University of California), **USA**
Brunilde Sansò (École Polytechnique de Montréal), **Canada**
Christian Schindelhauer (University of Paderborn), **Germany**
Nicolas Sklavos (Technological Educational Institute of Messolonghi), **Greece**
David Simplot-Ryl (Université de Lille 1, LIFL & INRIA), **France**
Otto Spaniol (RWTH Aachen University), **Germany**
Steve Uhlig (Université catholique de Louvain), **Belgium**
Pascal Urien (ENST Paris), **France**
Sung-Ming Yen (National Central University), **Taiwan**
Cui Yong (Tsinghua University), **China**
Bin Zhu (Microsoft Research Asia),**China**
Ahmad Fadlalla (ENST-Paris), **France**
Madiagne Diallo (IND PUC-Rio), **Brazil**

Sponsoring/Publicity Program Committee

Ouahiba Fouial (ESRGroups), **France**
Samer El Sawda (ENST Paris), **France**
David Perez (Metric), **France**
Romain Georgin (NRGIC), **France**

CONTENTS

Ad Hoc and Sensor Networks

Resource Management, QoS

CHAPTER 1

CRITICAL TRANSMISSION RANGE FOR CONNECTIVITY IN AD-HOC NETWORKS

HOSSEIN AJORLOO[1], S.–HASHEM MADDAH–HOSSEINI[1], NASSER YAZDANI[2], AND ABOLFAZL LAKDASHTI[3]

[1] *Iran Telecommunication Research Center, Tehran, Iran, {ajorloo, maddah}@itrc.ac.ir http://www.itrc.ac.ir*

[2] *Electrical and Computer Engineering Faculty, University of Tehran, Tehran, Iran,yazdani@ut.ac.ir*

[3] *Rouzbahan Institute of Higher Education, Sari, Mazandaran, Iran lakdashti@rouzbahan.ac.ir*

Abstract: One of the challenging problems in the ad hoc networks is how to determine the critical transmission range for each communicating node to achieve a connected network with minimum power consumption and communication interference. In this paper, an analytical approach is proposed to determine this parameter based on the number of nodes, physical dimensions of the network, and probability of connectivity. Our proposed approach resulted in Cumulative Distribution Functions (CDF) for the critical transmission range for various numbers of nodes

Keywords: Ad hoc networks, Cumulative distribution function, Critical transmission range

1. INTRODUCTION

One of the major challenging problems in ad hoc networks is the connectivity of the network. Reliability of connections depends on many factors, such as the transmission radius of each node, movement of nodes, environmental conditions, number of nodes, etc. In [1] an analytical procedure is proposed for the computation of the node isolation probability in an ad hoc network in the presence of channel randomness, with applications to shadowing and fading phenomena. However, in a simplistic model some authors have tried to bind together the probability of connectivity, the number of nodes, size and shape of the area in which nodes are located, and the transmission radius of each node, given a certain distribution of nodes.

H. Labiod and M. Badra (eds.), New Technologies, Mobility and Security, 1–12.
© 2007 *Springer.*

Santi and Blough [2] provided tight upper and lower bounds on the critical transmitting range for one-dimensional networks and non-tight bounds for two and three-dimensional networks.

Gupta and Kummar [3] have shown that if n nodes are placed in a disc of unit area in $\mathbb{R}^2$ and each node transmits at a power level so as to cover an area of $\pi r^2 = (\log n + c(n))/n$, then the resulting network is asymptotically connected with probability one if and only if $c(n) \to \infty$ for $n \to \infty$. This is a limit which does not help us to, *e.g.*, determine the number of nodes required to have a connected network with a certain probability say 95%.

Penrose [4] derived the distribution of the maximum of the edge lengths in a minimum spanning tree (MST), denoted M_n constructed from n points distributed uniformly in the unit square and proved that

$$\lim_{n\to\infty} P[n\pi_v M_n^v - \log n \leq \alpha] = \exp(-e^{-\alpha}), \quad \alpha \in \mathbb{R} \tag{1}$$

where π_v denotes the volume of the unit ball in v dimensions. As can be seen, here again we have an asymptotic relation for $n \to \infty$ which does not help us to determine n based on the size of the area, propagation radius of nodes and the probability of connectivity.

Tang *et al.* [5] have proposed a model for the probability of connectivity in ad hoc networks considering various values for the propagation radius using Monte-Carlo simulations. In this paper, we propose another model in the opposite direction: finding the transmission radius considering the probability of connectivity, but with an analytical procedure.

Some authors used a model with Poisson distribution of nodes (for example, see [2]). These models are appropriate for unlimited areas. But when we interest in finding models for limited areas, the distributions defined for limited areas such as the uniform distribution should be used instead.

The aim of this paper is to find a model that binds together three quantities, namely, the number of nodes in an ad hoc network, the maximum distance over which two nodes can communicate, and the area over which the nodes are scattered, in such a way that the resulting network is connected with a high probability when the nodes are assumed to be spatially uniformly distributed.

Designing power efficient protocols for ad hoc networks is a well documented topic in the literature (See [6–11] for some recently proposed solutions). In most of these protocols, it is required to determine the critical transmission power to achieve a connected network. One approach to determine this parameter is to use a message passing protocol such as one proposed in [12]. However, this approach suffers from the delay and communication load required for passing messages. On the other hand, if each node knows the approximate number of nodes in the network, it can determine the transmission power required to have a connected network with a certain probability using our proposed method. Although our method is not accurate as message passing approaches, it is faster and does not pose any communicating burden on the network.

The remainder sections of this paper are organized as follows: In section 2 we have discussed our analytical modeling. In section 3 the experimental results are presented. Finally, in section 4 we conclude the paper.

2. FINDING THE CRITICAL TRANSMISSION RANGE FOR CONNECTIVITY

In this section, we propose an analytical model for determining the critical transmission range given the probability of connectivity of ad hoc networks and the number of nodes. For the sake of simplicity, we consider these assumptions:
- The propagation radius is equal for all nodes.
- n nodes are distributed uniformly in the unit square.
- The x and y coordinates of the nodes are independent.
- The coordinates of nodes are independent.

Moreover, we have not assumed the mobility of nodes in our analytical approach. However, one can use the results for an instance, when assuming mobility in the network. By considering any mobility model, one can determine the critical transmission range to have a connected network for a given probability of connectivity in a certain time interval.

To begin, an important result given in [13,14] is presented:

Theorem 1: The critical transmission range for connectivity R_{crit} is equal to the longest link distance in the minimum spanning tree of the nodes.

From several algorithms proposed for finding the MST, we used the Prim algorithm [15]: starting with any single node, new nodes are added to the tree one by one, so that at each step the node closest to the nodes included so far is added [14]. One realization with 100 nodes as well as their MST is depicted in Fig. 1.

For finding the probability density function (PDF) of the maximum edge of the MST, we begin by the following equation:

$$f_{d_1,d_2,\ldots,d_N}(d_1, d_2, \ldots, d_N) = f_{d_1}(d_1) f_{d_2}(d_2) \ldots f_{d_N}(d_N). \tag{2}$$

where the left side denotes the joint PDF on N distances, and $f_{d_i}(d_i)$ is the marginal PDF of the distance d_i. This means that the joint PDF of N distances between nodes distributed independently in a square area equals to the product of the marginal PDFs of each of them.

In the next subsection, we derive the PDF and CDF of the distance between two uniformly distributed nodes.

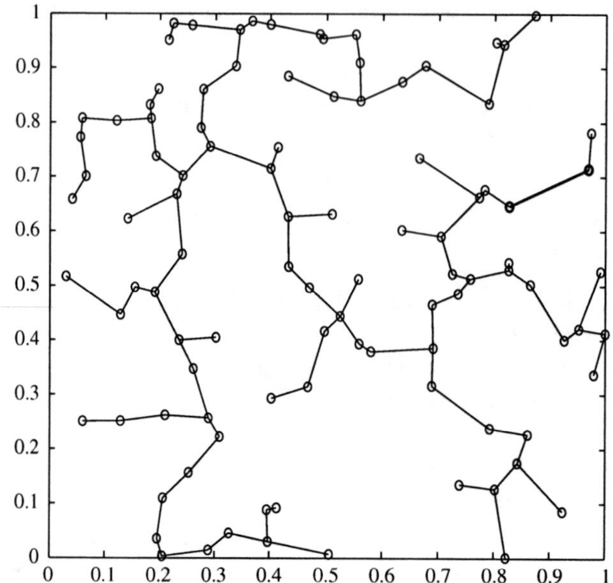

Figure 1. A sample set of 100 nodes and their MST. The longest link is shown with a dark line

2.1. PDF and CDF of the Distance Between Two Uniformly Distributed Nodes

We proceed by finding the PDF of the distance between two uniformly distributed nodes, d, in the unit square. In Appendix 4 we proved that this parameter has the following PDF

$$f_D(d) = \begin{cases} 2d^3 - 8d^2 + 2\pi d, & 0 \leqslant d < 1 \\ 8d\sqrt{d^2-1} + 8d\sin^{-1}(\frac{1}{d}) - 2d\pi - 4d - 2d^3, & 1 \leqslant d \leqslant \sqrt{2} \end{cases} \quad (3)$$

The probability distribution and density functions of d is shown in Fig. 2.

2.2. PDFs in the Prim Algorithm

In the Prim algorithm, at the first step, one node is chosen randomly. Then, the nearest node to this node is selected. In other steps, the nearest node to the set of selected nodes is chosen. We denote the edge chosen at the ith step by μ_i. Finally, $R_{crit} = \max\{\mu_1, \mu_2, \cdots, \mu_{N-1}\}$. If we denote the node chosen at step i by number i, then the PDFs of the random variables μ_1 to μ_{N-1} is

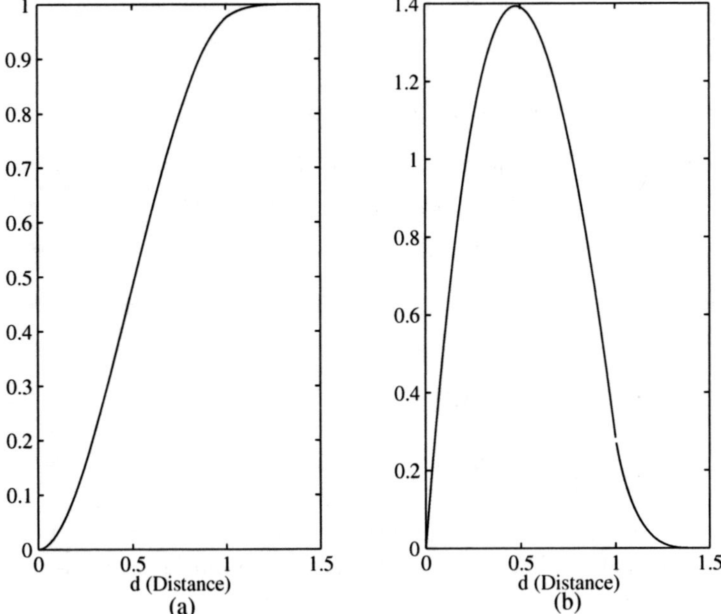

Figure 2. (a) Cumulative distribution and (b) Probability density functions of the distance between two uniformly distributed nodes in the unit square

$$
\begin{aligned}
f_{\mu_1}(\mu_1) &= P\left(\min_{j=2}^{N}\{d_{1j}\}\right),\\
f_{\mu_2}(\mu_2) &= P\left(\min_{\substack{i=1,2\\3\leqslant j\leqslant N}}\{d_{ij}\}\,\middle|\,d_{12}=\min_{j=2}^{N}\{d_{1j}\}\right),\\
f_{\mu_3}(\mu_3) &= P\left(\min_{\substack{1\leqslant i\leqslant 3\\4\leqslant j\leqslant N}}\{d_{ij}\}\,\middle|\,d_{12}=\min_{j=2}^{N}\{d_{1j}\},\,d_{13}=\min_{\substack{i=1,2\\3\leqslant j\leqslant N}}\{d_{ij}\}\right),\\
&\vdots\\
f_{\mu_{N-1}}(\mu_{N-1}) &= P\left(\min_{i=1}^{N-1}\{d_{iN}\}\,\middle|\,d_{12}=\min_{j=2}^{N}\{d_{1j}\},\right.\\
&\qquad \left. d_{13}=\min_{\substack{i=1,2\\3\leqslant j\leqslant N}}\{d_{ij}\},\cdots,d_{1(N-2)}=\min_{\substack{1\leqslant i\leqslant N-2\\j=N-1,N}}\{d_{ij}\}\right).
\end{aligned}
\tag{4}
$$

We should first determine the PDF of $f_{\mu_1}(\mu_1)$. Because of Theorem 1, the random variables d_{ij} are independent. Therefore, we should determine the joint PDF of

$N-1$ independent, identically distributed (i.i.d.) random variables. This is done in the next subsection.

2.3. Joint PDF of the Minimum of Independent Random Variables

We first find the joint pdf of the minimum of two i.i.d. random variables x and y. Defining $z = \min\{x, y\}$, we proved in Appendix 4 that the PDF of z is

$$f_Z(z) = 2f_X(z)\left(1 - F_X(z)\right) \tag{5}$$

For determining the joint PDF of N i.i.d. random variables, we use the following recursive formula

$$\min_{i=1}^{N}\{x_i\} = \min\{x_N, \min_{i=1}^{N-1}\{x_i\}\} \tag{6}$$

$$f(\min_{i=1}^{N}\{x_i\}) = f(\min_{i=1}^{N-1}\{x_i\})f(x_N) \tag{7}$$

Note that there does not exists any formula in an enclosed form for the joint PDF of the minimum of N i.i.d. random variables, and hence, we should use the above equations.

2.4. PDF of R_{crit}

As mentioned earlier, R_{crit} is the largest edge of MST of nodes, *i.e.*,

$$R_{crit} = \max_{i=1}^{N-1}\{\mu_i\} \tag{8}$$

Now, similar to the previous section we find the joint PDF of the maximum of $N-1$ independent random variables. Suppose that two random variables x and y are independent. Then, if $z = \max\{x, y\}$, according to Fig. 7–b the CDF of z is

$$\begin{aligned}
F_Z(z) &= P(\max\{x, y\} \leqslant z) \\
&= \int_0^z f_{XY}(z, y)dy + \int_0^z f_{XY}(x, z)dx \\
&= f_X(z)\int_0^z f_y(y)dy + f_Y(z)\int_0^z f_X(x)dx && (9) \\
&= f_X(z)F_Y(z) + f_Y(z)F_X(z) && (10)
\end{aligned}$$

Moreover, the following recursive formula is valid

$$\max_{i=1}^{N}\{x_i\} = \max\{x_N, \max_{i=1}^{N-1}\{x_i\}\} \tag{11}$$

$$f(\max_{i=1}^{N}\{x_i\}) = f(\max_{i=1}^{N-1}\{x_i\})f(x_N) \tag{12}$$

Again, there does not exists any formula in an enclosed form for the joint PDF of the maximum of N random variables.

3. EXPERIMENTAL RESULTS

In this section, we present our experimental results. We implemented Eqs. (5)–(12) using numerical methods. More precisely, we partitioned the range $d \in [0, \sqrt{2}]$ into infinitesimal sections of equal length, and for various numbers of nodes, we calculated the CDF of R_{crit} in a discrete form. The resultant CDFs for some selected values of n, the number of nodes, is shown in Fig. 3. In this figure, a horizontal line indicating 95 percentile line is sketched.

From the intersection of this line by each of the curves, one can determine the required critical transmission range (and consequently the required power) for a known number of nodes to have a network that is connected with the probability of 95%. For example, if we have 100 nodes distributed uniformly in a unit square, the critical transmission range for all nodes required to have a connected network with the probability of 95% is about 0.21.

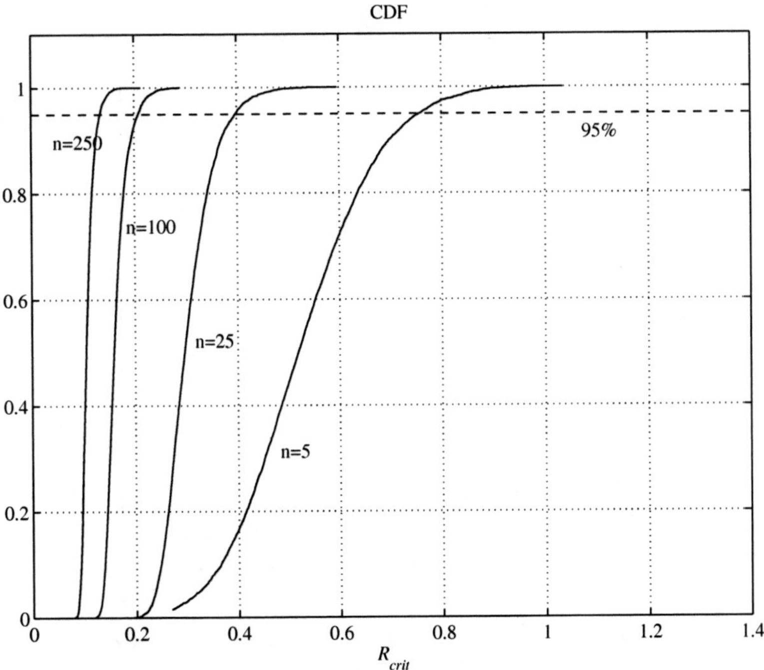

Figure 3. Computed CDFs for various numbers of nodes

Figure 4 shows the critical transmission range as a function of n for various values of percentiles. Using this figure, one can determine the number of nodes given the critical transmission range, required to have a connected network with a known probability.

For example, if the critical transmission range (resulted from a known power) for all nodes distributed uniformly in a unit square equals to 0.3, then the number of nodes required to have a connected network with the probability of 99% should be 63 nodes and 45 nodes for the probability of 95%.

Now, we are going to describe how the results can be used in potential applications toward more efficient networks. For example, consider one application in which scientists scatter some sensors in an area (such as around a volcano) which can communicate each other to convey their registered information. They may fix the transmission power for the nodes and desire in the number of nodes required to scatter in the area to have connected network with the probability of, say 95%. They can generate a plot similar to Fig. 4 using our proposed method and from which determine the number of nodes. The reverse situation may also be arisen: The number of nodes is fixed and the transmission power is of interest given a certain probability for connectivity.

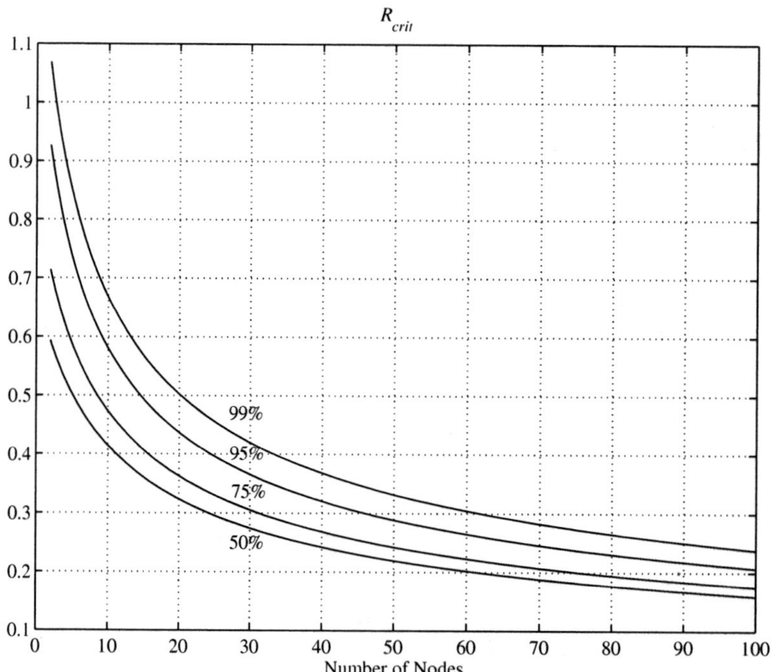

Figure 4. R_{crit} vs. number of nodes for various values of percentiles for connectivity

In design of power efficient routing protocols for ad hoc networks, the plots of Fig. 4 can be used, for example in the form of look-up tables, to adjust the power levels for the nodes to have a connected network with a certain probability and with least communication interferences. Using this approach, the nodes in the network only require to know the approximate value of the number of nodes in a certain area.

4. CONCLUSION

In this paper, an analytical approach is proposed to compute the CDF of the critical transmission range required to have a connected ad hoc network with a given probability as a function of the number of nodes and the physical dimensions. We obtained two recursive formulas to determine the mentioned CDF and used numerical methods to implement them. Our experiments resulted in curves of CDFs for a given number of nodes and critical transmission ranges as a function of the number of nodes. One can use these curves to determine the number of nodes required to have a connected network with a given radius of propagation for each node, or conversely, determine the required power for nodes to have a connected network with a given number of nodes in a square of known size and a certain probability of connectivity.

REFERENCES

1. Miorandi, D., Altman, E.: Coverage and connectivity of ad hoc networks in presence of channel randomness. Proceedings of IEEE INFOCOM 2005. 24th Annual Joint Conference of the IEEE Computer and Communications Societies, Vol. 1. 13–17 March (2005) 491–502
2. Santi, P., Blough, D.M.: The critical transmitting range for connectivity in sparse wireless ad hoc networks. IEEE Transactions on Mobile Computing, Vol. 2(1). Jan–March (2003) 25–39
3. Gupta, P., Kumar, P.R.: Critical power for asymptotic connectivity in wireless networks. Stochastic Analysis, Control, Optimization and Applications: A Volume in Honor of W.H. Fleming, W.M. McEneaney, G. Yin, and Q. Zhang (Eds.), Birkhauser, Boston (1998) 547–566
4. Penrose, M.D.: The longest edge of the random minimal spanning tree. The Annals of Applied Probability, Vol. 7(2). (1997) 340–361
5. Tang, A., Florens, C., Low, S.H.: An empirical study on the connectivity of ad hoc networks. Proceedings of IEEE Aerospace Conference, Vol. 3. March (2003) 89–98
6. Zhang, J., Zhang, Q., Li, B., Luo, X., Zhu, W.: Energy-efficient routing in mobile ad hoc networks: mobility-assisted case. IEEE Transactions on Vehicular Technology, Vol. 55(1). January (2006) 369–379
7. Li, D., Jia, X., Liu, H.: Energy efficient broadcast routing in static ad hoc wireless networks. IEEE Transactions on Mobile Computing, Vol. 3(2). April–June (2004) 144–151
8. Chen, K., Qin, Y., Jiang, F., Tang, Z.: A Probabilistic Energy-Efficient Routing (PEER) Scheme for Ad-hoc Sensor Networks. 3rd Annual IEEE Communications Society on Sensor and Ad Hoc Communications and Networks, 2006. SECON '06, Vol. 3. (2006) 964–970
9. Ping, Y., Yu, B., Hao, W.: A Multipath Energy-Efficient Routing Protocol for Ad hoc Networks. 2006 International Conference on Communications, Circuits and Systems Proceedings, Vol. 3. June (2006) 1462–1466
10. El-Hajj, W., Kountanis, D., Al-Fuqaha, A., Guizani, M.: A Fuzzy-Based Hierarchical Energy Efficient Routing Protocol for Large Scale Mobile Ad Hoc Networks (FEER). 2006 IEEE International Conference on Communications, Vol. 8. June (2006) 3585–3590

11. Li, F., Wu, K., Lippman, A.: Energy-efficient cooperative routing in multi-hop wireless ad hoc networks. 25th IEEE International Performance, Computing, and Communications Conference, 2006. IPCCC 2006, Vol. 8. April (2006) 10–12
12. Ovalle-Martinez, F. J., Stojmenovic, I., Nocetti, F. G., Solano-Gonzalez, J.: Finding minimum transmission radii for preserving connectivity and constructing minimal spanning trees in ad hoc and sensor networks. Journal of Parallel and Distributed Computing, Vol. 65(2). February (2005) 132–141
13. Sanchez, M., Manzoni, P., Haas, Z. J.: Determination of critical transmission range in Ad-Hoc Networks. Proceedings of Multiaccess Mobility and Teletraffic for Wireless Communications 1999 Workshop (MMT'99). October (1999)
14. Koskinen, H.: Connectivity and Reliability in Ad Hoc Networks. Master thesis, Helsinki University of Technology, Department of Electrical and Communications Engineering, February (2003)
15. Prim, R.C.: Shortest connection networks and some generalizations. Bell Systems Technology Journal, Vol. 36. (1957) 1389–1401

APPENDIX I: FINDING THE DISTRIBUTION OF THE DISTANCE BETWEEN TWO UNIFORMLY DISTRIBUTED NODES

Define z as $z = |x - y|$ where x and y have uniform distribution in the range (0, 1) and are independent (Fig. 5–(a)). Considering Fig. 5–(b), the CDF of z is

$$(13) \qquad F_Z(z) = P(Z \leqslant z) = 1 - (1 - z)^2 = 2z - z^2, \qquad 0 \leqslant z \leqslant 1$$

Taking the derivative of $F_Z(z)$ with respect to z, we get

$$(14) \qquad f_Z(z) = 2(1 - z), \qquad 0 \leqslant z \leqslant 1$$

Defining two new variables $z_1 = |x_1 - x_2|$ and $z_2 = |y_1 - y_2|$ with the PDF of (14), we observe that these variables are independent. Since, $d = \left((x_1 - x_2)^2 + (y_1 - y_2)^2\right)^{1/2}$, the CDF of d is

$$(15) \qquad F_D(d) = P(D \leqslant d) = P(\sqrt{z_1^2 + z_2^2} \leqslant d)$$

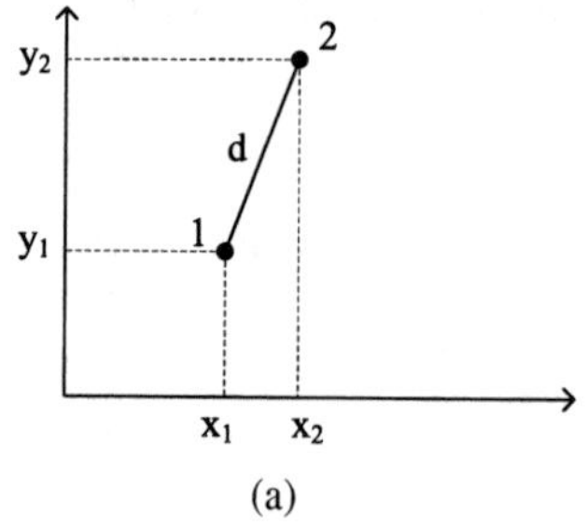

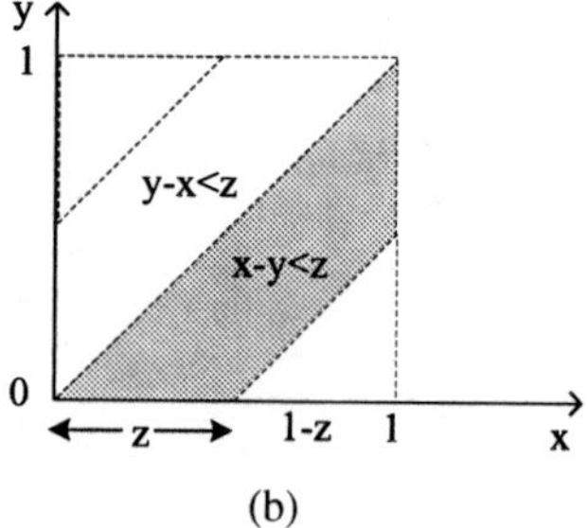

Figure 5. (a) The distance between two nodes having uniform distributions; (b) Computation of the distribution function of $z = |x - y|$

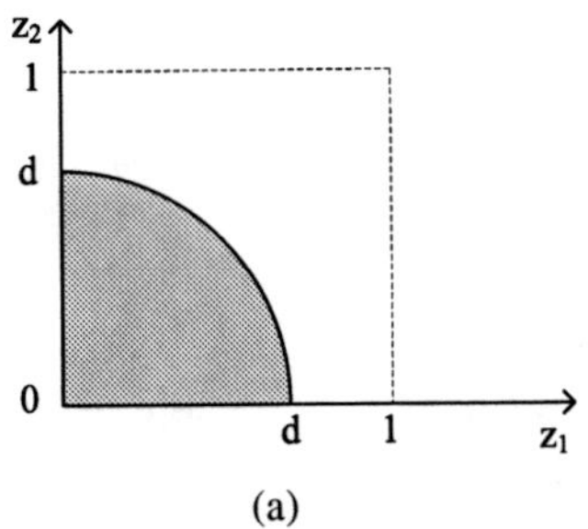

(a)

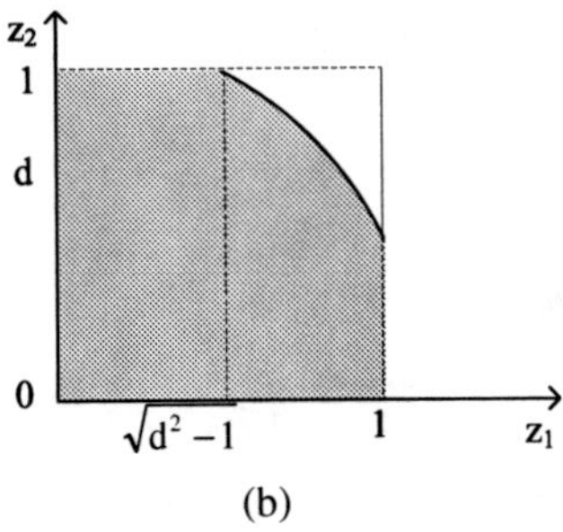

(b)

Figure 6. Computation of the CDF of d for (a) $0 \leqslant d < 1$; (b) $1 \leqslant d \leqslant \sqrt{2}$

To obtain this distribution, we should consider two cases:

Case 1: $0 \leqslant d < 1$, Considering Fig. 6–(a), we obtain

$$F_D(d) = \int_0^d \int_0^{\sqrt{d^2-z_1^2}} 4(1-z_1)(1-z_2)dz_1dz_2 \tag{16}$$

Case 2: $1 \leqslant d \leqslant \sqrt{2}$ (Fig. 6–(b)),

$$\begin{aligned} F_D(d) &= \int_0^1 \int_0^{\sqrt{d^2-1}} 4(1-z_1)(1-z_2)dz_1dz_2 \\ &\quad + \int_{\sqrt{d^2-1}}^1 \int_0^{\sqrt{d^2-z_1^2}} 4(1-z_1)(1-z_2)dz_2dz_1 \end{aligned} \tag{17}$$

Combining these two cases, we get

$$F_D(d) = \begin{cases} \frac{1}{2}d^4 - \frac{8}{3}d^3 + \pi d^2, & 0 \leqslant d < 1 \\ \frac{4}{3}\sqrt{d^2-1} + \frac{8}{3}d^2\sqrt{d^2-1} + 4d^2\sin^{-1}(\frac{1}{d}) \\ -d^2\pi - 2d^2 - \frac{1}{2}d^4 - \frac{1}{3}, & 1 \leqslant d \leqslant \sqrt{2} \end{cases} \tag{18}$$

Taking the derivative of (18), we obtain (3).

APPENDIX II: DETERMINING THE JOINT PDF OF THE MINIMUM OF TWO I.I.D. RANDOM VARIABLES

Defining $z = \min\{x, y\}$, according to Fig. 7–(a) the CDF of z equals to

$$\begin{aligned} F_Z(z) &= P(\min\{x, y\} \leqslant z) \\ &= \int_0^z \int_0^x f_{XY}(x, y)dydx + \int_z^{\sqrt{2}} \int_0^z f_{XY}(x, y)dydx \\ &\quad + \int_0^z \int_0^y f_{XY}(x, y)dxdy + \int_z^{\sqrt{2}} \int_0^z f_{XY}(x, y)dxdy \end{aligned} \tag{19}$$

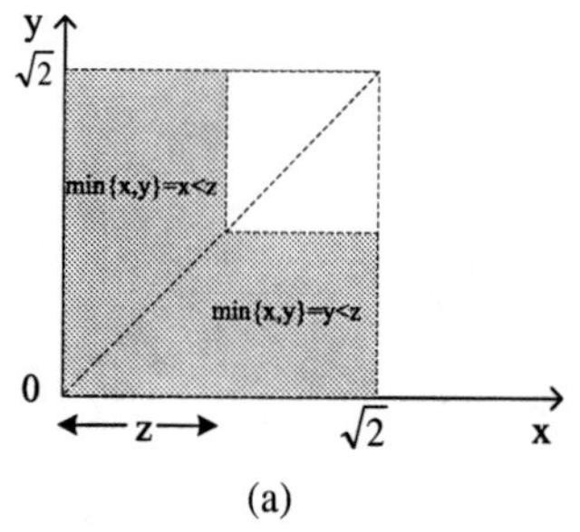

(a)

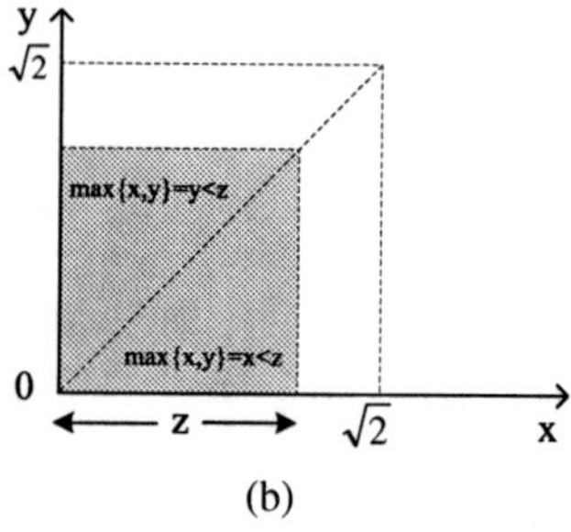

(b)

Figure 7. Computation of the CDF of (a) $z = \min\{x, y\}$; (b) $z = \max\{x, y\}$

According to Leibnitz theorem, if $F_Z(z) = \int_{a(z)}^{b(z)} f(x, z)dx$, then

$$f_Z(z) = \frac{db(z)}{dz} f(b(z), z) - \frac{da(z)}{dz} f(a(z), z) + \int_{a(z)}^{b(z)} \frac{\partial f(x, z)}{\partial z} dx \tag{20}$$

In our example, $z = \min\{x, y\}$ and x and y are i.i.d. random variables. Therefore,

$$\begin{aligned} f_Z(z) &= \int_z^{\sqrt{2}} f_{XY}(x, z)dx + \int_z^{\sqrt{2}} f_{XY}(z, y)dy \\ &= f_Y(z) \int_z^{\sqrt{2}} f_X(x)dx + f_X(z) \int_z^{\sqrt{2}} f_Y(y)dy \\ &= 2f_X(z) \int_z^{\sqrt{2}} f_X(x)dx \qquad (21) \end{aligned}$$

Determining (5) from (21) is straightforward.

CHAPTER 2

SCATTERFACTORY: AN ARCHITECTURE CENTRIC FRAMEWORK FOR WIRELESS SENSOR NETWORKS

MOHAMMAD AL SAAD, BENJAMIN HENTRICH, AND JOCHEN SCHILLER
Free University Berlin, Takustr. 9, 14159 Berlin, Germany

Abstract: ScatterFactory is a generative infrastructure for the model driven development of software for the Embedded Sensor Boards (ESB) of the WSN-Platform ScatterWeb. The chosen architecture centric approach represents an instance of the Model Driven Architecture. The goal is the furthermost automated and standardized production of software system families for the ScatterWeb ESBs. For this purpose, we developed a component meta model which builds a basis for a complete tool chain, from the model platform all the way to the deployment of the generated code onto the Sensor Boards. To model a ScatterWeb network, we developed a domain specific graphical editor on the basis of the Eclipse Modeling Framework and the Graphical Modeling Framework. For the examination of static model constraints, we integrated a real time validation into the editor. We used the openArchitectureWare framework for the transformation from models into code. The ScatterFactory framework was completed with additional components like flash-components for the automatic deployment of generated artifacts in an existing network. We realized our ScatterFactory tool with the Eclipse Framework as a basis

Keywords: model driven architecture (MDA), architecture centric model driven software development (AC-MDSD), software tools, wireless sensor networks (WSN), ScatterWeb

1. INTRODUCTION

This paper is organized as follows: The first chapter describes briefly the ScatterWeb Platform for which we developed the ScatterFactory. Since ScatterFactory follows an architecture centric model driven approach this paradigm is the topic of the second chapter. The several technologies and frameworks of which ScatterFactory is constructed are illustrated in the next chapter. The forth chapter shows the collaboration between the involved frameworks, whereby the architecture of ScatterFactory is explained. The following chapter deals with the achieved features of ScatterFactory, before we conclude the paper with a description of our experiences and future work.

H. Labiod and M. Badra (eds.), New Technologies, Mobility and Security, 13–30.
© 2007 *Springer.*

2. SCATTERWEB PLATFORM

ScatterWeb [6] is a platform for teaching and prototyping wireless sensor networks (WSN), which was developed by the Work Group Computer Systems and Telematics of the Free University Berlin. The hardware components of the ScatterWeb platform mainly consist of Embedded Sensor Boards (ESBs) and the sink (eGate/USB, see Fig. 1). The ESB has in addition to a controller and transceiver many functions at its disposal, such as a sensor for luminosity, vibration, temperature and IR movement detection, a beeper, LEDS, as well as a microphone. Thus a prototype of a comprehensive monitoring sensor is created, which makes studying the insertion of WSNs in various areas and scenarios – like environmental monitoring, intelligent buildings, Ad hoc process control, etc. – possible. The eGate is hooked up to the computer with a USB and functions as sink, which allows a connection to a ScatterWeb wireless sensor network. With this ability, various applications running on the computer can communicate with ScatterWeb ESBs via the eGate, and vice versa, which makes data-gathering, debugging, monitoring, over the air software updates, etc. possible.

The software that runs on the ESBs consists of two levels. The upper level is the application level. As the name suggests, this is the area, in which the application that runs on the ESB is implemented. The lower level is the firmware, software closely adapted to the hardware, created out of various (firmware-) modules, whereas each

Figure 1. ESB (above and eGate/USB (below) of ScatterWeb

module is responsible for a certain function. This division makes the software development for the ESBs easier. [7]

3. ARCHITECTURE CENTRIC MODEL DRIVEN SOFTWARE DEVELOPMENT (AC-MDSD)

While the main objectives of the OMG relative to the Model Driven Architecture (MDA) are the increase of the portability and interoperational ability of the software on a universal basis, the architecture centric model driven software development (AC-MDSD), as the name states, puts the focus on each an application domain. Instead of generating the same software for different platforms, the AC-MDSD has the goal of variations of software (software families) for a certain domain to automate as much as possible. This attempt is motivated with the observation that the (self repeating) infrastructure code has a considerable part of the entire code-basis in similar applications. With eBusiness applications, it lies around 70%, but with programming closer to the hardware, for instance with embedded systems, this share lies often between 90 and 100%. [4] Consequently it is naturally preferred to create the part automatically so that the actual application specific logic can be concentrated on. In this way, the concentration is set on an application domain for a model language, which would allow the concepts for the underlying platform to be domain related and precisely expressed.

Such a domain specific language (DSL) has as advantage over the usually more complex UML-based models used in the MDA, that the models created in it have a more complete knowledge of the domain. Since the model elements of DSL stand for concrete architectural concepts or aspects of the domain, a model written in DSL offers a higher abstraction level, but is concrete at the same time. The semantic gap between model and code becomes smaller. As a side-effect this simplifies the transformation of the models to code, because the step-by-step refinement of the models to code can often be skipped, since the underlying platform is known and clearly restricted. Overall, the objective target of the paradigm of the AC-MDSD can be compared to the use of modern product lines in the automobile industry.

At the beginning stands the prototype (*Reference Implementation*), in which the most important concepts are included. The prototype shows what the vehicle that is to be produced is supposed to look like. The construction plans (*Models*) serve as the starting basis for the end product (*Generated Artifact*) and point out which units (*Components*) are required.

In order to simplify the construction of the product line (*Generative Architecture*), as well as the later production (*Code-Generation*), logical coherent Components are summarized to production units. Production units, which are not automated or are too complicated to automate, have to be done by hand (*Manual Code*). To offer a wide production palette (*System Family*), the components, as a rule, have to be varied during the production process, while the production platform as such is left unchanged. Thus in context of the AC-MDSD, this approach is also called Product Line Engineering (PLE, see Fig. 2).

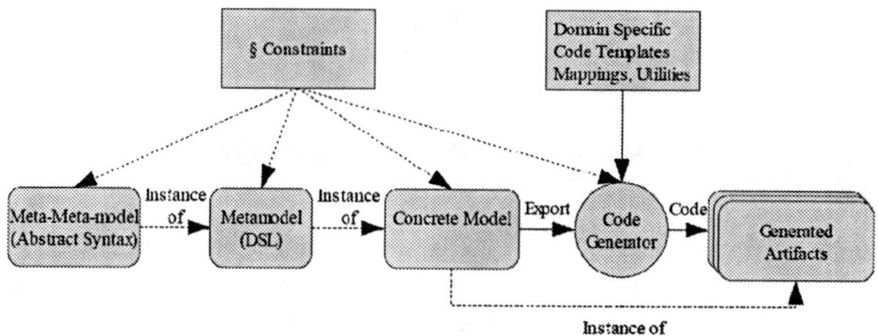

Figure 2. AC-MDSD as Product-Line

4. USED TECHNOLOGIES AND FRAMEWORKS FOR SCATTERFACTORY

For the development of the model driven, generative infrastructure according to the paradigm of AC-MDSD, many instruments were used for ScatterFactory, which we describe briefly in the following. The concrete realization will be explained in the next chapter.

4.1. The Eclipse-Framework

The Eclipse-Framework [1] serves as basis for the entire infrastructure. Since Version 3.0, the Eclipse-IDE consists of a run-time environment, in which the various plug-ins run. All together, there exist about 700 commercial and non-commercial plug-ins for Eclipse. Examples for open source plug-ins respectfully projects for Eclipse are all of the following instruments, of which the ScatterFactory architecture is constructed.

4.2. Eclipse Modeling Framework

The meta-meta model is the backbone of MDA-oriented activities. In the context of Eclipse, the task of meta-meta modeling is accomplished by the Project Eclipse Modeling Framework (EMF) [2]. The EMF allows itself to be divided into two levels: a model level and a generative level.

1) *model level*: EMF uses the language ECore for the description of meta-meta models, which is based on a subset of the MOF of the OMG, which has the name EMOF (Essential MOF), and which depicts the core of MOF. In ECore, written meta-meta models describe in turn concrete meta-meta models. Ecore models are about structural models, like the way they are known as UML-class diagrams or XML scheme definitions.
2) *generative level*: The platform independent model (PIM), in this case the Ecore model, must, analogue to the MDA, be changed into a platform specific model (PSM) before the code generation. This happens in EMF, almost completely automated, with the help of a wizard, whose input is the respective Ecore model.

The wizard delivers a generator-model as result. With the help of so-called Java Emitter Templates (JET), the generator model can create Code Artifacts directly. An important artifact, created from this process, is the Java implementation of the meta model, which is the input of the GMF created graphical editor.

4.3. Graphical Modeling Framework

The goal of the Graphical Modeling Framework (GMF) [3] is to provide a generative infrastructure for the creation of graphical editors for domain specific languages (DSL). The basic idea is to gain a domain specific model out of an ECore meta model so that graphical DSL-editors can be created.

4.4. OpenArchitectureWare

The openArchitectureWare framework (oAW) [5] is an open-source generator framework. Since January 2006, oAW is part of the Eclipse Technology Projects. oAW is able to be divided into two functional, coherent areas: The *Front End* reads the model and analyzes it for conformity to the meta-meta model as well as for fulfillment of the static constraints, which are called Expression Checks. As an outcome, the Front End delivers a model instance in the form of a Java Object Graph, which builds the input of the *Back End.* The heart of the Back End is built off of the template supported Code generator. The template language of oAW is called Xpand. Each model has its corresponded (root) template. Starting with this designated template all transformation regulations for the model must be directly or indirectly accessible in further templates. Also references to additional manual extensions in the form of Java classes are enclosed in the templates.

5. ARCHITECTURE OF SCATTERFACTORY

Figure 3 shows the architecture of ScatterFactory and the above mentioned frameworks used for realization, as well as their collaboration.

5.1. Meta Model

The meta model (see Fig. 4) was gradually adapted and furthered in the course of many iteration cycles in regard to the demands of ScatterFactory. In the following the single elements of the meta model and their meanings will be described. The root element of the ScatterFactory meta model is the class *Network*, which stands for a concrete network. The class *Connection* represents the generalization of all possible connections existing between model elements, which serve the modeling of software. The class *Sensor* builds the upper class for the classes *ESB* and *ECR*, which are inherited from it and helds the common attributes. To each Sensor a descriptive name is to be assigned. ECR is besides ESB another hardware component, which offers very limited functionalities compared to ESB, but functions in an identical

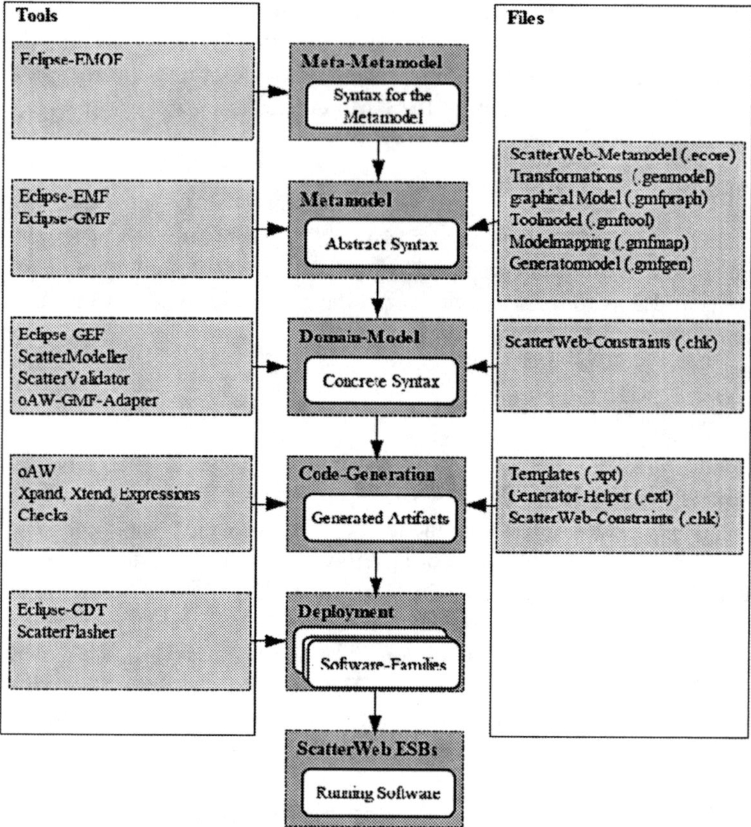

Figure 3. Architecture of ScatterFactory

way regarding the development of its software. Concerning ScatterFactory we are dealing only with ESBs, but we consider ECRs in the meta model, in order to demonstrate the ability to adapt the meta model to the dynamics of the domain, e.g. new coming hardware components. An object of the class *Container* represents the logic arrangement of several software components for the modeling of firmware, which can be composed of a multitude of containers as well. In order to make the most efficient use of the Container concept, it is vital to held up a high level of standardization, which increases the possibility of reusability and replaceability. Thus the productivity and robustness of the software development process is considerably raised. The class *Application* is the connection between Sensors and Containers. It is the region in which the developer inserts his manual code. Thus it is realized on templates level as a protected region in order not to be overwritten in the following generation process. The *Component* constitutes logically connected firmware modules of the software, which runs on an ESB. We want to point out that several alternatives (instances) for the same Component type can exist, since

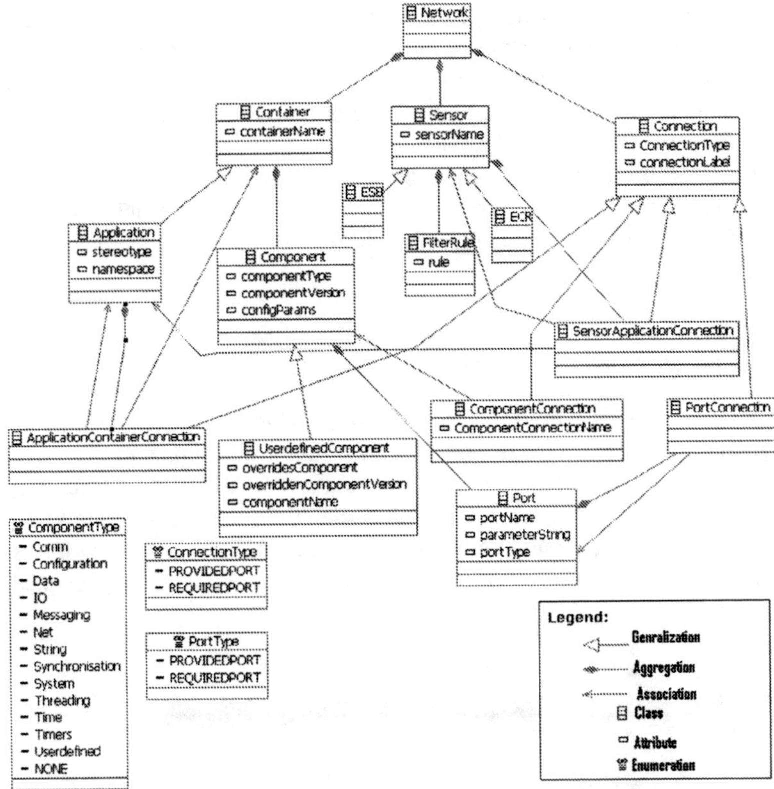

Figure 4. Ecore-based meta model of ScatterFactory

it is possible to have different implementations for one firmware module, as we will see in chapter VI. C. Furthermore it is possible to define for each component so called *Ports*, in order to address selectively subsets of a component – similar to UML. *Required Ports* specify which function is required for a component from another. *Provided Ports* define functions, which can be offered by a component.

5.2. Editor

The Screenshot (Fig. 5) shows the editor of ScatterFactory, in which some model elements have already been inscribed. The instrument palettes are found in 1. It is divided into four parts: General instruments for the selection of elements, to maximize or minimize the model view or to attach notes onto model elements, for example to comment on them, are found at the uppermost level. Underneath are the three groups of instruments to build model elements. The instruments for the ESBs are listed in the first group. The second group holds everything which is needed to model software, like containers, components and ports. In the lowest group, all the tools to build connections are listed. As shown in 2, elements can also be added directly over a context menu in the diagram. The menu appears as soon as the mouse moves over the diagram, as only the elements are offered in the menu, which can be added at that particular place. In properties view in 3, various attributes of a chosen element can be specified. In case an error is noticed by the live validation, it will be shown under problems view as shown in 4, as well as marking the concerned element in the Editor. The outline view in 5 helps to keep an overview of large models. A mouse click will lead to the corresponding place in the diagram.

5.3. Code Generation

The developer generates code in ScatterFactory from the model diagram by clicking with the right mouse button on the file *start_generator.oaw* (see Fig. 5, marked area at the left side) in the directory src, then in the context menu *Run As...* and after that, *oAW Workflow*. The generator is now started and will switch to *Console View*, where the progress of the generation process can be seen. Afterwards, a new folder named *src-gen* appears in the related Eclipse project of the model. The generated code is in this folder.

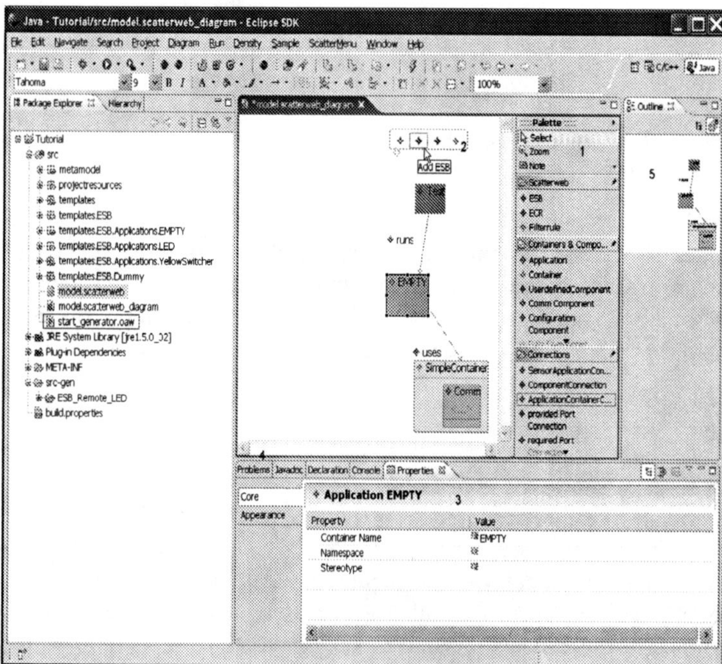

Figure 5. ScatterFactory Editor

In the code generation process we use two important techniques. The first technique is the usage of dummy templates in the generator while generating firmware modules. In this way, dummy templates for code files are emitted for many firmware modules that are not explicitly named in the model for an application, while the header files belonging to them are left unchanged. The dummy templates for code files hold merely empty procedure hulls without any function. This guarantees a compiled code, in any case. The modeler has to activate the usage of dummy templates explicitly by the specification of the appropriate filter rule called USE_DUMMIES for an ESB. This leads us to the second technique.

The second technique is the specification of global filter rules, which can be fixed for an ESB by the modeling. Depending on the specific filter rules, certain functions can be filtered on a sensor board. Specifically such parts are the functions of firmware, which can be found in different places of the code, like macros for console-commands. This is an example for a filter rule (NO_SERIAL) in a template:

```
<<IF !containsCommandFilter (sensor)>>
<<IF !sensor.filterRules.rule.exists (e l e.toString ()
  .matches("NO_SERIAL") ) >>
COMMAND (png, SERIALONLY) {
       packet_t packet;
       packet.to = String_parseInt(&str[4], NULL); packet.type =
  PING_PACKET;
       :
       :
```

Three examples – in addition to USE_DUMMIES – for such filter rules can be found: NO_COMMANDS: Using this rule various codes for console commands are left out. NO_SERIAL: Here various codes for console-commands, only used for the serial interfaces, are left out. MIN_COMMANDS: A large part of the available commands are filtered out. Only a few commands, like the one for setting an ID or for resetting, are left intact. To implement these rules to run on a template-basis, extend-expansions are called in places in the template source code, on which the commands of a firmware module are found, which evaluate whether the following code is to be emitted or not.

5.4. Compilation and Flashing

The Flashing of sensor boards with code image in ScatterFactory is possible in many ways. For this purpose a Makefile and an ANT-Build are generated with every generated sensor board project (see Fig. 6), in order to compile the various projects and flash the code image onto a sensor board. To create sensor boards in a series, ScatterFactory brings along a plug-in named ScatterFlasher. It supports not only the flashing of several sensor boards over a JTAG-Adapter, but also the flashing of entire groups of sensor boards over the eGate. In the case of over-the-air

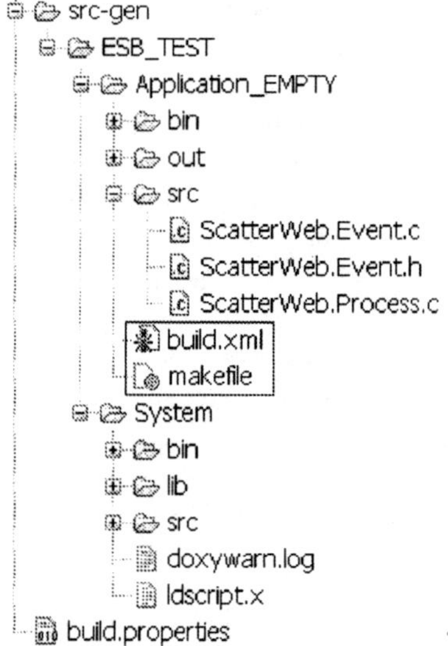

Figure 6. src-gen is the result of the generation process

flashing, sensor board IDs or even entire intervals of IDs can be chosen, to bring onto them the corresponding Software (see Fig. 7). The dialogue for the flashing over a JTAG-Adapter is built analogue, with the difference that instead of IDs, the number of sensor boards to be flashed per image is to be given.

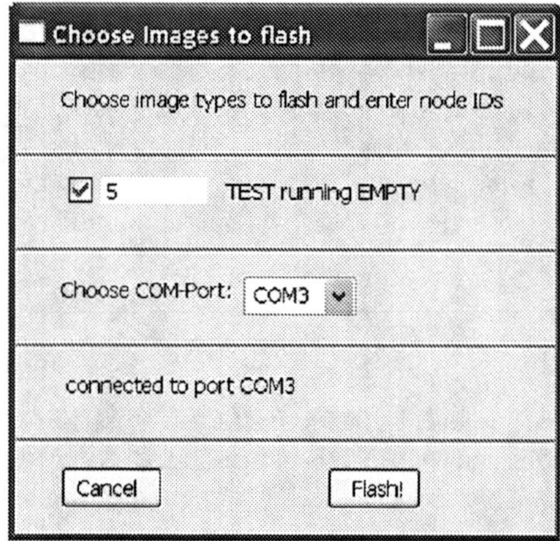

Figure 7. OTA-Flashing, TEST running EMPTY is the name of a certain generated application, for in the same model diagram different applications can be modeled. Thus the user can choose which image to flash on selected ESBs

The implementations behind the two forms of flashing differentiate from one another greatly. To flash over a JTAG-Adapter, the plug-in calls merely the corresponding software of the MSP430-Toolbox from Texas Instruments with the corresponding parameters. If more sensor boards are to be flashed with the same software, this happens iteratively again and again. If errors are caused, it will be shown through corresponding dialogue windows. The mass flashing over an eGate is realized with a bit more work. The logic needed for this is implemented in the class FlashHelper of ScatterFlasher plug-in. The OTA Flashing process consists of the following steps: First a serial connection to the corresponding eGate will be built via Java COM Ports. For this purpose we use the javax.comm package. Over this connection a binary image of the software (Hex-File) will be loaded line by line in the EEPROM of the eGate. Out of this, the image will be sent to various addressed ESBs and at the same time, the connection to the eGate will be monitored, to see if there are any errors. If errors occur (e.g. suddenly not responding addressed sensor), these will be made known through corresponding dialogue windows.

6. FEATURES OF SCATTERFACTORY

For the judging and illustration of the features of ScatterFactory, we consider three different scenarios in each of which a different role for the sensor board application can be found. In the first scenario the sensor board plays the role of a local node, onto which the software is initially developed and experimented. It should offer all the functionalities of a sensor board and should be able to be accessed not only over the serial interface port, but over an eGate as well. The *EMPTY-Application* serves as the replica of the role of a local development Node. The accompanying flash image includes various firmware modules, so that the entire functionality of the firmware is accessible on such a node. The EMPTY-Application serves as a reference for the evaluation regarding the amount of code, which for both of the other scenarios, can be brought onto an ESB. Figure 8 shows a simple model, which models the EMPTY-Application for an ESB.

The generated artifact (Hex-File) from this model corresponds directly with the "conventional" EMPTY-Application. Since there are no firmware modules left out, a dummy template does not need to be created. The second scenario basically depicts the parade example for the possible uses of a sensor board. ESBs can send and receive packets, react to Sensor Events and are accessible over an eGate (sink). Thus the Application *YellowSwitcher* (see Fig. 9) behaves simply as follows: By pressing the button on an ESB (event), a packet is sent from this ESB and all other ESBs that receive such a packet get their yellow-LED toggled. The third scenario (Fig. 10) is the *LED-Application.* It represents nodes that merely function as actors in a WSN. This scenario serves to examine what the minimal need of code is, to operate a sensor board.

This exemplifies also the use of ports. The LED-Application implements merely a periodical switch for a LED. That a second container for the module Timers was used, serves only for demonstrational purposes to show how dependencies between

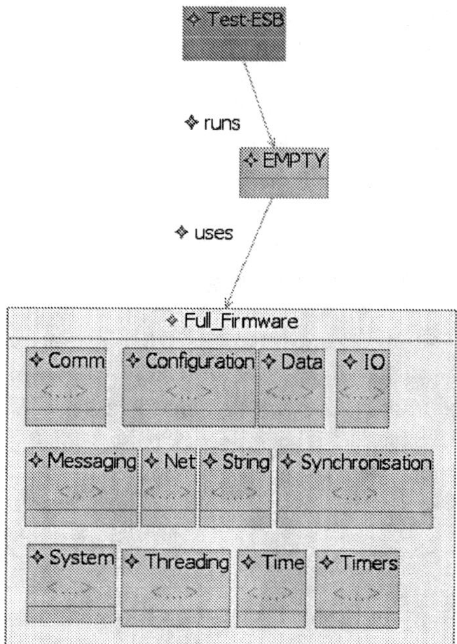

Figure 8. EMPTY-application with full firmware

components from different containers can be modeled. A provided port is defined for the module type *Data*. This shows that only the functions for passive behavior should be generated from the Data module (whereby it is implicitly specified that for all other functions of this module only dummy-code should be generated). The same applies for the defined port in the module *Timers*. It shows, that only the code should be generated that is needed to add a second timer. Through such a connection of the ports between the two containers it is ensured that the corresponding code-fragments are generated together.

6.1. Resource Consumption

One of the main goals of ScatterFactory is to bring as little code onto an ESB as possible. At this point it should be considered to what extend this is guaranteed. For the three pervious scenarios, we obtained the following code sizes of the generated applications without and with the use of several filter rules configurations:

As depicted in Table 1, the gaining of memory capacity already without the use of filter rules is noteworthy. If we consider that a typical task for a sensor board is the recording of sensor events and the processing of received packets, then already a simple model of saving code of about 27% is reached, in comparison to the

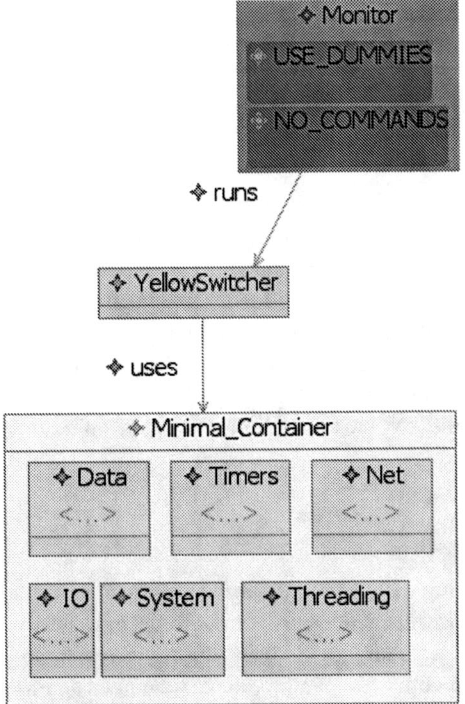

Figure 9. YellowSwitcher with two filter rules

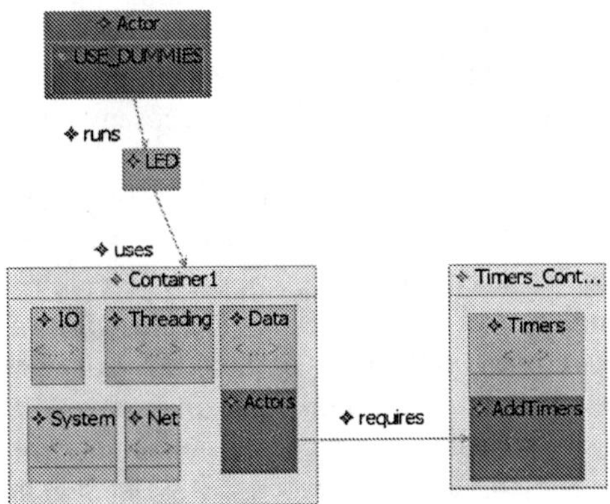

Figure 10. LED Application with two containers

Table 1. Size (in bytes) of the generated code for an ESB after the compilation with and without filter rules

Application	Without Filter	NO_SERIAL	MIN_COMMANDS	NO_COMMANDS
EMPTY	34050	32222	32215	31990
Yellow Switcher	24980	24230	24226	23950
LED	19846	19048	18944	18944

configuration of EMPTY-Application with the configuration of the YellowSwitcher Application. Through the use of filter rules, this percentage can practically be increased to near 30%. Even more clearly is the achieved result, in the case of LED Application, where additional ports would be used. Here code-sparing of almost 42% can be reached without any filter rules. The filter rules show in this case less effect, because already less complete modules are being generated. Nevertheless, in the extreme case (with the use of the rule "NO_COMMANDS") a goal of saving 45% can be reached.

6.2. Failure Detection

A model driven approach tries to notice logical design mistakes as early as possible in the development phase. This goal is reached easily through ScatterFactory, because the validation of a model in real time already happens during the modeling and before the code is generated and then compiled. If a model element goes against a model related constraint, this will be immediately shown in the Problems View of the Eclipse Workbench and the model element will be marked in the editor. A double-click on an entry in Problems-View selects the concerned model element in the editor. This is helpful, especially when many model elements are offending the same constraint. In Fig. 11, the experiment to add another connection from one ESB to an application is shown. This is, per the meta model, inadmissible, because the cardinality of a connection between a sensor board and an application can be one at the highest. The addition of a second connection will be stopped, in which the mouse pointer is signalized through a corresponding icon, that the chosen model element cannot be placed on the chosen destination.

The realization of our real time validation is based on an Eclipse plug-in named GMF-Adapter that the code generator (oAW) offers since the version 4.1. The GMF-Adapter provides an interface for the access onto the constraint checking system from oAW, as well as a Watchdog, which watches over the changing of the model. Plug-ins are allowed to be developed on the basis of the GMF-Adapter, which integrate the present GMF-Editor into the validation framework of oAW and the other way around. The necessity to care for redundant constraints for the Editor and generator is gone now and through the use of the watchdog, a real time validation during the editing of a model is possible. As mentioned, to connect the validation framework onto a GMF-Editor, a plug-in is needed that builds interfaces for this purpose. This approach was chosen for ScatterFactory and the

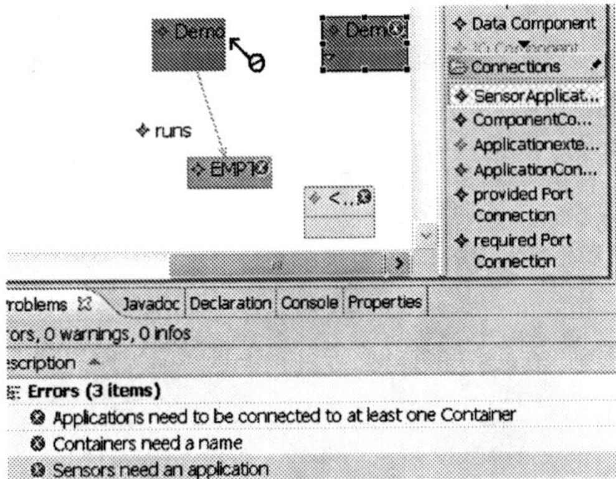

Figure 11. Real-time validation in ScatterFactory

plug-in ScatterValidator was developed. It extends the EMF extension point for the model validation (org.eclipse.emf.validation.constraint Providers) through a class ScatterWebValidateAction, which offers an implementation of the oAWConstraints-provider. This class also implements the necessary methods, in order to allow oAW access to the editor. Another class LiveConstraints loads the constraints that are to be used from another file, which holds all the constraints relevant for the model validation and makes them accessible to the validation frame work of the oAW. The constraints of ScatterFactory can be found in the file scatterweb.chk in the plug-in org.scatterweb.scattermodeller, which holds all the necessary models for SactterFactory as well as the implementation of the meta model generated through EMF.

6.3. Model Based Testing

The idea of *Model based Testing* has, on the one hand, the goal of being able to test different scenarios already on the level of a concrete model, but, on the other hand, a model can also specify a frame for testing, that should be accordingly generated. For testing on the model level, the validation mechanism from ScatterFactory, described previously, seizes control. In order to achieve a good result, which means that an error free model achieves robust code, all respective constraints must be defined to make sure that unreliable combinations of components and dependencies are already stated during the modeling. When different versions of components are maintained in the architecture, it is also important to watch with the help of the model relative constraints that possible incompatible combinations are noticed. In order to illustrate how various implementations of firmware modules in the generating of test phases can be combined in ScatterFactory, an alternative implementation of the Net-modules is used, which was nicely placed at disposal for this purpose by Markus Anwander from the University Bern. To be able to access this on the model level, the corresponding C-files must be transcribed into Xpand-templates and be made known to the generator. This happened simply by placing the relevant templates in the template, which steers the generation of components. This is the relevant piece of the template Component.xpt:

```
<<IF comp.componentType.toString( ) .matches ("Net")>>
  <<IF comp.componentVersion= =null>>
      <<EXPAND ESB : : ScatterWebNetc : : file(sensor, comp)
  FOR comp>>
      <<EXPAND ESB : : ScatterWebNeth : : file(sensor, comp)
  FOR comp>>
  << ENDIF>>
  <<IF comp.componentVersion= ="CSMACA">>
      <<EXPAND ESB : : ScatterWebNetCSMACAc : : file(sensor,
  comp) FOR comp>>
      <<EXPAND ESB : : ScatterWebNetCSMACAh : : file(sensor,
  comp) FOR comp>>
  <<ENDIF>>
<<ENDIF>>
```

If the attribute component Version is placed on "CSMACA" in the model for a component of the type Net with the help of the properties view, the alternative model will be generated.

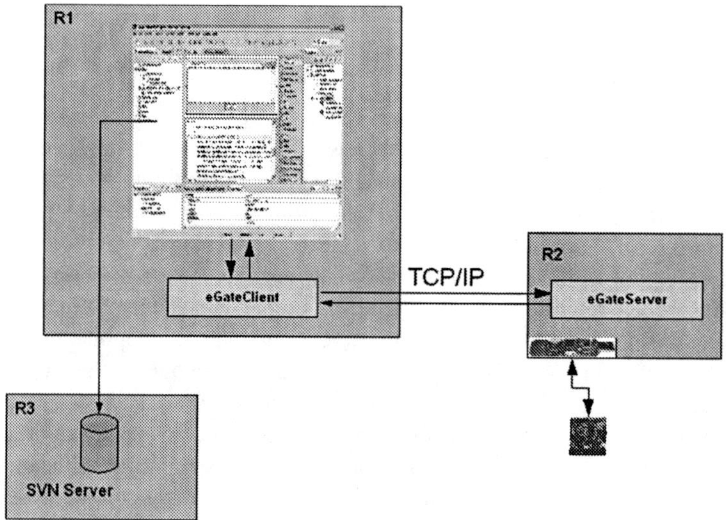

Figure 12. remote Scatter-flasher plug-in

6.4. Extensibility and Interoperability

Through the flexible plug-in structure of the Eclipse-frameworks, as well as the high number of disposable plug-ins, the functionality of ScatterFactory can be extended as much as wanted. The use of the standard already existing support in Eclipse from CVS-repositories definitely stands out as the first thing this offers. Here we want to point out that other self-developed plug-ins for ScatterFactory can be collaboratively combined with ScatterFactory, e.g. the use of our remote flashing plug-in developed independently from ScatterFactory. The remote flashing plug-in (as shown in Fig. 12) is able to flash a selected Hex-File (e.g. generated by ScatterFactory) through the internet on the basis of Java Remote Invocation Method (RMI) in one or more ESBs. R1 is the computer on which the plug-in runs. The eGate/USB is hooked up to R2. Thus R2 hosts the eGtaServer. R3 hosts the CVS Server. For the choice of ESBs, the plug-in –analogue to the flashing plug-in in ScatterFactory– gives the user the option of selecting the IDs of the ESBs as well as the IP address of the eGateServer of the WSN, in which the ESBs are deployed. Besides ScatterFactory, other Eclipse based frameworks are being developed for ScatterWeb, e.g. for planning, management, troubleshooting, etc. of the WSN. The idea is that ScatterFactory functions as a code deliverer in a comprehensive Eclipse based platform for ScatterWeb.

7. CONCLUSION

ScatterFactory produces a complete, model driven, architecture centric, generative infrastructure beginning from an essential MOF conformed meta-meta model all the way to the generating of C-Code for a MSP430-processor on an embedded sensor board. The creation of a domain specific tool chain can in turn itself be considered as a domain. Therefore many tools and frameworks have to be understood and mastered, to be able to install them and use them sensibly. Furthermore, you must currently be prepared for frequent migrations, since the individual tools are in the middle of a quickly moving development process and therefore cannot necessarily be seen as an "Out-of-the-box" solution. Here the GMF is to be named first, which started, at the beginning of the development of ScatterFactory (which began around 7 months ago), as the version 1.06M and now can be found in the current version 2.02M (to 13.01.2007). This is the same with other instruments, as noticed with oAW. For example, the aspect oriented templates were first efficient in the later phases of the development of ScatterFactory, because the mechanism functioned error free only since the version 4.1.1 (released on 30.11.06). Since the development continues on a wider front and the combinations out of EMF/GMF and oAW are becoming more popular, it is only a question of time before the mentioned childhood diseases are eliminated. Because of the development of ScatterFactory, we had to research and combine different new and self-developing technologies in the area of MDSD, which gave us great incentive to research further in this direction. We have a special interest in researching the integration of activity diagrams in ScatterFactory next, as well as the extension of the functionality of the real time validation.

REFERENCES

1. http://www.eclipse.org
2. http://www.eclipse.org/emf
3. http://www.eclipse.org/gmf
4. Eisenecker U, Czarnecki K (2000) Generative Programming. Addison-Wesley Longman, Amsterdam
5. http://www.openarchitectureware.org
6. http://scatterweb.mi.fu-berlin.de
7. Schiller J, Liers A., Ritter H (2005) ScatterWeb: A Wireless Sensornet Platform for Research and Teaching. Computer Communications 28: 1545–1551

CHAPTER 3

PERFORMANCE OF MULTI-HOP RELAYING SYSTEMS OVER WEIBULL FADING CHANNELS

SALAMA IKKI AND MOHAMED H. AHMED
Faculty of Engineering and Applied Science, Memorial University
St. John's, Newfoundland, Canada
E-mail: {ikki, mhahmed}@engr.mun.ca

Abstract: In this paper we present closed-form expressions for the lower bounds of the error and outage performance of multi-hop relaying with non-regenerative relays over independent and identically distributed Weibull fading channels. The end-to-end signal-to-noise ratio (*SNR*) is formulated and upper bounded. Novel closed-form expressions are derived for the moment generating function (*MGF*), the probability density function (*PDF*), and the cumulative distribution function (*CDF*) of the bounded *SNR*. These statistical results are then applied to study the outage probability and the average bit-error probability. Numerical examples compare analytical and simulation results, verifying the tightness of the proposed bounds

Keywords: bit-error probability, Weibull fading, multi-hop relaying, outage probability, quality of service

1. INTRODUCTION

Multi-hop relaying is a promising technique to achieve broader coverage and to mitigate wireless channels impairment. The basic idea is that communication is achieved by relaying the information from the source to the destination via many intermediate terminals. Relaying techniques enable network connectivity where traditional architectures are impractical due to location constraints. The relaying concept can be applied to cellular, wireless local area networks (WLAN*s*), hybrid networks (multi-hop cellular), ad-hoc networks, and mesh networks. More recently and depending on this concept, the idea of cooperative diversity is introduced [1–3]. The main idea is that a mobile terminal relays a signal between the base station and a nearby mobile terminal in addition to the direct link between the transmitter terminal and the receiver terminal.

H. Labiod and M. Badra (eds.), New Technologies, Mobility and Security, 31–38.
© 2007 *Springer.*

The performance analysis of multi-hop wireless relaying systems operating in fading channels has been an important area of research in the past few years. Hasna and Alouini have presented a useful and semi-analytical framework for the evaluation of the end-to-end outage probability and average error probability of multi-hop wireless relaying systems with non-regenerative channel-state-information-assisted relays over Rayleigh and Nakagami-m fading channels [4–7]. Karagiannidis has studied the performance bounds for multi-hop relayed transmissions with fixed-gain relays over Nakagami- (Rice), Nakagami- (Hoyt), and Nakagami-m fading channels using the moments-based approach [8–10]. Recently, Boyer *et al.* [11] have proposed and characterized four channel models for multi-hop wireless communications, and have also introduced the concept of multi-hop diversity.

In this paper, we focus on *amplify-and-forward* multi-hop systems and study their end-to-end performance over independent identically, Weibull fading channels. The reason of using the Weibull distribution is due to the fact that Weibull distribution is a flexible statistical model for describing multi-path fading channels for both indoor and outdoor propagation environments. Alouini and Simon [12] have presented an analysis for the evaluation of the generalized selection combining technique performance over independent Weibull fading channels. Recently, some other contributions dealing with switched and selection diversity, as well as second-order statistics over Weibull fading channels have been presented by Sagias *et al.* in [13–15]. In these works, useful performance criteria including the average output *SNR*, outage probability, and the bit-error rate performance have been studied.

A large body of literature has been devoted to the study of digital communications over Nakagami channels. However, to our knowledge, there exist very limited results discussing transmission over Weibull fading channels and especially in multi-hop relaying networks.

The main contribution of this paper is to derive a closed form expressions of the probability density function (*PDF*), cumulative distribution function (*CDF*), the moment generating function (*MGF*) of the end-to-end *SNR* of the multi-hop relaying system. By using these results we determine the error performance and outage probability of the multi-hop system.

The remainder of this letter is organized as follows. Section 2 introduces the system and channel models under investigation. Section 3 presents closed-form expression for the outage probability, average bit error rate and amount of fading of these systems. In section 4 we summarize the main results of this paper.

2. SYSTEM AND CHANNEL MODELS

We consider an N-hop wireless communication system (as shown in Fig. 1), which operates over i.i.d. Weibull fading channels. The source terminal S communicates with the destination terminal D through nodes terminals. These terminals relay the signal only from one hop to the next, acting as non-regenerative relays.

Assuming coherent detection at the relays and destination furthermore we assume that all the additive white Gaussian noise (*AWGN*) has equal variance N_0 the

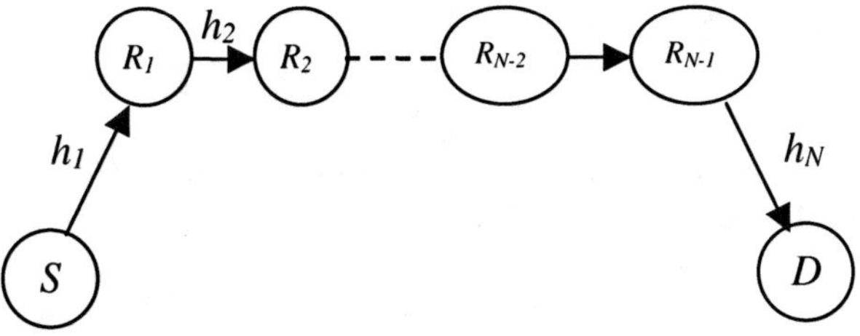

Figure 1. Illustration of a multi-hop relaying system where S is the source, D is the destination and R_i is the ith relay, $(i = 1, 2, \ldots N\text{-}1)$

end-to-end *SNR* can be written as [1,6]

$$\gamma_{eq} = \left[\prod_{n=1}^{N} \left(1 + \frac{1}{\gamma_n} \right) - 1 \right]^{-1} \quad (1)$$

where $\gamma_n = h_n^2 E_s / N_0$ is the *SNR* of the nth hop, h_n is the fading coefficient of the nth hop and E_s is the transmitted signal energy.

The equivalent *SNR* expression in (1) is not easily tractable due to the complexity in finding the statistics (i.e., the *PDF*, *CDF*, and *MGF*) associated with it. Fortunately, this form can be bounded by [16]

$$\gamma_{eq} < \gamma_b = \min(\gamma_1, \ldots, \gamma_N) \quad (2)$$

This form of γ_b has the advantage of the mathematical tractability over that in (1). It will be shown later in results section that this approximation can be used to derive a tight lower bound on the average bit-error rate and outage probability. The *PDF* of the fading amplitude h_n is Weibull distribution and it can be written as

$$f_{h_n}(x) = \frac{c}{\beta} x^{c-1} \exp\left(-\frac{x^c}{\beta} \right), \quad c > 0, \; x \geq 0, \; \mathrm{n} = 1, .., \mathrm{N} \quad (3)$$

where c and β are the fading and scaling parameters, respectively. As c increases the severity of fading decreases, while in the special case when $c = 1$, the Weibull distribution becomes an exponential distribution while for $c = 2$, (3) reduces to the well-known Rayleigh *PDF*. The cumulative distribution function (*CDF*) is given by

$$F_{h_n}(x) = 1 - \exp\left(-\frac{x^c}{\beta} \right), \quad c > 0, \; x \geq 0 \quad (4)$$

The corresponding *PDF* of the instantaneous *SNR* γ_n is also Weibull distributed with β and $c/2$ are the Weibull parameters. It should be noted that average *SNR* $(\bar{\gamma}_n)$ of each hop can be written as

$$\bar{\gamma}_n = \beta^{2/c} \Gamma\left(1 + \frac{2}{c} \right) \frac{E_s}{N_0} \quad (5)$$

where $\Gamma(.)$ is the gamma function defined in [17, eq. (8.310.1)]

3. PERFORMANCE ANALYSIS

3.1. Error Performance

In order to calculate the *MGF* of γ_b $\left(M_{\gamma_b}(s) = E\left(e^{-s\gamma_b}\right)\right)$, we need the *PDF* of γ_b. The *PDF* of γ_b can be found by determining first the *CDF* of γ_b which it can be written as

$$
\begin{aligned}
(6) \qquad F_{\gamma_b}(\gamma) &= 1 - \left(P(\gamma_{h_i} > \gamma)\right)^N \\
&= 1 - \exp\left(-\frac{\gamma^{c/2}}{\beta/N}\right)
\end{aligned}
$$

Then *PDF* can be found by taking the derivative of (6) with respect to γ , yielding

$$
(7) \qquad f_{\gamma_b}(\gamma) = \frac{c/2}{\beta/N}\gamma^{c/2-1}\exp\left(-\frac{\gamma^{c/2}}{\beta/N}\right)
$$

The interesting result in (7) that γ_b is also Weibull distributed with parameters $c/2$ and β/N. The *MGF* of γ_b can be written as

$$
(8) \qquad M_{\gamma_b}(s) = \int_0^\infty \frac{c/2}{\beta/N}x^{c/2-1}\exp\left(-\frac{x^{c/2}}{\beta/N}\right)\exp(-sx)dx
$$

To evaluate the integral in (8), we will first express the exponential function for any arbitrary function $g(x)$ by its Meijer's function defined in [17, equ. 9.301] as

$$
(9) \qquad \exp(-g(x)) = G_{0,1}^{1,0}\left[g(x)\middle|\begin{matrix}-\\0\end{matrix}\right]
$$

Then *MGF* of γ_b in (8) can be written as

$$
(10) \qquad M_{\gamma_b}(s) = \frac{c/2}{\beta/N}\int_0^\infty x^{c/2-1}G_{0,1}^{1,0}\left[\frac{x^{c/2}}{\beta/N}\middle|\begin{matrix}-\\0\end{matrix}\right]G_{0,1}^{1,0}\left[sx\middle|\begin{matrix}-\\0\end{matrix}\right]dx
$$

With the help of [18, equ. 21] the integral in (10) can be written in closed form as

$$
(11) \qquad M_{\gamma_b}(s) = \frac{c/2}{\beta/N}\frac{\left(\frac{k}{l}\right)^{\frac{1}{2}}\left(\frac{l}{s}\right)^{\frac{c}{2}}}{(2\pi)^{\frac{k+l}{2}-1}}G_{l,k}^{k,l}
$$

$$
\left[\frac{(\beta/N)^{-k}}{s^l}\frac{l^l}{k^k}\middle|\begin{matrix}\frac{1-c/2}{l},\frac{2-c/2}{l},\ldots,\frac{l-c/2}{l}\\0,\frac{1}{k},\frac{2}{k},\ldots\ldots\ldots\ldots,\frac{k-1}{k}\end{matrix}\right]
$$

where k and l are positive integers. In order for (11) to be valid, k and l should be chosen as the smallest integers satisfying the condition ($l/k = c/2$; e.g., for $c = 3.5$ we should choose $l = 7$ and $k = 4$). The Meijer's function is widely available in many scientific software packages, such as Mathematica and Maple.

With the aid of *MGF* the performance of a great variety of modulation schemes can be easily obtained [19]. Examples of such modulation schemes include M-ary quadrature amplitude modulation (MQAM), M-ary phase-shift keying (M-PSK) and non-coherent binary schemes such as binary frequency-shift keying (BFSK) and differential phase-shift keying (DPSK). For example, the average error probability of binary DPSK can be given by $0.5M_{\gamma_b}$ (1).

By obtaining the *MGF* in a closed form, the analytical expression for the moments can be easily found and these moments are very useful since it can be used to directly obtain the average end-to-end *SNR* or the amount of fading (*AF*) defined in [19]. It must be noted here, that (11) can be used to study several other quality measures, such as the kurtosis and the skewness [20].

3.2. Outage Probability

The probability of outage is defined as the probability that the instantaneous *SNR* falls below a specified threshold. This threshold is a minimum value of the *SNR*, above which the quality of service is satisfactory. In the case of the multi-hop system under consideration, the use of upper bound γ_b leads to lower bounds for the outage probability at the destination terminal D, expressed as $P_{out} > F_{\gamma b}(\gamma_{th})$.

4. SIMULATIONS AND NUMERICAL RESULTS

In this Section, we show numerical results of the analytical error rate and outage probability found above for differential binary phase shift keying (DBPSK). We also show the results of the computer simulations used for verification. It is assumed that the channel fading coefficients of different links are mutually independent.

In Fig. 2, lower bounds for the *BER* of a multi-hop system are plotted for $N = 2, 3$, and 5 and for $c = 1$. We chose to plot the *BER* at $c = 1$ only to avoid entanglement. It is evident that the proposed bounds are tight particularly at medium and high *SNR*. Also, it is clear that the *BER* deteriorates with the increase of the number of hops (N). This is because we use here a symmetric network (all links have the same path-loss regardless of the number of hops). If asymmetric network is assumed, this trend might be reversed.

Figure 3 shows the error performance is studied for different fading parameters c. It should be noted that for both Fig. 2 and Fig. 3 the tightness of the error performance increase as *SNR* increased; however, the proposed bounds lose their tightness in the low *SNR* region as N increases. This happens due to the fact that the accuracy of the γ_{eq}upper bound increases as the *SNR* increases (see (3) and (4)).

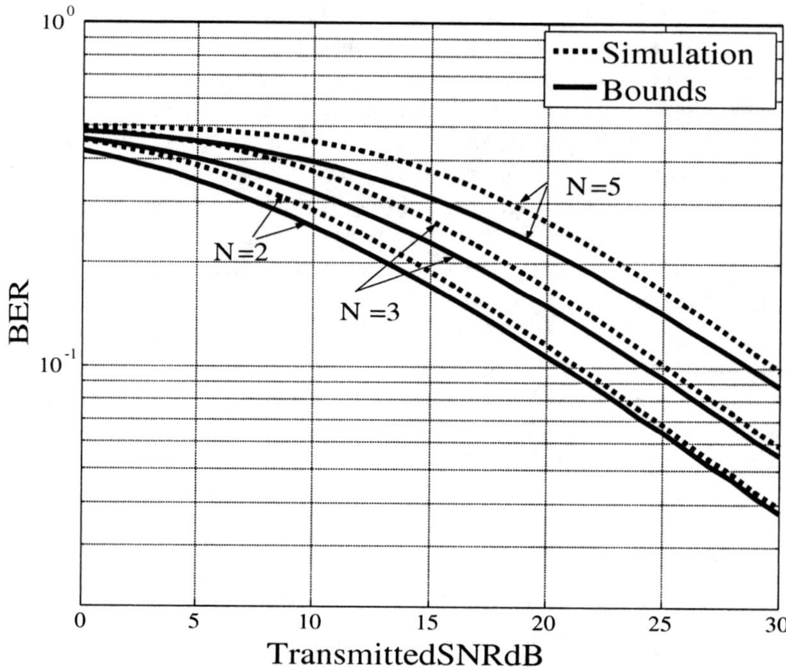

Figure 2. DBPSK error bounds for a multi-hop system in i.i.d. Weibull fading channels ($c = 1$)

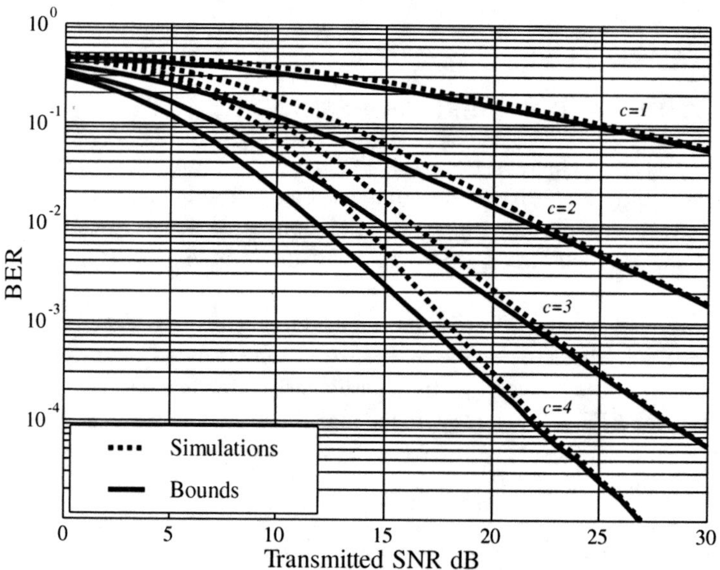

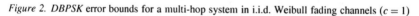

Figure 3. DBPSK error bounds for a multi-hop system in i.i.d. Weibull fading channels for different values of fading parameters

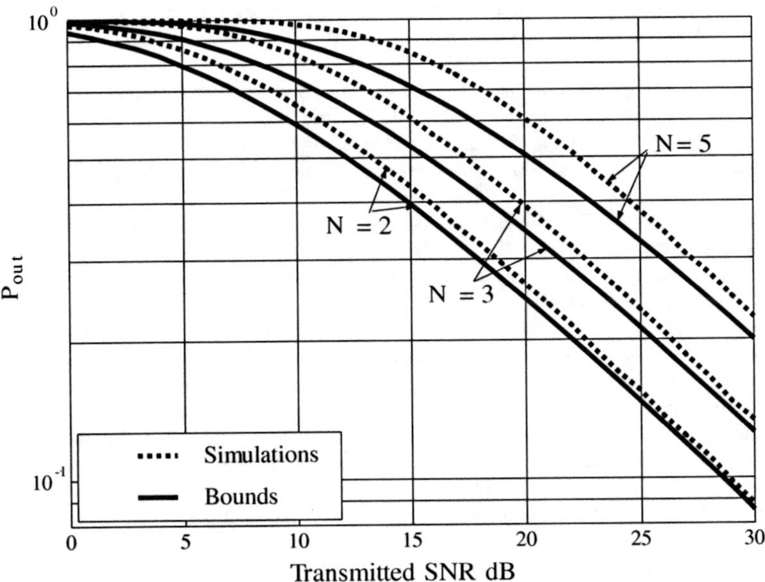

Figure 4. Outage probability bound for a multi-hop system in i.i.d. Weibull fading channels ($c = 1$)

Lower bounds for the outage probability are plotted in Fig. 4 for $\gamma_{th} = 0\,\text{dB}$. The obtained results show that the obtained bound gets tighter with the increase of the *SNR*. Also, it is obvious that the outage bound gets tighter as N gets smaller.

5. CONCLUSION

Performance bounds for multi-hop transmissions with channel-state-information-assisted relays operating over i.i.d. Weibull fading channels have been investigated. The end-to-end *SNR* is stated and upper bounded and novel closed-form expressions for the *MGF*, *PDF*, and *CDF* of this upper bounded *SNR* have been derived. Additionally, tight lower bounds for the outage probability, the average error probability, and amount of fading have been presented. Our numerical results show that this approximation is either tight lower bound or exact on the performance. Finally, we consider extending this work in future publications to address the following issues: Performance study under non-identical Weibull fading amplitudes and the existence of an optimal number of hops for asymmetric network.

REFERENCES

1. J. N. Laneman, D. N. C. Tse, and G. W. Wornell, "Cooperative diversity in wireless networks efficient protocols and outage behaviour," *IEEE Trans. Inform. Theory*, vol. 50, no. 12, pp. 3062–3080, Dec. 2004.

2. ___ "Distributed space-time coded protocols for exploiting cooperative diversity in wireless networks," in *Proc. Global Telecommunications Conf.*, vol. 1, Taipei, Taiwan, R.O.C., Nov. 2002, pp. 77–81.
3. A. Sendonaris, E. Erkrip, and B. Aazhang, "User cooperation diversity–Part I: System description," *IEEE Trans. Commun.*, vol. 51, no. 11, pp. 1927–1938, Nov. 2003.
4. M. O. Hasna and M.-S. Alouini, "End-to-end performance of transmission systems with relays over Rayleigh fading channels," *IEEE Trans. Wireless Commun.*, vol. 2, no. 6, pp. 1126–1131, Nov. 2003.
5. ___ "Harmonic mean and end-to-end performance of transmission systems with relays," *IEEE Trans. Commun.*, vol. 52, no. 1, pp. 130–135, Jan. 2004.
6. ___ "Outage probability of multihop transmission over Nakagami fading channels," *IEEE Commun. Lett.*, vol. 7, no. 5, pp. 216–218, May 2003.
7. ___ "A performance study of dual-hop transmissions with fixed gain relays," *IEEE Trans. Wireless Commun.*, vol. 3, no. 6, pp. 1963–1968, Nov. 2004.
8. G. K. Karagiannidis, "Performance bounds of multihop wireless communications with blind relays over generalized fading channels," *IEEE Trans. Wireless Commun.*, to be published.
9. G. K. Karagiannidis, "Moments-based approach to the performance analysis of equal-gain diversity in Nakagami-*m* fading," *IEEE Trans. Commun.*, vol. 52, no. 5, pp. 685–690, May 2004.
10. Karagiannidis, G.K.; Tsiftsis, T.A.; Mallik, R.K.;" Bounds for multihop relayed communications in Nakagami-*m* fading" *IEEE Trans. Commun*, vol. 54, no. 1, pp. 18–22, Jan 2006
11. J. Boyer, D. D. Falconer, and H. Yanikomeroglu, "Multihop diversity in wireless relaying channels," *IEEE Trans. Commun.*, vol. 52, no. 10, pp. 1820–1830, Oct. 2004.
12. M.-S. Alouini and M. K. Simon, "Performance of generalized selection combining over Weibull fading channels," in *Proc. Vehicular Technology Conf.*, vol. 3, Atlantic City, NJ, 2001, pp. 1735–1739.
13. N. C. Sagias, D. A. Zogas, G. K. Karagiannidis, and G. S. Tombras, "Performance analysis of switched diversity receivers in Weibull fading," *IEE Electron. Lett.*, vol. 39, no. 20, pp. 1472–1474, Oct. 2003.
14. N. C. Sagias, P. T. Mathiopoulos, and G. S. Tombras, "Selection diversity receivers in Weibull fading: Outage probability and average signal-tonoise ratio," *Electron. Lett.*, vol. 39, no. 25, pp. 1859–1860, Dec. 2003.
15. N. C. Sagias, G. K. Karagiannidis, D. A. Zogas, P. T. Mathiopoulos, and G. S. Tombras, "Performance analysis of dual selection diversity in correlatedWeibull fading channels," *IEEE Trans. Commun.*, vol. 52, no. 7, pp. 1063–1067, Jul. 2004.
16. Salama Ikki and Mohamed. H. Ahmed "Performance analysis of wireless cooperative diversity networks over Nakagami-m channels" *IEEE commun. Letter* accepted to be published, 2007
17. I. S. Gradshteyn and I. M. Ryzhik, table of integrals, series and products San Diego, CA: Academic Press, 6th edition, 1994
18. V. S. Adamchik and O. I. Marichev "the algorithm for calculating integrals of hypergeometric type functions and its realization to REDUCE system" in *proc. International Conference on Symbolic and Algebraic Computation*, Tokyo, Japan, 1990 pp 212–224.
19. M. K. Simon and M.-S. Alouini, Digital Communication over Fading Channels. New York: Wiley, 2000.
20. Zogas, D.A., Karagiannidis, G.K., Sagias, N.C., Tsiftsis, T.A., Mathiopoulos, P.T., Kotsopoulos, S.A., "Dual hop wireless communications over Nakagami fading" *IEEE Vehicular Technology Conference*, 2004. VTC 2004-Spring.

CHAPTER 4

PERFORMANCE STUDY OF A NEW MAC ALGORITHM TO REDUCE ENERGY AND DELAY IN WIRELESS SENSOR NETWORKS

IOANNIS GRAGOPOULOS[1], IOANNIS TSETSINAS[1], EIRINI KARAPISTOLI[2], AND FOTINI-NIOVI PAULIDOU[2]

[1]*Informatics and Telematics Institute, 1st Km Thermi-Panorama Road, Thessaloniki, GR-57001, Greece, {grag,tsetsinas}@iti.gr*

[2]*Department of Electrical and Computer Engineering, Aristotle University of Thessaloniki, GR-54124, Greece {ikarapis,niovi}@auth.gr*

Abstract: In this paper we focus on the problems of high latency and low throughput arising from the periodic operation of MAC protocols for wireless sensor networks. In order to meet both design criteria we propose an energy-efficient, low-delay, fast-periodic MAC algorithm based on the IEEE 802.15.4 physical layer. Our proposal relies on the short periodic communication operation of the nodes comprising the WSN. This is achieved by decreasing the actions that a node needs to perform at the beginning of every communication period and by incorporating a variable radio-on operation. Moreover, the algorithm introduces differences in nodes' scheduling to further reduce delay. The proposed MAC is evaluated and compared to S-MAC and T-MAC through extensive simulations, showing a significant improvement in terms of low energy consumption and average MAC delay

Keywords: Wireless Sensor Networks, MAC protocol, energy efficiency

1. INTRODUCTION

Wireless sensor networking is an emerging technology that has a wide range of potential applications [1]. A typical wireless sensor network (WSN) is a large set of wireless nodes, with sensing, monitoring and processing capabilities, deployed in an *ad hoc* fashion. These wireless nodes are autonomous, battery-operated devices, with limited energy capacity, mostly designed for unattended operation. Therefore, they require mechanisms to minimize energy consumption, in order to ensure a

H. Labiod and M. Badra (eds.), New Technologies, Mobility and Security, 39–51.
© 2007 *Springer.*

long-lasting operation without the need for replacing/recharging their batteries. These mechanisms need to be implemented in all the network layers, from physical to application. In this study, we concentrate on medium access control (MAC) protocols since this gives a fine-grained control to switch the wireless radio on and off and therefore to effectively prolong the network lifetime.

As already mentioned, energy conservation is the primary and most important challenge to meet since it determines the lifespan of a sensor network. Since the power consumption of a transceiver is remarkably high during channel listening, the best way to achieve energy conservation is to turn off the radio electronics on every network node for as long as possible. The crucial challenge is to keep the nodes' radios on only for the time necessary to exchange data. The only solution that approaches this, since there is no "Wake-Up Radio" in the market developed yet, is to periodically switch on and off a node's radio, ultimately to keep the radio on only when communication is needed.

Derived from the above, the key factor for energy conservation appeared to be the *periodic operation*. Periodicity can be implemented either in a straightforward way (i.e. in schedule-based MAC protocols) or not (i.e. in contention-based MAC protocols). Yet, whatever of the access technique in use, the selection of a proper *duty cycle*, which is the percentage of on-time with respect to total period duration, is mandatory. Smaller duty cycle values improve power consumption but also lead to higher end-to-end delays which can be a serious drawback, especially in multi-hop systems. This tradeoff between consumption and delay with respect to period selection was the major motivation for this paper.

In this paper, we propose a new MAC algorithm for wireless sensor networks that is capable of using small period values by decreasing the actions that a node needs to perform at the beginning of every communication period and by incorporating a variable radio-on operation. The algorithm also incorporates the idea of scheduling the listening times of the nodes, while using contention over the transmission of data during each node's on period with a simple back-off mechanism. The above features together with the algorithm's fast periodic operation lead to an energy-efficient MAC algorithm that reduces delay in wireless sensor networks.

The remainder of the paper is organized as follows: in Section 2 we describe existing widely accepted access methods. In Section 3, we present an analytical description of the proposed algorithm. Section 4 illustrates the obtained simulation results, followed by detailed reports. Finally, conclusions are given in Section 5.

2. RELATED WORK

Piconet [2] is one of the first contention-based MAC protocols that introduced periodic sleep for energy conservation. However, there is no coordination and synchronization among nodes about their sleep and listen time. The scheme to enable communication among neighboring nodes is to let a node broadcast its

ID when it wakes up from sleeping. If a sender wants to talk to a neighbor, it must keep listening until it receives the neighbor's broadcast. Although ID reception secures that the destination node is alive, there is no information about its schedule. This results in energy waste, since the source node may have to wait for unnecessary long time with its receiver on, until the reception of neighbor's ID.

One of the famous contention-based access protocols is the Sensor-MAC protocol or S-MAC [3]. S-MAC uses a coordinated sleeping mechanism, similar to the IEEE 802.11 DCF power saving (PS) mode, and in-channel signaling to avoid overhearing. In S-MAC, data is exchanged using the RTS-CTS handshake procedure. Although RTS-CTS can alleviate the hidden terminal problem, it incurs high overhead because data packets are typically very small in sensor networks. Exchanging control messages in the order of the actual data packets leads to bandwidth waste and increased protocol overhead. Moreover, although S-MAC achieves low power operation, it accomplishes this by trading off energy for latency. In a follow up paper [4], the increased latency caused by the periodic sleep of each node, is improved by utilizing an adaptive listening mechanism.

Several S-MAC variations have been proposed since then. T-MAC [5] is one of them and improves S-MAC's energy usage by using a very short listening window at the beginning of each active period. After the SYNC section of the active period, there is a short window to send or receive RTS and CTS packets. If no activity occurs in that period, the node returns to sleep. The adaptive duty cycling in T-MAC saves more energy under variable traffic, but, unlike our proposal, it achieves lower throughput due to TMAC's identified early sleeping problem. DMAC [6], on the other hand, lowers the latency in convergecast communications in an energy efficient way. This is achieved by assigning subsequent slots to nodes that are successive in the data transmission path. Though, collision avoidance methods are not utilized; hence, when a number of nodes that have the same schedule try to send to the same parent node, collisions will occur.

TDMA-based protocols, such as [7], are naturally energy conserving, because they have a built-in duty cycle, and do not suffer from collisions. However, maintaining a TDMA schedule in an ad-hoc network is not an easy task and requires much complexity in the nodes. TDMA-based protocols usually require the nodes to form communication clusters. Thus, when the number of nodes within a cluster changes, due to addition or removal of sensor nodes, it is not easy for them to dynamically change its frame length and its time slot assignment. So, its scalability is not as good as that of a contention-based protocol.

In all the above proposals, the main effort has been focused on energy consumption minimization, disregarding in a degree other critical system parameters such as delay and throughput. In our proposal, we address both energy, latency and throughput issues by introducing a novel MAC algorithm where no synchronization and topology information is needed.

3. PROPOSED ALGORITHM DESCRIPTION

3.1. The Concept

In a contention-based MAC algorithm, the key element in optimizing the energy consumption is to minimize the duration of receiver's on time before the actual data exchange can take place. This time interval, which we refer to as *minimum active time* (t_m), consists of possible signaling, handshaking, collision avoidance mechanisms, plus the necessary Tx/Rx turnaround and calibration time and poses the lower bound on the radio on time. This bound varies in different MAC implementations. An implementation that needs more time to accomplish the above procedures clearly requires to operate in longer periods so as to achieve the same duty cycle d. This is shown in Figure 1, where the sum $(t_m + t_{on})$ corresponds to the radio-on time and the fraction $(t_m + t_{on})/d$ to the period length (p) given the duty cycle's definition. Furthermore, in those systems that operate in a periodic manner, data transfer latency naturally depends on the frequency of the operation cycle (namely period) with faster cycle iterations yielding better performance. Summing up, the ability of a MAC algorithm to shrink the period and therefore reduce latency, emanates from its potential to operate at a very short minimum active time, otherwise either there will be no useful time interval to exchange data, or the duty cycle will be increased in such values that practically no energy conservation will be finally achieved.

In S-MAC and its variations, the synchronization phase plus the RTS-CTS handshaking procedure implies a relatively long minimum active time, restricting their ability to absorb the traffic fluctuations in a wireless sensor network while keeping the average radio-on time, the delay and throughput in acceptable levels. In a typical WSN, a sensor node spends most of its operation time without communicating, so it has to operate with a minimum duty cycle that will extend network lifetime. On the other hand, the same duty-cycle should guarantee an acceptable delay and throughput when traffic exists. As it will be shown in Section 4 , S-MAC and T-MAC do not fulfill the above requirements.

In this study, we attempt to remedy these issues by spreading neighbor nodes' wake-up schedules within a period (in contrast to S-MAC where nodes synchronize their schedules), and by replacing the RTS-CTS handshaking with a simple back-off algorithm. These techniques allow us to considerably reduce the minimum active time, improving the overall performance.

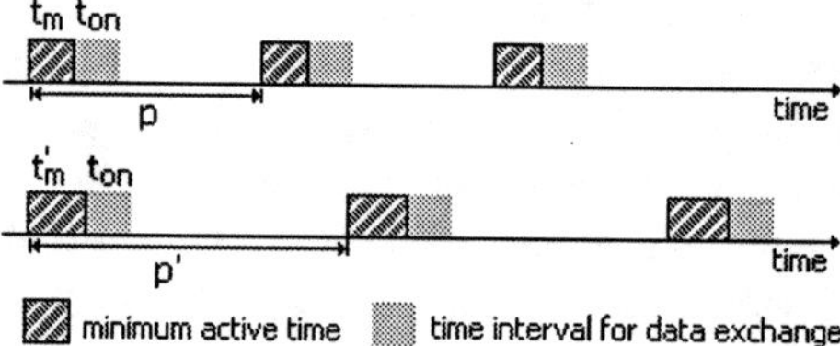

Figure 1. Minimum active time with respect to the operation period

3.2. Detailed Description

The core of the proposed algorithm relies on the periodic sleep/wake-up operation of the wireless nodes comprising the WSN. At the beginning of every period there is a radio-on time (T_L) which consists of the broadcasting of a synchronization frame and a minimum idle listen time that a node requires in order to identify possible transmissions (Figure 2).

The synchronization frames contain essential parameters of the transmitting node, such as its id, its period and its oscillator's drift. A node that receives a synchronization frame, stores its information locally and consequently learns the consecutive moments that the corresponding neighbor node will be able to receive data. Considering that 802.15.4 radio transmitters consume about as the same energy

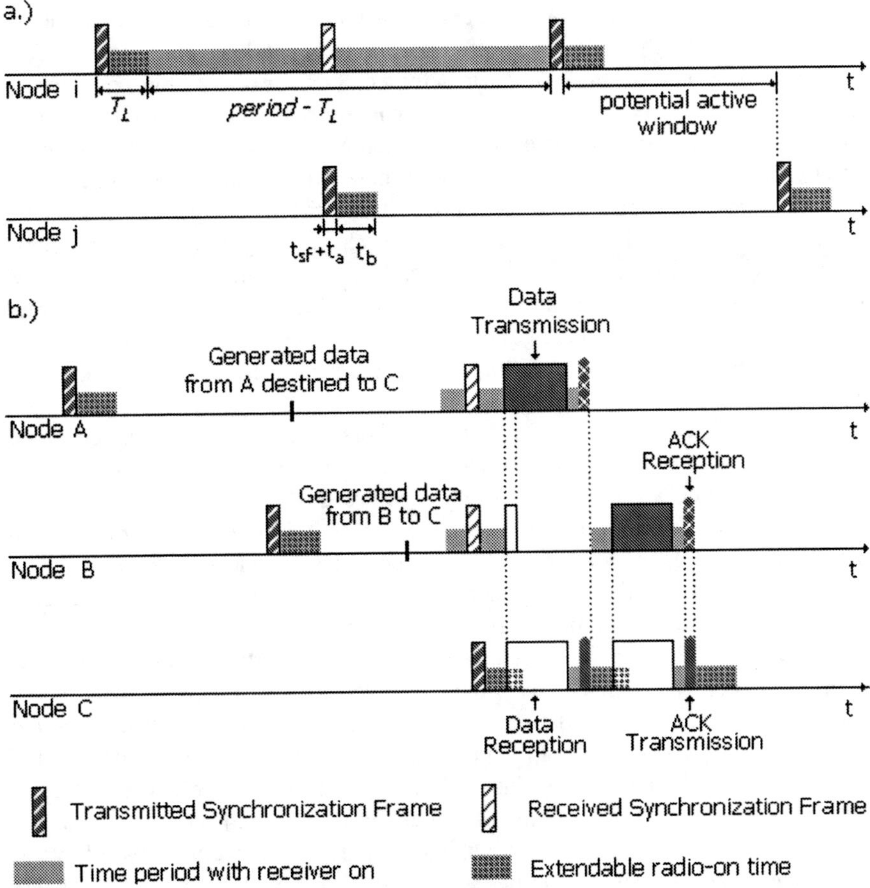

Figure 2. a) Channel activity while node i is searching for neighbors nodes and b) Channel activity for receive and transmit processes between node A, B and C

as its receivers [8], there is no extra power consumption during the transmission of these frames. Immediately after broadcasting the synchronization frame, the node turns over from Tx to Rx and listens for potential data. If no channel activity occurs after time T_L, the node turns its radio off. If data is detected, the node extends its active time interval in order to complete the reception procedure, which includes the acknowledgment mechanism. After the reception of a complete packet, the node continues to listen for an additional time, equal to t_b (Figure 2), for further possible transmissions. This procedure can be repeated until the next successive synchronization frame of any other neighbor node. This time duration is called the *potential active window* and it depends both on the length of the operation period and the number of nodes comprising the network. As the number of nodes within vicinity increases the length of the operation period should be long enough to provide the nodes with sufficient potential active window for data transmission. The remaining data, if any, will be delayed until the next cycle of the receiving node.

During the network setup stage and before each node starts its periodic operation, it needs to run a neighbor discovery protocol and to schedule its on-period in a *free* time window. Hence, the node first listens for a certain amount of time, equal to *two* periods. This constraint gives the receiver enough time to collect synchronization frames in the neighborhood. If during the listening time it does not hear channel activity, it immediately broadcasts its synchronization frame. On the other hand, if a node receives synchronization frames during the listening time, it chooses a time window that will not overlap with the already scheduled ones and starts broadcasting its own synchronization frame. The neighborhood discovery process described above during which data reception is disabled, is essential since it enables nodes to adapt to network changes and to compensate timing errors attributed to oscillators' drift and can be a periodic or on demand operation according to the application requirements.

When a node wishes to transmit to its neighbors it needs to wait their scheduled wake-up period. Thus, the node first checks the corresponding stored timing information and switches on its radio slightly before the expected broadcast of the neighbor's synchronization frame in order to receive it. The reception of this frame is mandatory in order to confirm that the destination node is still alive. Upon reception of this frame the transmission procedure begins with a randomly selected back-off time. After this time elapses, carrier sense is performed at the physical layer and if no radio activity is detected, transmission starts. During the back-off time, all the contending nodes have their receivers on, so when one acquires the channel and starts data transmission, the rest of the contenders are able to listen the transmission and consequently to learn the packet length in order to calculate the end of the transmission. During this time interval, they switch off their receivers (since another node has gained the channel and transmits data) and schedule a new back-off contention period, if the destination's potential active window allows it, immediately after the completion of transmission which includes the acknowledge mechanism. Note that in 802.15.4-like networks [9] the addressing scheme used allows this trick to be easily implemented, since the packet length resides at the beginning of each packet, immediately after the preamble sequence and before the address bytes. So each node that overhears a transmission has the ability to learn its length, even if it is destined to another node, before address recognition mechanism fails. Thus, a contenting node overhears just six bytes of the total packet (4 bytes preamble, 1 byte start frame delimiter and 1 byte packet length for 802.15.4 implementation), which are actually useful information in order to safely schedule the next contention interval in the same active window.

The benefits of the described procedure are a significant decrease of overhearing energy waste and elimination of retransmissions due to lost acknowledgement packets. An acknowledgement packet can be destroyed because the time interval between the end of reception and start of acknowledge transmission (Rx/Tx turnaround time) is enough for a contending node to assess clear channel and begin transmission of a new packet. This actually destroys two packets, and in our case is effectively avoided.

The reliability of the algorithm is based on each node's uninterrupted process of transmitting synchronization frames. Keeping the correct timing for broadcasting synchronization frames is the highest priority task for each node and guarantees the algorithm's reliable operation. Whatever a node is doing, it will be interrupted for the accurate on-time transmission of the synchronization frame. Although CSMA/CA algorithm is used for every transmission, there is still probability for collisions in data as well as in synchronization frames. In the case of data frames, collisions can easily be identified and coped with retransmissions. However, in synchronization frames, which are broadcast packets, it is impossible to detect collisions due to lack of an acknowledgement mechanism. In the case that a node (say α) does not succeed in getting a schedule (due to a collision), a neighboring node (say β) that receives a packet destined from α (note that node α does not appear in node's β neighboring list), could inform the latter for a possible collision via enabling a flag at the acknowledgement packet. In turn, node α will re-run the neighbor discovery protocol to obtain a non-colliding schedule.

3.3. Performance Considerations

The crucial parameter in our approach is the determination of an effective minimum active time T_L (Figure 2). The T_L time duration is the sum of three elements:

(1) $$T_L = t_{sf} + t_a + t_b$$

where t_{sf} is the required time to broadcast the synchronization frame, t_a is the Tx/Rx turnaround time and t_b is the maximum time required to implement one back-off. The values of t_{sf} and t_a are quite deterministic. Time t_{sf} depends on the transmission rate (250 kbps for 802.15.4 networks) and the number of bytes transmitted within the synchronization frame. Time t_a depends on the transceiver's PLL lock and calibration time. Thus, the only parameter that significantly affects T_L and can be adjusted, is the time required to implement one back-off, t_b. The back-off is performed as follows. Time is divided in slots of 160μs, i.e. the minimum time for a 802.15.4 receiver to assess clear channel. A node randomly selects a number of slots with uniform distribution within the range of $[0, B_{max}]$. The value of B_{max} depends on the possible medium contenders, namely neighbor nodes. It should be small enough to maintain a short T_L (otherwise it will lead to energy waste), while keeping collisions in acceptable levels. Assuming n is the number of contending nodes, then the probability P_s that a node will acquire the channel in a single contention period, has been analytically calculated as:

$$P_s = n \sum_{i=1}^{B_{max}} \frac{1}{B_{max}} \left(\frac{B_{max} - i}{B_{max}} \right)^{n-1} \tag{2}$$

The choice of B_{max} affects both the radio on operation and latency metric. After simulations, it was found that any B_{max} value between 10 and 20 is close to optimal for both metrics and for a number of contenders ranging from 2 to 14. The probability P_s exceeded the 70% for this choice.

4. SIMULATION RESULTS

In order to test the robustness of the proposed algorithm, we conducted a series of simulation tests. The simulation environment which resembles a static wireless network was created using the OMNeT++ discrete event simulator. The protocols that have been chosen for comparison were S-MAC [3] and T-MAC [5], with a fixed duty-cycle and an adaptive one respectively. The two metrics we consider to evaluate the performance of the three protocols are:

1. Radio-ON time: is the percentage of time that the node's receiver is ON. It is the only fair metric to evaluate the performance of each MAC algorithm in terms of energy consumption, since it is independent from factors such as current technology's features and the characteristics of different chipsets whose energy consumption varies. It is obvious that, the smaller the radio-ON time percentage, the higher the node's lifetime extension.
2. Average MAC delay: is the average end-to-end delay of a packet from its birth up until correct reception at its destination.

All graphical presentations that follow show overall network performance with a constraint of zero lost probability (no packet drops), meaning that all packets are successfully delivered to their destinations. If a protocol starts dropping packets under certain conditions, the corresponding result entry is empty. We used the following testing scenarios for simulations:

4.1. One-Hop Network

The network setup consists of 8 sensor nodes with overlapping radio ranges, meaning that all nodes are placed within a one-hop neighborhood. All wireless nodes generate sensing data to random destinations at different rates varying from 0.25 packets/sec to 4 packets/sec. Each data packet has a size of 37 bytes including a preamble of 6 bytes, a header of 11 bytes and data payload of 20 bytes. In each simulation set, all three protocols share the same network parameters.

Initially, extensive simulations were realized so as to see how our proposal behaves at very short periods. For that reason, we varied the period length from 60 ms to 250 ms. Figure 3a depicts the percentage of time radio receiver is on. The proposed algorithm operates effectively in all cases and regardless of the traffic variations. Its radio-on operation never surpasses 7.8 percent showing that the idle listening is highly reduced. Next, the average MAC delay depicted in Figure 3b shows that the shorter the period, the lower the obtained MAC delay following the analysis illustrated in Section 3.1. Our algorithm successfully serves the heavy traffic conditions keeping the average MAC delay at very low levels that do not surpass 0.25 sec. The interesting performance characteristic of the proposed algorithm is that it succeeds in keeping both metrics low at the same time without the need of making a tradeoff between them. Moreover, its traffic-independent characteristic, results from its fast periodic operation and the fact that the wake-up schedules of the nodes are spread within the period and are not synchronized.

Following on, we conducted comparative simulations over the three protocols. We tested the S-MAC protocol with a period of 200 ms and a fixed duty cycle of 10% and 15%. For the T-MAC protocol, we used the same period and an interval TA equal to 2.5% of each period as suggested in [5]. Overhearing avoidance, full-buffer priority and FRTS features that aim at addressing the early sleeping problem and therefore at increasing the algorithm's achieved maximum throughput, were not enabled when T-MAC was simulated, since T-MAC achieves the 100% throughput limit we set in all testing scenarios. Finally, the period in our algorithm is chosen to be equal to the S-MAC's and T-MAC's periods.

From Figure 4a we can clearly see that regardless the traffic conditions, the proposed algorithm results in very low radio-on operation, which is significantly lower than the one of the two other protocols. As expected, S-MAC with a 10% duty-cycle fails to serve traffic conditions higher that 3 packets/sec. The results expose one main drawback that algorithms based on a fixed schedule have; the inability to adapt to the traffic variations met in a sensor network. A fixed duty

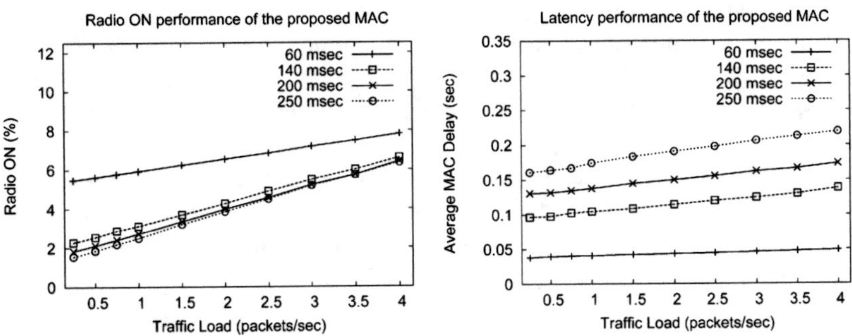

Figure 3. a) Radio-ON time and b) Average MAC delay over traffic load variations

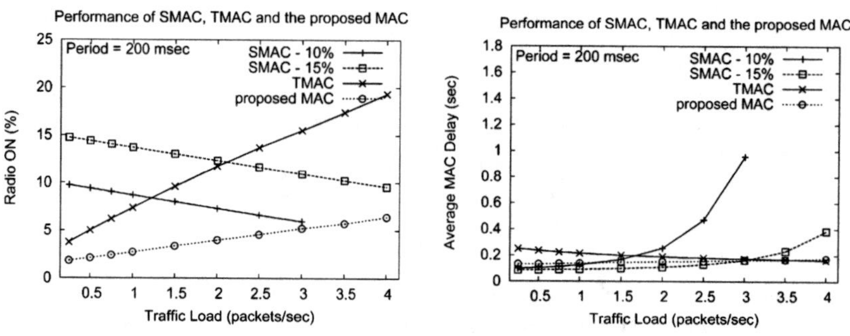

Figure 4. a) Radio-ON time and b) Average MAC delay over traffic load variations

cycle adjusted for high traffic load results in significant energy waste when traffic is low, while a duty cycle for low traffic load results in low message delivery and long queuing delay. On the other hand, T-MAC for light traffic conditions remains active for a short percentage of time, and, as traffic increases, it has to stay on for a long time. Simulation results concerning the measured message latency are illustrated in Figure 4b. From these, we derive that S-MAC due to its fixed sleep/listen schedule, increases the queuing delay and therefore the average MAC delay. T-MAC presents lower levels of latency compared to S-MAC in all traffic conditions. However, our algorithm shows the lowest latency levels, which remain stable regardless of traffic.

We also ran the simulations with higher period lengths, such as 500 ms and 1000 ms, and we obtained very similar results. The proposed MAC, at these high period lengths where its fast periodic operation advantages are hidden, operates in a very energy efficient way compared to S-MAC's and T-MAC's performance.

4.2. Two-Hop Network

The two-hop topology consists of two clusters of 4 nodes each and a sink node. The two clusters' radio range overlap on the sink node, and thus act as *hidden terminals* for each other. For the nodes-to-sink communication pattern all wireless nodes generate sensing data at different rates varying from 0.25 packets/sec to 4 packets/sec. Moreover, S-MAC's and T-MAC's period length was chosen equal to 200 ms, whereas a smaller period (60 ms) that better reflects a fast periodic operation was chosen for our algorithm.

Figure 5a shows that our algorithm consumes less energy than S-MAC and T-MAC in almost all traffic conditions. S-MAC with a 10% duty cycle achieves slightly better radio-on performance. A closer look at Figure 5b though, shows that this reduction is traded off with a higher average MAC delay. T-MAC on the other hand, exhibits the optimal latency performance especially when the traffic load is high (above 3 packets/sec) showing that its adaptive behavior positively results in that direction. However, once again, our algorithm succeeds in keeping both

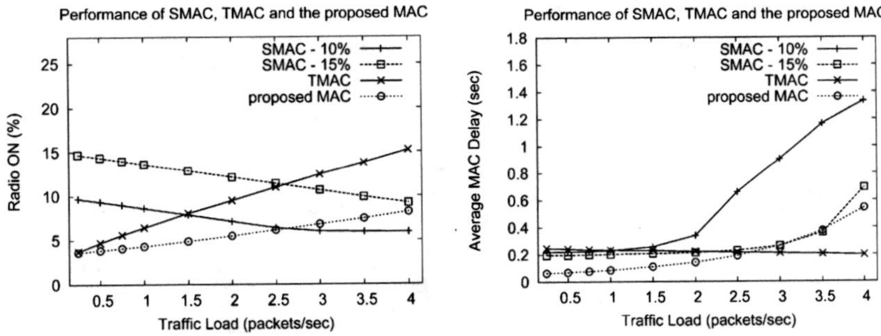

Figure 5. a) Radio-ON time and b) Average MAC delay over traffic load variations

metrics low at the same time without the need of making a tradeoff between them. Moreover, the simple back-off algorithm that it implements, proved to be sufficient to handle the hidden terminal problem observed in this testing scenario.

4.3. Multi-Hop Network

The multi-hop topology consists of 15 nodes placed in a 3 by 5 grid. We have chosen a radio range so that all non-edge nodes have 4 neighbors. Though a small numbered multi-hop setting, the conclusions that we obtain are indicative of the algorithms' performance. In this testing scenario we also applied a nodes-to-sink communication pattern, where nodes send packets to a single sink node at the corner of the network. A randomized shortest path routing scheme was used where next hop nodes are eligible if they have fewer hops to the destination. From these next hops, a random one is chosen. Thus packets flow in the correct direction, but do not use the same path every time. S-MAC was tested with and without adaptive listening and a 10% duty cycle as suggested in [4]. Period lengths were kept the same as in the two-hop topology.

From Figure 6a we can see that both S-MAC variations succeeds in keeping the radio-on operation at very low levels. The comparatively increased radio-on percentage that the proposed algorithm presents, can be interpreted by the increase in collided packets that are retransmitted. Algorithm's overall performance though is very satisfying if we also consider its achieved average MAC delay. Figure 6b shows that S-MAC at 10% duty cycle without adaptive listening presents the highest latency and has about twice the average latency than that of S-MAC with adaptive listening. However, even with the adaptive listening mechanism, S-MAC does not reach T-MAC's and our algorithm's latency performance. Since the adaptive listening mechanism cannot guarantee the immediate transmission at each hop, if a node fails to receive a CTS from the intended receiver after a RTS transmission, it has to wait for one sleep cycle which increases the overall end-to-end delay. On the contrary, the fast periodic operation of our algorithm maintains the end-to-end delay at the lowest levels.

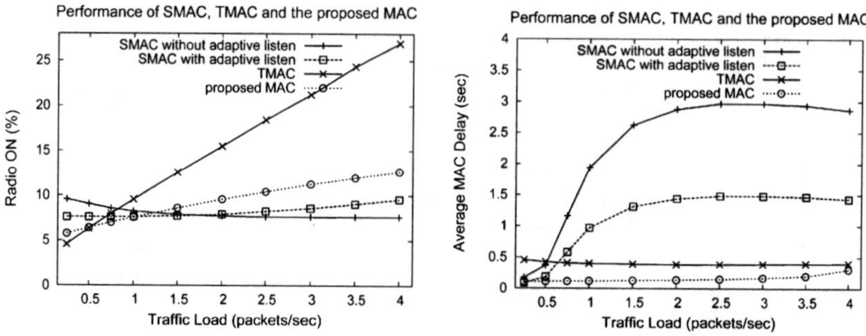

Figure 6. a) Radio-ON time and b) Average MAC delay over traffic load variations

The above simulation results in all testing scenarios clearly indicate the superior performance of our proposal in both energy efficiency and average MAC delay. The examination of its traffic-independent characteristic under more complex network topologies and traffic scenarios remain for further studies.

5. CONCLUSIONS

In this study, a new MAC algorithm based on the IEEE 802.15.4 physical layer is presented that enhances energy conservation and reduces delay in wireless sensor networks. The algorithm is capable of using relatively small period values by decreasing the actions that a node needs to perform at the beginning of every communication period and by incorporating a variable radio-on operation. Moreover, the introduced differences in nodes' scheduling positively acted upon reducing the delay. The obtained benefits from our proposal are the maintenance of low radio-on time and low mean delay in a wide range of traffic variations. This stable network performance, in terms of the above metrics, makes feasible the efficient implementation of our algorithm in a wide range of applications.

6. ACKNOWLEDGMENTS

This work was performed within the framework of the project PENED 2003 (Grant No. 636), funded by the European Union-European Social Fund, and from the GSRT of the Hellenic Ministry of Development.

REFERENCES

1. I. F. Akyildiz, W. Su, Y. Sankarasubramaniam, and E. Cayirci, "Wireless sensor networks: a survey," *Computer Networks*, vol. 38, no. 4, pp. 393–422, Mar. 2002.
2. F. Bennett, D. Clarke, J. B. Evans, A. Hopper, A. Jones, and D. Leask, "Piconet: Embedded mobile networking," *IEEE Personal Communications Magazine*, vol. 4, no. 5, pp. 8–15, Oct. 1997.
3. W. Ye, J. Heidemann, and D. Estrin, "An energy-efficient mac protocol for wireless sensor networks," in *Proceedings of the 21st International Annual Joint Conference of the IEEE Computer*

and Communications Societies (INFOCOM 2002), vol. 3, New York, NY, USA, June 2002, pp. 1567–1576.
4. W. Ye, J. Heidemann, and D. Estrin, "Medium access control with coordinated, adaptive sleeping for wireless sensor networks," *IEEE/ACM Transactions on Networking*, vol. 12, no. 3, pp. 493–506, June 2004.
5. T. V. Dam and K. Langendoen, "An adaptive energy-efficient MAC protocol for wireless sensor networks," in *The 1st ACM Conference on Embedded Networked Sensor Systems (Sensys03)*, Los Angeles, USA, Nov. 2003, pp. 171–180.
6. G. Lu, B. Krishnamachari, and C. S. Raghavendra, "An adaptive energy-efficient and low-latency MAC for data gathering in wireless sensor networks," in *Proceedings of 18th International Parallel and Distributed Processing Symposium*, Apr. 2004.
7. A. Sridharan and B. Krishnamachari, "Max-min fair collision-free scheduling for wireless sensor networks," in *IEEE International Conference on Performance, Computing, and Communications*, 2004, pp. 585–590.
8. "Chipcon AS SmartRF© CC2420, preliminary datasheet (rev. 1.2)," June 2004.
9. "IEEE std 802.15.4™-2003: Wireless Medium Access Control (MAC) and Physical Layer (PHY) Specifications for Low-Rate Wireless Personal Area Networks."

CHAPTER 5

A FAST AND EFFICIENT SOURCE AUTHENTICATION SOLUTION FOR BROADCASTING IN WIRELESS SENSOR NETWORKS

TAOJUN WU, YI CUI, BRANO KUSY, AKOS LEDECZI, JANOS SALLAI, NATHAN SKIRVIN, JAN WERNER, AND YUAN XUE[1]

Institute for Software Integrated Systems (ISIS) and EECS, Vanderbilt University
Email: {taojun.wu, yuan.xue, nathanael.skirvin, jan.werner, branislav.kusy, janos.sallai, akos.ledeczi}@vanderbilt.edu

Abstract: Wireless sensor network has drawn increasing attentions in recent years due to its wide range of applications. Often deployed in hostile environments, wireless sensor network is particularly vulnerable to malicious attacks. Thus security becomes a critical issue. This paper studies the security support for source authentication for broadcasting in wireless sensor networks. Our problem is motivated by a real sensor network application scenario – Dirty Bomb Detection and Localization, which requires efficient broadcast source authentication service in real-time. Although there exist broadcast source authentication solutions developed for wireless sensor networks, they either require significate latency in key release from a one-way hash key chain, or need a large memory space and/or involve high communication overhead. None of these solutions could meet the strict requirements from real-time communication and the limited memory space in our application

To address this issue, we present a broadcast source authentication mechanism based on multiple message authentication codes (MultiMAC). The novel contribution of this work is that it proposes a deterministic combinatorial key distribution scheme that provides scalable authentication service with limited key storage need. This authentication service is implemented as a security component in TinyOS as part of the Dirty Bomb Detection and Localization application, where its performance is validated

[1]This work was supported in part by TRUST (The Team for Research in Ubiquitous Secure Technology), which receives support from the National Science Foundation (NSF award number CCF-0424422) and the following organizations: Cisco, ESCHER, HP, IBM, Intel, Microsoft, ORNL, Pirelli, Qualcomm, Sun, Symantec, Telecom Italia and United Technologies.

H. Labiod and M. Badra (eds.), New Technologies, Mobility and Security, 53–63.
© 2007 *Springer.*

1. INTRODUCTION

The convergence of micro-eletro-mechanical system technology, wireless communication and digital electronics leads to the emergence of wireless sensor networks [1], which are capable of sensing, data processing, and communicating. Sensor networks can be readily deployed in diverse environments to collect and process useful information in an autonomous manner. Thus, they have a wide range of applications in the areas of health care, military, and disaster detection.

Often deployed in hostile environments, wireless sensor networks are vulnerable to a variety of attacks [2,3]. Thus security becomes a critical issue to ensure safe operation of wireless sensor network. Existing research has provided a variety of security supports (e.g., protection of data confidentiality and integrity) for wireless sensor networks. Representative works include link layer security architecture [4], secure routing [2,3], and key management and distribution mechanisms [5,6].

In this paper, we study the problem of source authentication for broadcast[2] traffic in wireless sensor networks. Our problem is motivated by the need from a real sensor network application scenario – Dirty Bomb Detection and Localization [7]. In this application, the master sensor, carried by a moving policeman, will broadcast the localization command to the rest of the sensors via multi-hop wireless communication so that they could start the synchronization and localization operation simultaneously. In this application, every round of localization operation, including communication delay, needs to finish in less than 3 seconds. This application also requires each receiver sensor node to be able to authenticate the broadcast message (i.e., localization command) from the source (the master sensor). Acceptance of false command will trigger unnecessary synchronization and localization operations, which waste scarce battery energy and cause operation confusion among sensors in the worst case.

There exist two catagories of solutions for broadcast source authentication: asymmetric-key-based and symmetric-key-based mechanisms. In wired network, asymmetric-key-based solution is a popular approach due to its convenience. However, its high computation overhead becomes a huge obstacle for its application in wireless sensor networks. The work of [8,9] have studied the fitness of the most popular asymmetric key algorithms including RSA and ECC in wireless sensor networks. Through advanced implementation technologies, their study shows that the cryptographic operations of ECC algorithm are viable when the computational power and space of the sensor node are dedicated to security operations. However, such an asymmetric-key-based solution could not be applied to our application scenario, as most storage space and computation power of the sensors are used for the detection and localization functions, leaving little space for security functions.

---

[2]In this paper, broadcast refers to the flooding messages that may be forwarded through multiple hops, instead of the local broadcast message.

In $\mu TESLA$ [10], the authors present a symmetric-key-based broadcast source authentication scheme. This scheme is further extended in BABRA [11]. The basic idea of these two schemes is to create an *asymmetry* in time among the broadcasting source and the receivers through the delayed disclosure of key. In particular, the source node will generate a one-way key chain by applying a hash function iteratively from the last key. The keys will be applied to generate message authentication codes (MAC) sequentially, but will be released with certain delay after the packets are received. As a consequence, the receivers are unable to authenticate the messages they receive immediately. This approach introduces time latency in authentication, thus is not applicable in real-time systems.

Pair-wise keys and their establishment have received extensive research [12–16 6,17]. Once established, pair-wise key could also be used to provide broadcast authentication service either through generated session key or unique key-path. Yet these two approaches will either involve high delay and computational overhead in repeatedly generating and verifying MAC, or high communication overhead in generation of session keys that prevents it to scale to large networks.

To address the above issues and meet the real-time and efficient authentication need from our application, we propose a novel broadcast source authentication mechanism based on multiple message authentication codes (MultiMAC). This mechanism requires the sensor nodes to have different yet overlapping set of keys (so-called key ring). To authenticate a message, the source node will generate a list of MACs based on its keys, and append them to the message. The receiver node will verify the message based on the MACs which are generated using the keys that are shared with the source. To support such an authentication service over the sensors with limited storage space, the key ring has to be designed to satisfy a set of conditions. We formulate this problem as a combinatorial problem and present a deterministic combinatorial key distribution scheme that supports scalable authentication service for wireless sensor networks.

The major contributions of this work are summarized as follows. Theoretically, it integrates a hierarchical key structure with the deterministic combinatorial key distribution schemes to support large-scale sensor network broadcast authentication services. Practically, it implements the proposed authentication solution as a security module in TinyOS as part of the Dirty Bomb Detection and Localization system and validates its performance through the integrated system experiment. It is also worth noting that although our MultiMAC scheme is designed for the Dirty Bomb Detection and Localization system, it could be applied to general broadcast scenarios in wireless sensor networks.

The rest of the paper is organized as follows: Section 2 explains our symmetric-key-based broadcast source authentication approach. A key pre-distribution scheme is next presented in Section 3 that scales to large sensor networks in a deterministic way. The detailed protocol implementation of our approach as well as field measurement results are included in Section 4. Section 5 concludes the paper and points out some future work.

2. SYMMETRIC-KEY-BASED BROADCAST SOURCE AUTHENTICATION MODEL

We investigate the source authentication problem for broadcast in wireless sensor networks based on symmetric key mechanisms. Here are two benchmark scenarios of broadcast: one scenario involves only a single sender; the other scenario allows every node to be the source of broadcast. Our research focuses on the second scenario while the first scenario could be straightforwardly addressed based on our approach.

Inspired by the work of [18,19], the basic idea of broadcast source authentication in wireless sensor networks works as follows. Each sensor node in the network $i \in N$ has a *different* set of keys S_i called *key ring*. To authenticate message M, the sender node i will generate a *message authentication code* $MAC(K_i^j, M)$ using each key in its key ring $K_i^j \in S_i$. The full collection of MACs $MAC(K_i^1, M)||MAC(K_i^2, M)||\ldots||MAC(K_i^l, M)$ will be transmitted together with the message M. Each recipient node $r \in N$ verifies all the MACs that are created using the common keys which it shares with the sender, *i.e.*, the keys in set $S_i \cap S_r$ where S_r is the key ring of r. If any of these MACs is incorrect, then r rejects the message. Fig. 1 illustrates a simple example of this idea, where sensors A, B, C, D have different combinations of overlapping keys, and sensor A tries to imposter C. When sensor A sends out a message M, it attaches $MAC(1, M)||MAC(4, M)$ to it. The other recipients of the message then verify MACs generated with any common keys they are sharing with the sender. However, this key pre-distribution does not assure sufficient authentication and we would discuss later the conditions necessary to achieve authentication with shared keys.

In comparison with the existing solutions for wireless sensor network broadcast authentication such as $\mu TESLA$ [10] and [11], our approach offers latency-free

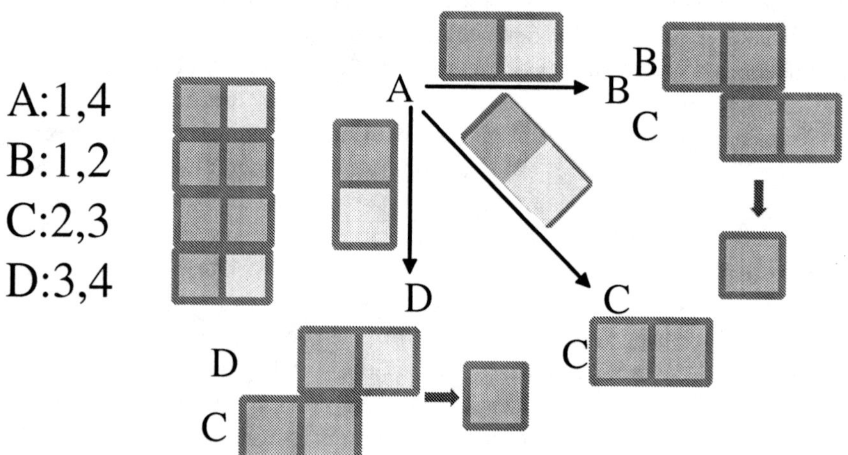

Figure 1. Simple example of shared key authentication

authentication – the recipient nodes are able to verify the message immediately without waiting for key disclosure. Further without using delayed key disclosure, this mechanism does not require any clock synchronization among sensor nodes, which itself is a hard problem [20] and can be susceptible to attacks.

The major challenge to apply the presented mechanism to broadcast authentication is the key pre-distribution. There are two major issues involved: (1) whether the key assignment scheme supports authentication of any sensor node as source of broadcast; (2) whether the scheme is scalable to larger systems. These issues become even more challenging when the application system enforces a strict extra delay bound. Our solution to this problem is one of the major contribution of this paper. Next in this section we will present some conditions on key distribution.

Generally speaking, the key grouping scheme for any network needs to fulfill the following **Baseline Grouping** conditions.

1. *Non-empty subset condition*: any sensor N_i has at least one key, i.e. $\forall S_i, |S_i| = k_i \geq 1$.
2. *Unique subset condition*: no two sensors have the same subset of keys, i.e., $\forall i, \forall j, j \neq i$ implies $S_i \neq S_j$. Otherwise, the i-th sensor can easily imposter the j-th node.

Furthermore, the grouping of sensor nodes in a network needs to satisfy **Authentication Feasibility** conditions below in order for the grouping scheme to provide multi-hop broadcast source authentication.

1. *Subset overlapping condition*: any subset of keys S_i assigned to sensor N_i overlaps with at least one other subset of keys S_j of some sensor N_j, i.e., $\forall i, \exists j \neq i,\ s.t.\ |S_i \cap S_j| = d_{ij} \geq 1 (S_i \cap S_j \neq \Phi)$. The *subset overlapping condition* ensures any sensor can be authenticated by at least one other sensor. In other words, for any possible identity spoofing, there will exist some sensor in the system to detect the activity.
2. *Overlapping distinction condition*: the sets of overlapping keys of any subset of keys S_i with any two other key subsets S_j, S_k are distinguishable, i.e., $\forall i, \forall j, \forall k \neq j, S_i \cap S_j \not\subseteq S_i \cap S_k$.

In order for this group-based authentication scheme to work, the grouping parameters of sensors mentioned above need to satisfy the following constraints:

1. *Key number bound condition*: the number of keys in any subset is bounded by total number of keys in key pool. $d_{ij} \leq k_i, k_j \leq m$;

Table 1. Notations used in Sec. 2

m	number of distinct keys
n	number of wireless sensors
N_i	the i-th sensor
S_i	the i-th subset of keys, assigned to N_i
Key_j	the j-th key
$Group_j$	the j-th group of sensors, sharing the j-th secret key
k_i	$\|S_i\|$, number of keys in S_i
d_{ij}	$\|S_i \cap S_j\|$, number of common keys between S_i and S_j

2. *Sensor number condition*: the number of obtained different key subsets needs to exceed the number of sensors. In many occasions, it is convenient to have equal sized subsets, i.e. $\forall i, k_i = k^*$. In this case, a bottom-line version of this constraint becomes $\binom{m}{k^*} \geq n$, which ensures the number of possible combinations of k^* keys larger than sensor number n.

Additionally, some similar works have assumed stronger constraints to achieve their goals of establishing single secret key for any pair of sensors. Some typical constraints are:

1. *Equal-size overlapping condition*: any sensor N_k has equal sized overlapping set of keys with all other sensors. $\forall i, \forall j \neq i, d_{ij} = d^*$;
2. *Unique overlapping condition*: any sensor N_k has unique overlapping set of keys with all other sensors. This can be expressed as: $\forall i, j, \nexists k,\ s.t.\ S_i \cap S_k = S_k \cap S_j$. An even stronger version of this condition requires no two overlapping combination of keys are the same, in other words: $\forall i, j, k, l,\ s.t.\ S_i \cap S_j \neq S_k \cap S_l$;

3. KEY PRE-DISTRIBUTION SCHEME

In this section we present the key pre-distribution scheme for our symmetric-key-based broadcast source authentication. Our solution starts with a base scheme and it is afterwards extended in a hierarchical way to scale up to larger sensor networks.

3.1. Base Key Distribution Scheme

Our key pre-distribution scheme takes the advantage of existing research in combinatorics. Generally speaking, a working key distribution scheme needs to satisfy **Authentication Feasibility** conditions listed in Sec. 2. These define a combinatorial problem. According to Colbourn et al. [21], schemes satisfying these conditions are known for most $n \leq 50$.

The complexity in finding satisfying schemes grows rapidly [21] when n increases. To support large-scale sensor networks, we choose following grouping scheme ($n = 7$) as our *BaseScheme* and extend it with a hierarchical structure.
$BaseScheme = \{(1,2,4,7), (1,2,5,6), (1,3,4,6), (1,3,5,7), (2,3,4,5), (2,3,6,7), (4,5,6,7)\}$.

3.2. Extension of Base Scheme

For ease of presentation, before explaining how to extend the base scheme, we denote $KeyScheme = \{a, b, c, d, e, f, g\}$, where $a, b, \ldots, g$ are called $KeyBundle$s which are key combinations following *BaseScheme*. For example, if we have a set of keys $KeySet = \{Key_1, Key_2, \ldots, Key_7\}$, we can map element a to the first $KeyBundle_1 = (Key_1, Key_2, Key_4, Key_7)$. Therefore, if we have two distinct key sets (KS_1, KS_2), and $KS_1 \cap KS_2 = \Phi$, we can have two different key schemes $KeyScheme_1 = \{a, b, \ldots, g\}$ and $KeyScheme_2 = \{A, B, \ldots, G\}$.

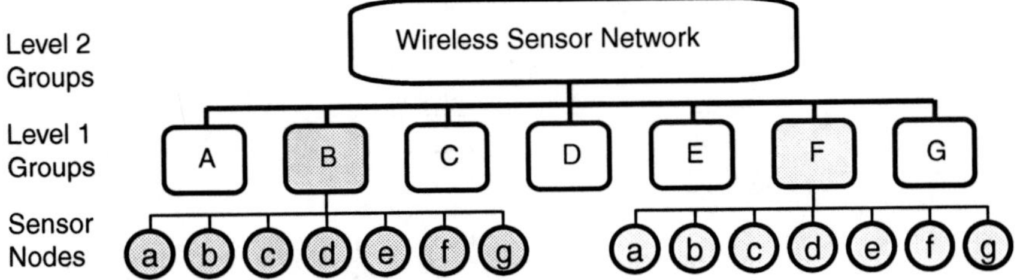

Figure 2. Illustration of extending base scheme to support $n = 49$

First of all, the n sensors of the network is divided into $GroupNum$ groups of size 7. Within each of these groups, the nodes are assigned one of the seven key bundles (a combination of keys chosen from seven keys in the key pool) out of $BaseScheme$. In this way, any nodes within the same group can authenticate mutually. However, to enable broadcast authentication among nodes from different groups, more efforts are needed. In the next stage, seven new keys are introduced. All the nodes in the same group are treated as one single parent node, and the series of groups generated in the first step is regarded as a higher level network consisted of parent nodes. These parent nodes are again separated into parent groups, each containing 7 parent nodes, or 49 sensor nodes. In a similar way, key bundles organized according to $BaseScheme$, consisting of new keys introduced during this stage, are assigned to parent nodes within these parent groups. Consequently, all sensor nodes of the same parent group now have four additional keys, while maintaining their four old keys obtained during stage one. The previous two steps would recursively repeat until $GroupNum \leq 7$.

The Exclusion Basic System (EBS) presented in [22] also provides a key management scheme for multicast in wireless sensor networks. The major difference of our combinatorial deterministic approach and EBS is that our scheme satisfies the **Authentication Feasibility** condition. This eliminates the need of establishing session keys before actual authentication for EBS case. The relatively small and predictable size of key ring also distinguishes our scheme. For example, when $n = 49$, our scheme only needs 8 keys per sensor ($4log_7 n$).

4. PROTOCOL IMPLEMENTATION AND PERFORMANCE EVALUATION

We implement and evaluate our protocol as a nesC component (MultiMAC) under TinyOS 1.1.15 in the real system "Dirty Bomb Detection and Localization" [7]. The system tracks a radiation detector as it traverses an area where a network of sensor nodes (Mica2 motes) has been deployed.

The key grouping scheme we use is illustrated in Table 2, following discussions of Sec. 2. For compatible reason and to save storage space, we use the SkipJack implementation provided in TinySec [4] as our symmetric cipher to compute and verify

Table 2. Key distribution scheme used for implementation

Node #	Assigned Key Ring
0	[1, 2, 4, 7, 8]
1	[1, 2, 5, 6, 8]
2	[1, 3, 4, 6, 8]
3	[1, 3, 5, 7, 8]
4	[2, 3, 4, 5, 8]
5	[2, 3, 6, 7, 8]
6	[4, 5, 6, 7, 8]
7	[1, 2, 4, 7, 9]
8	[1, 2, 5, 6, 9]
9	[1, 3, 4, 6, 9]
10	[1, 3, 5, 7, 9]
11	[2, 3, 4, 5, 9]
12	[2, 3, 6, 7, 9]
13	[4, 5, 6, 7, 9]

MACs. As shown in Table 2, every sensor node in the system stores a different key ring in its ROM. Multiple MACs of every outgoing message are calculated, using the key ring assigned to the sending mote. The receiving mote authenticates the message by recomputing MACs using its shared keys with the sender. Our protocol fits into contingent real-time constraints of the system and is discussed in detail next.

4.1. Protocol Procedure

Our broadcast authentication scheme operates in following steps:

1. The key ring for every sensor N_i is initialized to one out of n subset of keys, S_i;
2. A key mapping function (or structure) exists in every sensor N_i so that the detailed key grouping is available locally and N_i can get to know subset of keys S_j for any sensor N_j;
3. When a sensor N_i sends out messages, it appends multiple MACs to the message. Each MAC is computed with a key in S_i, hence there are k^* MACs in total;
4. When a sensor N_j receives a message from N_i, it checks to find its common keys with N_i and then verifies if the corresponding MACs are correct. N_j will "reject" the message if the provided multiple MACs contain any wrong MAC. Otherwise, it will "accept" it;

4.2. Reducing Length of Multiple MACs

One obvious problem with our protocol design is that our scheme requires k^* times as many bytes for single message authentication. The *Bloom filter* structure can serve as a possible solution to this problem. In [23], Ye et al. proposed to use several hash functions to transform a bunch of MACs from various sensors to a combinated Bloom filter of equal size. This bit string then serves as one MAC.

Similarly, we transform individual MAC to a smaller space and obtain a shorter MAC. In this way, each key will still have a designated portion in the final MACs, although weakened.

4.3. Broadcast Authentication Overhead

We will now present the measurement results from the system. Our MultiMAC component requires a reasonable amount of memory storage. Table 3 shows that the storage requirements scale gracefully with larger key rings, as larger key rings simply use more memory to store additional keys. The memory requirements imposed by MultiMAC are acceptable for the "Dirty Bomb Detection and Localization" system.

We next discuss overhead of computating single MAC of different length. Table 4 shows the time (in milliseconds) required to compute a single MAC for various MAC lengths (in bytes). As the result shows, generating MACs of different lengths does not affect processing time much.

As a receiving sensor can share varying number of keys with the sending mote, we finally measure the time (in milliseconds) of verifying different number of MACs for a single message. This is shown in Table 5. In this table, the number of MACs represents the number of shared keys between the broadcasting source node and the receiving node. The verification time is almost linear to the number of MACs needing verification, as expected.

Table 3. Memory usage

# of Keys	ROM (bytes)	RAM (bytes)
2	2778	105
3	2814	153
4	2824	201
8	2894	393
16	3022	777
32	3278	1545

Table 4. Single MAC processing time

MAC Length(bytes)	Compute Time(ms)
2	3.42
3	3.45
4	3.51

Table 5. Multiple MACs processing time

Number of MACs	Verify Time (ms)
0	<0.1
1	1.3
2	2.5
3	3.7

5. CONCLUSIONS AND FUTURE WORK

In this paper we have presented a fast and efficient broadcast source authentication protocol in a real-time system: "Dirty Bomb Detection and Localization". We address scalability issue of existing deterministic approaches by applying hierarchical structure to combinatorial results. The resulting key pre-distribution scheme ensures efficient authentication and the measured results confirm us of the efficiency and low overhead of our approach. Our nesC implementation of MultiMAC for wireless sensor networks under TinyOS fulfils its security demands and satisfies the contingent time constraints.

For future work we are interested in adding re-keying and key revocation mechanisms. These are essential for the system to operate properly if we want to add new sensor nodes and remove faulty ones after deploying it. Solution to this problem might require adding extra information to current key mapping function mentioned in second step of protocol procedure (Sec. 4).

REFERENCES

1. I. F. Akyildiz, W. Su, Y. Sankarasubramaniam and E. Cyirci, "Wireless Sensor Networks: A Survey," *Computer Networks*, vol. 38, no. 4, pp. 393–422, 2002.
2. C. Karlof and D. Wagner, "Secure routing in wireless sensor networks: Attacks and countermeasures," *Elsevier's AdHoc Networks Journal, Special Issue on Sensor Network Applications and Protocols*, vol. 1, no. 2–3, pp. 293–315, September 2003.
3. Y. C. Hu, A. Perrig, and D. B. Johnson, "Packet leashes: a defense against wormhole attacks in wireless networks," in *INFOCOM 2003. Twenty-Second Annual Joint Conference of the IEEE Computer and Communications Societies. IEEE*, vol. 3, 2003, pp. 1976–1986 vol.3. [Online]. Available: http://ieeexplore.ieee.org/xpls/abs_all.jsp?arnumber=1209219
4. C. Karlof, N. Sastry, and D. Wagner, "Tinysec: a link layer security architecture for wireless sensor networks," in *SenSys '04: Proceedings of the 2nd international conference on Embedded networked sensor systems*. New York, NY, USA: ACM Press, 2004, pp. 162–175.
5. W. Du, J. Deng, Y. S. Han, and P. K. Varshney, "A pairwise key pre-distribution scheme for wireless sensor networks," in *CCS '03: Proceedings of the 10th ACM conference on Computer and communications security*. New York, NY, USA: ACM Press, 2003, pp. 42–51.
6. H. Chan, A. Perrig, and D. Song, "Random key predistribution schemes for sensor networks," in *SP '03: Proceedings of the 2003 IEEE Symposium on Security and Privacy*. Washington, DC, USA: IEEE Computer Society, 2003, p. 197.
7. "Dirty bomb detection and localization," http://www.isis.vanderbilt.edu/Projects/rips/.
8. D. J. Malan, M. Welsh, and M. D. Smith, "A public-key infrastructure for key distribution in tinyos based on elliptic curve cryptography," in *SECON 2004. First IEEE International Conference on Sensor and Ad Hoc Communications and Networks*, October 2004.

9. N. Gura, A. Patel, A. Wander, H. Eberle, and S. Shantz, "Comparing elliptic curve cryptography and rsa on 8-bit cpus," in *CHES's 2004 Workshop on Cryptographic Hardware and Embedded Systems*, Aug 2004, pp. 119–132.
10. A. Perrig, R. Szewczyk, J. D. Tygar, V. Wen, and D. E. Culler, "Spins: security protocols for sensor networks," *Wirel. Netw.*, vol. 8, no. 5, pp. 521–534, 2002.
11. Y. Zhou and Y. Fang, "Babra: Batch-based broadcast authentication in wireless sensor networks," in *Proc. of IEEE GLOBECOM*, Nov 2006.
12. R. Blom, "An optimal class of symmetric key generation systems," in *Proc. of the EUROCRYPT 84 workshop on Advances in cryptology: theory and application of cryptographic techniques.* New York, NY, USA: Springer-Verlag New York, Inc., 1985, pp. 335–338.
13. L. Zhou, J. Ni, and C. V. Ravishankar, "Efficient key establishment for group-based wireless sensor deployments," in *WiSe '05: Proceedings of the 4th ACM workshop on Wireless security*, 2005.
14. J. Lee and D. R. Stinson, "Deterministic key predistribution schemes for distributed sensor networks," in *Selected Areas in Cryptography*, 2004, pp. 294–307.
15. D. Liu and P. Ning, "Location-based pairwise key establishments for static sensor networks," in *SASN '03: Proceedings of the 1st ACM workshop on Security of ad hoc and sensor networks.* New York, NY, USA: ACM Press, 2003, pp. 72–82.
16. L. Eschenauer and V. D. Gligor, "A key-management scheme for distributed sensor networks," in *CCS '02: Proceedings of the 9th ACM conference on Computer and communications security.* New York, NY, USA: ACM Press, 2002, pp. 41–47.
17. W. Du, J. Deng, Y. S. Han, S. Chen, and P. K. Varshney, "A key management scheme for wireless sensor networks using deployment knowledge," in *INFOCOM 2004. Twenty-third AnnualJoint Conference of the IEEE Computer and Communications Societies*, vol. 1, 2004.
18. R. Canetti, J. Garay, G. Itkis, D. Micciancio, M. Naor, and B. Pinkas, "Multicast security: A taxonomy and some efficient constructions," in *Proc. IEEE INFOCOM'99*, vol. 2. New York, NY: IEEE, Mar. 1999, pp. 708–716.
19. P. Rohatgi, "A compact and fast hybrid signature scheme for multicast packet authentication," in *CCS '99: Proceedings of the 6th ACM conference on Computer and communications security.* New York, NY, USA: ACM Press, 1999, pp. 93–100.
20. M. Manzo, T. Roosta, and S. Sastry, "Time synchronization attacks in sensor networks," in *SASN '05: Proceedings of the 3rd ACM workshop on Security of ad hoc and sensor networks.* New York, NY, USA: ACM Press, 2005, pp. 107–116.
21. C. J. Colbourn and J. H. Dinitz, "Graphical designs," in *The CRC Handbook of Combinatorial Designs.* Boca Raton: CRC Press, 1996, pp. 367–369.
22. L. Morales, I. H. Sudborough, M. Eltoweissy, and M. H. Heydari, "Combinatorial optimization of multicast key management," in *HICSS '03: Proceedings of the 36th Annual Hawaii International Conference on System Sciences (HICSS'03) - Track 9.* Washington, DC, USA: IEEE Computer Society, 2003, p. 332.2.
23. F. Ye, H. Luo, S. Lu, and L. Zhang, "Statistical en-route filtering of injected false data in sensor networks," in *INFOCOM 2004. Twenty-third Annual Joint Conference of the IEEE Computer and Communications Societies*, vol. 4, 2004, pp. 2446–2457.

CHAPTER 6

TMN MANAGEMENT SYSTEMS USING A GRID BASED AGNOSTIC MIDDLEWARE

P. DONADIO [1], A. CIMMINO [2], A. PAPARELLA[3], AND B. BERDE[4]

[1] *Alcatel Italia - Optical Network Division (OND) - pasquale.donadio@alcatel-lucent.it*

[2] *Alcatel Italia - MS & T Technology Program - antonio.cimmino@alcatel-lucent.it*

[3] *Alcatel Italia - Optical Network Division (OND) - andrea.paparella@alcatel-lucent.it*

[4] *Alcatel Research & Innovation (R&I) - bela.berde@alcatel-lucent.fr*

Abstract: In this paper we propose a new paradigm for Telecommunication Management Network (TMN) based on Service Oriented Architecture. Web-based Grid technologies, indeed, allow for combined flexibility at the level of information structuring and distribution, while leverage existing systems for managing both information and system complexity. On one hand, the Web technology frees human users from complicated management system interfaces, introducing a common and univocal management interface: the browser! On the other hand, the Grid technology enables to create cooperation between sophisticated management systems in order to manage their complexity. The improvement to TMN Management System proposed is a Grid based framework that transforms a layered TMN architecture into a simply manageable "flat" network. The framework provides a generalized TMN interface, based on Web Services that integrates protocol-dependent gateway systems. By using this interface, the TMN management application can be easily constructed to provide complex TMN management services to user via user-friendly Web browsers. On the basis of this framework, we have designed and implemented a prototype of TMN Manager system that validates our framework

Keywords: Telecommunication Network Management, Grid Computing, Web based Management, Service Oriented Architectures and Web Services

1. INTRODUCTION

Recent growth in TMN networking technologies has created much and more complex and heterogeneous networking environments. To manage network devices, partitions, and services in such environments, various organizations have developed standard managements platforms based on management protocols (e.g., SNMP [1], CMIP [2], TL1 [3]). This increasingly poses the requirement on integrating

H. Labiod and M. Badra (eds.), New Technologies, Mobility and Security, 65–73.

© 2007 *Springer.*

many different network management schemes; research efforts have concentrated on how to harmonize legacy systems using new technologies [4] [5]. The widespread Internet is the living environment of Grid computing technologies. A basic premise of Open Grid Services Architecture (OGSA) is that everything is represented as a service, which is a network-enabled entity that provides capability to users. In addition there is a network paradigm shift dictated by the need of rapid and autonomic service creation, deployment, activation, and management combined with context customization and customer personalization. Such a motivation can be traced in different organizations and research activities as well as market forces. Since Grid provides an infrastructure for the interoperability of legacy applications in distributed environments, research people have started using Grid concepts to integrate and enhance network management systems.

In this paper we propose a new paradigm for Telecommunication Management Network (TMN) [6] [7] based on Service Oriented Architecture (SOA), using Web Services and Grid technologies. The tool used to add programmability to Grid-enabled network management is extension to the widely used Grid supporting tool, Globus, which is also a powerful supporting tool for (OGSA) [8].

The organization of the paper is as follows. In section 2 we position and analyze the benefits of the Grid Computing for TMN, while focusing on differences with CORBA with reference to related works on Grid technology. In section 3 we propose a Grid based framework that transforms a layered TMN architecture into a simply manageable "flattened" network. The framework provides a generalized TMN interface, based on Web Services that integrates protocol-dependent gateway systems. Section 4 presents the Globus agnostic middleware that enables the TMN infrastructure on the Grid and Web infrastructure. On the basis of this infrastructure, in Section 5 we show the designed and implemented prototype of TMN Manager system that validates the framework.

2. BACKGROUND AND RELATED WORK

Grid computing for TMN can be considered as the natural evolution of CORBA based TMN architecture and services [9]. A key distinction between CORBA and Grid computing is that CORBA assumes object orientation, while Grid computing does not. In CORBA, every entity is an object and supports mechanisms such as inheritance and polymorphism [10]. In Grid computing, there are similarities to some object concepts, but there isn't a presumption of object-oriented implementation in the architecture. However, the use of a formal definition language (such as Web Services Definition Language (WSDL)) means that interfaces and interactions are just as precisely defined as in CORBA, sharing one of the major software engineering benefits also exhibited by object-oriented design. Another distinction is that TMN Grid architecture is built on a Web Services foundation. TMN CORBA architectures integrate with and interoperate with Web Services. TMN architectures based on CORBA are hierarchical. TMN architectures based Grid computing is "flat". One of the problems with CORBA was that it assumed too much of the "endpoints," which are basically all the machines (clients and servers) participating in a CORBA environment. There are also issues of interoperability between vendors' CORBA implementations, how CORBA nodes are able to interoperate on the Internet, and how endpoints are named. This means that all of the machines in the cohort had to conform to certain rules and to a certain way of doing things for CORBA to work. Grid Computing solves this kind of limitations in a native way, allowing aggregated computing resources to be harnessed to improve TMN performance services and usability.

In last years research on Grid computing for TMN was originated to solve the lacks of CORBA and create a "glue" technology able to manage the large varieties of underlying native TMN platforms. A valid example of this tentative is the Lambda Grid Project [11].

The added value of our research is to propose a complete architecture able to realize a glue technology for TMN applications (legacy or not) and capable to manage the large variety of TMN services simply using a standard web browser.

3. THE ARCHITECTURE: A GRID BASED TMN SYSTEM

In this section, we propose a TMN architecture based on Grid concepts and Web technologies. Figure 1 shows the functional blocks of a Grid aligned TMN architecture. Gateway proxies, proxy coordinator, system management functions, TMN Services, high-level service functions and legacy applications are virtualized and connected to the Grid as depicted in Figure 1. Every module is implemented as a Web Service and can be processed through open Internet standards or modified (if legacy) to support the Simple Object Access Protocol (SOAP) channel for application exchanges. In this way each component attached to the TMN architecture is automatically attached to the other components and reachable via Web using a single browser interface.

At the lowest level, the **Gateway proxies** provide methods for interoperations between the Web Service technology and different network management protocols, such as SNMP, CMIP, and other proprietary protocols. The related protocol dependent information structure and interaction functionality are transformed here to be accessed as Web Services and vice-versa.

The **Proxy coordinator** is a module to introduce modularity and generalizing the different protocol dependent management interfaces in a way the generalized Grid supportive interfaces can be exposed to the upper level.

On top of the proxy coordinator, there are **Systems Management Services** implemented using Web Services. These general TMN systems management functions are thus exposed as callable services. Web Services enables these components to be easily interconnected to form further sophisticated TMN services.

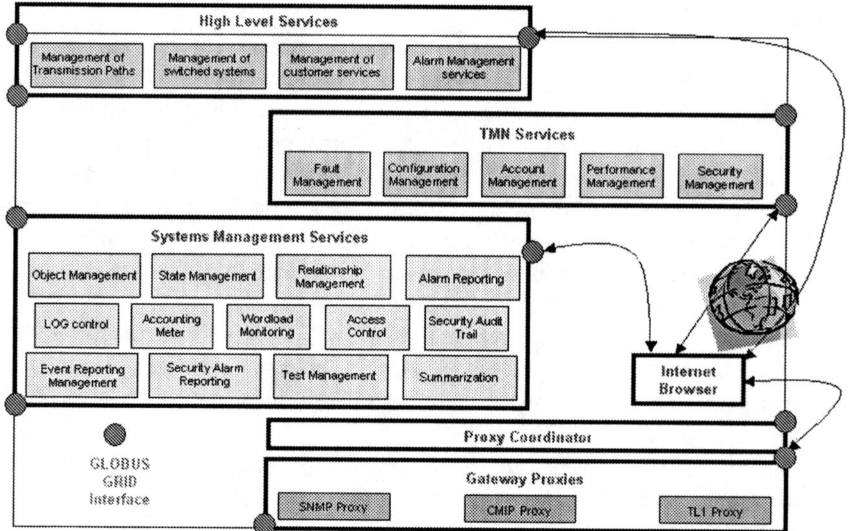

Figure 1. Grid-wise service structuring for TMN system

The **TMN Services** are the five general management functionality in the TMN framework: fault, configuration, account, performance, and security. Each management function can be realized combining available Web Services in a standard way from the systems management services layer.

The **High-level Services** are specific task-oriented management services for end users. Users can define any kind of services based on TMN services. For instance, the management of transmission paths, switched systems, customer services, or alarm surveillance systems can be built here.

Using the layered service architecture in Figure 1, Web Services aligned Grid extensions easily enables software developers to implement distributed business applications for TMN management systems, where Graphical User Interfaces are managed starting from the same application: the browser.

4. THE TECHNOLOGY: GRID COMPUTING USING THE GLOBUS FRAMEWORK

The architecture for Grid Computing is defined in OGSA standard that describes the overall structure and services to be provided in Grid environments [12] Figure 2.

The companion implementation standard, OGSI, is a formal specification of the concepts; it specifies a set of service primitives that define a nucleus of behavior common to all Grid services. OGSI, in effect, is the base infrastructure on which the OGSA is built.

As just noted, OGSA is a distributed interaction and computing architecture that is based around the Grid service concept, assuring interoperability on hetero-

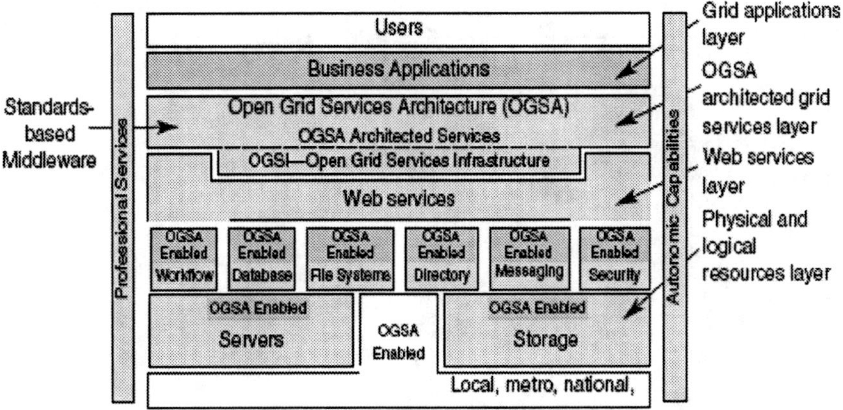

Figure 2. OGSA built on web services basis

geneous systems. As a result, different types of systems can communicate and share information. OGSA allows system composition to perform a specific task, or solve a challenging problem, by using distributed resources over the interconnecting network [13][14].

The related architecture is now being developed based on Internet protocols (for example, communication, routing, file transfer, name resolution, etc.) and services. Building on concepts and technologies from the Grid and Web Services communities, OGSA architecture defines uniform exposed service semantics (the Grid service) and standard mechanisms for creating, naming, and discovering transient Grid service instances; it also provides location transparency and multiple protocol bindings for service instances and supports integration with underlying native platform facilities [8].

Due to its many benefits, OGSA and Globus [15] [16] are adopted in this paper as a guideline for Grid services. This architecture aims to provide mechanisms automatically adapting Grid network elements to different Grid services and the management of the Grid system itself.

5. THE PROTOTYPE: A TMN MANAGER CONNECTED TO THE TMN GRID

The essential open source components include XML User interface markup Language (XUL) [17][18], NetSNMP, Zend, and Globus toolkit. The network management paradigm transformed by Grid technology permits the communication between actors of the TMN network in a simple way (e.g., Network Element - NE, Regional Manager - RM, Broadband Manager - BM and Craft Terminal - CT).

Each legacy system manages its own internal database using the traditional paradigm, and communicates over the Grid using one or more XML links, also

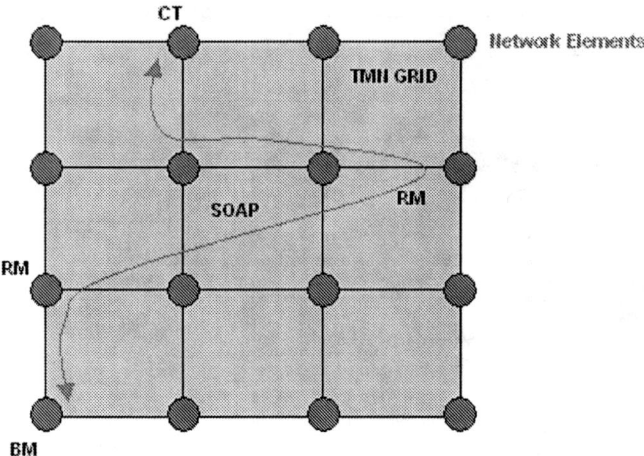

Figure 3. SOAP channels between TMN components

named SOAP channels - Figure 3. Figure 4 shows the use of TMN Grid architecture with the sequence of logical data flow through the Grid system built around managed Network Elements (NE), where marked lines are incoming requests and outlines are replies. In incoming direction browser get-requests (marked lines) are traduced in NE monitoring or management actions, accessing various network elements in the following way:

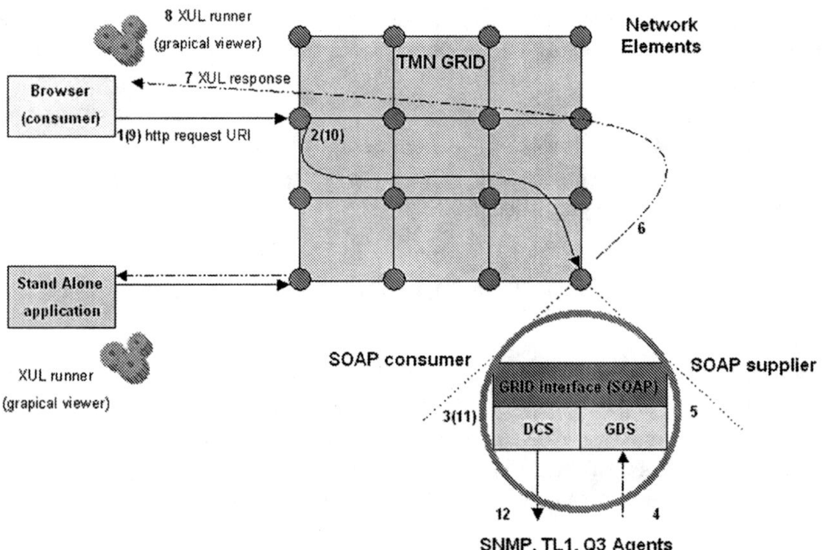

Figure 4. A complete logical data flow of TMN grid

1. The browser invokes a well-defined http Uniform Resource Identifier (URI), containing the address of the NE to be reached and the resources to be managed (e.g., SNMP get-request of an Ethernet interface on an NE with intranet address 151.98.22.22);
2. The first available NE acts as "access point" of the Grid, capturing the URI and translating it into a SOAP request to the target NE. The manager is then connected to the NE in indirect way via Grid middleware (based on Globus framework);
3. The target NE consumes the SOAP packet and translates the operation in a request on the remote DB (e.g., it recovers the SNMP ifTable on the NE and translates the SOAP request in SNMP "get request" on ifTable).

In outgoing direction (outlines), responses from the NE are managed in the following way:

4. The database daemon returns the data to the Globus middleware (e.g., NetSNMP returns the SNMP ifTable data);
5. The Graphical Decorator Service (GDS) encapsulates data in XUL format (e.g., if Table data are encapsulated in XUL frames with a well defined graphical meaning) and supplies the data on the Grid interface;
6. The supplier publishes the XUL data on the Grid formatting it in SOAP responses;
7. The grid "access point" consumes the SOAP responses to build the XUL data responses;
8. Finally the XUL data response is first parsed by XUL runner and then transformed in Graphical User Interfaces populating the browser.

In incoming direction browser set-requests are traduced in NE monitoring/management actions, accessing the Network Elements in the following way:

9. The browser invokes a well-defined URI, containing the address of the Network Element to be reached and the resources to be managed (e.g., SNMP set on MTU value of Ethernet Interface on NE with intranet address 151.98.22.22);
10. The first available NE acts as "access point" of the Grid capturing the URI and translating it in a SOAP request to the target NE. The manager is then connected to the Network Element in indirect way via Globus middleware;
11. The target NE consumes the SOAP packet and translates the operation in a request on the remote DB (e.g. it sets the MTU value inside the SNMP ifTable on the NE and traduces the SOAP request in SNMP "set request" on if Table);
12. The Distributed Constraint Service (DCS) first locks the set-operation, in order to verify the coherency of attributes referenced inside the set-request; then unlocks the set-operation and executes it.

An example of manageability of the TMN Grid system is given in Figure 5. Beyond the craft terminal main view for a NE supporting Ethernet, TDM, and

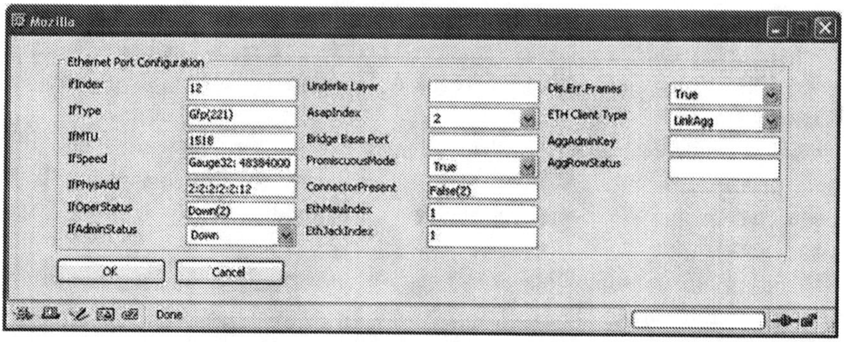

Figure 5. Ethernet port configuration status via a browser

WDM, due to Web Services, the user through a browser can visualize the current status of an Ethernet port configuration.

6. LONG TERM EVOLUTION

Competitions in the telecommunications industry have modified the operational needs of both companies and service providers while driving the technology base of the TMN systems. Since years, global, end-to-end management solutions have never been adopted despite the standardisation effort and innovation in TMN protocols. Probably, in the past years, one of the motifs for complexity was the fact that the resource management is designed and assigned to NEs even though transport resources are mainly inhomogeneous. One of the main arguing by operators, however, in selecting a particular network solution is the efficiency of the management platform and its easy integration in existing operation centres.

The convergence to IP-based networks and beyond, the relevance of optical layer should reverse this last point instead. Therefore, an architecture for TMN, at least for transport and network layers (sub-IP area), based on Grid technology might be one of the key technological enablers towards single global management framework given that it allows not only a common approach in resource management but equally to meet important requirements to the management platform such as plug-in (Grid Nodes), distribution of management function, service oriented and service aware, and adaptability to the evolution of NEs.

7. CONCLUSIONS AND FUTURE WORK

In this paper we have proposed a Grid based TMN architecture by using Web Services and Grid technologies. For validation we have implemented a TMN surveillance system based on the proposed system architecture. Web-based integration has many advantages, especially in constructing distributed systems and simple manageability using a Web interface.

Various research groups around the world are using Grid to realize network applications. The TMN system provides users with user-friendly, easy-to-use, integrated interfaces, and provide developers the reuse of legacy applications with a generalized implementation architecture on top of an abstraction layer hiding all protocol-dependent details. The proposed architecture can be easily extended so that other TMN services can be built within it. In the future we plan to develop the complete TMN architecture in a Service Oriented vision. This implies that each TMN Service could be attached to the TMN Grid in a very simple way, using only one integrated management interface: the browser. The tool, developed by the Alcatel Italia Laboratories and University of Naples Federico II, is now in beta-test beside the Laboratories of the same University.

REFERENCES

1. RFC3416 Version 2 of the Protocol Operations for the Simple Network Management Protocol (SNMP) - R. Presuhn, Ed.
2. CMIP/CMIS are defined in ISO (www.iso.org) documents 9595, 9596 and ITU (www.itu.org) X.700, X.711.
3. TR-NWT-000831, OTGR Section 12:1: Operations Application Messages - Language for Operations Application Messages – 1984 Bellcore (now Telecordia).
4. W.W. Agresti, "New Paradigms for Software Development", IEEE Computer Society Press, 1986.
5. T. Bray, J. Paoli, C. M. Sperberg-McQueen eds. "Extensible Markup Language (XML) 1.0" W3C Recommendation 10 February 1998.
6. www.itu.int/TMN/
7. www.tmn.org/
8. Ian Foster, Carl Kesselman, Jeffrey M. Nick and Steven Tueckel. The Physiology of the Grid: An Open Grid Services Architecture for Distributed Systems Integration. Open Grid Service Infrastructure Working Group (OGSI-WG) of Global Grid Forum. Available at: http://www.ggf.org/ogsiwg/drafts/ogsa_draft2.9_2002-06-22.pdf
9. ITU recommendation Q816: CORBA-based TMN services
10. ITU recommendation X780: TMN guideline for defining CORBA managed objects. It specifies guidelines for defining CORBA-based interfaces to software objects
11. The Lambda Grid Project – University of Amsterdam. www.science.uva.nl/~deλlaat
12. Daniel Minoli - A Network Approach to GRID Computing - Wiley Interscience A John Wiley & Sons,INC.,Publication.
13. "draft-ggf-ghpn-netservices-1.0", http://forge.gridforum.org/projetcs/ghpn-rg, February 2004.
14. V. Sander (Ed.) et al."draft-ggf-ghpn-netissues-3", http://forge.gridforum.org/projects/ghpn-rg, May 2004
15. The Globus Alliance http://www.globus.org/
16. Foster I. and Kesselman C. "Globus: A Toolkit-Based Grid Architecture". Foster, I. and Kesselman, C. (eds.), The Grid: Blueprint for a New Computing Infrastructure. Morgan Kaufmann, 1999. Pages 259–278.
17. XML User interface Language (XUL) http://www.mozilla.org/projects/xul/
18. XUL Tutorial http://xulplanet.com/tutorial/xultu/

CHAPTER 7

DIRECT RF CONVERSION TRANSCEIVERS AS A BASE FOR DESIGNING DYNAMIC SPECTRUM ALLOCATION SYSTEMS

OLEG PANFILOV, ANTONIO TURGEON, RON HICKLING, IGOR ALEXANDROV, AND KELLY McCLELLAN

TechnoConcepts, Inc., 14945 Ventura, Sherman Oaks, CA 91403, USA {panfilov, tony, ronh, igorsa, kpm}@technoconcepts.com

Abstract: A new solution for implementing dynamic spectrum allocation (DSA) principles is described. It is based on using the Technoconcepts' direct conversion from RF to baseband transceiver chips. Such RF/D^{TM} chips convert the received signals into digital form immediately after an antenna making it possible to provide all required signal processing in digital form. It yields systems capable of operating in different dynamically allocatable frequency bands and allowing integration of different types of services including voice and multimedia. These systems may be not only frequency agile and protocol independent, but configuration agnostic as well

Keywords: Direct RF conversion, dynamic spectrum allocation, software radio

1. INTRODUCTION

Farther progress in the spectacular growth of communication services is hindered by the serious limitation of available spectrum. The acute shortages of available frequencies slow adaptation of new wireless applications and services convergence of different mix of voice, data and video. The major culprit in the current spectrum shortage is the nature of frequency allocation. Being static in its origin it restricts efficiency of frequency bands utilization. The necessity to review current state of affairs in spectrum utilization was emphasized in [2], [3] suggesting to rely more on market forces rather than on administrative regulations.

Dynamic channel frequency assignment is proposed as the main mechanism that has to be implemented in the design of physical and link layer design of the seven layers of OSI network model. That idea attracted a lot of attention in finding solution

H. Labiod and M. Badra (eds.), New Technologies, Mobility and Security, 75–88.
© 2007 *Springer*.

to the current spectrum shortage. For example, adaptive, dynamic mechanisms of spectrum reuse based on the smart radio are considered in [4], [5] as a viable solution to the current static frequency allocation. The importance of adaptive principles in DSA including priority classes of different services for individual or multiple operators in the multi -vendor environment is described in [6–12]. Current paper picks up the torch of ideas in [4–12] and makes one additional step further focusing on DSA implementation. Its main focus is on the practical realization of spectrum management principles through utilizing the direct conversion RF/D^{TM} chips. These chips are able to provide a universal network access to a broad range of frequencies covering the major wireless communication protocols including CDMA, GSM, WiMax, WiFi and any new ones that are still on the drawing boards. The paper emphasizes using the dynamically downloadable software to operate in frequency agile environment that is central in dynamic frequency management. A given range of possible scenarios of implementing RF/D^{TM} chips illustrates benefits of frequency agility and protocol independence in solving DSA problems.

As with everything in life, the solution of one problem, particularly dynamic spectrum allocation, brings other problems asking for adequate attention. This paper shows the benefits and challenges of broadband frequency agile chips in light of specifics in operating in broad frequency ranges. That specifics boils down to a necessity of cancelling powerful interferers at the receiver front end while these interferers might be outside of spectrum width of particular desirable signals as it can be seen from Fig. 1 borrowed from [1]. The spectral occupancy measurements for that paper were done in New York City during 2004 republican convention when traffic intensity was substantially above the average level. It can be seen from Fig. 1 that a lot of interference can be picked up by the broadband receiver.

Proper addressing of inband and outband interference will preserve the dynamic range of a receiver front end and avoid the possibility of its operation in the

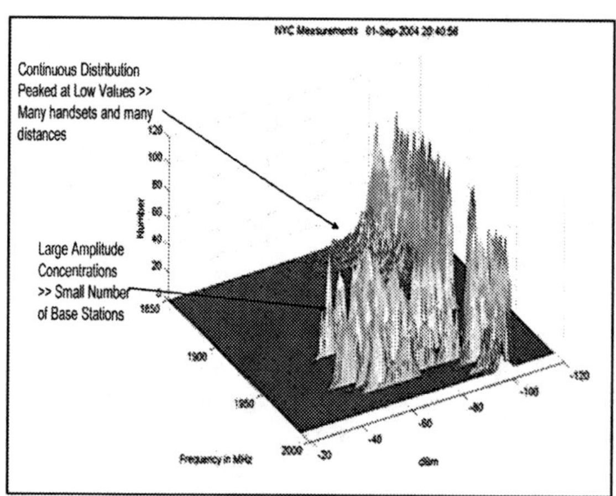

Figure 1. Amplitude histogram of PCS band (courtesy of [1])

nonlinear region with all negative consequences of it. It has to be noted that since interference by its nature is not predictable it has to be cancelled by using adaptive means. A portion of this paper will be devoted to that issue.

Dynamic spectrum allocation can be viewed currently as an area where we can provide a second wind to the industry that is in a strait jacket of static spectrum allocation principles. Such outdated allocation principles result in gross inefficiencies of allocated frequencies on one side and acute shortage of these frequencies on the other. Successful solution of the spectrum availability problem will create highly sought after win-win situation for both parties - customers as well as service providers.

The main body of the paper has the following structure. Section two describes the main DSA challenges. Section three shows the main features and technical realization specifics of the frequency agile receiver. Section four does the same for a transmitter. Section five is devoted to the chip test results. The results will show the level of chips maturity, the areas where it can be utilized right away as well as it will show of what kind of improvements we may expect in the future chip generations. Section six concludes the paper by summarizing obtained results and offering a glimpse to the enhanced features of future chip generations.

2. DYNAMIC SPECTRUM ALLOCATION: HARDWARE CHALLENGE

While the current regulatory regime results in large chunks of unused spectrum ([1], [16]), there are still nevertheless challenges associated with developing a new (and somewhat ad hoc) regime for spectrum re-use. Any dynamic spectrum allocation scheme must be practical from the standpoint of realizability as well as power dissipation overhead (which will be limited inevitably by the capabilities of the mobile device).

While the overall utilization of spectrum is generally low from both a spatial and temporal point of view, the peak to average ratio of the usage of a chunk of spectrum can be very large. Stated another way, most crowded bands are characterized by occasional "bursts" of extremely high usage, which drives a requirement of being able to do extremely quick sensing in order to take advantage of potential transmission opportunities.

Santivanez, et al. [16] focuses primarily on the secondary (and opportunistic) use of licensed spectrum on an "as available" basis. While this paradigm does not apply to all possible dynamic spectrum allocation scenarios, the paper articulates some key concepts that are useful in evaluating hardware technologies with respect to their applicability to spectrum allocation.

In accordance with [16], communications using of dynamic spectrum allocation is decomposed into three sub-problems:

- *Opportunity Awareness* by which communication nodes determine the number and nature of transmission opportunities;
- *Opportunity Allocation*, which is the medium access control in this context; and

- *Opportunity Use*, which is the process of efficiently communicating how the (possibly non-contiguous) channels are to be used.

The opportunity awareness will require quite different mode of operation compared to the opportunity use. In the first case a system has to analyze the operating environment by continuously scanning or sequentially analyzing segments of the entire spectrum width using the efficient automatic gain control system (AGC) to provide operation of receiver in the linear mode. The scaling coefficients of AGC for each analyzed frequency sub-band are used in opportunity allocation procedure in spotting an opportunity. The results of the opportunity allocation can be substantially widened if the adaptive interference cancelling can be used.

The opportunity use mode of operation may utilize the adaptive interference cancellation to improve the signal/(noise + interference) ratio (SNIR) ([13]-[15]) to accommodate substantially larger number of users compared to otherwise.

There is a need for two separate and distinct types of programs necessary to solve the sub-problems above:

- *Spectrum Agility* or the ability to identify transmission opportunities and to use them over a broad range of frequencies.
- *Policy Agility* or the ability to accomplish opportunistic communications in accordance with a variety of frequency use policies or "etiquettes".

In the case of spectrum reuse in a licensed band, it must be assumed that some kind of coordination must take place in order for the first priority (licensed) traffic to operate seamlessly. Yet in the general case (for example using an unlicensed band or a combination of many bands, some of which are licensed and some of which are not) it may be necessary for the transmitting node to observe transmission etiquettes without the aid of coordinating information from other nodes.

Of course, there are many aspects to the above paradigm that must be resolved in order to develop a successful opportunistic communication system. However, many of these (e.g., policy agility) are primarily within the realm of the baseband processing hardware. Consequently, this paper will concentrate on aspects of opportunistic communication that place constraints on, or are limited by the performance of RF hardware. Thus the thrust of this paper will be on the aspects of spectrum agility.

3. RECEIVER REQUIREMENTS

Santivanez, et al. [16] breaks down *opportunity awareness* into three separate functions: sensing, identification, and dissemination. The RF portion of an SDR receiver is involved primarily in sensing and to a smaller extent in identification. Most of the remaining aspects of the opportunistic communication problem are handled by baseband processors at the physical or media access layer of the network.

By far, these functions (as opposed to those corresponding to the transmitter) pose the greatest challenge because of some fundamental physical limitations. The first of these is the tradeoff between frequency resolution and spectrum sampling period.

According to the Heisenberg-Gabor inequality:

$$\Delta t \Delta f \geq \frac{1}{4\pi}. \tag{1}$$

That is to say, the user must trade off the resolution of the spectrum sensed with the length of time over which information about the spectrum is collected. Within limits, one can have either a very accurate spectrogram that takes a relatively long while to construct (a relatively slow update rate) or a quickly constructed (fast update rate) spectrogram that is relatively coarse.

The above equation is, of course, a theoretical limit and does not address issues associated with further "overhead" imposed by the sampling system used to construct the spectrogram. Results obtained by simply "chopping up" the time record into finite increments (a rectangular window) results in spectral anomalies that limit the resolution. Remedies for these anomalies require more complex windows (for example, the Hanning window), which places an overhead on the number of computations required to compute the spectrogram. Finally, the limitations placed by "real" logic circuits adds further overhead to the computation of the spectrogram, resulting in actual realizable update rates that are far slower than those predicted by Heisenberg-Gabor.

For widely frequency agile transceivers some mechanism for frequency sweeping must be implemented resulting in further overhead to the frequency resolution versus sample time tradeoff. If this span is fairly wide (say, 400 MHz to 6 GHz) the time required to construct a kHz resolution spectrogram could be slow enough to be useless in high speed packet-switched environment. Clever methods may be required to conserve computing resources such as a two-pass approach wherein "no opportunity" bands are identified in the first pass and "potential opportunities" are scrutinized in the second pass. In any event, the overall challenge in opportunistic transceiver designs will be minimizing the overhead required in finding opportunities for transmission compared to exploiting these opportunities.

Another major challenge in designing opportunistic networks is the fact that it is impossible to constrain geographies in the same way that they are constrained in, for example, in Ethernet. Although Ethernet is an opportunistic network, its non-segmented wired implementation ensures that all nodes can "hear" all traffic. Thus when interference occurs (or a "collision") all nodes can sense this collision and take corrective action. In contrast, in the general case of networks without coordination, it is conceivable that even without multipath effects a radiating node may not be capable of sensing its potential interference with a competing transmitter. This is apparent by considering the simple one- dimensional problem shown in Fig. 2.

Using the Friis transmission equation we can write

$$\begin{aligned} \frac{P_{r1,2}}{P_t} &= G_t G_{r1} \left(\frac{\lambda}{4\pi R_{1,2}} \right)^2 \quad \text{or} \\ \frac{P_{r2}}{P_{r1}} &= \frac{G_{r2}}{G_{r1}} \left(\frac{R_1}{R_2} \right)^2 \end{aligned} \tag{2}$$

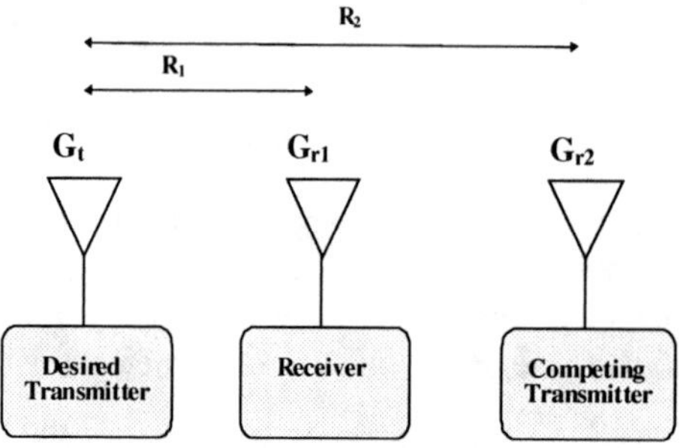

Figure 2. Simple one-dimensional problem demonstrating the challenge of "listening" for competing traffic

where $P_{r1,2}$ is the received power at the 1st or 2nd receiver; P_t – transmit power, λ- wavelength.

The above equation (2) clearly shows that if either the gain of the sensing antenna on the competing transceiver G_{r2} is significantly less than that of the receiver G_{r1} or the distance R_2 is much greater than R_1, the sensed signal at the competing transmitter will be significantly smaller than that at the receiver. This could lead to an erroneous assessment of an available opportunity particularly if the competing signal is small enough to fail at being sensed. Naturally, the presence of obstacles that would cause multipath effects could further exacerbate this effect. This is particularly true when the competing transmitter is a secondary user and is required not to interfere with the "desired transmitter", a primary user.

The foregoing discussion suggests some traits that are necessary for acceptable opportunistic transceiver performance:

- *Wide Frequency Agility*. Particularly for the case of secondary use when the availability of bandwidth is not guaranteed by virtue of a dedicated channel. Consequently, the availability of uninterrupted service can only be ensured through the availability of a wide diversity of available frequencies.
- *Sensitivity to Low Level Signals*. The probability of successful detection of potentially competing primary users requires receivers that can detect low level signals. In any event, common courtesy would dictate that any receiver used for sensing must be capable of detecting any signal that could be detected viably by a user with whom we are avoiding interference. This is particularly important as the FCC considers "interference temperature" [17] approach to interference management and spectrum re-use.
- *Large Spurious Free Dynamic Range (SFDR)*. As the frequency range over which sensing is performed is widened, the requirement for the "native" spurious free dynamic range of a receiver increases. For narrow band applications, the SFDR can be relaxed if a filter is used to partially attenuate *outband* blockers. Yet wide frequency agility requires that sensing be performed quickly, which requires a large bandwidth.

TechnoConcepts' TSR receiver technology is particularly well-suited to opportunistic communications because of the following traits inherent to its architectural approach:

- *Wide Frequency Agility.* Because its architecture is based on high speed RF sampling, and avoids the use of tuned passive circuits for filtering, it is capable of changing its tuning parameters over a wide range of frequencies (less than 400 MHz to 6 GHz). With the capability to shift the IF frequency along with its inherent linearity, it is capable of avoiding many of the spurious product problems encountered in conventional superheterodyne radios.
- *Programmability to Trade off Resolution versus Bandwidth.* The core architecture is based upon an RF sampled delta-sigma A/D converter. This permits widening the detection bandwidth in order to obtain a fast sweep of frequencies with low resolution, or a slow sweep of frequencies with much higher resolution. Both of these options are important since a radio that is capable of dynamic spectrum allocation must constantly be on the lookout for transmission opportunities.
- *Planned introduction of adaptive interference cancellers [15].* They will allow to substantially improve signal/(noise+interference) ratio increasing substantially the number of usable frequency sub-bands as it is shown in Fig. 3.

Most of interference to received signals is narrowband. Quantitative analysis of the interference suppression shown on Fig. 3 indicates that quality of interference suppression depends on the difference in the spectrum width of desirable and interfering signals. That is why interference suppression for the WCDMA systems may be expected about 5 dB better than for CDMA systems.

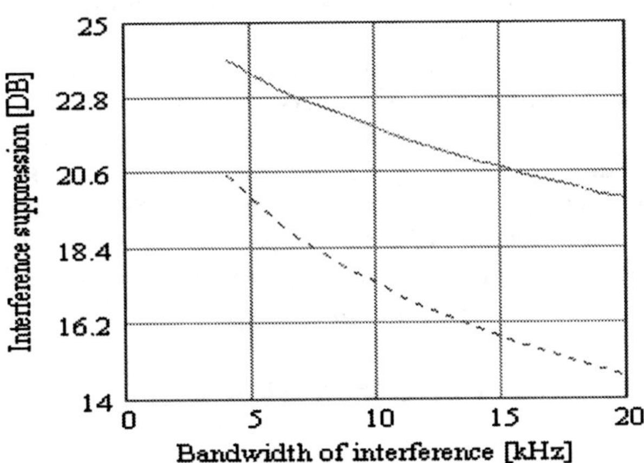

Figure 3. Interference suppression from one interferer vs. its bandwidth Legend: _____ WCDMA system; - - - - - CDMA system

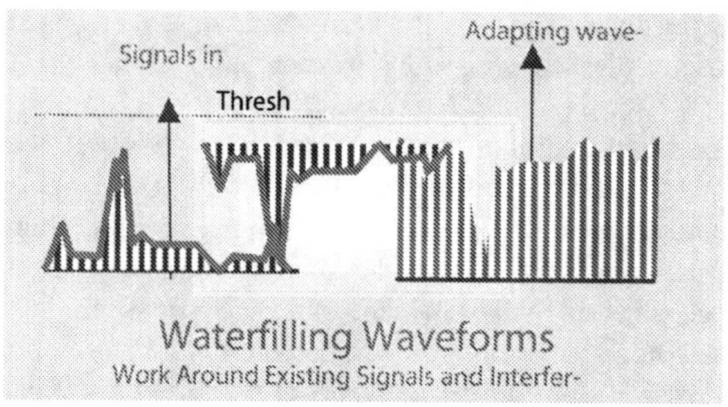

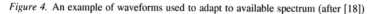

Figure 4. An example of waveforms used to adapt to available spectrum (after [18])

4. TRANSMITTER REQUIREMENTS

While the process of sensing is critical to finding transmission opportunities, a frequency agile transmitter is a key to exploiting these opportunities.

Analogous to the receiver, a widely frequency agile transmitter is crucial to maximize the number of opportunities for transmission particularly if the traffic requires a relatively large amount of bandwidth (for example, video or data applications).

Recently, techniques for shaping the transmission spectrum in order to match the "shape" of the available spectrum have been proposed (for example, [18]). It is important to note that in such transmission schemes, the permissible transmission power to avoid interfering with primary users may (and most likely would) vary as a function of frequency. Thus it is not adequate to determine which channels are "open" and to distribute the total radiated power evenly among this non-contiguous spectrum. It is instead necessary to determine the optimum (or at least acceptable) spectrum shape and to make the transmitter comply with this shape. An example of such a spectrum (after [18]) is shown in Fig. 4.

Needless to say, the generation of such complex analog waveforms requires the use of a DAC in conjunction with digital signal processing technology, since conventional analog circuits are incapable of generating these waveforms with the required fidelity and without severe accompanying distortion. TechnoConcepts' TSR transmitter technology is capable of solving this problem. Since its RF frequency delta-sigma technology is capable of quantizing wide bandwidth signals with high resolution, it is ideally suited to the producing the analog waveforms associated with prescribed spectrum generating principles.

5. TECHNOCONCEPTS' DIRECT RF CONVERSION CHIPS TESTING

Direct RF conversion transmitter and receiver chips were designed by Technoconcepts and fabricated on the Jazz Semiconductor 0.18 μm SiGe BiCMOS process. Spurious free dynamic range (SFDR) and sensitivity were measured in the broad range of frequencies. These parameters allow making judgment of the signal distortion and receiver noise, respectively.

Shown results are obtained for the first silicon of TechnoConcepts' SiGe chip. Other production version chips are in development now with added features and improved performance. We show some typical results for key test methods to illustrate the range of amplitude and bandwidth for the RF/D and D/RF conversion processes. The receiver results are taken at approximately 2.6GHz – the chip performance is similar over a continuously tunable range from less than 800MHz to greater than 5GHz. The receiver results are presented first, then the transmitter results.

Receiver results include:

- Noise shaping characteristic of ADC
- Single tone test result – to show noise floor
- Two tone test result - intermodulation
- Application waveform test – wide bandwidth and effect of linearity on noise & distortion floor

Noise shaping characteristic reduces the in-band noise compared to the out-of-band noise. The out-of-band (high frequency) noise is normally removed by a digital filter.

Figure 5 shows the spectrum of the ADC digital output without filtering. Digital filtering removes the high frequency noise and limits the bandwidth to the application requirements. The following receiver test results include a baseband low pass filter with a corner frequency of approximately 18MHz.

The measured noise spectral density for the initial receiver chip design is approximately −132 dBFS per Hz. This translates into a SNR of 82 dB for a 100 kHz bandwidth signal, 72 dB for a 1 MHz bandwidth signal, and 62 dB for a 10 MHz bandwidth signal. Furthermore, (and as expected), this noise floor (as referred to the input) is far below the thermal noise of the front end and is thus easily competitive

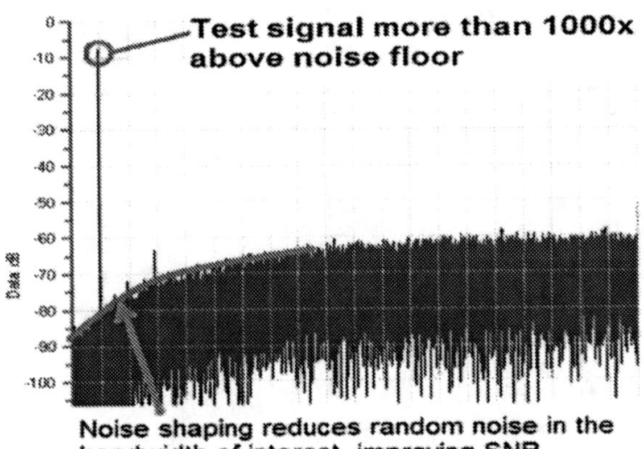

Figure 5. ADC noise shaping

with conventional architectures in performance with the added benefit of frequency agility. These specifications already exhibit more than adequate performance for wideband applications (such as WiFi and WiMax) and performance fairly close to that required for narrowband applications (such as GSM). We expect that with the improvement of the analog front end circuitry in our current redesign GSM requirements can also be met.

A key feature of this technology for DSA is the capability of digital filtering to tune quickly to any frequency and bandwidth without affecting the analog signal path.

Another key performance parameter for any receiver is the distortion, particularly with respect to limiting the spurious free dynamic range (SFDR). The importance of SFDR is demonstrated in Fig. 6. The spurious free dynamic range gives us an indication of what the maximum ratio between a strongest interferer and a weak signal of interest. In the described first design of the receiver chip, a raw spurious free dynamic range of approximately 60 dB was achieved. This is an important result since it comprises the cumulative performance of the front end circuitry combined with the A/D converter without the use of passive filtering.

While avoiding the use of passive filtering is a challenge, it is ultimately required if a truly frequency agile solution is to be achieved. A single tone test results on Fig. 7 show a noise floor and linearity.

The two tone test results on Fig. 8 show intermodulation characteristics obtained. Fig. 9 illustrates a wideband modulation test that shows the power ratio between occupied channels and adjacent spectrum. A CDMA2000 modulation is used with non-standard carrier frequencies. Please note that the spur at 5MHz is due to test instrument limitations.

Transmitter results include:

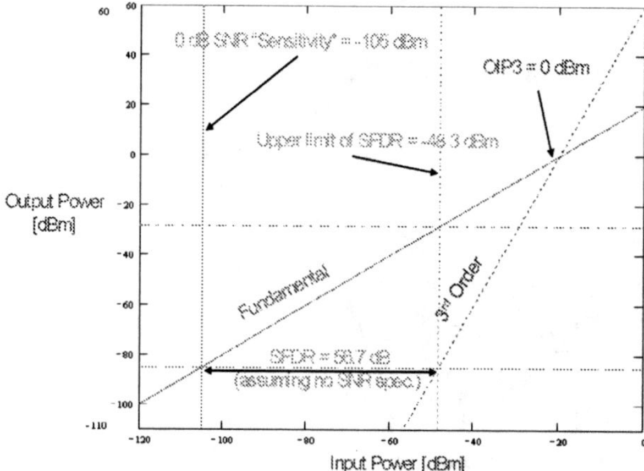

Figure 6. A diagram showing the relationship between spurious free dynamic range (SFDR), 3rd order intercept, gain and noise floor

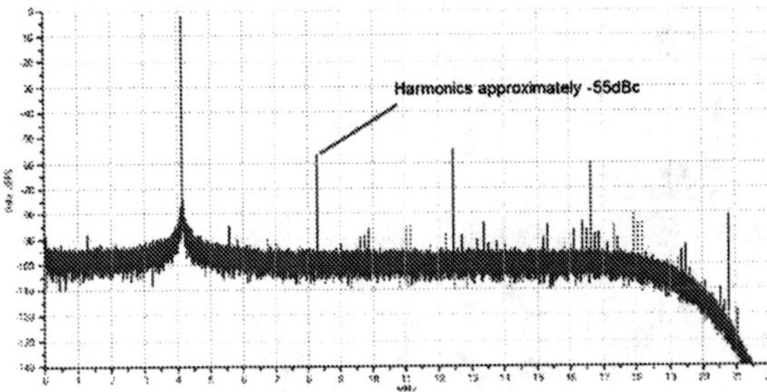

Figure 7. ADC single tone test

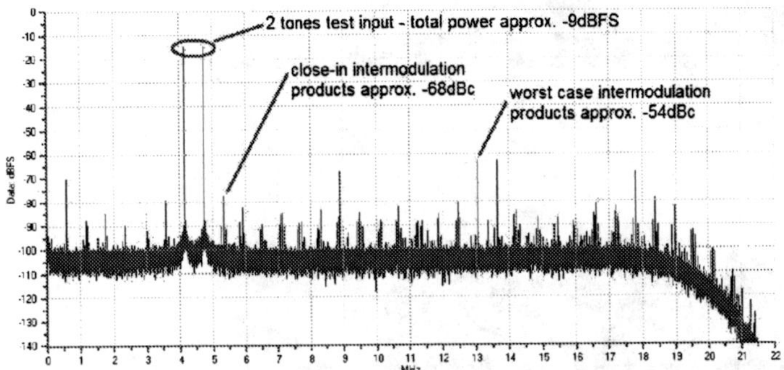

Figure 8. ADC two tone test

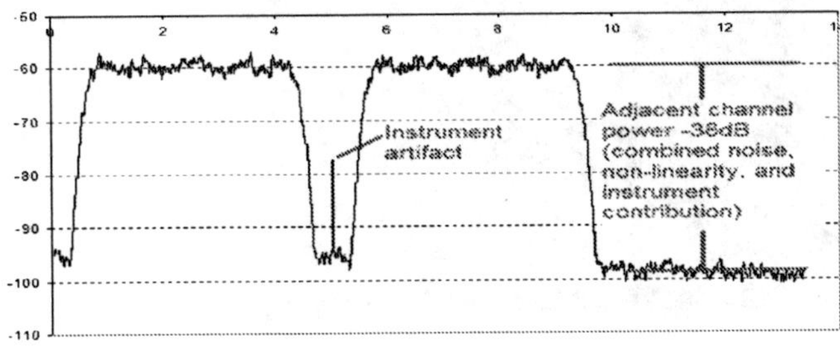

Figure 9. ADC wideband modulation test

- Narrow bandwidth noise and linearity
- Wide bandwidth noise and linearity

The transmitter measurements are taken through a special test port to access the DAC output directly. Fig. 10 shows is a narrow bandwidth test results for a carrier frequency of 1.8GHz and a baseband modulation frequency of 500 kHz.

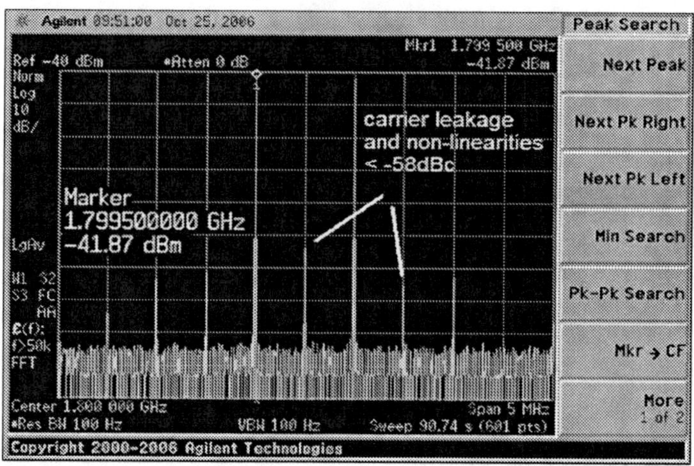

Figure 10. DAC narrowband test

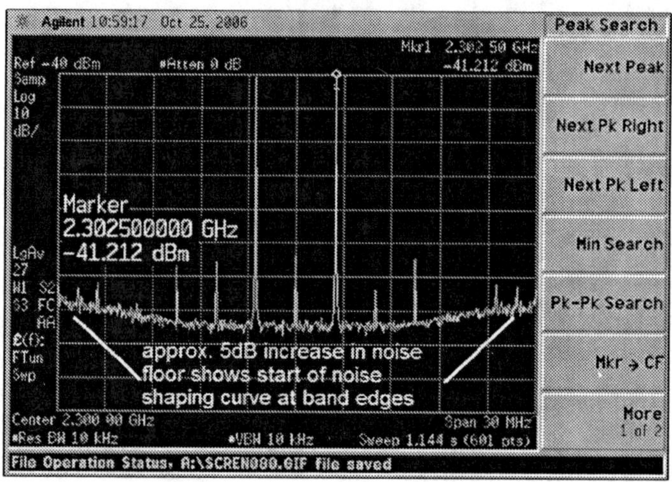

Figure 11. DAC wideband test

Fig. 11 is a wide bandwidth test with a carrier frequency of 2.3GHz and a baseband modulation frequency of 2.5MHz. In this test, please note that the wider bandwidth (30MHz span in the measurement) shows the start of the noise shaping curve characteristic, with an increase of approximately 5dB in the noise floor at the band edges.

6. CONCLUSIONS

Direct conversion homodyne receivers and its matching transmitters are well suited for realization of opportunistic DSA principles. For example, the Technoconcepts RF/D^{TM} transmitter chips operating in broad frequency range from 0.4 to 6 GHz allow fast analysis of broad frequency swath with good resolution due to its inherent capability of high speed RF sampling without using tuned passive circuits for filtering. As a result it may change its tuning parameters over a wide range of frequencies. With the capability to shift the IF frequency along with its inherent linearity, it avoids many spurious product problems encountered in conventional superheterodyne radios. The fast search of available frequencies may be facilitated by its inherent ability to change resolution on the fly. It is achieved by using delta-sigma A/D converter for signal sampling. That permits widening the detection bandwidth in order to obtain a fast sweep of frequencies with low resolution, or a slow sweep of frequencies with much higher resolution. Both of these options are important since a radio that is capable of dynamic spectrum allocation must constantly be on the search for transmission opportunities.

Being a new approach in spectrum allocation, DSA technical solutions will be done in phases with each subsequent phase offering higher performance and corresponding better spectrum utilization. The chips making it possible will go in turn through several generations addressing different needs of the promising DSA implementation.

REFERENCES

1. McHenry M. "NSF Spectrum Occupancy Measurements", New York, 2004
2. Hoffmeyer, J.A"Regulatory and standardization aspects of DSA technologies – Global Requirements and Perspectives", DySPAN 2005, Pages: 700–705
3. Pawelczak, P.; Prasad, R.V.; Liang Xia; Niemegeers, I.G"Cognitive radio emergency networks – requirements and design" DySPAN 2005, Pages: 601–606
4. Maldonado, D.; Bin Le; Hugine, A.; Rondeau, T.W.; Bostian, C.W. "Cognitive radio applications to dynamic spectrum allocation: a discussion and an illustrative example", New Frontiers in Dynamic Spectrum Access Networks, 2005. DySPAN 2005. 2005 First IEEE International Symposium, Nov. 2005, Pages: 597–600
5. Cowen-Hirsch, R.; Shrum, D.; Davis, B.; Stewart, D.; Kontson, K. "Software radio: evolution or revolution in spectrum management", MILCOM 2000. 21st Century Military Com. Conference Proceedings, Volume 1, Date: 2000, Pages: 8–14 vol.1
6. Lin Xu; Tonjes, R.; Paila, T.; Hansmann, W.; Frank, M.; Albrecht, M. "DRiVE-ing to the Internet: Dynamic Radio for IP services in Vehicular Environments" LCN 2000. Proceedings. 25th Annual IEEE Conference on Local Computer Networks, 2000.
7. Hoon Kim; Yeonwoo Lee; Sangboh Yun "A dynamic spectrum allocation between network operators with priority-based sharing and negotiation" Personal, Indoor and Mobile Communications, 2005. PIMRC 2005. IEEE 16th International Symposium, Volume 2, Date: 11–14, Sept. 2005
8. R. Kulkarni and S. Zekavat "Traffic - Aware Inter-vendor Dynamic Spectrum Allocation: Performance in Multivendor Environment", IWCMC'06 July 3–6, 2006, Vancouver, British Colambia, Canada
9. Oner, M.; Jondral, F.; "Cyclostationarity-based methods for the extraction of the channel allocation information in a spectrum pooling system", Radio and Wireless Conference, 2004 IEEE, 19–22 Sept. 2004

10. Leaves, P.; Moessner, K.; Tafazolli, R.; Grandblaise, D.; Bourse, D.; Tonjes, R.; Breveglieri, M.;"Dynamic spectrum allocation in composite reconfigurable wireless networks", Communications Magazine, IEEE Volume 42, Issue 5, May 2004
11. Grandblaise, D.; Moessner, K.; Leaves, P.; Bourse, D.; "Reconfigurability support for dynamic spectrum allocation: from the DSA concept to implementation", Mobile Future and Symposium on Trends in Communications, 2003. SympoTIC '03. Joint First Workshop on 26–28 Oct.
12. Keller, R.; Lohmar, T.; Tonjes, R.; Thielecke, J.; "Convergence of cellular and broadcast networks from a multi-radio perspective", IEEE Wireless Communications, Volume 8, Issue 2, April 2001
13. Jeffry Andrews "Interference Cancellation for Cellular Systems", University of Texas, Austin, February 8, 2005
14. Xianing Lu "Novel Adaptive Methods for Narrow Band Interference Cancellation in CDMA Multiuser Detection", ICASSP 05, 2005
15. Oleg Panfilov, Ron Hickling, Tony Turgeon, Kelly McClellan "Direct Conversion Software Radio – Its Benefits and Challenges in Solving System Interoperability Problems", Mobility 2006 conference, Bangkok, Thailand, October 2006
16. C. Santivanez, R. Ramanathan, C. Partridge, R. Krishnan, M. Condell, S. Polit, "Opportunistic Spectrum Access: Challenges, Architecture, Protocols", WiCon '06, Boston MA, August 2–5, 2006
17. Laura Van Wazer (FCC), "Spectrum Access and the Promise of Cognitive Radio Technology", presented at the FCC Cognitive Radio Workshop, May 19, 2003.
18. Preston Marshall (DARPA), "Beyond the Outer Limits – XG Next Generation Communications", presented at the FCC Cognitive Radio Workshop, May 19, 2003

CHAPTER 8

SCALABLE COMMUNICATION FOR HIGH PERFORMANCE AND INEXPENSIVE RELIABLE STRICT QOS

I. CHEN AND M. R. ITO

{ingwherc, mito}@ece.ubc.ca, Electrical and Computer Engineering, University of British Columbia, Vancouver, B. C., Canada V6T 2G9

Abstract: Traffic that requires high service quality often uses a single path with reserved resources for data transmission. The use of a single dedicated path suffers from single link failures. This paper provides a solution (TPmax-S) that uses scalable routing information to find alternate paths in the event of a single link failure in the network. When such a failure occurs, a flow that requires a path with reserved resources has a high rate of successfully finding enough resources on the backup path computed by TPmax-S, and thereby maintains continuous, high quality service to the user. This reliable high quality service can be provided without reserving more resources prior to the failure, and hence is an inexpensive solution to offer a better service. In this paper, initial tests are performed on TPmax-S and other similar backup path computation methods. These initial results show that TPmax-S provides very good backup paths while using scalable communication overhead

Keywords: quality of service, routing, traffic engineering

1. INTRODUCTION

Traffic that requires high quality-of-service (QoS), such as a requirement that data be delivered with a small delay and jitter, often uses a single path with reserved resources for data transmission [4,5,13]. The use of a single dedicated path suffers from single link failures in the network, and a high rate of service disruption can occur without advanced planning [2,3,12]. To provide continuous high quality of service before and after the failure, this paper provides a solution, TPmax-S, to compute an appropriate alternate path to backup the primary reserved path. This backup path remains unused and resources on the backup path are not reserved before the failure. When a failure occurs, resource reservation on the backup path is

H. Labiod and M. Badra (eds.), New Technologies, Mobility and Security, 89–98.
© 2007 *Springer.*

attempted so that when data transmission is switched to the backup path, the same strict QoS can be maintained. Without reserving extra resources on the backup path before the failure, TPmax-S ensures that there are no resources that are reserved but unused. While waiting until a failure occurs to reserve resources on the backup paths may risk failure of resource reservation resulting in service disruption, TPmax-S uses extra routing information to compute good backup paths that result in a high success rate in resource reservation. Furthermore, TPmax-S also limits the amount of extra routing information to a constant size in a fixed network.

In the rest of this paper, Section 2 describes some related work that also focuses on providing solutions to re-route traffic in the event of a single link failure. Section 3 describes the how TPmax-S computes the backup paths. Section 4 provides some initial analyses of TPmax-S compared to other similar backup path computation methods. Sections 5 and 6 provide some concluding thoughts and future work.

2. RELATED WORK

Other studies have also provided backup path computation methods that aim to find backup paths that have higher chances of successful resource reservation in the event of a link failure in the network [9,10]. In these cases, each flow also has a reserved primary path and an unreserved backup path, and resource reservation is only attempted on the backup path when the primary path fails. However, these backup path computation methods require routing information of every single flow to be exchanged, which is extremely costly and not scalable.

As mentioned in Section 1, waiting until a failure occurs to reserve resources may lead to service disruption because of unavailability of resources despite rigorous planning. Thus, another approach to recover from a single link failure is to reserve both the primary and the backup paths for the duration of the flow [1,6,7,8,11]. However, one major drawback of this approach is the waste of resources that are reserved on the backup paths. Because backup paths are rarely activated, the resources reserved on the backup paths are usually wasted.

3. BACKUP PATH COMPUTATION USING TPMAX-S

3.1. System Assumptions

TPmax-S is designed to be used within an autonomous system that runs a link state routing protocol such as OSPF. It takes advantage of the security measures in OSPF to prevent malicious use of sensitive extra routing information that is exchanged among nodes. TPmax-S also takes advantage of existing technology such as MPLS to transmit data on a specific path.

3.2. Extra Routing Information

Because data of a flow is forwarded along a specific primary path when there are no failures and along a specific backup path when a failure occurs, each source node is required to store the primary and backup path information of its flows. However, because the number of flows can increase indefinitely, path information of individual flows is not sent to other nodes. Instead, the condensed information, *PTP* and *BTP*, is periodically sent out to other nodes. Based on refreshed routing information, newer and more appropriate backup paths are also re-computed periodically. The refresh rates can be constant and pre-determined, but future work will attempt to find the best update strategies.

Assume that $\mathcal{L}$ is the number of links in the network. Suppose that flow f is a flow originated from N_i and that the links are represented as l_i, l_j, etc. Based on the path information of each flow originated at a source node N_i, N_i constructs two $\mathcal{L}xL$ vectors PTP^{N_i} and BTP^{N_i} defined as follows. Let the sets, $\mathcal{P}_i$, $\mathcal{B}_i$, $\mathcal{A}P_{j,k}$, and $\mathcal{A}B_{j,k}$, be defined as in Table 1. Then, the values in the vectors PTP^{N_i} and BTP^{N_i} are defined in Table 2.

3.3. Backup Path Computation

Assume that a flow f originates at the source node N_i and that it already has a primary path pp_f. Before computing the backup path bp_f, extra weights are added to undesirable links in the network (graph). Link weights are adjusted based on the stored path information and the condensed information, *PTP* and *BTP*, received from other source nodes. Then, the actual backup path computation is performed by calling Dijkstra's single source, shortest path algorithm on the graph with the adjusted link weights. Section 3.4 describes how to use the stored path

Table 1. A summery of the sets that group the flows originating at a source node

Set	Definition	Value
$\mathcal{P}_i$	The set of flows whose primary paths use link l_i.	
$\mathcal{B}_i$	The set of flows whose backup paths use link l_i.	
$\mathcal{AP}_{j,k}$	The set of flows that continue to use their primary paths when l_j fails and these primary paths all use l_k.	$\{f \mid f \notin \mathcal{P}_j \wedge f \in \mathcal{P}_k\}$
$\mathcal{BP}_{j,k}$	The set of flows that has a failed primary path when l_j fails and their corresponding backup paths all use l_k.	$\{f \mid f \in \mathcal{P}_j \wedge f \in \mathcal{B}_k\}$

Table 2. The definitions and values in the condensed routing information

Data	Description	Value
$PTP^{N_i}_{j,k}$	The number of flows originated at N_i that use their primary paths when l_j fails and these primary paths all use l_k.	$\mid \mathcal{A}P_{j,k} \mid$
$BTP^{N_i}_{j,k}$	The number of flows originated at N_i that use their backup paths when l_j fails and these backup paths all use l_k.	$\mid \mathcal{A}B_{j,k} \mid$

information of flows that also originate at N_i to modify link weights. Section 3.5 describes how to use *PTP* and *BTP* received from other nodes to modify the link weights. Section 3.6 emphasizes that the backup path needs to be disjoint from its corresponding primary path.

3.4. Using Stored Path Information

As mentioned in Section 1, backup paths are not reserved before the failure. Thus, to increase the chances of successfully reserving resources on the backup path bp_f when it is needed, extra weight should be added to links that may be used by other flows when bp_f is activated. To find the undesirable links and add extra weight to them, the source node N_i goes through each pair of primary and backup path information it has stored. For each flow f_i with primary-backup pair $pp_{f_i} - bp_{f_i}$, N_i calculates the number of shared links between pp_f and pp_{f_i} and modifies link weights in the following order.

The first step is to modify the link weights so that bp_f avoids conflicting primary paths. If there are no shared links between pp_f and pp_{f_i}, extra weight can be added to links on pp_{f_i} to prevent the eventual bp_f from conflicting with pp_{f_i}. This is needed because, if pp_f and pp_{f_i} do not share any links, pp_f may fail while pp_{f_i} does not in the event of a single link failure. Consequently, bp_f and pp_{f_i} may be active simultaneously, and should avoid using the same links. Thus, for bp_f to be successful, it should avoid links used by pp_{f_i}, especially if all those links are already full. To avoid links on pp_{f_i}, their new link weights can be calculated by adding extra weight proportional to the path length of pp_{f_i}.

bp_f should then avoid conflicting backup paths. If pp_f and pp_{f_i} share at least one link, then bp_f and bp_{f_i} may potentially be activated simultaneously. To increase the chance of successfully reserving resources on both bp_f and bp_{f_i} at the same time, between bp_f and bp_{f_i}, one of the paths should avoid the other path. Thus, if bp_f should avoid bp_{f_i}, possibly because flow f was started later than flow f_i, then link weights of links used by bp_{f_i} should be updated to reflect their undesirable status. (Since flow f_i and flow f both originated at the same node, their source node can easily determine the priority of its own flows without violating its commitments to these flows.) The new links weights can be calculated by adding an extra weight proportional to the number of shared links.

3.5. Using *PTP* and *BTP* Received from Other Nodes

Similar to avoiding traffic generated by N_i itself, the backup path bp_f should also avoid traffic originated by other nodes N_j, $j \neq i$. The same rules in Section 3.4 are also applied in this case. Links that carry primary path traffic at the same time that bp_f may be activated should be avoided. Links that carry backup path traffic at the same time that bp_f may be activated should also be avoided, although in this case, there should be a rule to determine whether the other backup path traffic should avoid the backup path traffic on bp_f or vice versa.

As before, bp_f should first avoid conflicting primary path traffic. As defined in Table 2, $PTP_{m,n}^{N_j}$ stores the number of primary path flows that link l_n carries when link l_m fails. Assume that link l_m is a link in pp_f ($l_m \in pp_f$). Then $PTP_{m,n}^{N_j}$ is also the number of primary path flows that l_n carries when pp_f fails and bp_f should be activated. Thus, to improve the success rate of reserving resources on bp_f, extra weight should be added on l_n. To avoid adding too much weight to l_n, for each node N_j in the network where $N_j \neq N_i$, only weight proportional to $\max_{l_m \in pp_f} \{PTP_{m,n}^{N_j}\}$ should be added.

Similarly, the second step is for bp_f to also avoid conflicting backup paths that do not already try to avoid bp_f. Because backup paths are unreserved before activation, a particular backup path does not have priority of unreserved network resources over another backup path. If bp_f and another backup path choose to use a particular link, bp_f and the other backup path both have the same priority over the link. As a result, there should be a rule that all the nodes agree upon on which backup paths should actively avoid other conflicting backup paths. If the rule is for backup paths of nodes with higher IP addresses (e.g., 216.x.x.x) to avoid backup path traffic from nodes with lower IP addresses (e.g., 128.x.x.x), then $BTP_{m,n}^{N_g}$ should be considered for nodes N_g with lower IP addresses than N_i. As defined in Table 2, $BTP_{m,n}^{N_g}$ stores the number of backup path flows that l_n carries when l_m fails. Assume that link l_m is a link in pp_f ($l_m \in pp_f$). Then $BTP_{m,n}^{N_g}$ is also the number of backup path flows that link l_n carries when pp_f fails and bp_f should be activated. To improve the success rate of reserving resources on bp_f, extra weight should be added on l_n. To avoid adding too much weight to link l_n, for each N_g with lower IP address than N_i, only weight proportional to $\max_{l_m \in pp_f} \{BTP_{m,n}^{N_g}\}$ should be added.

3.6. Backup Path should be Disjoint from Primary Path

Finally, the backup path bp_f should also avoid its corresponding primary path pp_f. Thus, enough weight, more than any weight added in Sections 3.4 and 3.5, should also be added to the links in pp_f.

4. EVALUATION

4.1. Overview

In the rest of this section, TPmax-S will be evaluated against three other backup path computation methods that also take the approach of finding good backup paths before a failure but reserving resources on the backup paths only when the corresponding primary paths fail. These three methods are BV + APV (a method that uses BV combined with APV [9]), APLV Norm (a method that uses APLV norm [10]), and CV (a method that uses CV [10]). Additionally, a simple method that uses the shortest paths disjoint from the primary paths (disjoint method) is also used as the least expensive backup path computation method. In Section 4.2 BV + APV,

APLV Norm, CV, and the disjoint method are compared. In Sections 4.3 and 4.4, simulations are performed on these five methods to compare the performance of their backup paths.

4.2. Communication Cost

Because TPmax-S, BV+APV, APLV Norm, and CV all take the approach of only reserving resources on backup paths when the corresponding primary paths fail, the backup paths have to be carefully planned to increase the success rate of reserving resources on the backup paths when needed. As a result, TPmax-S, BV+APV, APLV Norm, and CV all use extra routing information to find more suitable alternate paths as backup paths. As described in Section 3.2, TPmax-S nodes exchange link-based extra routing information that is fixed sized within a fixed network. On the other hand, BV+APV, APLV Norm, and CV all require flow-based routing information where path information of each individual flow to be communicated to other nodes in the network. This need to communicate path information of individual flows causes the communication overhead to be not scalable and impractical for deployment.

Assume that N is the number of nodes in the network, L is the number of links in the network, and F is the number of flows in the network. For simplicity, assume that communicating information to all nodes (broadcast) requires sending the information out all the links in the network, and hence requires $O(L)$ cost. Table 3 lists the extra routing information each method uses and how the information is communicated. The upper bounds to communicate the extra routing information are also listed. As expected, the communication overhead of TPmax-S is fixed in a fixed network, and the disjoint method does not require any extra routing information. However, the communication overhead of BV+APV, APLV Norm, and CV is ultimately bounded by F ($F >> N$, $F >> L$), which is not fixed even in a fixed network and can increase indefinitely.

Table 3. Communication strategies and upper bounds

Algorithm	Communication Overhead	Upper-bound
TPmax-S (Section 3)	• Placement of traffic on each link in the network originated by each node is sent to all nodes.	• $O\ (NL)$
BV + APV [9]	• Primary and backup path information of each flow is sent along its primary path.	• $O\ (FN)$
	• Information to compute backup path of each flow is sent along its primary path.	• $O\ (FN)$
APLV Norm, CV [10]	• Primary and backup path information of each flow is sent along its backup path.	• $O\ (FN)$
	• Condensed path information of each link is sent to all nodes.	• $O\ (NL)$
Disjoint	• No extra information is needed.	• $O\ (1)$

4.3. Simulation Overview

In Section 4.4, a simulation is performed on TPmax, BV+APV, APLV Norm, CV, and the disjoint method to test how many backup paths can be successfully activated when a single link failure occurs. In the simulations, traffic flows using reserved primary paths are placed in the network initially. For simplicity, the QoS requirement for each flow is that one unit of resource should be reserved on each link in the path it takes. Thus, at the beginning of the simulation, one unit of resource is reserved on the links of the primary path for each flow. Backup paths are computed and then single link failures are injected into the network. At this time, for each flow with a failed primary path, resource reservation on the corresponding backup path is attempted. To maintain the same strict QoS requirement after the failure, one unit of resource needs to be reserved on each link of a backup path whose corresponding primary path has failed. The traffic of a flow is only re-routed and considered successfully recovered from the failure if the required resources can be reserved on its pre-planned backup path. For comparison, the resulting recovery rate and success rate are computed. Assume that $\mathcal{F}$ is the set of flows started at the beginning of the simulation; $\mathcal{F}P$ is the set of flows with failed primary paths; $\mathcal{S}RR$ is the set of flows $f \in \mathcal{F}P$ that are successfully re-routed on backup paths; $\mathcal{S}F$ is the set of flows that are successful even after a single link failure, routed either on the primary path or backup path. Then, the recovery rate and the success rate are defined as follows.

Recovery rate—the percentage of flows successfully re-routed:

$$\frac{|\mathcal{SRR}|}{|\mathcal{FP}|} * 100\%.$$

Success rate—the percentage of successful flows after failure:

$$\frac{|\mathcal{SF}|}{|\mathcal{F}|} * 100\%.$$

4.4. Simulation on a Random Network

The simulation described in Section 4.3 is run on a randomly generated network of 20 nodes, each with a degree of four. Each simplex link has 20 units of resources. Because the QoS requirement of each traffic flow requires one unit of resource on every link in the path it takes (Section 4.3), each simplex link can carry up to 20 traffic flows. 100 distinct source-destination node pairs are randomly generated, and the shortest path between each pair is designated as the primary path of the traffic between that particular pair of nodes. 200, 300, 400, and 500 more flows between these initial 100 pairs of nodes are randomly chosen and added into the network. The resulting scenario is that traffic flows are concentrated in 100 paths, leaving free resources spread out throughout the network in an un-uniform manner.

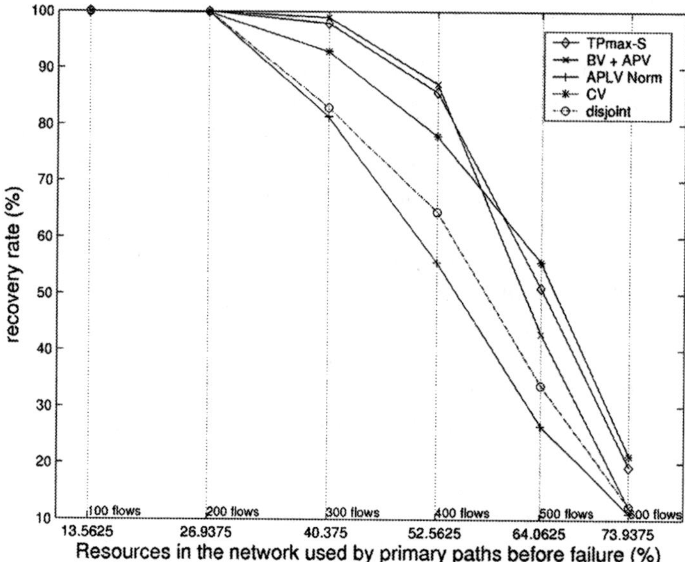

Figure 1. Recovery rate in the random network

Fig. 1 shows that the recovery rate of TPmax-S is very close to the best performance. When the network loads are roughly at most half full, the recovery rate of TPmax-S is only slightly lower than that of BV+APV. When the network loads are more than half full, the recovery rate of TPmax-S is only slightly lower than CV. Thus, while there is no method that provides the best recovery rate under all circumstances, TPmax-S provides a very close second-best recovery rate for all network loads. Furthermore, Fig. 2 shows that when the overall success rate of the traffic is considered, TPmax-S always performs the best, or performs nearly indistinguishable from the best. Taking both the communication cost and performance into consideration, TPmax-S is also the best solution among the five methods compared under this simulation setup.

In terms of the unreserved backup path approach that TPmax-S takes, Fig. 2 shows that when a single link fails in the network, at worst 98% of the traffic flows can continue data transmission at the same strict QoS requirement as before. Thus, overall, using TPmax-S, reliability at a strict QoS requirement can be provided to the majority of the traffic at scalable communication cost and at no extra resource reservation cost to the network.

5. CONCLUSIONS

This paper provides a scalable and inexpensive solution, TPmax-S, so that strict QoS requirements can be provided both before and after a single link failure in the network. When a flow uses a single reserved path for data transmission to

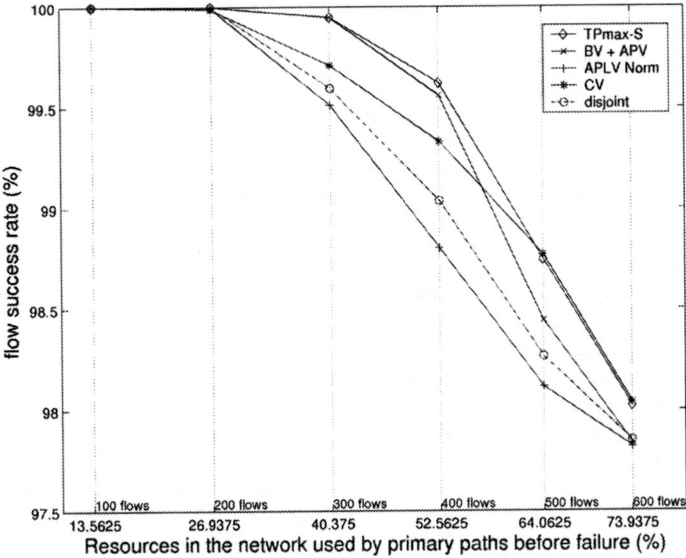

Figure 2. Success rate in the random network

obtain the strict QoS it needs, the single path easily suffers from single link failures in the network, which results in complete disconnection of the flow. To maintain continuous data transmission at the strict QoS, TPmax-S takes the approach that plans but does not reserve backup paths in advance. Using scalable and less routing information than other similar methods, TPmax-S provides the best performance in the 5×5 mesh network setup and provides the best overall performance in the random network setup. Despite not reserving the backup paths in advance, the majority of the traffic flows (at least 98%) can recover from a single link failure in the network, which may be adequate guarantee for some applications. Overall, TPmax-S provides a scalable and inexpensive solution for flows to overcome a single link failure in the network using unreserved backup paths paired with reserved primary paths.

6. FUTURE WORK

The simulations presented in this paper are the initial tests to show the potential of a new algorithm. Future simulations will be run on larger networks, heavier traffic volumes, and different traffic patterns. Randomization and variation of information propagation, link failure timing, bandwidth requirement, and QoS guarantee will also be introduced into future simulations. Future work is also planned on finding alternate recovery processes to overcome a single link failure in the network. In addition, other single point failures, such as single node failures, will also be studied.

REFERENCES

1. Anderson, J., Doshi, B. T., Dravida, S., and Harshavardhana, P. Fast Restoration of ATM Networks. IEEE Journal on Selected Areas in Communications, 12(1): 128–138, Jan. 1994.
2. Banerjea, A. Fault Recovery for Guaranteed Performance Communications Connections. IEEE/ACM Transactions on Networking, 7(5): 653–668, Oct. 1999.
3. Banerjea, A., Parris, C., and Ferrari, D. Recovering Guaranteed Performance Service Connections from Single and Multiple Faults. Technical report TR-93-066, University of California, Berkeley, 1993.
4. Blake, S., Black, D., Carlson, M., Davies, E., Wang Z., and Weiss, W. An Architecture for Differentiated Service. RFC 2475, Dec. 1998.
5. Braden, R., Clark, D., and Shenker, S. Integrated Services in the Internet Architecture: an Overview. RFC 1633, Jun. 1994.
6. Gummadi, K. P., Pradeep, M. J., and Murthy, C. S. R. An Efficient Primary-Segmented Backup Scheme for Dependable Real-Time Communication in Multihop Networks. IEEE/ACM Transactions on Networking, 11(1): 81–94, Feb. 2003.
7. Gupta, A., Gupta, A., Jain, B. N., and Tripathi, S. QoS Aware Path Protection Schemes for MPLS Networks. Proceedings of the 15th International Conference of Computer Communication, 2002.
8. Kawamura, R., Sato, K., and Tokizawa, I. Self-Healing ATM Networks Based on Virtual Path Concept. IEEE Journal on Selected Areas in Communications, 12(1): 120–127, Jan. 1994.
9. Kim, S. and Shin, K. G. Improving Dependability of Real-Time Communication with Preplanned Backup Routes and Spare Resource Pool. Proceedings of the 11th International Workshop on Quality of Service (IWQoS), 2003.
10. Kim, S., Qiao, D., Kodase, S., and Shin, K. G. Design and Evaluation of Routing Schemes for Dependable Real-Time Connections. Proceedings of IEEE International Conference on Dependable Systems and Networks (DNS), 2001.
11. Kodialam, M. and Lakshman, T. V. Dynamic Routing of Locally Restorable Bandwidth Guaranteed Tunnels using Aggregated Link Usage Information. IEEE Infocom, 2001.
12. Parris, C. and Banerjea, A. An Investigation into Fault Recovery in Guaranteed Performance Service Connections. Technical report TR-93-054, International Computer Science Institute, University of California, Berkeley, Oct. 1993.
13. Zhang, L., Deering, S., Estrin, D., Shenker, S., and Zappala, D. RSVP: A New Resource Reser-Vation Protocol. IEEE Network, 7(5), Sep. 1993.

CHAPTER 9

A USER INTERFACE FOR RESOURCE MANAGEMENT IN A MOBILE ENVIRONMENT

BADR BENMAMMAR [1], ZEINA JRAD [2], AND FRANCINE KRIEF [3]

[1] *GET / Télécom Paris CNRS LTCI, 46 rue Barrault - 75634 Paris Cedex 13*
benmamma@enst.fr, http://www.ltci.enst.fr/
[2] *INRIA Rocquencourt, Domaine de Voluceau, B.P. 105, 78153, Le Chesnay, France*
zeina.jrad@inria.fr, http://www.inria.fr/rocquencourt/
[3] *LaBRI Laboratory, Bordeaux1 University, 33400 Talence, France*
francine.krief@labri.fr, http://w5.labri.fr/

Abstract: This paper describes a user interface for QoS management in mobile IP networks. The paper context is built in conformance with the generic signaling environment, which is standardized by the NSIS IETF working group. In this work, we investigate the use of some techniques of the AI (Artificial Intelligence) domain to implement a user interface called NIA (Negotiation Individual Assistant) in order to determine the QoS profile and negotiate the QoS parameters in the new domain after the handover. Therefore, we use the connectionist learning in the management of the negotiation profiles and the agent technology to help the user to choose the best service provider, dynamically negotiate the QoS on the user's behalf, follow the user's behaviour to be able to anticipate the negotiation and manage the re-negotiation. The resource management, presented in this work, provides to mobile terminals the required QoS based on user's mobility and QoS profile. This QoS profile is determined by the NIA

1. INTRODUCTION

The development of real time applications as well as multimedia applications has witnessed an exponential increase. The real time constraints of these applications present a big challenge for their integration. That's why we need services adapted to specific application needs with a guaranteed Quality of Service (QoS) [1].

However, the implementation of QoS mechanisms is a very heavy task. It is difficult to manually configure all the network devices because of the abundance of QoS information and because of the dynamic nature of QoS configurations. The operator must control the attribution of network resources according to applications and users characteristics. Using management tools adapted to QoS quickly proves

H. Labiod and M. Badra (eds.), New Technologies, Mobility and Security, 99–110.
© 2007 *Springer.*

essential. In order to simplify the router's configuration by permitting its automation, the IETF proposed a general framework called policy-based networking [2] for the control and management of these IP networks. The Internet New Generation is based on the policy-based networking management.

Most applications cannot dynamically express their QoS requirements to obtain the adapted level of service. For each application, the customer and the provider have to agree on rules of assignment of service levels. They sign a contract called SLA (Service Level Agreement) which is then translated into high-level policies. These policies are not directly executable by the network devices. They must be translated into intermediate and then into low level policies which are understandable by network devices. The SLS (Service Level Specification) is the technical version of the SLA [3] and the QoS parameters (also called performance metrics) are a part of the SLS parameters. The negotiation of these parameters is a difficult process because the user needs to identify himself their values according to his context. The user interface, called NIA (Negotiation Individual Assistant), presented in this paper is supposed to determine using artificial intelligent techniques, a context-aware profile for the QoS negotiation.

On the other hand, the IETF has launched the Next Steps In Signaling working group (NSIS). The initial objective of this group was to unify all the existing solutions of IP signaling or to make them coexist. With the emergence of IP networks and the increasing number of applications requiring a high level of QoS, the signaling problem became increasingly critical. Provide a universal signaling, which takes into account the QoS as well as the security and the mobility is a very difficult task. Initially, the NSIS working group aimed the QoS, and proposed the QoS NSLP [4] signaling application.

The objective of our work is to propose a signaling environment for QoS negotiation and advance resource reservation in mobile IP networks in conformance with the generic signaling environment standardized by the NSIS IETF working group. The QoS negotiation and resource reservation is based on the profile determined by the NIA (Negotiation Individual Assistant).

This paper is organized as follows. The first section presents a synthesis of the research relating to resources reservation in an IP mobile environment. The second section defines the NSIS environment in which we specify the advance resource reservation protocol. The third section presents our user's mobility and QoS profile. Then, we present the user interface for profile determination. The last section is a description of an utilization case of the signaling environment.

2. QOS IN MOBILE IP NETWORKS

Recent research takes an interest in advance resource reservation to provide the necessary QoS to the mobile terminals.

The authors in [5] proposed a new protocol of resource reservation in a mobile environment called MRSVP (Mobile RSVP). In this model of reservation, the mobile terminal can make advance reservations in a set of cells named MSPEC (Mobility Specification). The MSPEC is not very clear, it only indicates the future locations of the mobile terminal but the MSPEC is not described. Authors proposed new RSVP messages in order to treat the user's mobility. This technique requires additional classes of service, major changes of RSVP, and a lot of signaling.

Min-Sun Kim et al. [6] proposed a resource reservation protocol in a mobile environment. The proposed protocol introduces the *RSVP agent* concept in order to guarantee the necessary QoS through an anticipation of the resource reservation. In this protocol, there are 3 classes of resource reservation to obtain a better use of resources:

- *The Free class:* it represents the resources used by the best effort flows.
- *The Reserved class:* it represents the reserved resources for a specific flow, which are currently used.
- *The Prepared class:* it represents the reserved resources for a specific flow, which are not currently used.

Ferrari et al. [7] described a distributed mechanism in order to make reservations in advance for the real time connections. In this mechanism, the reservation demand is classified according to two types: *immediate* and *in advance*.

- An Immediate reservation is activated at the moment of the demand; its length is not specified.
- A reservation in advance is associated with two parameters: *starting time* (the time of the reservation activation) and *duration* (the reservation period).

Another way to obtain a better use of resources is to determine the future locations of the mobile terminal.

3. SIGNALING ENVIRONMENT

The IETF decided in 2002 to launch the NSIS working group to try to unify all solutions of signaling or to make them coexist. This group standardized two-layer architecture: the NSLP layer to generate signaling flows for different purposes and the NTLP layer to transport those flows in a path coupled way.

3.1. GIST

GIST (General Internet Signaling Transport) is the protocol that NSIS has adopted as a standard for the NTLP layer [8]. GIST is conceived for an in band transport of signaling flows generated by the NSLP layer, i.e. signaling flows follow the same path as the data flows. Besides, it only treats unicast signaling. Finally, GIST collaborates with the underlying transport and security layers to assure the good routing of signaling flows.

3.2. QoS NSLP

Whereas the NTLP layer has for essential goal the transport of signaling, the NSLP layer assures the generation of this signaling in accordance with of user needs. QoS NSLP is the first NSLP layer protocol to be elaborated in NSIS: it permits to generate a signaling to provide a certain level of QoS by making reservations on the data path independently of the QoS models (Diffserv, Intserv...) adopted by the different domains.

NTLP is independent of the NSLP layer signaling application and it is through the intermediary of one API that parameters asked by one layer are obtained.

QoS NSLP generates 4 messages types:

- *Reserve*: the only message, which handles the reservation state (refresh, create, remove).
- *Response:* using this message, a response is sent to a message received.
- *Query:* this message is used to require information concerning the nodes, which are on the data path, for example: the available resources.
- *Notify*: using this message, it possible to inform a node without preliminary request.

3.3. MQoS NSLP

We name MQoS NSLP, the procedure of resources reservation in advance using the QoS NSLP messages in a mobile environment. This procedure of reservation is applied in HMIPv6 architecture. The MAP (Mobility Anchor Point) plays a significant role to reserve the resources in advance on behalf of the mobile terminal.

QoS NSLP operates according to the two following modes: *Sender Initiated Reservation* and *Receiver Initiated Reservation.* In the first mode, the sender of the flow initiates the reservation (he generates the RESERVE message). In the second mode, the reservation is initiated by the receiver of the flow.

4. MOBILITY AND QOS PROFILE

The user's mobility profile is built on the basis of its behaviour / movement after **m** associations with the system.

The system model is based on the Continuous Time Markov Chain (CTMC).

Our system can evolve between N states defined by the following set: $C = (C_1, C_2, \dots\dots\dots C_i \dots\dots\dots C_n)$.

The system is in the state i = the terminal mobile is in the cell C_i.

P_{ij}: the probability of transition from the cell C_i to the cell C_j.

P_i (t_r): the probability, which defines the location of the mobile terminal in the cell C_i at the time t_r.

The user's mobility profile contains the following information:

- The user's identifier;
- User Preferences: User_P;

This attribute represents the set of the user's preferences and is determined by the NIA (Negotiation Individual Assistant).

The proposed format for the User_P is as follows:

User_P = <Preference ID> <Duration_P> <Cell_P> < QoS_level>

- <Preference ID>: it identifies the preference (the system can detect several preferences for the user).

- <Duration_P> : <start_P> <end_P>: it determines the period of time in which the user's preference is satisfied.
- <Cell_P>: it determines the cell in which the user's preference is satisfied.
- <QoS_level>: it is the QoS level needed by the user for the preference.
- M = [Pij] [N*N]: The Matrix of transition, which contains the P_{ij}, before the **m** associations, the P_{ij} are random.
  We note:

t [i, j]: the number of transition from the cell i to the cell j during the **m** associations with the system.

g (i): the number of transition outgoing from the cell i during the **m** associations with the system. We calculate it as follows:

$$g(i) = \sum_{j=1}^{n} t[i, j].$$

After the **m** associations, the probability of transition from the cell i to the cell j is calculated as follows: $P_{ij} = t[i, j]/g(i)$.

- V = [Pi(to)] [N]: This Vector contains the $P_i(t_o)$.
  We note:

$P_i(t_o)$: this probability defines the location of the mobile terminal in the cell C_i at the time t_o.

k (i): the number of association with the cell i during the **m** associations at the time t_o.

We have: $\sum_{i=1}^{n} k(i) = m$ and $P_i(t_o) = k(i)/m$.

- The MSpec (Mobility Specification): The MSpec determines the future locations of the mobile terminal.
  The proposed format for the MSpec is as follows:
  MSpec = <MSpec ID> <Duration> <Cell ID>.
- *MSpec ID* is the identifier of the MSpec.
- *Duration* is the interval of time (<start time>, <end time>) during which the future locations of the mobile terminal can be determined.
- *Cell ID* : <cell ID1>, <cell ID2>, <cell ID3>, ..., <cell IDn> is a set of cells identifiers. We suppose that each cell is identified by a single identifier.

We have $P_j(t_{r+1})$: the probability of the mobile terminal's location in the cell C_j at the time t_{r+1}.

We can calculate this probability by the following formula:

$$P_j(t_{r+1}) = \sum_{i=1}^{n} P_i(t_r)^* P_{ij}.$$

We define $\theta (0 \leq \theta \leq 1)$, which is a fixed or variable threshold. It is used to select the cells according to their probabilities. The MSpec is defined as follows:

$$MSpec(t_r) = \{C_j / P_j(t_{r+1}) \geq \theta\}.$$

5. USER INTERFACE FOR QOS NEGOTIATION

The user interface proposed in this paper is placed on the user terminal and is called NIA (Negotiation Individual Assistant). The NIA negotiates the QoS between the user and the service provider, from one side, and between the user and the network, from the other side (figure 1). The main purpose of the interface is the representation of the user in requesting and negotiating the desired QoS in a dynamic environment.

As shown in figure 1, the proposed interface contains different layers. The first one is the profile management layer. This is the layer of direct contact with the user. Once connected, the principal task of the modules of this layer is to react autonomously in order to follow the user's work. User and terminal contexts are saved along with the used applications and their requirements in the knowledge base of the system. These data are modified systematically according to any change in the user choices and actions. The reasoning modules will also use them in order to deduce a general profile that represents the user.

The second layer is for the control. Once the user's needs and preferences are identified, the next step consists of verifying that these preferences are converted into the appropriate SLS values.

The third layer is the Negotiation layer. This layer manages the service publication, subscription, selection and negotiation.

In this paper, we will focus on the first layer, further details concerning the second and the third layers can be found in [17,18].

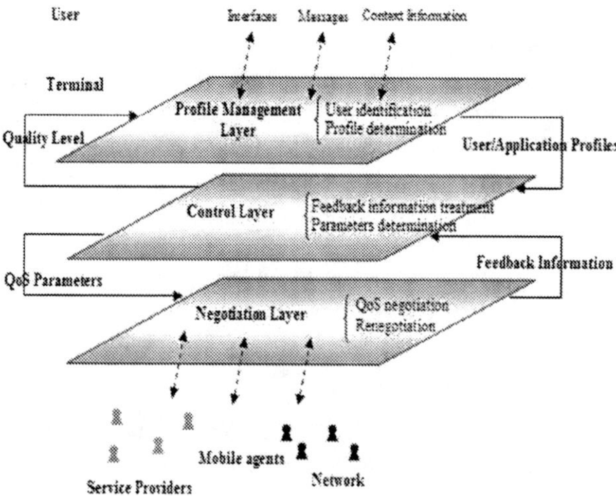

Figure 1. The layers of the proposed framework

5.1. Profile Management Layer

This layer is introduced to identify the user and to analyze his work. User preferences and application requirements are saved in the knowledge base of the system. These data are modified systematically according to any change in the user choices and actions. Applications can be classified in many categories according to their needs (delay, jitter…) and to the type of the supported information (data, voice, image…). The profile of the application will then be determined according to these categories and to the requirements of the user.

As an input to this layer, we can identify two types of data:

- Information collected through a communication with the user via graphical interfaces or messages.
- Information collected through an observation of the user's behavior.

Once the information is analyzed, the result is a user profile that represents the user's preferences in terms of quality and an application profile that describes the needs of each application.

Our approach, represented in figure 2, consists in recovering, first of all, data that represent traces of use (i.e. log files [9]). These data will be cleaned and re-coded in a numerical or binary format to be easily treated. We build then a Self Organizing Map from the re-coded file in order to extract profiles [10]. Finally, we carry out a classification to better see the clusters structure of the map, followed by a segmentation of the SOM in order to separate the different profiles.

The steps of our approach represented in figure 2 are detailed in the following subsections.

5.1.1. *The Unsupervised Connectionist Learning*

The unsupervised numerical learning, or automatic classification, consists in determining a partition of an instances space from a given set of observations, called training set. It aims to identify potential trend of data to be gathered into classes. This kind of learning approach, called clustering, seeks for regularities from a

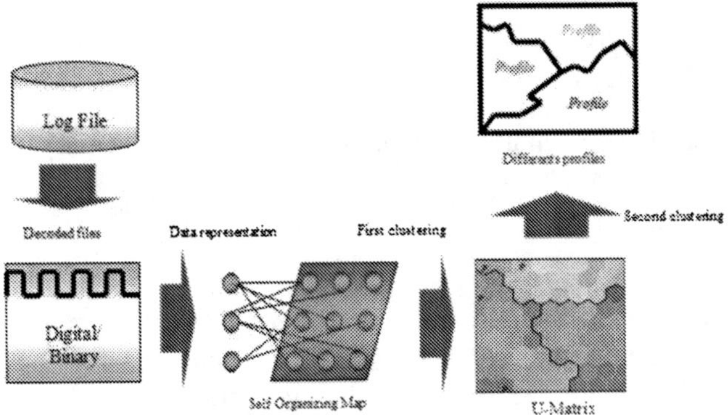

Figure 2. Profile management procedure

sample set without being driven by the use of the discovered knowledge. Euclidian distance is usually used by clustering algorithms to measure similarities between observations.

Self-Organizing Maps (SOM) implement a particular form of competitive artificial neural networks [10]; when an observation is recognized, activation of an output cell – competition layer – leads to inhibit activation of other neurons and reinforces itself. It is said that it follows the so called "Winner Takes All" rule. Actually, neurons are specialized in the recognition of one kind of observations. The learning is unsupervised because neither the classes nor their numbers are fixed a priori.

The training set is used to organize these maps under topological constraints of the input space. Thus, a mapping between the input space and the network space is constructed; closed observations in the input space would activate closed units of the SOM.

An optimal spatial organization is determined by information received from the neural networks. When the dimension of the input space is lower than three, both of the position of weights vectors and the direct neighborhood relations between cells can be visually represented. Thus, a visual inspection of the map provides qualitative information on its architecture.

The connectionist learning is often presented as a minimization of a risk function [11]. In our case, it will be carried out by the minimization of the distance between the input samples and the map prototypes (referents), weighted by a neighborhood function h_{ij}. To do that, we use a gradient algorithm. The criterion to be minimized is defined by:

$$E_{SOM} = \frac{1}{N}\sum_{k=1}^{N}\sum_{j=1}^{M} h_{j\,NN(x^{(k)})} \left\| w_{.j} - x^{(k)} \right\|^2$$

Where N represents the number of learning samples, M the number of neurons in the map, $NN(x^k)$ is the neuron having the closest referent to the input form x^k, and h the neighborhood function. The neighborhood function h can be defined as:

$$h_{rs} = \frac{1}{\lambda(t)} \exp\left(-\frac{d_1^2(r,s)}{\lambda^2(t)}\right)$$

$\lambda(t)$ is the temperature function modeling the neighborhood extent, defined as:

$$\lambda(t) = \lambda_i \left(\frac{\lambda_f}{\lambda_i}\right)^{\frac{t}{t_{\max}}}$$

λ_i and λ_f are respectively initial and the final temperature (for example $\lambda_i = 2$, $\lambda_f = 0.5$). t_{max} is the maximum number allotted to the time (number of iterations for the x learning sample). $d_1(r, s)$ is the Manhattan distance defined between two neurons r and s on the map grid, with the coordinates (k,m) and (i,j) respectively:

The learning algorithm of this model proceeds essentially in three phases:

- Initialization phase where random values are assigned to the connections weights (referents or prototypes) of each neuron of the map grid.
- Competition phase during which, for any input form $x^{(k)}$, a neuron $NN(x^k)$, with neighborhood, $V_{NN(x^{(k)})}$ is selected like a winner. This neuron has the nearest weight vector by using Euclidean distance:

$$d_1(r,s) = |i-k| + |j-m|$$

$$NN(x^{(k)}) = \underset{1 \le i \le M}{\arg\min} \left\| w_{.i} - x^{(k)} \right\|^2$$

- Adaptation phase where the weights of all the neurons are updated according to the following adaptation rules:

If $w_{.j} \in V_{NN(x^{(k)})}$ then adjust the weights using:

$$w_{.j}(t+1) = w_{.j}(t) - \varepsilon(t) h_{j\,NN(x^{(k)})} \left(w_{.j}(t) - x^{(k)} \right)$$

Else

$$w_{.j}(t+1) = w_{.j}(t)$$

Repeat this adjustment until the SOM stabilization.

5.1.2. SOM Map Segmentation

We segment the SOM using the K-means method (Figure 3). It is another clustering method that consists in arbitrarily choosing a partition; the samples are then treated one by one. If one of them becomes closer to the center of another class, it is moved into this new class. We calculate the centers of new classes and we reallocate the samples to the partitions. We repeat this procedure until having a stable partition.

The criterion to be minimized in this case is defined by:

$$E_{K-means} = \frac{1}{C} \sum_{k=1}^{C} \sum_{x \in Q_k} \|x - c_k\|^2$$

Where C represents the number of clusters, Q_k is the cluster k, C_k is the center of the cluster Q_k or the referent.

The basic algorithm requires fixing K, the number of wished clusters. However, there is an algorithm to calculate the best value for K assuring an optimal clustering. It is based principally on the minimization of Davies-Bouldin index [12], defined as follows:

$$I_{DB} = \frac{1}{C} \sum_{k=1}^{C} \max_{l \ne k} \left\{ \frac{S_c(Q_k) + S_c(Q_l)}{d_{ce}(Q_k, Q_l)} \right\}$$

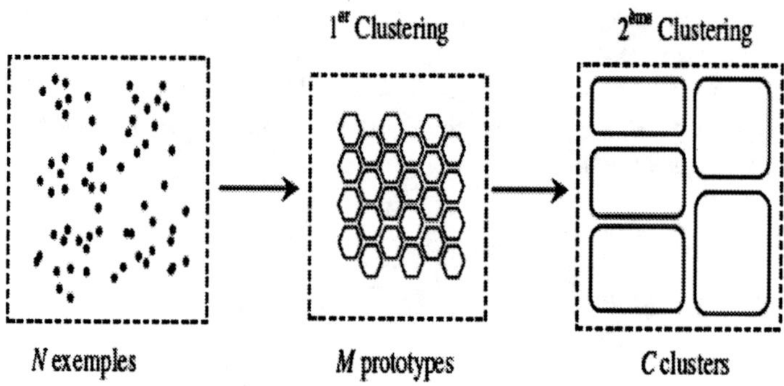

Figure 3. Two successive clusterings: SOM followed by K-means

With,

$$S_c(Q_k) = \frac{\sum_i \|x_i - c_k\|^2}{|Q_k|}$$

$$d_{cl}(Q_k, Q_l) = \|c_k - c_l\|^2$$

C is the number of clusters, S_c is the intra-cluster dispersion, and d_{cl} is the distance (centroid linkage) between the clusters centers k and l. This clustering procedure aims to find internally compact spherical clusters which are widely separated.

There are several methods to segment the SOMs [13]. Usually, they are based on the visual observations and the manual assignment of the map cells to the clusters. Several methods use the K-means algorithm with given ranges for K value. Our work is based on the approach of Davies-Bouldin index minimization.

We note that the K-means approach can be directly applied to the data instead of SOMs approach. In our work, we applied it to the SOMs results. The idea is to use SOMs as a preliminary phase in order to set a sort of data pretreatment (dimension reduction, regrouping, visualization…). This pretreatment has the advantage to reduce the clusters calculation complexity and also ensures a better visualization of the automatic classification results.

5.1.3. Simulations Results

We applied the two algorithms described above on our data (log files describing different traces of use) in order to determine the negotiation profiles. In the simulations, we used the *SomToolbox* proposed by the researchers of the HUT (Helsinki University of Technology) of the T. Kohonen team [14]. The results obtained are very promising (Figure 4).

Figure 4.a is a representation of a SOM map seen as "*Component Planes*" that allows the visualization of the partition of the different variable values. The highest values of the variables are in red and the lowest values are in blue.

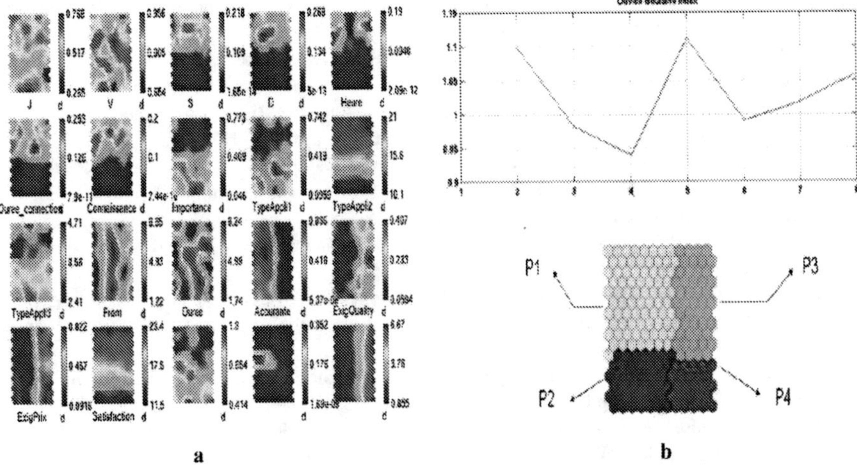

Figure 4a. SOM clustering, b. Resulting profiles

Figure 4.b represents the neurons segmentation of SOM map (second clustering). The curve shows that the minimal value of the index of Davies-Boudin corresponds to an optimal clustering resulting in four profiles. The various colors represent the different clusters or the identified negotiation profiles.

6. CONCLUSION

In this paper, we propose a user interface for QoS management in mobile environment. We have presented a mobility and QoS profile used for QoS negotiation and advance resource reservation in a mobile IP network. This reservation is made according to the MSpec object which determines the future locations of the mobile terminal. Our objective through this approach is to minimize the degradation of services during the handover. The Negotiation Individual Assistant (NIA) presented in this paper constitutes an interface between the user and the network in the context of the Internet of new generation. This interface integrates two interesting techniques of the Intelligence Artificial domain: The connectionist learning and the agent technology. Our NIA has been implemented and tested in two French national research projects: ARCADE [15] and IPSIG [16].

REFERENCES

1. QoSforum, "QoS protocols & architectures", White Paper, July 1999.
2. Yavatkar R., D. Pendarakis, R. Guerin, "A Framework for Policy Based Admission Control", RFC 2753, January 2000.
3. D. Goderis and al., Service Level Specification Semantics and Parameters, draft-tequila-sls-02.txt, Internet draft, January, 2002.

4. S. Van den Bosch, G. Karagiannis, A. McDonald, NSLP for Quality-of-Service signaling, <draft-ietf-nsis-qos-nslp- 10.txt>, March, 2006.
5. A. K. Talukdar, B. R. Badrinath, A. Acharya, MRSVP: A resource reservation protocol for an integrated services network with mobile hosts, ACM Journal of Wireless Networks, vol. 7, 2001.
6. K. Min-Sun, S. Young-Joo, K. Young-Jae, C. Young, A Resource Reservation Protocol in Wireless Mobile Networks. ICPP Workshops, Valencia, Spain September 03–07, 2001.
7. D. Ferrari, A. Gupta, G. Ventre, Distributed Advance Reservation of Real-Time Connections. Proceedings of the 5th International Workshop on Network and Operating System Support for Digital Audio and Video. P.16–27. April 19–21, 1995.
8. H. Schulzrinne, R. Hancock, GIST: General Internet Signaling Transport, draft-ietf-nsis-ntlp-09, February, 2006.
9. K. Benabdeslem, Approches Connexionnistes pour la visualisation et la classification des données issues d'usages de l'Internet, thesis in computer science, LIPN, University of Paris13, France, December, 2003.
10. T. Kohonen and S. Kaski and H. Lappalainen, Self-Organized Formation of Various Invariant-Feature Filters in the Adaptive-Subspace SOM, pages 1321–1344, Neural Computation, 1997.
11. Y. Bennani, Réseaux de neurones artificiels, Chapter in Encyclopedie d'Informatique et Science de l'information, Edition Vuibert, 2005.
12. Davies, D., L., Bouldin, D., W., A Cluster Separation Measure, IEEE Transactions on Pattern Analysis and Machine Intelligence, PAMI-1(2): pp. 224–227. (1979).
13. Juha, A. , Esa, A., Clustering of the Self-Organizing Map, IEEE Tractions On Neural Networks volume 11, n° 3, (2000).
14. E. Alhoniemi and J. Himberg and J. Parhankangas and J. Vesanto, SOM Toolbox, 2000, Copyright (C),http://www.cis.hut.fi/projects/somtoolbox/.
15. Projet Arcade, Architecture de Control Adaptative des Environments IP, Web Site: http://www-rp.lip6.fr/arcade/, 2002.
16. Projet IpSig, Signalisation générique du monde IP, http://www.telecom.gouv.fr/rnrt/rnrt/projets/res_02_85.htm, 2004.
17. Zeina JRAD, Thesis report, Apports des techniques de l'Intelligence Artificielle dans la négociation dynamique de la qualité de service : Proposition d'un assistant à l'utilisateur dans les réseaux IP de nouvelle génération, LIPN, University of Paris 13, France, May 2006.
18. Z. Jrad., B. Benmammar, J. Correa, F. Krief , N. Mbarek, A user assistant for QoS negotiation in a dynamic environment using agent technology. Second IFIP International Conference on Wireless and Optical Communications Networks WOCN 2005. Dubai, United Arab Emirates UAE, March 2005.

CHAPTER 10

APPLICATION OF REUSE PARTITIONING CONCEPT TO THE IEEE 802.16 SYSTEM: DESIGN AND ANALYSIS

S. HAMOUDA[(1)], P. GODLEWSKI[(2)], S. TABBANE[(1)], AND SENIOR MEMBER IEEE

[(1)]*Ecole Supérieure des Communications de Tunis (Sup'Com). Parc Technologique des Communications El-Ghazela, Ariana, 2083, Tunisia (e-mail: {soumaya.hamouda, sami.tabbane}@supcom.rnu.tn)*

[(2)]*Ecole Nationale Supérieure des Télécommunications (ENST), 46 rue Barrault, 75634 Paris Cedex, France (e-mail: godlewski@enst.fr)*

Abstract: This paper is a continuation of a previous work on resource management in fixed/portable Orthogonal Frequency Division Multiple Access (OFDMA) systems [1]. In [1], we gave efficient subcarrier assignment and power control algorithms to minimize the outage probability in such systems. Since the IEEE 802.16 standard is an OFDMA-based system, it is worth implementing our scheme and evaluating the achievable performances while respecting the particular specifications of the standard. Interesting results were found in terms of capacity. We showed that for a low outage probability, the number of simultaneous users in one sector requiring a given QoS can be doubled as compared to a classical subchannel assignment based on a universal reuse

1. INTRODUCTION

The WiMAX technology, based on the IEEE 802.16d-2004 air interface standard, was primarily designed for fixed Point-to-Multipoint (PMP) broadband applications (e.g. a backhaul for 801.11 hotspots) [2]. But it rapidly became very attractive as it proved to be a cost-effective wireless alternative to cabled access networks [2], [3]. More recently, a new amendment of the fixed/portable IEEE 802.16d-2004 standard was ratified in December 2005. The IEEE 802.16e-2005 standard, referred to as Mobile WiMAX, supports mobility and could therefore compete with the next generation of radio mobile networks. One of the major reasons for the success of the IEEE 802.16 is its Orthogonal Frequency Division Multiple Access (OFDMA) based multiple access technique. In fact, OFDMA is well-known for offering high data rates [4]. It also provides flexible resource allocation to benefit from

H. Labiod and M. Badra (eds.), New Technologies, Mobility and Security, 111–121.
© 2007 *Springer*.

multi-user diversity. The IEEE 802.16 air interface standard gives specifications regarding OFDMA symbol structure and sub-channelization [5]. However, it does not precisely specify any subchannel assignment algorithm leaving the designers free to apply their own strategies.

On the other hand, several studies were independently carried out to optimize subcarrier assignment in downlink OFDMA networks (e.g. [4], [6]–[10]). Different strategies were proposed: some (e.g. [6]–[8]) consisted of a so-called RA (Rate Adaptive) optimization which aims to maximize the total achieved data rate in the system; others (e.g. [4],[9],[10]) addressed an MA (Margin Adaptive) optimization which corresponds to a minimization of the total transmit power at the base station. These two strategies led to an efficient use of multi-user diversity and have thus improved the spectral efficiency. However, they failed to ensure fairness among the users as they only benefit users closer to the base station. The fairness issue was dealt in [11]–[13]. The authors proposed algorithms to maximize the user's minimum data rate [11], or to add user rate proportional constraints to the RA optimization [12],[13]. In [14], in order to achieve load balancing within the system and ensure a certain fairness, the authors minimized the maximum value of the QoS violation ratios. In a previous study [1], we proposed a centralized reuse partitioning scheme that also aimed to minimize the outage probability in the entire system. This scheme has lessened the impact of Co-Channel Interference (CCI) particularly for users at the cell borders.

Like most of the previous studies, the approach in [1] has proved its efficiency with theoretical system models. In this paper, we extend our approach to the IEEE 802.16 system. We evaluate to what extent this scheme can be effective in terms of capacity while respecting the IEEE 802.16 standard. A former paper has applied a fairness optimization strategy to the IEEE 802.16 system [15]. The authors showed that their scheme can lead to a fairness index of 0.96% if all subchannels are assigned in the cell. However, they assign one single subchannel to each user. In the case of a small number of users, they would only use part of the total available subchannels, which limits the total throughput in the system. In our strategy, a user may be assigned several subchannels according to the minimum required data rate and the available total bandwidth. The total throughput can therefore be higher when few users requiring high data rate are in the system.

This paper is organized as follows: in section 2, we present the subchannelization as specified in IEEE 802.16d-2004 [5]. We specially consider the subchannelization in DL-FUSC mode. Section 3 contains details about our system model referring to IEEE 802.16 architecture network and propagation channel models [16]. In section 4, we state the optimization problem and go over the main steps of our approach in [1]. Then, we present and discuss our simulation results in section 4. Finally, section 6 concludes this paper.

2. OFDMA SUB-CHANNELIZATION IN IEEE 802.16 SYSTEM (DL-FUSC MODE)

The OFDMA symbol structure, as defined in IEEE 802.16, consists of three types of subcarriers: data subcarriers for data transmission, pilot subcarriers for estimation and synchronization purposes and null subcarriers which are kept for guard bands and DC carriers[1]. Active (data and pilot) subcarriers are grouped into subsets of sub-carriers called subchannels [5]. Since the IEEE 802.16 system is based on OFDMA, each user in a given cell occupies a subset of sub-channels exclusively assigned to him at any given time.

The subchannel allocation, however, can be performed in partial usage of subchannels (PUSC) where some of the subchannels are allocated to the transmitter, or full usage of subchannels (FUSC) where all subchannels are allocated to the transmitter [5]. In this paper, we consider the FUSC mode in downlink (DL-FUSC) with a Fast Fourier Transform (FFT) size of 2048 subcarriers.

Among the total 2048 subcarriers, 1536 subcarriers are dedicated to data subcarriers and mapped into 32 subchannels, each with 48 subcarriers ($32 \times 48 = 1536$). In order to handle varying channel conditions, the subcarriers assigned to each subchannel are pseudo-randomly distributed across the available subcarrier set. To do so, the latter is subdivided into 32 contiguous groups. Then, each subchannel consists of one subcarrier from each of these groups (Fig. 1) [3][5]. A permutation formula is specified in [5] for an exact partitioning into data subchannels. Fig. 2 shows an example of the mapping of 3 subchannels. One can notice that subcarriers belonging to each subchannel are distributed all over the available active subcarrier space.

2.1. System Model

2.1.1. System Layout

In the downlink, the IEEE 802.16-2004 standard operates on a PMP basis with a central base station (BS) and a sectorized antenna that is capable of handling multiple independent sectors simultaneously [5]. The system supports a frequency reuse of FRF = 1 i.e. all the base stations in the system are using the same RF frequency [5]. The IEEE 802.16-2004 standard, nevertheless, proposes three options of operation in the reuse of 1: asynchronous, synchronous and coordinated synchronous configurations. In our study, we consider the coordinated synchronous configuration in which all the

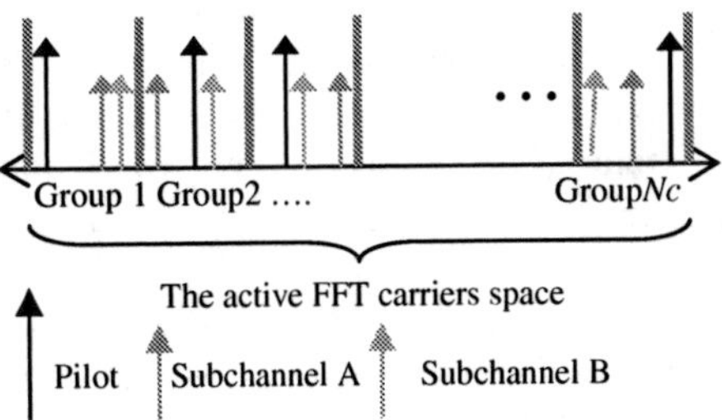

Figure 1. The principle of subchannelization in DL-FUSC [3]

[1] DC carriers are dedicated to the RF center frequency of the station.

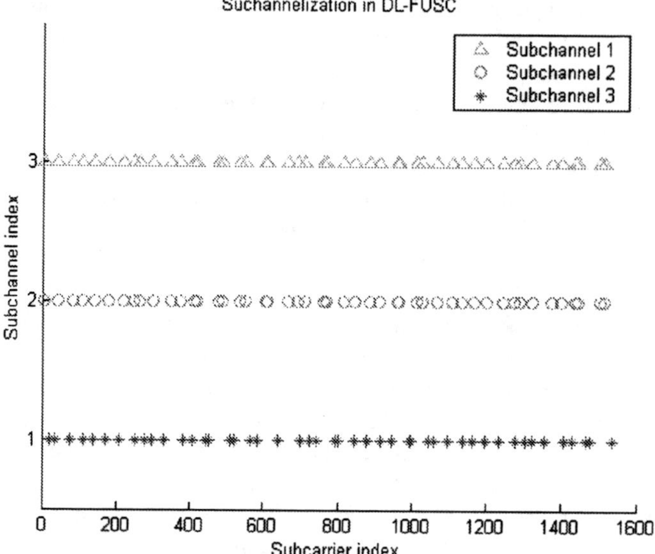

Figure 2. Example of 3 subchannels formed in DL-FUSC mode

base stations work in the synchronous mode but also use the same permutations [5]. This means that the subcarriers allocated to a given subchannel in a certain cell are the same as those allocated to that same subchannel in another cell. This configuration clearly introduces CCI in the system. The standard stipulates that an upper layer is responsible for the handling of subchannel allocations within the sectors of the base station and the system in order to achieve the best load balance between the sectors and within the system. However, it does not specify the manner to do so leaving the designers free to apply their own algorithms. Our proposed subchannel assignment scheme, based on a reuse partitioning concept, can easily be implemented in such a system to improve its performances.

The considered system model is therefore the outcome of three adjacent sectors from three adjacent hexagonal cells (Fig. 3). For users located in a given sector, CCI comes from only the two neighboring sectors. We assume that each BS b serves the same number U_b of uniformly distributed users. The location of the users is, however, different in each sector (Fig. 3). We denote by B the number of sectors ($B = 3$), and U the total number of users in the entire system (i.e. $U = U_b \times B$).

Furthermore, we suppose that all base stations have perfect knowledge of all propagation channel gains and that the maximum transmit power P_{max} is the same at each BS. On the other hand, fixed and equal transmit power are applied on each subcarrier whereas the transmission rate is variable using adaptive coding/modulation. Transmit power per subcarrier is obviously chosen such that the sum of the transmitted power per BS over all subcarriers is inferior to the maximum

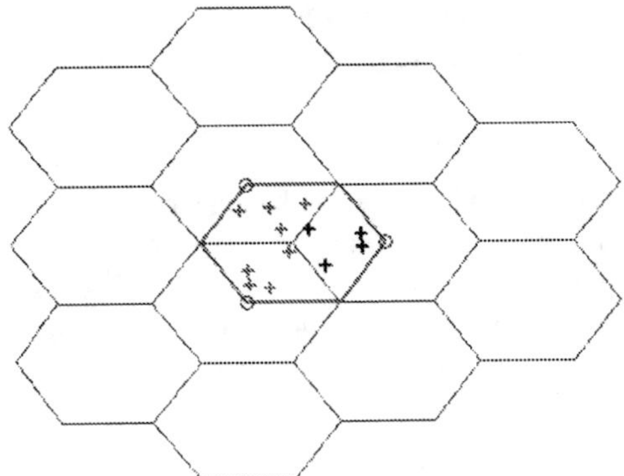

Figure 3. System layout and example of a user distribution

BS transmit power. We also assume the same background noise power over all subcarriers.

2.2. Channel Model

We consider propagation channel models provided by the IEEE 802.16 Broadband Wireless Access Working Group for fixed wireless applications [16]. This working group sets six typical Stanford University Interim (SUI) channel models for three types of terrains. We will use SUI-3 channel model in our simulations, which corresponds to a terrain with moderated propagation conditions (category B). This channel is also for low mobility with small delay spread (Tables 1). Other simulation parameters are presented in Table 2.

Table 1. SUI-3 channel

	Tap1	Tap2	Tap3	Units
Delay	0	0.4	0.9	μs
Power (omni. ant)	0	–5	–10	dB
K factor	1	0	0	
Maximum Doppler frequency	0.4	0.3	0.5	Hz

Table 2. Other Simulation Parameters

	Values	Units
Total bandwidth *W*	10	MHz
RF frequency	3.3	GHz
Cell radius	3.5	Km
Maximum transmit power at the base station	4	Watts
Power noise per Hz	–174	dBm/Hz

3. OPTIMIZATION PROBLEM AND PROPOSED REUSE PARTITIONING FOR THE IEEE 802.16 SYSTEM

3.1. Optimization Problem

Our objective is to optimize frequency planning and reuse such that data rate requirements for a maximum number of users is met. We define the outage probability P_{out} as the set of users u whose QoS requirements have *not* been satisfied. We also assume that the QoS relative to a certain service is determined by the minimum required data rate by that user $r°_{b,u}$. The outage probability is therefore:

$$(1) \qquad P_{out} = Prob\{u \text{ in all sectors } b/r_{b,u} < r°_{b,u}\}$$

where $r_{b,u}$ is the achieved data rate by user u in the sector b.
The optimization problem is thus expressed by:

$$(2) \qquad \begin{array}{c} \min P_{out} \\ \text{subject to } \Sigma_{u=1,..,Ub} N_{b,u} \le N \text{ for } b = 1, \ldots, B \end{array}$$

where $N_{b,u}$ is the total number of subchannels assigned to the user u in the BS b and N is the total number of subchannels.

3.2. Reuse Partitioning Concept for an IEEE 802.16 System

3.2.1. Principle

For a user u to whom is assigned a subchannel n, we assume that the Signal-to-Interference plus Noise Ratio (SINR) estimated at the BS b is $\gamma_{b,u,n}$. This ratio is the average of all the SINR $\gamma_{b,u,m}$ evaluated over the relative subcarriers m of the subchannel n. It is then equal to $\gamma_{b,u,n} = 1/N_c \sum_{m=1,..,Nc} \gamma_{b,u,m}$ where N_c is the number of subcarriers allocated to the channel n (according to the subchannelization previously described, $N_c = 32$ for all n). Then, according to Shannon's channel capacity formula, the data rate $r_{b,u}$ achieved by that user u is given by:

$$(3) \qquad r_{b,u} = \frac{W}{N} \sum_{n=1,..,N_{b,u}} \log_2(1 + \gamma_{b,u,n})$$

where W is the total bandwidth.

Consequently, to achieve a given data rate $r°_{b,u}$, the larger the SINR estimated by a user over the entire subchannels, the smaller the bandwidth allocated to him (i.e.: $N_{b,u}$). In cellular networks, assuming an equal transmit power and noise background per subchannel, smaller bandwidth is needed for users near the base stations, whereas users next to the cell edges require greater bandwidth as they experience more significant CCI.

The basic idea of our approach is to reduce the impact of CCI for users next to cell borders by increasing the frequency reuse factor (FRF) for subcarriers assigned to them. Therefore, the entire set of subchannels is partitioned into two subsets: in the first subset subchannels are reused with an FRF_1 equal to the one stipulated by the IEEE 802.16 standard, and in the second with an FRF_2 such that $FRF_1 > FRF_2$. We fix FRF_2 to $1/B$, B is the number of sectors considered in the system model. Subchannels with FRF_1 are assigned to the so-called *near* users (near to their serving BS) and can be reassigned in the entire system. Respectively, those with FRF_2 are attributed to *far* users (far from their serving BS) and are assigned in only one sector. Our subchannel assignment approach consists of a centralized two-step algorithm: first, we determine the number of subchannels with the respective FRF, that means classifying users into *far* and *near* and establishing for each one of them the number of subchannels needed. Then, we assign particular subchannels for each user. More details about these two steps can be found in [1]. We present hereafter the main points.

3.2.2. *Bandwidth Allocation*

In this step, we determine the number of subchannels used with each FRF. For each user u in the system, we estimate the SINR $\gamma_{b,u,n}$ at the serving BS b over each subchannel n assuming that the latter is reused with FRF_1 ($FRF_1 = 1$). The bandwidth $w_u{}^1$ which would be needed for the user u in that case is then deduced by:

$$(4) \qquad W_{b,u}{}^1 = r^{\circ}_{b,u}/\log_2(1+\bar{\gamma}_{b,u})$$

where $\bar{\gamma}_{b,u} = 1/N \sum\limits_{n=1,..,N} \gamma_{b,u,n}$ is the average of $\gamma_{b,u,n}$ over all subchannels.

In the same way, we estimate the bandwidth $w_{b,u}{}^{\infty}$ required for each user u at the BS b over each subchannel n reused with FRF_2 ($FRF_2 = B$):

$$(5) \qquad w_{b,u}{}^{\infty} = r^{\circ}{}_{b,u}/\log_2(1+\bar{\gamma}_{b,u})$$

where $\bar{\gamma}_{b,u} = 1/N \sum\limits_{n=1,..,N} \gamma_{b,u,n}$. Note that in this case $\gamma_{b,u,n}$ corresponds simply to the signal-to-noise ratios.

Then, for each user u in the cell/sector b, we evaluate the ratio $\alpha_{b,u} = r_{b,u}{}^1/r_{b,u}{}^{\infty}$ where $r_{b,u}{}^I$ and $r_{b,u}{}^{\infty}$ are the achievable data rate if only one FRF is used in the entire system, respectively FRF_I and FRF_2. Let $\bar{\alpha}$ be the average of all $\alpha_{b,u}$ ratios. For each user u, if $\alpha_{b,u}$ is superior to $\bar{\alpha}$ then the bandwidth allocated to u is equal to $w_{b,u}{}^{\infty}$ and the user u is considered a far user. Otherwise, a bandwidth of $w_{b,u}{}^I$ is allocated to u.

3.2.3. *Fixed Subchannel Assignment*

In this paper we only address fixed subchannel assignment algorithm. In fact, according to the subchannelization in DL-FUSC previously described, the subcarriers belonging to each subchannel are distributed over the entire FFT space. Consequently, dynamic subchannel assignment based on subchannel quality reordering won't add any significant improvement. Therefore, our subchannel assignment algorithm consists of assigning particular subchannels in a fixed manner while respecting the bandwidths determined earlier and the correspondent FRFs. We first assign subchannels to *near* users in each cell. Then, we assign subchannels to *far* users starting from the user that requires the smallest bandwidth. As we assumed a centralized scheme, starting by *far* users requiring the least bandwidth clearly reduces the outage probability in the entire system and provides better spectrum utilization.

4. SIMULATION RESULTS AND ANALYSIS

To evaluate the effectiveness of our subchannel assignment scheme based on the reuse repartitioning concept, we compare our approach to more classical subchannel assignment strategies which can be applied to an IEEE 802.16 system. We call FSA2 (Fixed Subchannel Assignment 2), the reuse partitioning scheme we propose ("2" refers to the deployment of two different FRF); and FSA1 (Fixed Subchannel Assignment 1) and EqBw (Equal-Bandwidth strategy) the schemes used for comparison purposes. The scheme FSA1 as well as EqBw are based on fixed subchannel assignment and operates with universal FRF (FRF = 1) in the entire system. In the FSA1 scheme, each user u in a sector b is allocated the bandwidth of $wb, u = wb, u^1$. The EqBw scheme merely subdivides the total bandwidth into contiguous sets of subchannels. There is as many of these sets as there are users in each sector. In both cases, for each sector b, subchannels are assigned to users successively one after the other.

We have conducted simulations over 1000 snapshots assuming that all users require the same data bit rate r° i.e.: $r^\circ = r_{b,u}$. Fig. 4 (upper simulations) gives the outage probability vs. r° for a given number of simultaneous users in the sector U_b ($U_b = 9$) in case of FSA1, FSA2 and EqBw. We observe that the FSA2 scheme offers higher capacity than FSA1 and BqBw ones at a given P_{out}. For instance at $P_{out} = 7\%$, FSA1 and FSA2 strategies respectively provide a minimum data rate of 2.25 Mbps and 4 Mbps. In fact, because of the large number of subcarriers inside a given subchannel, the average signal-to-noise ratios become much superior to the average signal-to-interference-plus-noise ratios when the FRF is decreased. The achieved data rates are thus enhanced specifically for users at the cell border.

Figure 4 (lower simulations) illustrates the outage probability vs. the number of users per sector for $r^\circ = 4$ Mbps and for the three strategies FSA1, FSA2 and EqBw. Also in this case, the reuse partitioning scheme considerably improves the outage probability in the system as it ensures a better fairness than FSA1. As for the EqBw scheme, it is obviously not efficient (in upper and lower simulations) because it doesn't take into account the user's requirement.

Fig. 5 gives the minimum required data rate r° vs. the number of simultaneous connected users per sector for an outage probability less than 7%. This figure shows for instance that with the reuse partitioning concept, the IEEE 802.16-2004 can offer a minimum data rate per user of $r^\circ = 3$ Mbps with a P_{out} less than 7%

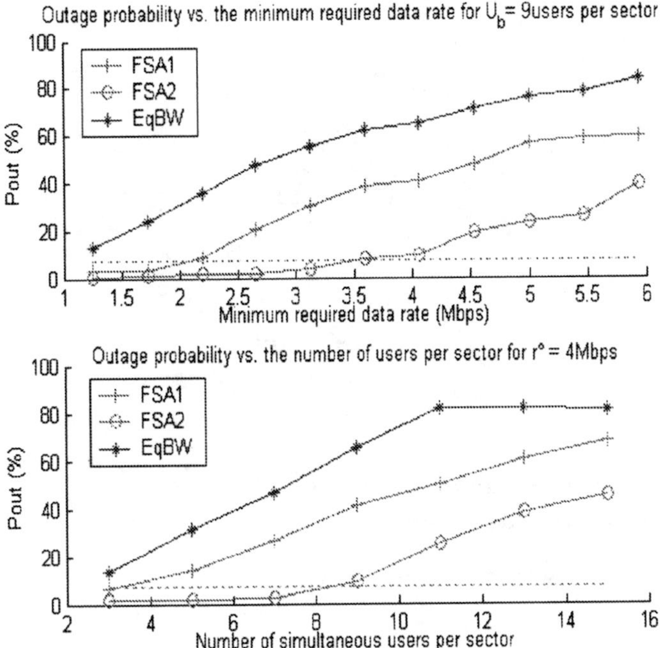

Figure 4. Comparison of the outage probability in the entire system

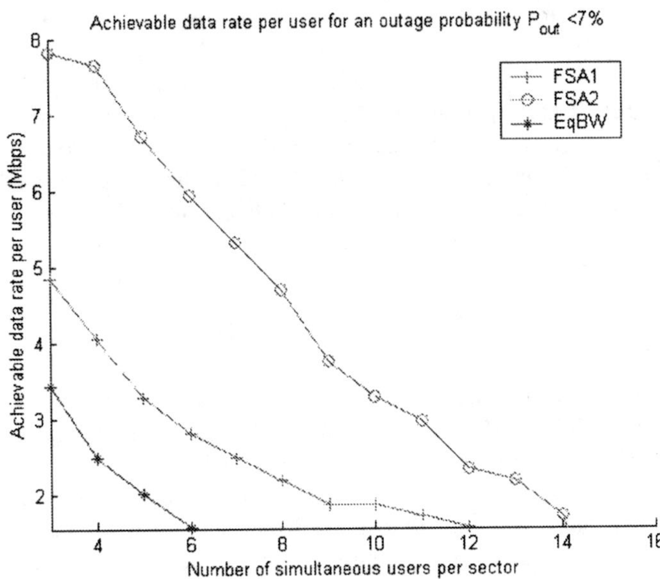

Figure 5. Achievable data rate per user vs. the total number of users per sector U_b at a outage probability $P_{out} < 7\%$

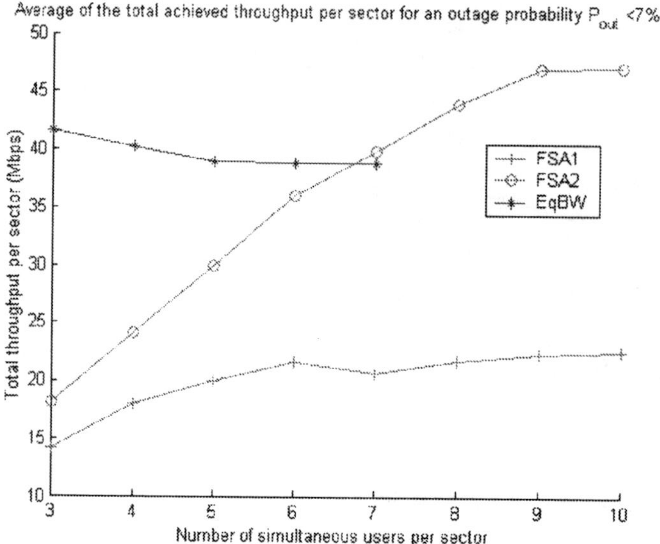

Figure 6. Average of the achieved capacity per sector vs. U_b for $P_{out} < 7\%$

for 11 simultaneous users per sector; whereas this rate can only be offered to 6 users with the FSA1 strategy. This enhancement represents about 100% of the simultaneous users in the sector as compared to FSA1 and 150% as compared to EqBw.

Finally, Fig. 6 gives the average total throughput per sector at each r° for $P_{out} < 7\%$ vs. the number of users per sector. Our approach achieves an average total throughput much higher than both of FSA1 and EqBw schemes for a loaded system. When the number of users per sector is small, EqBw clearly achieves a better total throughput as it allocates the entire available bandwidth. However, it doesn't satisfy the QoS user's requirement as shown in Fig. 5.

5. CONCLUSION

In this paper, we applied a subchannel assignment scheme based on the reuse partitioning concept, to the IEEE 802.16 in the DL-FUSC mode. We investigated its impact on the outage probability and the total throughput in the system. We showed that our approach is efficient in DL-FUSC in terms of fairness as well as system capacity. As it helps users at the cell border achieve their minimum required data rate, the outage probability in the entire system significantly decreased. Simulations conducted while respecting the IEEE 802.16 subchannelization showed that the number of simultaneous users in one sector can be doubled for a given QoS requirement at a low outage probability.

REFERENCES

1. Hamouda, S; Godlewski, P; Tabbane, S; "Enhanced Capacity for Multi-cell OFDMA Systems with Efficient Power Control and Reuse Partitioning", published in IEEE International Conference on Communication Systems 2006 (ICCS'06) proceeding, November 2006.
2. Ghosh, A., Wolter, D.R., Andrews, J.G., Chen, R., "Broadband wireless access with WiMax/802.16: current performance benchmarks and future potential Communications Magazine, IEEE Volume 43, Issue 2, Feb 2005 Page(s):129–136.
3. I. Koffman and V. Roman, "Broadband wireless access solutions based on OFDM access in IEEE 802.16," *IEEE Communications Magazine*, pp. 96–103, Apr. 2002.
4. C.Y. Wong, R.S. Cheng, K.B. Letaief and R.D. Murch, "Multi-user Subcarrier Allocation for OFDM Transmission using Adaptive Modulation," Proc. VTC, May 1999.
5. IEEE 802.16 Standard for Local and Metropolitan Area Networks Part 16: Air Interface for Fixed Broadband Wireless Access Systems – October 2004.
6. Chi-Hsiao Yih; Geraniotis, E., "Adaptive modulation, power allocation and control for OFDM wireless networks", Personal, Indoor and Mobile Radio Communications, 2000. PIMRC 2000. The 11th IEEE International Symposium on Volume 2, 18–21 Sept. 2000 Page(s):809–813 vol.2.
7. Guoqing Li; Hui Liu, "Downlink dynamic resource allocation for multi-cell OFDMA system", Vehicular Technology Conference, 2003. VTC 2003-Fall. 2003 IEEE 58th, Volume 3, 6–9 Oct. 2003 Page(s):1698–1702 Vol.3.
8. Jiho Jang, Kwang Bok Lee, "Transmit power adaptation for multiuser OFDM systems", IEEE Journ. on SAC, Vol. 21, Issue 2, Feb. 2003, pp. 171–178.
9. Pietrzyk, S.; Janssen, G.J.M., "Radio resource allocation for cellular networks based on OFDMA with QoS guarantees", Global Telecommunications Conference, 2004. GLOBECOM '04. IEEE Volume 4, 29 Nov.-3 Dec. 2004 Page(s):2694–2699 Vol.4.
10. Kivanc, D.; Guoqing Li; Hui Liu, "Computationally efficient bandwidth allocation and power control for OFDMA", Wireless Communications, IEEE Transactions on, Volume 2, Issue 6, Nov. 2003 Page(s):1150–1158.
11. Rhee, W.; Cioffi, J.M.; "Increase in capacity of multiuser OFDM system using dynamic subchannel allocation",Vehicular Technology Conference Proceedings, 2000. VTC 2000-Spring Tokyo. 2000 IEEE 51st; Volume 2, 15–18 May 2000 Page(s):1085–1089 vol.2.
12. Wong, I.C.; Zukang Shen; Evans, B.L.; Andrews, J.G.; "A low complexity algorithm for proportional resource allocation in OFDMA systems", Signal Processing Systems, 2004. SIPS 2004. IEEE Workshop on 2004 Page(s):1–6.
13. Hanbyul Seo; Byeong Gi Lee; "Proportional-fair power allocation with CDF-based scheduling for fair and efficient multiuser OFDM systems"; Wireless Communications, IEEE Transactions on Volume 5, Issue 5, May 2006 Page(s):978–983.
14. Hojoong Kwon, Won-Ick Lee, and Byeong Gi Lee , "Low-Overhead Resource Allocation with Load Balancing in Multi-cell OFDMA Systems", Vehicular Technology Conference, 2005. VTC 2005-Spring. 2005 IEEE 61st Volume 5, 30 May-1 June 2005 Page(s):3063–3067 Vol. 5.
15. Li-Chun Wang; Wei-Jun Lin; "Throughput and fairness enhancement for OFDMA broadband wireless access systems using the maximum C/I scheduling"; Vehicular Technology Conference, 2004. VTC2004-Fall. 2004 IEEE 60th Volume 7, 26–29 Sept. 2004 Page(s):4696–4700 Vol. 7.
16. IEEE 802.16, Broadband Wireless Access Working Group, Channel Models for Fixed Wireless Applications, July. 2001.

CHAPTER 11

EVALUATION OF THE PERFORMANCE OF THE SLOPS: AVAILABLE BANDWIDTH ESTIMATION TECHNIQUE IN IEEE 802.11B WIRELESS NETWORKS

AMAMRA ABDELAZIZ, HOU KUN MEAN, AND CHANET JEAN-PIERRE

LIMOS Laboratory, UMR 6158 CNRS, University of Blaise Pascal
BP 1025 24 av des Landais 63173 Aubière
France

Abstract: Over the past few years, to actively measure the end-to-end available bandwidth of a network path several algorithms have been created (TOPP, SLoPS, Spurce, VPS ...). These algorithms are dedicated to wired Ethernet network; however the bandwidth estimation in wireless network field remains a challenge. In the present work we evaluate the performance of Self-Loading Periodic Stream (SLoPS) method to estimate the available bandwidth in the IEEE 802.11b wireless network. By using different probe packet sizes and different Cross-Traffics, we show that the available bandwidth measurements are affected by the varying Cross-Traffic in both Ethernet and wireless IEEE 802.11b network. The variation of probe packet size affects only the available bandwidth measurements in the wireless network and not in Ethernet network case. Also SLoPS technique provides less accurate measurements in the IEEE 802.11b wireless network. The simulation is built in NS-2. The results are analyzed by MatLab software

Keywords: IEEE 802.11b, SLoPS, Available Bandwidth, Cross-Traffic, Quality of Service

1. INTRODUCTION

1.1. Overview

Until now, Quality of Service (QoS) in wireless network is still an open problem, due to the dynamic topology of wireless nodes and the brittleness of wireless link [12]. In this paper we study one aspect of QoS, which is the guarantee of a high available bandwidth and capacity that a link or path can deliver. The capacity is defined as the maximum IP-layer throughput, which the link can provide to a flow when there is no competing traffic load (Cross-Traffic) [14]. The link capacity is

H. Labiod and M. Badra (eds.), New Technologies, Mobility and Security, 123–132.
© 2007 *Springer.*

determined by its physical characteristics. Furthermore, the available bandwidth is defined as the maximum IP-layer throughput, which the link can provide to a flow in the presence of Cross-Traffic [14]. The available bandwidth depends on Cross-Traffic and link capacity. The available bandwidth is continually varying over time. So, it is important to measure it promptly. This is especially true for specific applications that use available bandwidth measurement to adapt their transmission rate [10,16].

1.2. Related Work

To estimate the capacity and the available bandwidth in wired network different types of techniques are developed [15,14,11]. The well-known ones is the Dispersion-based techniques [4]. Theses techniques use the Round Trip Time (RTT) or the One Way Delay (OWD) to estimate the available bandwidth. They are based on self-induced congestion by injecting probe packets into the network. The probe packets temporarily induce network congestion, if and only if the probing bit-rate exceeds the path available bandwidth, thus causing a noticeable increase in queuing delay. Consequently, the minimum probing bit-rate that causes network congestion gives an estimate of the available bandwidth. The well-known techniques are: Packet Pair/Train Dispersion (PPTD) [4], Self-Loading Periodic Streams (SLoPS) [10,11] and Trains of Packet Pairs (TOPP) [1,14,13].

PPTD technique is proposed in [9,2,7], measure the end-to-end capacity of a path. It consists of sending two same packets back-to-back from source to sink. The capacity of the link is the ratio of packet size by the time dispersion between two packets. This technique assumes that the link is empty of any other traffic. It uses very heavy and long statistical methods to filter out erroneous measurements.

TOPP technique is an extension of the packet pair probing technique. It has been studied by a number of researchers [13,9,3]. TOPP sends many trains of packet pairs from the source to the sink, by increasing the sending rate gradually to reach the available bandwidth bound.

SLoPS technique proposed in [10], measures the end-to-end available bandwidth. The sender probes the link with periodic packet stream of different rates. The receiver analyzes the sequence of OWD variations of the probing packets stream. The sender attempts to bring the stream rate close to the available bandwidth. The measurement algorithm is iterative and requires a cooperation of both sender and receiver. This technique will be described in detail in the section 2.

In our previous work [1], we have evaluated the TOPP technique in wireless network field. We have shown that in Ethernet LAN, the capacity is always stable but the available bandwidth fluctuates when the Cross-Traffic varies. However, both the link capacity and the available bandwidth of wireless network fluctuate with the Cross-Traffic and packet size change.

The SLoPS technique showed particular performance in the wired network field. In IEEE 802.11b wireless network, the available bandwidth estimation depends on the end-to-end delay measurement. The transmission error rate is high and packet loss can occur even when there is no congestion. Moreover the link rate variability is also due to Cross-Traffic using the same radio frequency channel within the ad-hoc network. Therefore, the bandwidth measurement in IEEE 802.11b wireless network is a challenging task. In our work, we apply SLoPS to estimate the available bandwidth in IEEE 802.11b wireless network. Then, the obtained results will be compared with the Ethernet ones.

This paper is organized as follows. In Section 2, we will describe the SLoPS technique in detail. Section 3 presents the simulation scenarios implemented by using NS-2, taking into account the network topology, the Cross-Traffic and the packet size. The obtained results are shown and explained in Section 4. Finally, some conclusions and the ongoing work will be presented in section 5.

2. SELF-LOADING PERIODIC STREAMS (SLOPS)

In this section, we describe the basic idea of SLoPS measurement technique. More information on definitions and theory can be obtained from [11]. SLoPS is an active measurement technique based on the dispersion of the probe packets. To measure the available bandwidth, the sender transmits a periodic probe packets stream to the receiver. Each probe stream consists of **K** packets (e.g. $K \approx \mathbf{100}$). All packets have equal size of **L** bits. The stream transmission rate is $\mathbf{R} = \mathbf{L/T}$ bit per second, where **T** (seconds) is the dispersion between packets (i.e. time interval between the first bits of two successive packets). Before their transmission, the packets are time stamped by the sender (Si). The receiver time stamps each arrival packet (Ri). Hence, the receiver can compute OWD of each packet as $\mathbf{Di} = \mathbf{Si} - \mathbf{Ri}$. **Di** is the (absolute) OWD. We are only interested in the relative magnitude of OWD. Thus, the measurement methodology does not need synchronized clocks. When the probe packets stream has been received, the receiver inspects the sequence of OWDs, and check whether the transmission rate **R** is larger than the available bandwidth **A**. The way we examine the relation between **R** and **A,** is the key idea in the measurement methodology, it will be detailed later.

The methodology involves monitoring OWDs variations of the probe packets. If the stream rate **R** is greater than the path's available bandwidth **A**, the stream will cause a *short-term overload* in the tight link queue of the path. During that overload period, the tight link receives more traffic than what it can transmit, so its queue size is gradually increased. Thus, when $\mathbf{R} > \mathbf{A}$, the OWDs of the probe packets stream are expected to have an increasing trend (Fig. 2.a) [11]. Moreover, if the stream rate **R** is less than the available bandwidth **A**, the probing packets stream will go through the path without causing an overload at the tight link. Thus, the backlog of that link will not keep increasing with every new stream packet. So, when $\mathbf{R} < \mathbf{A}$, the OWD of the stream packets is expected to have a non-increasing trend (Fig. 2.b) [11].

The goal of the SLoPS technique is to bring the stream rate **R** close to the available bandwidth **A**, following an iterative algorithm similar to binary search. The result is an interval with lower and upper bounds, shows as the Min and Max values of available bandwidth.

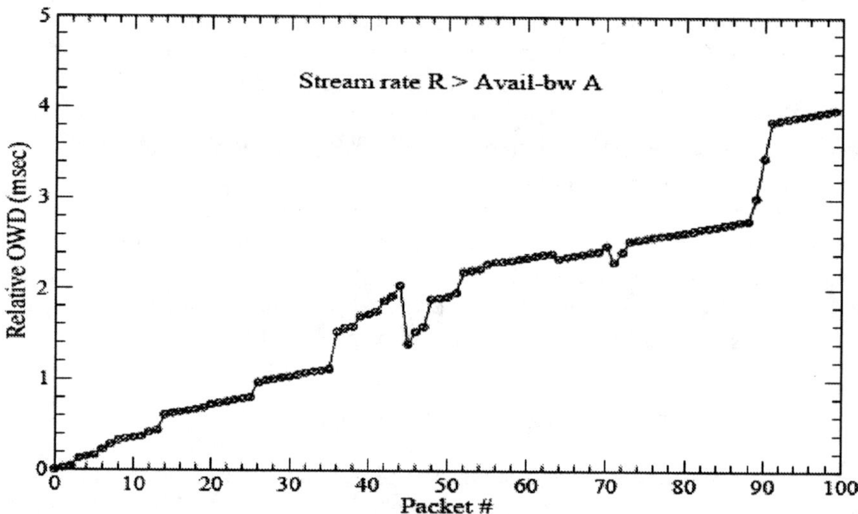

Figure 2.a. Increasing trend of OWD when R > A [11]

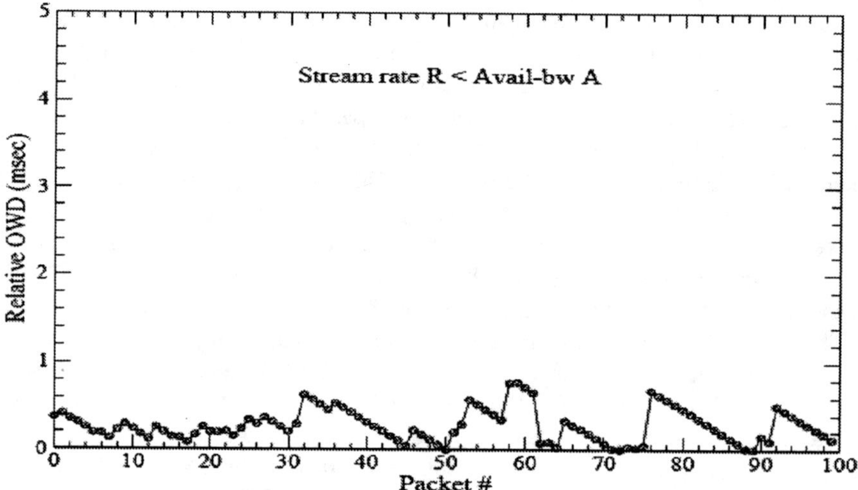

Figure 2.b. Decreasing trend of OWD when R < A [11]

3. SIMULATION

SLoPS technique is adapted to Ethernet network field, it provided a good performances. By applying SLoPS technique in IEEE 802.11b wireless network, we would like to explore its behavior and compare the results of WLAN with the LAN ones. In future work, we will be able to introduce a possible improvement in measurement technique. To achieve this goal, we carry out a simulation in NS-2 on two topologies networks, showed on Fig. 3.a and Fig. 3.b

Two simulation scenarios are implemented. In each scenario, two nodes exchange a constant traffic, (Cross-Traffic). To respect the principles of SLoPS technique, two other nodes probe the network to estimate the available bandwidth by sending Probe-Traffic. Different Cross-Traffic rate and different packet size are applied. The simulation results are analyzed using MATLAB.

3.1. Topology

In the first experimentation, the Ethernet nodes are used. As shown in the Fig. 3.a nodes A, B, C and D are equipped with 10 Mbps Ethernet network card (raw theoretical capacity). The Probe-Traffic transits over the link **A** →X-Y→ **D**. The Cross-Traffic transits over the link B → X-Y → C. We note that the packets of

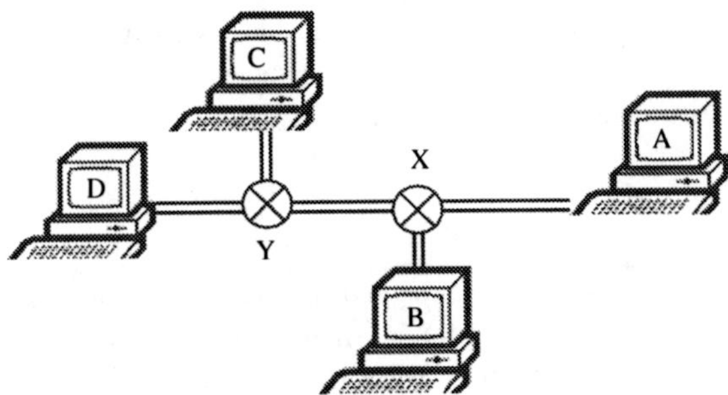

Figure 3.a. Ethernet network topology. Probe-Traffic A-x-y-D, Cross-Traffic B-x-y-C

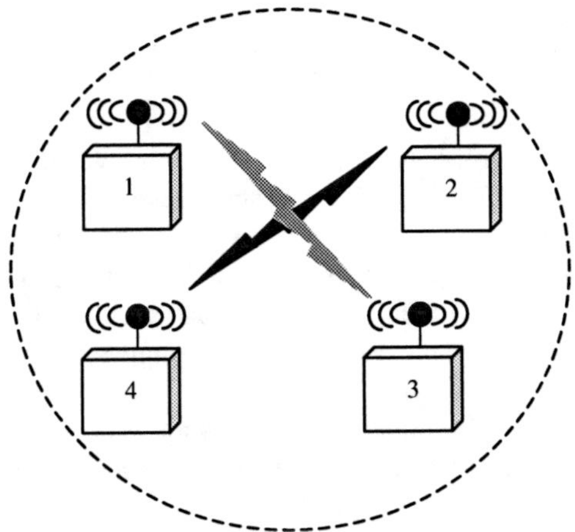

Figure 3.b. Wireless network topology. Probe -Traffic 2–4, Cross-Traffic 1–3

both traffics have the common link X $\longleftrightarrow$ Y. The theoretical capacity of this link is 10 Mbps. In our experimentation, we estimate the available bandwidth of this link.

In the second experimentation (Fig.3.b), we use the wireless nodes. Nodes 1, 2, 3 and 4 are equipped with IEEE 802.11b wireless network card (11Mbps of rough theoretical capacity). We choose this rate because it is similar to the wired network rate one (10 Mbps). The node locations are in a diameter that does not exceed 300m. Each of which would be within communication range of others. Then, wireless link is common between the nodes. The Cross-Traffics are sent from the node 1 towards the node 3. Nodes 2 and 4 try to probe the network to estimate the available bandwidth, by sending a Probe-Traffic.

3.2. Packet Size

For the packet size, the MTU *(Maximum Transfer Unit)* is taken into account [8,12], and it equals to 1518 bytes before packet fragmentation (of which 1500 bytes is available for the payload). Frames larger than 1518 bytes are normally fragmented to comply with the standard. The packet sizes used in the scenarios are 128, 512, 1000 and 1500 bytes. But the Cross-Traffic is fixed at 2 Mbps.

3.3. Cross-Traffic

In all scenarios, we use a CBR (Constant Bit Rate) to generate Cross-Traffic. The same rate of Cross-Traffic is maintained for all experimentations. Furthermore, we realized other experimentations with different CBR Cross-Traffic (0, 2, 3 and 4 Mbps), to estimate the available bandwidth of the link in presence of Cross-Traffic perturbation.

4. RESULTS AND DISCUSSION

4.1. Packet Size Effect

The diagram in Fig. 4.1, illustrates the probe packet size effect in available bandwidth measurements when Cross-Traffic is 2 Mbps. Probe packet has the following size: 128, 512, 1024 and 1500 bytes, shown on the x-axis. The y-axis shows the measurement in Mbps. The maximum values of available bandwidth measurements are plotted in solid line, the minimum values in dashed line. The following subsections will show that the available bandwidth measurements are depended on the packet size.

In Fig. 4.1, the two upper curves show the measured available bandwidth obtained in the Ethernet scenario. Note that, the results of the measurement of the link available bandwidth are between 7.51 Mbps and 7.70 Mbps. It is quite stable throughout the different packet size values. Thus, we can deduce that in Ethernet network, the available bandwidth link is independent of the packet size variation.

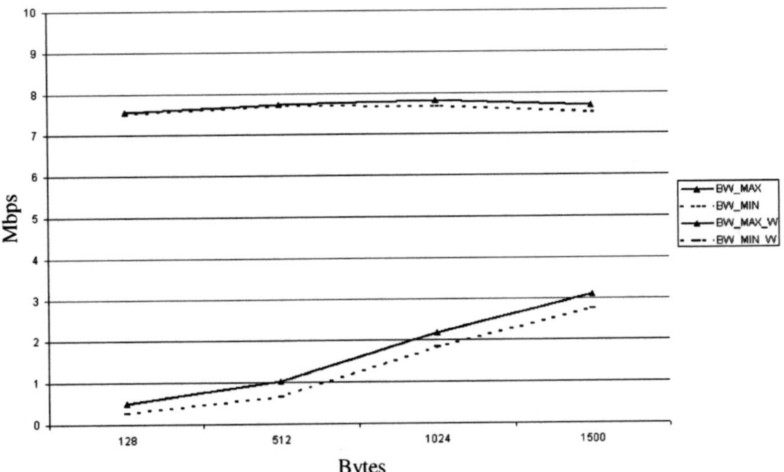

Figure 4.1. The available bandwidth in function of packet size in both Ethernet and IEEE 802.11b network scenarios Cross-Traffic is 2 Mbps

Also, the estimate interval obtained from the SLoPS technique is small. Then, the measurements are more or less accurate.

Nevertheless, the two lower curves in Fig. 4.1 prove that the variation of the packet size from 128 bytes up to 1500 bytes, gives increasing values of available bandwidth measurement, in the wireless scenario. The maximum value is 3.09 Mbps when packet size is equal to MTU. The available bandwidth depends on the packet size variation. It should be observed that the difference between the maximum and the minimum of available bandwidth measurements are huge, because the estimate interval obtained from the SLoPS technique is large. Then the wireless scenario's measurements are more inaccurate than the wired scenario one. (We explain the causes in detail in the subsection C).

4.2. Cross-Traffic Effect

The effect of the Cross-Traffic on available bandwidth measurements is illustrated in Fig. 4.2. The packet size is 1024 bytes. The CBR Cross-Traffic rate is shown on the x-axis, and varying from 0 to 4 Mbps. The y-axis shows the measurement in Mbps. Solid lines represents maximum values of available bandwidth measurements and dashed lines represent minimum values.

The two upper curves in Fig. 4.2 show the available bandwidth measurements, obtained in the Ethernet scenario. We can observe that varying CBR Cross-Traffic from 0 Mbps up to 4 Mbps, gives decreasing values of the link available bandwidth measurements. This explains that in Ethernet network the available bandwidth is depending on the CBR Cross-Traffic variation. The estimate interval of SLoPS method remains small.

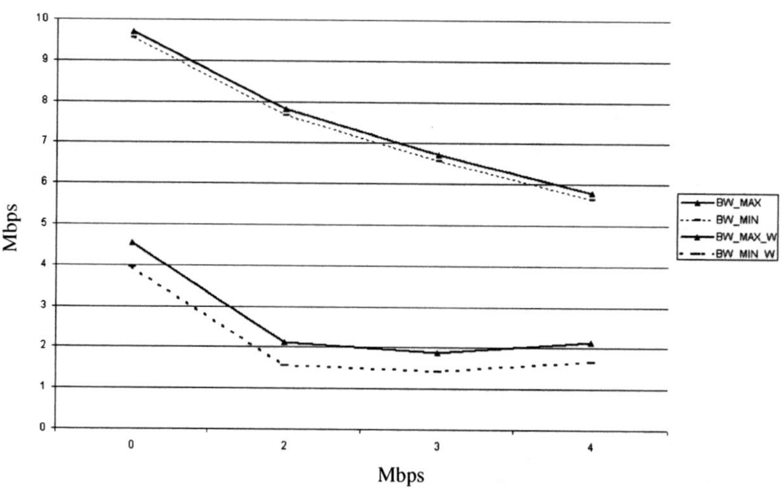

Figure 4.2. The available bandwidth in function of CBR Cross-Traffic in both Ethernet and IEEE 802.11b network scenarios. Probe packet size is 1024 bytes

The two lower curves in Fig. 4.2 prove that the variation of CBR Cross-Traffic from 0Mbps to 4Mbps, gives decreasing values of available bandwidth measurement, in wireless scenario. Then the available bandwidth depends on the packet size variation. But the estimate interval is larger than in upper curves. Therefore, the wireless scenario's measurements are more inaccurate than the Ethernet scenario ones.

4.3. Discussion

For Ethernet measurement (Fig.4.1), the available bandwidth values are around [7.51Mbps–7.70 Mbps] with a precision of 0.09 Mbps, the Cross-Traffic is 2Mbps. This estimation is very reasonable and found its principal explanation in characteristic of Ethernet medium access method. In the Ethernet networks, the medium access is made by the means of CSMA with *Collusion Detection* (CD). This method is known by a very low ratio of collision.

Furthermore, in IEEE 802.3 protocol, the data overhead (MAC header, CRC, Inter Frame Gap...etc.) is 38 bytes. Comparing with the total size of the packet 1500 bytes, the CSMA\CD is efficient with ~97%[6], which less affects the available bandwidth. That explains the precision of Bandwidth measurements. Thus, distance interval between BW_MIN and BW_MAX is small. Contrary to measurements in the wireless network, the distance between BW_MIN_W and BW_MAX_W is large; this is due to characteristics of access method CSMA with *Collision Avoidance* (CA) and the high generated collision ratio. The explanation is detailed in the following subsection.

According to the curves obtained in previous subsections, in Fig.4.2, the maximum values of the available bandwidth measurements are around [4.00 Mbps, 4.50 Mbps] with a precision of 0.25 Mbps, in case of no Cross-Traffic. This estimation is around the half of the theoretical capacity of link (11 Mbps). This is due in particular to data overhead necessary to CSMA\CA for medium access management. In IEEE 802.11b, there are three distinctly different types of frames: control, management and data. These frames carry information related to their names. In all cases they are sent and received in basic rate (1 Mbps). To send each packet, IEEE 802.11b needs to send 178 bytes (RTS, CTS, ACK and MAC header, Phy header, with DSSS physical layer), and wait a minimum 100μs (SIFS, DIFS, Slot time. . .etc corresponds to send $\sim$ 138 byes). This is a very important comparing to data payload. In IEEE 802.11b, data overhead is approximately [45% – 55%] of link capacity use [12]. These justify the difference between the estimate values and the theoretical values.

In wireless network, the available bandwidth values change with the variation of CBR Cross-Traffic and packet size. We find the explanation of this phenomenon in the characteristics of MAC and Physical layers of wireless network. In IEEE 802.11b, for each sent packet, MAC layer wait for contention period, send RTS\CTS\ACK packets and wait for three SIFS delays between packets. Hence if the packets number increases, the contentions period number, the RTS\CTS\ACK packets and SIFS delays increase. Consequently, the link available bandwidth decreases.

Furthermore, the reason of the decreasing measurement values of the available bandwidth can be explained by the amount of the induced overhead data. If the probe packet size is small, the overhead induced by CSMA\CA is more important comparing with the large probe packet size one. Hence, we come to the conclusion that large probe packets will measure a large available bandwidth than small probe packet.

A final remark is that in all cases, the IEEE 802.11b wireless network is more sensitive to Cross-Traffic variation than Ethernet wired network due to the vulnerability of wireless link and a different undesirable phenomena, Doppler effect, multi-path etc.

5. CONCLUSIONS

In this article, we have shown the effect of the variation of the packet size and the Cross-Traffic on available bandwidth measurements. To realize these measurements, the SLoPS technique is applied to wireless field. For comparison and validity we have implemented with NS-2 simulation two similar scenarios (Ethernet and Wireless) having the same topology, and we have shown that the available bandwidth estimations are not the same in both fields.

The available bandwidth of Ethernet network depends only on Cross-Traffic but not on packet size. However, the wireless available bandwidth depends on Cross-Traffic and on probe packet size.

Our conclusions are that the estimation measurement in the wireless field are difficult and require a specific technique which takes into account the difference between the Ethernet and the wireless field, in particular the MAC and the physical layers.

In the future research, we will propose a new available bandwidth estimation technique dedicated to wireless sensor network (WSN) by combining the SLoPS and the TOPP techniques. This new technique takes into account the loss packet ratio and the variations of OWD. Moreover, all these work will be implemented on the LivePlatform [5] and the obtained results will be compared with the simulation ones. Finally, the best available bandwidth estimation technique will be implemented to improve the QoS in WSN.

REFERENCES

1. Amamra A, Hou KM, Chanet JP (2006) Wireless Available Bandwidth Estimation: TOPP. 5th edn, international I2TS, Cuiabà, MT-Brazil
2. Bolot JC (1993) Characterizing End-to-End Packet Delay and Loss in the Internet. ACM SIGCOMM, pp 289–298
3. Carter, Crovella ME (1996) Measuring bottleneck link speed in packet-switched networks. Technical Report TR-96-006, Boston University Computer Science Department, Boston, MA, USA
4. Dovrolis C, Ramanathanm P, Moore D (2001) What Do Packet Dispersion Techniques Measure?. In: IEEE INFOCOM'01, Anchorage, AK, USA, vol 2, pp 905–914
5. Hou KM et al (2006) LiveNode: LIMOS versatile embedded wireless sensor node. Technical Report, LIMOS, France, pp 1–6
6. Http://www.erg.abdn.ac.uk/users/gorry/course/lan-pages/enet-calc.html
7. Jacobson V (1988) Congestion Avoidance and Control. ACM SIGCOMM, pp 314–329
8. Johnsson A, Melander B, Björkman M (2005) Bandwidth Measurement in Wireless Networks. Mediterranean Ad Hoc Networking Workshop, Porquerolles, France
9. Keshav S (1991) A Control-Theoretic Approach to Flow Control. ACM SIGCOMM, pp 3–15
10. Manish J, Dovrolis C (2002) End-to-end available bandwidth: Measurement methodology, dynamics, and relation with TCP throughput. ACM SIGCOMM, Pittsburg, PA, USA, pp 295–308
11. Manish J, Dovrolis C (2002) Pathload: A Measurement Tool for End-to-End Available Bandwidth. Passive and Active Measurements, Fort Collins, Colorado, USA, pp 14–25
12. McGraw Hill Osborne (2003) CWNA Certified Wireless Network Administrator: Official Study Guide (Exam PW0-100). Snd Edn. ed Planet3 Wireless
13. Melander B, Bjorkman M, Gunningberg P (2000) A New End-to-End Probing and Analysis Method for Estimating Bandwidth Bottlenecks. Global Internet Symposium
14. Prasad RS, Murray M, Dovrolis C, Claffy K (2003) Bandwidth estimation: Metrics, measurement techniques and tools. IEEE Network vol 17, 6:27–35
15. Ribeiro VJ, Riedi RH, Baraniuk RG, Navratil J, Cottrell L (2003) PathChirp: Efficient Available Bandwidth Estimation for Network Paths. Passive and Active Measurement Workshop, San Diego
16. Strauss J, Katabi D, Kaashoek F (2003) A Measurement Study of Available Bandwidth Estimation Tools. 3rd edn, ACM SIGCOMM Internet Measurement Workshop, Miami Breach, FL, USA, pp 39–44

CHAPTER 12

QUALITY OF SERVICE PROVISIONING ISSUE OF ACCESSING IP MULTIMEDIA SUBSYSTEM VIA WIRELESS LANS

ASMA A. ELMANGOSH, MAJDI A. ASHIBANI, AND FATHI BEN SHATWAN

Electrical Engineering Department, The Higher Institute of Industry
{a_elmangosh, mashibani, ashatwan}@hii.edu.ly

Abstract: The Third Generation Partnership Project (3GPP) has specified the IP Multimedia subsystem (IMS) as a service control platform that allows creation of new multimedia and multi-session applications utilizing wireless and wireline transport capabilities. Supporting many types of communications such as instant messaging (IM), push-to-talk cellular walkie-talkie service, and multimedia telephony; the IMS should provide mechanisms to negotiate various levels of Quality of Service (QoS) that should be established at session set-up and maintained throughout the life of the session. In this paper we will cover the accessing of IMS service using Wireless LAN (WLAN) in the transport plane. We will investigate the interworking architecture specified by 3GPP, and propose a framework of the QoS provisioning required in establishing IMS sessions from/to WLAN terminals

1. INTRODUCTION

The IP Multimedia Subsystem (IMS) standard by the 3rd Generation Partnership Projects (3GPP and 3GPP2) represents the global service delivery platform standard for providing multimedia applications in Next Generation Networks (NGN) [2]. The IMS can be defined as "a global, access-independent and standard based IP connectivity and service control architecture that enables various types of multimedia services to end-users using common Internet-based protocols".

IMS allows convergence of different transport networks by employing a standardized architecture independent of the access network technology, this way IMS services can be provided over any IP connectivity access network (IP-CAN), such as Universal Mobile Telecommunications System (UMTS), Digital Subscriber Line (xDSL), or Wireless local access network (WLAN). Although Release 5

H. Labiod and M. Badra (eds.), New Technologies, Mobility and Security, 133–143.
© 2007 *Springer.*

IMS specifications contain some UMTS-specific features, in Release 6 access specific issues were separated from the core IMS description. IMS employs Internet Engineering Task Force (IETF) protocols to realize standardized reference architecture, and provides a common signaling framework for end user registration, session control, security, profile management, accounting, and end-to-end QoS management. The Session Initiation Protocol (SIP) is the main signaling protocol used in IMS to establish, modify and terminate multimedia sessions between participants. Other protocols are used for specific tasks, such as Session Description Protocol (SDP) to describe the media characteristics to be supported within the sessions, Common Object Policy Service (COPS) to perform local policy-based admission control in different IP-CAN gateways, and Diameter to query HSS registry and to perform charging actions.

3GPP has recently developed a cellular-WLAN interworking architecture, as an add-on to the existing 3GPP cellular system specifications published within 3GPP Release 6 specifications [3], the main driver is to enable 3GPP system operators to provide public WLAN access as an integral component of their total service offering to their cellular subscribers. The subscribers benefit from the interworking service through the greater coverage, higher data-rate, and lower overall cost. The desired interworking between WLAN and IMS can be seen from different points of view. However, the most important aspects are the session negotiation level; which provide service continuity from the user perspective, and the Quality of Service (QoS) provisioning, which guarantee QoS consistency across different access networks. This paper aims to address the problem of how QoS can be provisioned for the end user willing to establish a multimedia IMS session using WLAN as his/her IP-CAN.

The remainder of this paper is organized as follows. Section 2 presents the interworking architecture of 3GPP-WLAN as specified by the 3GPP. Section 3 provides a brief view of the related work concerning the subject. Section 4 discusses the QoS aspects of 3GPP IMS-WLAN interworking. Finally the paper is concluded in Section 5.

2. INTERWORKING ARCHITECTURE OF 3G AND WLAN

3GPP-WLAN interworking architecture, as specified by 3GPP, is focused on the interworking functionality between 3GPP based core network and 802.11 technology based WLAN systems. The loose coupling approach has been chosen in this architecture since it provides greater flexibility and scalability than the tight coupling approach. The WLAN is directly connected to the Internet backbone and the 3GPP core network. The typical 3GPP-WLAN interworking system, as specified in 3GPP [3] is shown in Fig. 1.

The WLAN User Equipment (WLAN UE) is equipment that uses UICC (Universal Integrated Circuit Card) utilized by a 3GPP subscriber to access the WLAN AN for 3GPP interworking purpose. The WLAN UE may be capable of WLAN access only, or it may be capable of both WLAN and 3GPP radio

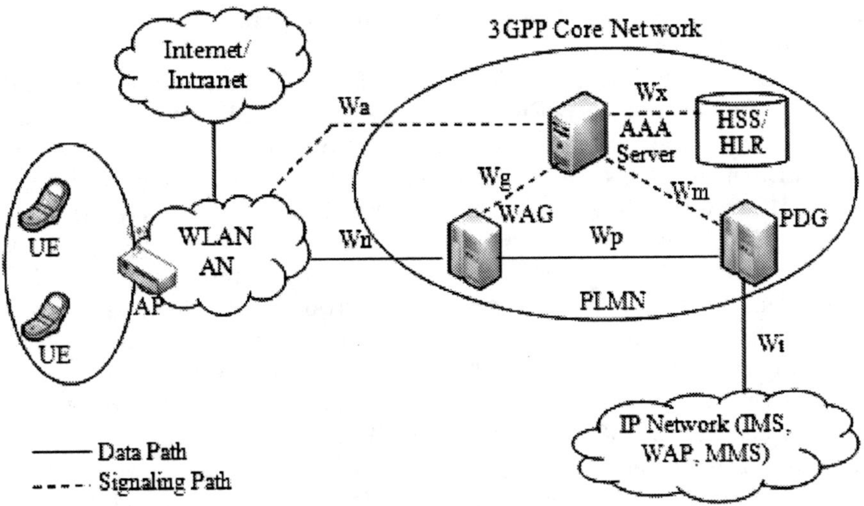

Figure 1. 3GPP-WLAN Interworking System Architecture

access. For enabling interworking with WLANs, the 3GPP Packet Switched (PS) core network incorporates three new functional elements: the 3G Authentication, Authorization and Accounting (AAA) server, the WLAN Access Gateway (WAG), and the Packet Data Gateway (PDG). The WLAN must also support similar interworking functionality to meet the access control and routing enforcement requirements.

Authentication, Authorization and Accounting (AAA) is one basic prerequisite for providing IP connectivity and other services by the 3G network via a WLAN system. The 3G AAA server in the 3GPP network terminates all AAA signaling originated in the WLAN that pertains to the WLAN UE. This signaling is securely transferred across the Wa interface, which is based on Extensible Authentication Protocol (EAP) [7].

The 3G AAA server interfaces with other 3G components, such as the WAG, PDG, and home subscriber server (HSS), which stores information defining the subscription profiles of 3G subscribers. The 3G AAA server can also route AAA signaling to/from another 3G networks, in which case it serves as a proxy and is referred to as a 3G AAA proxy. PDGs are responsible for charging data generation, IP address management, tunnel termination and QoS handling. The PDG role in QoS handling will be further discussed in this paper.

A WAG is a 3GPP network's gateway via which the data to and from the WLAN Access Network (WLAN AN) is routed via a Public Land Mobile Network (PLMN) through a selected PDG in order to provide a WLAN UE with 3G PS based services. WAGs are responsible for charging data generation and routing enforcement. HSS is located within the 3GPP subscriber's Home Network, which maintains required authentication and subscription data to access the 3GPP-WLAN interworking service.

To access 3G services, the subscriber needs to first authenticate and get authorized from the 3GPP network, and then WLAN access network allows the UE to access the 3G Core network, and assign him/her a local IP address, which identifies the WLAN UE in the WLAN AN. An end to end IP tunnel is then set up between a UE and a PDG to direct all UE packets through the 3G operator network. The home operator may also wish that all subscriber data were routed via the home network to collect independent charging information and to apply operator's policies. The requirements to remotely connect to private IP networks and to route traffic via a certain point in the network imply the need to tunnel data packets, as per 3GPP specification. According to the 3GPP specification [5], IPSec [8] tunnel using Encapsulated Security Protocol (ESP) [9] is used between the UE (within WLAN AN) and the PDG of 3G core network. In IPSec tunneling, a tunnel endpoint encrypts and encapsulates the data packet within a new packet destined to the other endpoint of the tunnel. The packet is then decapsulated, decrypted and delivered to the indicated original destination. The Remote IP address is used to transport the encapsulated data packet over the IPSec tunnel. It represents the identity of the UE in the network and is used by UE in accessing the 3G service. The Remote IP address is used for the inner packet (containing encapsulated data payload) of the IPSec tunnel.

3. RELATED WORK

A feasibility study carried out within the 3GPP introduced six interworking scenarios at [1], each scenario realizes an additional step in integrating WLAN in the 3GPP service offering and naturally includes the previous level of integration of the previous scenario. The third scenario "Access to 3GPP system packet switch based services" is the most related scenario to our paper. This scenario extends the 3GPP system PS-based services (e.g. IMS, MBMS) to the WLAN. However, it is a matter of implementation whether all services or a subset of the services is provided. Scenario 3 does not provide service continuity. When a user changes access (e.g. from 3G to WLAN), the sessions need to be re-established. The interworking of WLANs and cellular networks will maximize user satisfaction in terms of service availability and high data rates at rational cost, and further boost the penetration of the two networks. Categories and examples of future services that could be provided using the 3G-WLAN interworking are discussed in [10].

The 3GPP-WLAN interworking has been discussed by Marquez et al. [11] from the session negotiation point of view, analyzing the SIP interworking for each interworking level as defined by 3GPP [1]. The authors also provided a comparison between basic SIP as defined by IETF and the SIP extensions contained in the 3GPP specification.

Early WLAN standards can be considered as a wireless version of Ethernet by virtue of supporting a best-effort service based on Ethernet like medium access control (MAC) protocol and up to 54Mbps transmissions rates. A new extension of IEEE 802.11 is emerging to enhance the MAC to support QoS in WLANs named IEEE 802.11e [6], which provides DiffServ type aggregate QoS classes to the air interface.

The limitations of legacy 802.11 MAC layer in QoS supporting have been described in [12], the article also investigates the features of 802.11e by evaluating the performance of the Enhanced Distributed Channel Access (EDCA) and the HCF Controlled Channel Access (HCCA) mechanisms using NS-2 simulator.

Wallenius et al. [13] analyze how well 802.11e's QoS properties perform under the different kind of simulation scenarios when packet sizes and channel error rates are varied. The paper concluded that EDCF can be implemented with 3G to support QoS at the situations when 3G devices will use fast WLAN access to the core network; however their simulation work focused only on the wireless medium.

The technical specification in [4] was introduced to investigate the necessity and reliability of the applicable QoS mechanism between the WLAN UE and PDG, and the possible impacts on the 3GPP-WLAN interworking entities, also to ensure that the architecture for 3GPP/WLAN Interworking described in Section 2 is supported by other QoS-related mechanisms being developed in 3GPP. In this paper we provide further investigation of the QoS provision procedures to be supported in the interworking framework.

4. QOS PROVISIONING USING WLAN ACCESS NETWORKS

QoS refers to the ability to deliver network services according to the parameters specified in a Service Level Agreement (SLA). At a network resource level; QoS refers to a set of capabilities that allow a service provider to prioritize traffic, control bandwidth, and network latency. Typically different applications have different requirements on end-to-end delay characteristics, throughput, and reliability. While video streaming requires higher bandwidth, it can tolerate slightly more delay and loss rates than does voice over IP (VoIP). As QoS perceived by the user is as weak as the weakest link, the list of addressed network elements must cover the full end-to-end chain of the radio channel, IP core network and IMS domains.

A policy-based QoS solution is adopted by the 3GPP for ensuring that sufficient QoS resources are provided to the authorized users in the network [14]. This way the QoS policy is separated from its implementation on various devices. The policy rules are stored in the policy repository from which the Policy Decision Function (PDF) retrieve the appropriate policy rules in response to policy events that are triggered by the contracted IP QoS services by the Policy Enforcement Function (PEF). The PDF translates the acquired policy rules into a set of QoS mechanism configuration actions based on the capabilities of the PEF and the current network conditions.

The PEF then executes these PDF-supplied actions to handle the triggering policy events in accordance with the requested IP QoS services.

In 3GPP-IMS Release 6 QoS provisioning for IMS over interworking WLAN was not considered. In this paper we try to address this issue and investigate the possible QoS parameters mapping producers using different implementations of the PDF in the 3GPP core network.

The end-to-end service should provide transport of user data and signaling between the WLAN UE and another terminal Equipment (TE) over different bearer services of the network. The considered QoS architecture for WLAN-3GPP IP access consists of 3GPP IP access bearer service and external bearer service, as shown in Fig. 2. The external bearer service may implement several network services such as UMTS bearer service. The 3GPP IP access bearer service provides transport of signaling and data for the IP domain between the UE and the PDG. WLAN bearer service supports WLAN access network specific bearer capability between WLAN UE and the access network. The WLAN support of QoS in layer 2 bases is being addressed by the IEEE 802.11e Work group [6].

Most WLAN technologies do not provide support for QoS, except of IEEE 802.11e. When the WLAN technology used to access IMS service, does not offer sufficient support for QoS (e.g. IEEE 802.11a/b), best effort approach may be used to approximate the service, however this will provide limitations on the services that can be offered to the end user in non-real time services only. We propose that the QoS capability of the WLAN access network should be send along with other capabilities to the 3GPP AAA server in the access request. This information can be used later on the session initiation procedure to prevent provisioning a high-priority QoS at core network for an end user attached to a QoS support-less access network.

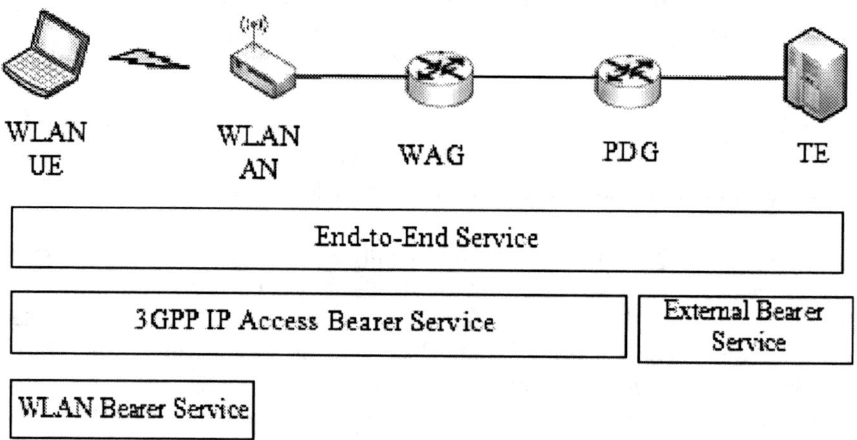

Figure 2. QoS architecture for WLAN 3GPP IP Access

According to the 3GPP's proposed architecture described in Section 2, a tunnel is established between the WLAN UE and the PDG carrying traffic of different QoS levels. Since the data within this tunnel is encrypted including the IP header, the intermediate nodes may not distinguish the individual IP flows and their QoS requirements. For this situation be believe that the best option is using the Differentiated Service (DiffServ) QoS architecture by both UE and PDG. DiffServ outperforms other QoS architectures; such as Integrated Service (IntServ) by providing scalability, flexibility and efficient resource allocation. We expect DiffServ to become a key QoS assurance technique in the upcoming All-IP NGN.

Basically, the IP DiffServ octet contains a Differentiated Services Code Point (DSCP) to identify and select the particular Per-Hop Behavior (PHB) that an IP packet will receive at a given network node. A DSCP is set in the IP header of each packet. As for the resource admission control within a DiffServ domain, the PEP is the node that handles resource and policy enforcement (typically the edge router), while the PDP is the server that handles resource allocation and policy decisions (a bandwidth broker in DiffServ terminology).

The user's QoS profile will be downloaded from the HSS to the AAA server at the QoS provisioning phase. According to our purposed freamwork, the QoS profile will include the maximum bandwidth and the maximum DCSP per service allowed to the user.

In 3GPP specification in [3] suggests that the implementation of PDG by reusing the gateway GPRS support node (GGSN) deployments of the UMTS. This implementation could be done via using a subset of the Gn reference point, in order to allow re-use of existing GGSN functionality without upgrading them. In this configuration the end-to-end tunnel between WLAN UE and PDG is terminated by the Tunnel Termination Gateway (TTG) of the PDG. A setup of a GTP (GPRS tunneling Protocol) tunnel is triggered towards the GGSN part of the PDG. In the following subsections we compare and analyze the possible QoS parameters mapping process using different ways of implementing the PDG.

4.1. Using a Stand-alone PDG

In this implementation the resource reservation could be based on DiffServ QoS mechanism mainly, without the need of the Packet Data Protocol (PDP) context activation used in the UMTS network. Fig. 3 shows our proposed QoS parameter translation process for this PDG implementation. When the Proxy Call Session Control Function (P-CSCF) gets the negotiated SDP parameters of each multimedia component of the desired session from the terminating side through SIP signaling interaction, the P-CSCF maps the SDP parameters to service information per application, and passes it to the PDF over the Gq interface.

The PDF authorizes every component negotiated for the session, generates an authorization token and sends it to the P-CSCF (using Gq Interface). The PDF

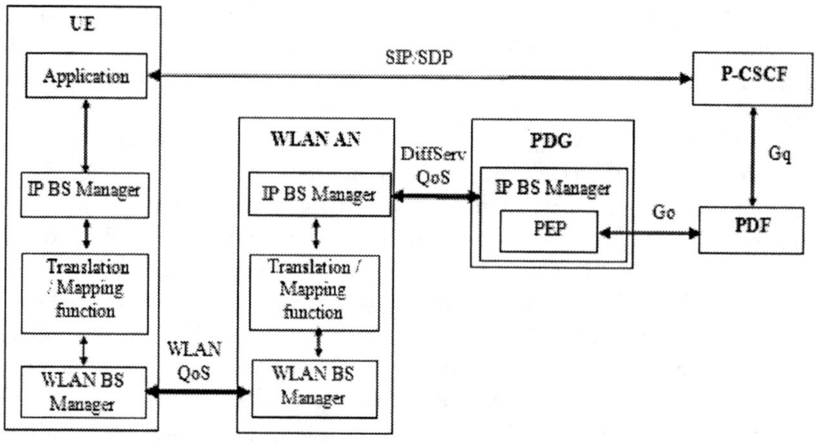

Figure 3. QoS parameters translation using a stand-alone PDF

shall map from the service information to the Authorized IP QoS parameters (data rate and DiffServ Code-point (DSCP)) per flow, which shall be passed to the PEP in the PDG via the Go interface. The UE have to derive the IP QoS parameters (data-rate and DSCP). The mapping/translation function in the UE maps the IP QoS parameters to the WLAN QoS parameters (EDCA Access Category (AC)) for each flow identifier. It is also required that the Access point (AP) in the WLAN access network (WLAN AN) supports a mapping function in order to translate the WLAN QoS parameters to/from IP QoS parameters. We also propose that the support of DiffServ QoS mechanism has to be added to the WAG to handle the routed packets according to their DSCP.

In our proposed framework, the WLAN UE performs packet classification and conditioning in the IP layer before forwarding the IP packets to the AP. the translation/mapping function in the UE has to map IP layer QoS (DiffServ) parameters to the 802.11e MAC layer parameters, by mapping the DSCP placed in the IP header to a suitable EDCA AC, thus IP packets are encapsulated into EDCA 802.11e MAC frames, and forwarded to the proper priority queue. The PDG in this implementation has to support DiffServ and acts as an edge router. The PEP element in the PDG

Table 1. QoS mapping between DiffServ PHBs and EDCA ACs

Traffic Class	DiffServ PHB	EDCA AC	IMS QoSClass
Voice	EF	AC_VO	A
Video	AF12	AC_VI	B
Best Effort	AF22	AC_BE	C
Best Effort	AF23	AC_BE	D
Best Effort	AF33	AC_BE	E
Background	BE	AC_BK	F

will be responsible for the resource and policy based authorization, such as the comparison between the requested QoS parameters by the UE and the authorized QoS parameters, and it could degrade it if necessary. Table 1 presents the proposed mapping scheme between DiffServ PHBs and EDCA 802.11e ACs. This mapping scheme has been evaluated in our previous work [15].

4.2. Implementing PDG by Reusing GGSN

This implementation can be used by mobile operators those willing to reuse their existing infrastructure and functionality to support users accessing from a WLAN UE. Fig. 4 depicts the required QoS parameters translation according to this implementation of the PDF. The same mapping processes in the P-CSCF, the PDF, and the UE explained in the previous subsection have to be done without any notable changes. Using the UMTS GGSN the resource reservation means PDP context activation. Thus the TTG element in the PDG has to perform a mapping of the QoS parameters from IP level to UMTS level in order to establish a GTP tunnel with the GGSN for every IP tunnel with the UE. The GTP tunnel is established by sending a PDP context activation request from the TTG towards the GGSN.

Using the Go Interface the PEP element in the GGSN sends a COPS request to the PDF, which in response sends the configuration parameters to be enforced by the PEP through the Go Interface. The GGSN shall compare the UMTS QoS parameters of the PDP context (set by the TTG) against the derived authorized UMTS QoS parameters from mapping of authorized IP QoS parameters. Table 2 presents the proposed mapping scheme between DiffServ PHBs, UMTS QoS classes and EDCA 802.11e ACs.

We believe that this configuration will be costly in term of the required signaling and translation/mapping process, because of the PDP Context Activation procedure which has to be held between the TTG and the GGSN in the PDF. In addition

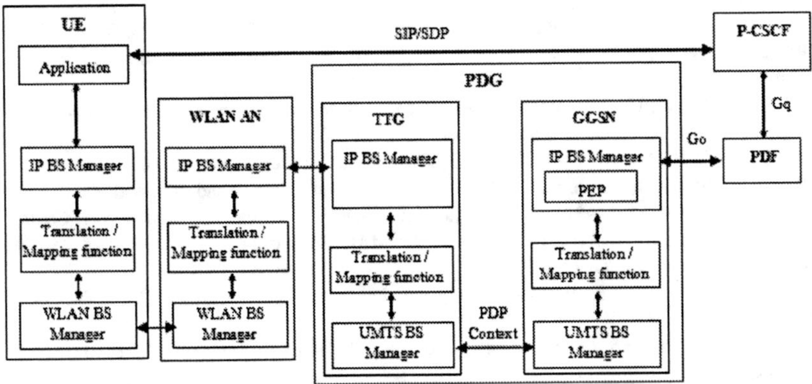

Figure 4. QoS parameters translation re-using GGSN

Table 2. QoS mapping between DiffServ PHBs, UMTS QoS class and EDCA ACs

Traffic Class	EDCA AC	DiffServ PHB	UMTS QoS Class	IMS QoS Class
Voice	AC_VO	EF	Conversational	A
Video	AC_VI	AF12	Streaming	B
Best Effort	AC_BE	AF22	Interactive	C
Best Effort	AC_BE	AF23	Interactive	D
Best Effort	AC_BE	AF33	Interactive	E
Background	AC_BK	BE	Background	F

the multi-mapping process in this implementation could cause inconsistency in the delivered QoS level to the end user.

5. CONCLUSION

In this paper we have discussed the access of IMS services by users connected to WLAN, with focus on the QoS provisioning issue. We propose and analyze the mapping processes that are required for end-to-end resource reservation for users willing to access IMS services using their IEEE 802.11e equipment terminals.

We have proposed DiffServ as a promising approach for QoS provisioning over the end-to-end IMS-WLAN interworking in order to achieve scalability. We expect that DiffServ will become a key QoS assurance technique in the upcoming All-IP NGN. Some further work should be undertaken to evaluate both ways of the PDG implementations in providing consistency in the end-to-end QoS delivered to the end WLAN user.

REFERENCES

1. 3GPP TR 22.934 v6.2.0, "Feasibility Study on 3GPP System to WLAN Interworking (Release 6)," Sept. 2003.
2. 3GPP TS 23.228 v6.b.0, "IP Multimedia Subsystem (IMS)," TS 23.228, Release 6, Oct. 2005.
3. 3GPP TS 23.234 v6.7.0, "3GPP System to WLAN Interworking; System Description (Release 6)," December 2005.
4. 3GPP TS 23.836 v1.0.0: "Quality of Service (QoS) and policy aspects of 3GPP-Wireless Local Area Network (WLAN) interworking (Release 7)", Nov 2005.
5. 3GPP, 3GPP TS 33.234 V 6.1.0, "3G Security, Interworking Security (Release 6)".
6. IEEE 802.11 WG, Draft Supplement to Standard for Telecommunication and Information Exchange between Systems – LAN/MAN Specific Requirements – Part II: Wireless Medium Access Control (MAC) and Physical Layer (PHY) Specifications: MAC Enhancements for Quality of Service, IEEE 802.11e Draft 11.0, October 2004.
7. L. Blunk, J. Vollbrecht, "PPP Extensible Authentication Protocol (EAP)", IETF RFC 2284, March 1998.
8. S. Kent and R. Atkinson, "Security Architecture for the Internet Protocol", IETF RFC 2401, November 1998.
9. S. Kent and R. Atkinson, "IP Encapsulating Security Payload (ESP)", IETF RFC 2406, November 1998.

10. D. Axiotis et al., "Services in interworking 3G and WLAN environments", IEEE wireless Communication, October 2004.
11. F. Marquez et al., "Interworking of IP multimedia core networks between 3GPP and WLAN", IEEE Wireless Communications, June 2005.
12. Q. Ni. "Performance Analysis and Enhancements for IEEE 802.11e Wireless Networks". IEEE Network, July/August 2005.
13. E. Wallenius, T. Hämäläinen, T. Nihtilä, J. Joutsensalo , K. Luostarinen, "3G Interworking with WLAN QoS 802.11e", IEEE VTC 2003.
14. W. Zhuang, Y. Gan, K. Loh, K. Chua, "Policy-Based QoS Architecture in the IP Multimedia Subsystem of UMTS", IEEE Network, May/June 2003.
15. A. Elmangosh, M. Ashibani, F. Ben-Shatwan, "The Interworking between EDCA 802.11e and DiffServ", IEEE eSCO-Wi 2007.

CHAPTER 13

THREE-COLOR MARKING WITH MLCN FOR CROSS-LAYER TCP CONGESTION CONTROL IN MULTIHOP MOBILE AD-HOC NETWORKS

YI-JEN LU, CHI-JEN HUANG, AND TSANG-LING SHEU

Department of Electrical Engineering
National Sun Yat-Sen University, Kaohsiung, Taiwan
E-mail: d9131811@student.nsysu.edu.tw and sheu@ee.nsysu.edu.tw

Abstract: This paper presents a three-color marking strategy with MAC layer contention notification (MLCN) for cross-layer TCP congestion control to achieve high throughput and low latency in a multi-hop mobile ad-hoc network (MANET). Unlike most of the existing works, the proposed method features by a three-color marking and an estimation of backlog packets at a TCP sender. Adopting from the FECN (Forward Explicitly congestion control) technique, a router with three-color marking can timely report contention status to a TCP sender, which then estimates the number of in-flight packets to determine a proper congestion window size. Performance evaluation performed on NS-2 is mainly based on a dynamic topology. From the simulation results, we show that the proposed cross-layer scheme can adaptively control TCP congestion window size to provide outstanding benefits in throughput and propagation delay as compared to the previous works

Keywords: Cross-layer, Three-color marking, TCP, MAC, Congestion window, and Mobile ad-hoc network

1. INTRODUCTION

One of the major attractions of wireless communication is the ease of deployment, which enables an effective connectivity for those where wired infrastructures are not possible. In order to deploy a wireless ad hoc network more effectively, copious amounts of research works have been proposed with the techniques of how to increase TCP/IP throughput [3][5][9,10,11] and balance its fairness [4][8] being of particular interest. The primitive spirit of TCP transmission is to fill the network pile gradually with slow start, until congestion occurs in the queue of some wireless

H. Labiod and M. Badra (eds.), New Technologies, Mobility and Security, 145–157.
© 2007 *Springer.*

nodes. When a TCP sender detects the congestion based on the information that some packets are lost, it begins to cut the congestion window (*cwnd*) to reduce the transmission rate. Consequently, the key to achieving high throughput is the effectiveness of controlling *cwnd* at a TCP transmitter. In the 1990's, TCP/IP has been widely used in wired networks, and hence a number of algorithms, such as Tahoe, Reno, Vegas, and NewReno, were developed to avoid congestion so that the transmission can be boosted. Nevertheless, though the algorithms can perform well in wired networks, they are not suitable for a wireless environment. For this reason, recent research works [3,4,5][7,8,9][11] have been focused on how to avoid the congestion in mobile ad hoc networks (MANET) to achieve a better TCP layer throughput.

Besides the TCP/IP layer, the linking operations of wireless networks in the physical/MAC layer are significantly different from those of wired networks. Compared to wireless channels, a wired line, e.g., cable or fiber, can provide much lower bit-error rate and a colossal bandwidth for transmitting data. Thus, the effect of bit error can almost be ignored, and only the congested queues in some nodes need to be dealt with. Yet in wireless communications, the problem of detecting the congestion becomes more complicated by the fact that the inevitable interferences from inter-bits and contending resources (e.g., media, time slot, carrier frequency, etc.) may affect TCP handler in detecting the congestion resulting from the jamming queue in some routers or from the unsuccessful bit-delivery in some wireless links. Fig. 1 shows how the congestion occurs in a wireless environment after a TCP session is established. We can see from this figure that, once a MAC frame cannot be successfully received at a downstream node, its upstream node has to re-transmit this frame within a specific time; otherwise, this frame must be discarded when the retransmission time is larger than a predefined value. Therefore, in a MANET, we need to address not only the congestion issue, but also the signal interference and media contention issues.

In addition, recognizing the high frame error rates in a wireless network, the MAC layer must provide a relatively reliable link to its neighbor nodes. Schedule-based link (e.g., FDMA, TDMA, and CDMA) and contention-based link (e.g., CSMA/CA) are the two most dominant linking methods used in the MAC layer. Consider a schedule-based link. There are two factors that may result in the re-transmission of erroneous frames: the strength of transmitting power and the interference from noise or adjacent slots. When the transmitting power or SNR (Signal-to-Noise Ratio) is low in a link, a receiver may not receive an accurate frame. In this case, time consumption in re-sending the frame may cause the queue in IP layer to get crowded because new packets can continuously arrive. To better handle the resending process, a recent research work [1] has developed an efficient energy-conserving algorithm to achieve low frame-error rates in a wireless network. On the other hand, media occupation to send frames is another impact on the throughput for a contention-based link, such as Wireless LANs (802.11-based networks). As a result, TCP cannot perform well in terms of throughput and latency if the media contention effect is not taken into account. To enhance the TCP throughput and

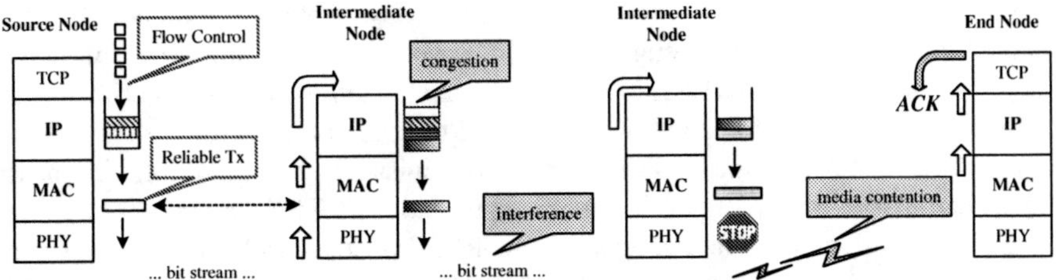

Figure 1. Congestion, interference, and media contention in a wireless environment

reduce the propagation delay in an 802.11-based network, this study proposes a cross-layer framework, which enables a TCP sender to quickly aware network congestion status such that it can adjust its *cwnd* accordingly in advance of serious performance degradation. The required information is to estimate the number of in-flight packets, thereby assisting the TCP sender to determine an appropriate *cwnd*. In summary, our cross-layer design approach consists of a MAC-layer contention notification based on a three-color marking strategy and an estimation of backlog packets at a TCP sender. Using a practical topology for performance evaluation, our experiments show that the proposed scheme is more suitable for a large-scale MANET and time-sensitive applications, as compared to TCP-NewReno and a previous work in [3].

2. RELATED WORKS

We begin the discussion of enhancing the TCP throughput in wireless networks by reviewing some previous works. These works involve related concepts on restricting congestion window, adjusting RTS/CTS in the MAC layer, and trimming TCP operation, which serve as the bases of this study to design a cross-layer approach to improve the TCP-layer throughput.

To solve the problem in fairness and throughput for a wireless network, a previous work [4] proposed limiting TCP transmission rate by estimating the overall collision probability along the paths. Using this probability, the receiver sets a proper advertised window (*adw*) in the ACK packet. Thus the sender can confine the congestion window to an acceptable range. Furthermore, adjusting congestion window (*cwnd*) is another effective approach to increase TCP throughput. In [3], they have proved that the upper bound of congestion window for a chain topology must be limited to one fifth of the round-trip hop count. However, the hop-count distribution with a frequent re-routing is not readily to implement in a dynamic environment, and hence influences the performance in throughput. In addition, Kim [5] proposed a scheme to compute an adaptive advertised window by estimating the ECN (Explicit Congestion Notification) rate, and thereby set the congestion window limit. It is shown that appropriate restriction in congestion window can improve performance in a mobile network.

With regard to a wireless network in the MAC layer, many researchers [6,7,8] devoted to improving TCP throughput. Fu [6] proposed two techniques, Link-layer Random Early Detection (LRED) algorithm and adaptive pacing algorithm, for multi-hop flows. The LRED policy is designed to compute the drop probability in terms of the number of retries in the frame transmission. When the retry count is larger than a specific threshold, it issues a notification to TCP senders by marking the ECN bit in IP header. In analogous to the RED concepts, in [8] they proposed a Neighborhood Random Early Detection (NRED) scheme to operate the distributed neighborhood queues, thereby enhancing TCP fairness. This distributed algorithm is suitable for ad hoc wireless networks, because it extends RED design to estimate queue length of neighbor nodes from the view of an overall drop probability. Meanwhile, Mesut [7] introduced a dynamic short retry to adapt RTS/CTS retransmission in an 802.11 WLAN. This adaptive limit is determined by network density, which means the number of wireless nodes in a defined area. These works suggest that an effective transmission can be achieved by better handling collision and retry in wireless networks.

Viewed from the angle of modifying TCP operation, many works [9,10,11] focus on promoting TCP efficiency. In [9] they presented a TCP-ADA mechanism, whereby the TCP sinker returns one delayed acknowledge to its sender when receiving several packets. This achieves the minimization of the contentions between data and ACK packets. In [10], Lilakiatsakun demonstrates that the delayed ACK might also provide high utilization on bandwidth and reduce the consumption on a node's energy. In addition to enhancing TCP retransmission technique, Kim [11] presents an early scheme to re-send the lost packet right after the first duplicate ACK in mobile ad hoc networks is received, instead of waiting for the arrive of the third duplicate ACK packet. This suggests that TCP performance could be improved by handling ACK packets systematically.

3. THE CROSS-LAYER DESIGN

This section describes our cross-layer algorithm embedded in the TCP New Reno. To enhance the TCP throughput in MANET and collect detailed information of network congestion, we propose a feedback mechanism, whereby a TCP sender is alerted with network condition. The mechanism is derived from the concept of Forward Explicit Congestion Notification (FECN) in ATM network, where the switch being congested marks an alert in the cell header, thereby notifying its destination by RM cells to reduce transmission rate. As an analogy, in wireless networks, when frame error rate continuously increases or media access cannot be permitted promptly, a wireless node has to alert a TCP sender to adjust its flow. To this end, an intermediate node marks a three-color notification in the IP header if the above-mentioned conditions are detected. After extracting the notification from the IP packet, a TCP receiver returns an ACK packet with the status alert of wireless links. Then, a TCP sender can adjust congestion window, based on the prompt information, which is more efficiently than simply detecting the signal of packet loss. Our cross-layer algorithm is thus named as MAC layer contention notification (MLCN).

In order for the cross-layer scheme to operate in MANET practically, we define the following new variables and equations, which are further divided into three groups by the role of a node.

At a TCP sender,

- Congestion window, *cwnd*, which represents the parameter that we will determine in this paper;
- Round-trip time, *RTT*, which can be periodically measured by sending one packet and then receiving its ACK;
- *Min_RTT*, which indicates a minimum round-trip time in the TCP session;
- Backlog packets, *N*, which represents how many aggregated packets are flying in the networks when a new *RTT* is measured.

In addition, we compute an average backlog, called *avg_N*, during the TCP session, and define *prev_N* as the value of *N* in receiving the previous *RTT*.

At an intermediate node,

- Current transmission count, *CRC*, which represents how many times one frame can be successfully transmitted to neighbor nodes in a contention link.
- Average retransmission count, *ARC*, which represents a moving average of all *CRC*s computed.

At a TCP sinker,

- Collision notification, *CN*, which indicates the collision information presented by three-color markers and to be sent back along with ACK.

Note that although our algorithm is mainly designed for an 802.11-based network, it can also be applied to other wireless networks with no contention in a link.

3.1. Contention Notification

This subsection describes how a wireless node chooses an appropriate hue as the indication in an IP header. In an 802.11 link, each node puts an RTS signal into the shared media to own the resource of delivering frames. A node is permitted to transmit frames only when it receives a CTS signal; otherwise, it has to send an RTS signal again. Here, we define the number of retrying RTS signals as the transmission count. In an 802.11 DCF mode, RTS/CTS retry limit is set to 7 as default. When the transmission count hits the retry limit, frames pending to send out will be discarded. Thus, if too many packets pending to forward are aggregated at some nodes in a queue, a valid retry may become difficult to fulfill. After TCP senders detect the condition that a packet has been lost and then decrease *cwnd*, it is, however, too late to relief the burden of the queues. Therefore, our goal is to effectively predict the trend of RTS retransmission and to rapidly notify the TCP sender to adjust an appropriate *cwnd*.

Fig. 2 shows the range of RTS retransmission to classify how a node contends with others. In our scheme for contention notification, the IP header in the transmitted frame will be marked with a color according to the range where the RTS

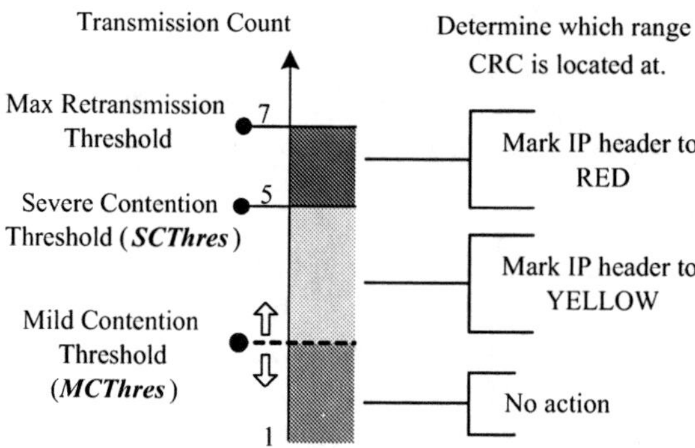

Figure 2. Three-color marking strategy for sending a frame

retry is located. In this figure, an IP header will be marked with RED if the number of retrying exceeds *SCThres* (Severe Contention Threshold). Similarly, an IP header will be marked with YELLOW if the number of retrying is between *SCThres* and *MCThres* (Mild Contention Threshold). Here the *SCThres* is a fixed threshold, while the *MCThres* is a dynamic threshold determined by all of the past *CRC*s. The dynamic design helps to avoid global synchronization problem on multiple TCP flows. In the range below *MCThres*, we preserve the original TCP functionality without inserting any additional alert.

In the three-color marking strategy, we determine *MCThres* in terms of average retransmission count, *ARC*, given by Eq. 1. This equation demonstrates either long-term or short-term tendency of media contention by choosing the value of α. The bigger the value is selected, the longer we observe a contention tendency.

$$ARC = \alpha \times ARC + (1 - \alpha) \times CRC, \quad (1)$$

where $0 < \alpha < 1$. Note that the dynamic range of *MCThres* must be limited to $1 < MCThres < SCThres - 1$. The following represents the pseudo-code of computing mild contention threshold with *ARC* in each intermediate node.

```
//Intermediate node computes marking threshold.
  if ( ARC > MCThres ) //Contention is getting worse.
    MCThres–;
  else if ( ARC < MCThres ) //Contention is being relieved.
    MCThres++;
  end
  if ( MCThres >= SCThres )
    MCThres = SCThres – 1; //It must be less than SCThres.
  else if ( MCThres < 2 )
    MCThres = 2; //The threshold is at least 2.
  end
```

To implement the three-color marking strategy into an IP header, we adopt one field of two bits, ECN, which has been defined in [12] as congestion notification. As a result, the indication extracted from the IP header will be embedded into the TCP header of an ACK packet at a TCP sinker. The main reason is that its IP header could be modified at some nodes during the returning route.

3.2. Computation of Congestion Window

As the above-mentioned notification provides a prompt alert for a TCP sender to adjust congestion window, the flying packets, called backlog packets, are another additional indication to precisely compute a proper *cwnd* in our paper. The hidden terminal problem in a real ad hoc environment may cause a high rate of RTS/CTS retry even though few packets are waiting for delivery. Thus the estimation of in flight packets assists the TCP sender in judging a collision notification from delivering excess packets or having hidden terminal. To this end, the aggregated packets in the network pile were mentioned in [2], which gives

$$\text{Actual\_Rate} = cwnd/RTT, \tag{2}$$

$$\text{Expected\_Rate} = cwnd/Min\_RTT, \tag{3}$$

$$\text{Diff} = \text{Expected\_Rate} - \text{Actual\_Rate}. \tag{4}$$

Here the TCP sender can obtain the difference in transmission rate to estimate the in-flight packets. Therefore, we have

$$\begin{aligned} N &= Min\_RTT \times \text{Diff} \\ &= \text{Actual\_Rate} \times (RTT - Min\_RTT). \end{aligned} \tag{5}$$

The following represents the pseudo-code of calculating the backlog packets and *avg_N* in a TCP sender.

```
//Calculate N, prev_N, and avg_N, when packets are sent out.
TCPSender_Compute_N (void)
{
  count++; //Count the number of packets

  if ( Min_RTT = 0 or Min_RTT > RTT )
    Min_RTT = RTT;
  end

  prev_N = N;
  N = Min_RTT * (Expected_Rate – Actual_Rate);
  avg_N = ( avg_N * (count−1) + N ) / count;
}
```

Accordingly, associating the notification with backlog, we have several scenarios to determine an effective value for *cwnd*. In case of a green alert, no extra process to *cwnd* is needed in our algorithm just as to operate the original flow control. In case of a yellow alert, we adjust *cwnd* by proportional decrease when current backlog packets are over average. In case of a red alert, we must cut *cwnd* down to one if *N* is more than *avg_N*. Otherwise, we can adjust *cwnd* by proportional decrease. The detailed processes represent as follows. Here is the pseudo-code of processing congestion window according to the relationship of *CN*, *N*, *prev_N*, and *avg_N* in a TCP sender.

```
//When TCP sender receives a CN, it needs to process as follows.
  If ( CN = Green ) //Green alert in wireless links.
     Process cwnd according to original TCP control;
  else if ( CN = Yellow ) //Yellow alert in wireless links.
     If ( N > avg_N ) //Contention is getting worse.
        Cwnd = cwnd × avg_N / N ;
     else
        cwnd = cwnd; //keep static
     end
  else if ( CN = Red ) //Red alert in wireless links.
     If ( N > avg_N ) //Contention is the worst.
        Cwnd = 1;
     else if ( N > prev_N )
        cwnd = cwnd × prev_N / N ;
     else
        cwnd = cwnd; //keep static
     end
  end
```

4. PERFORMANCE EVALUATION

To investigate the performance of the proposed model, we create a MANET topology, where 50 mobile nodes are randomly located in a square of 1500m × 1500m. In addition, the initial position of the source and the destination are at (350, 750) and at (1150, 750), respectively. One observed TCP session from S node to D node will be established by adopting AODV (Ad-hoc On-demand Distance Vector) routing protocol. Assume one packet size is 1000 bytes. Moreover, selecting two nodes as a connection pair, we will randomly create 8 pairs to generate flows of 100 Kbps as background traffic. About power radiation of antenna, the maximum transmission range at each node is 150 meters. In our simulation, each node has the capability of moving at any direction with the speed of 0, 1, 5, and 10 m/s. To prevent the distance between S and D from being too close, we restrict the horizontal range of S moving from 0 to 400m, whereas the range of D moving

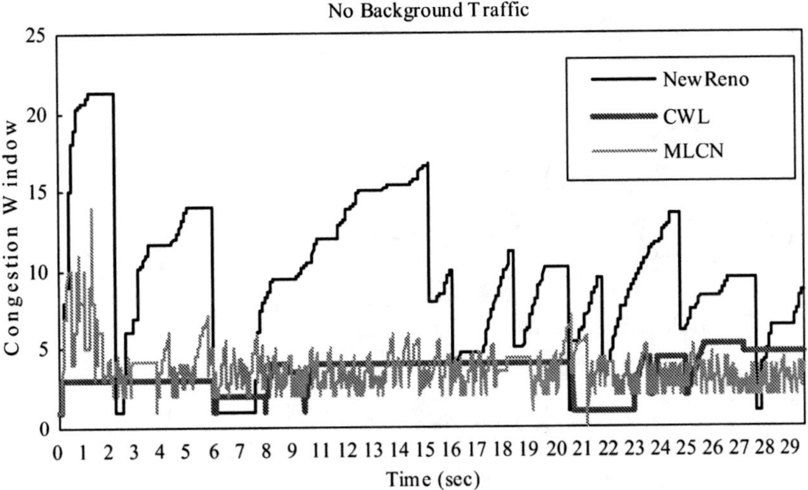

Figure 3. Comparisons of the congestion window in three mechanisms

from 1100m to 1500m. Here $\alpha = 0.4$ is chosen for Eq. 1. According to the above configuration, ns-2 simulator [13] was used to observe the efficiency of MLCN.

According to the above topology, the proposed algorithm, MLCN, was compared with TCP New Reno and CWL (Congestion Window Limit) [3] in our simulation. Fig. 3 shows the congestion window in each mechanism without feeding any background traffic. Although the congestion window may increase in New Reno until detecting packet loss, its performance in throughput is not comparable. The further study, CWL, provides an upper limit to restrict *cwnd* in static MANET to avoid excess packets stuck in queues. However, the limitation method to *cwnd*, which was derived from a chain topology, cannot effectively react to a dynamic topology. To this end, MLCN gives an adaptive *cwnd* to MANET, especially fast dynamic networks. Besides seeing the plot of congestion window, the other effects on throughput, node speed, and hop count are our major benchmark in the following investigations.

Figure 4 indicates the relationship between throughput on the observed session and moving speed of nodes. Without generating background traffic, CWL can perform relatively higher throughput in a static and low contention environment. Unfortunately, it is not suitable in a practical network as shown in the bottom of this figure, where the background has been generated. As a result, MLCN can provide better throughput than CWL by 6%–24% and than New Reno by 13%–53%. Furthermore, the three lines in the bottom on throughput represent to converge at a high speed since the frequent re-routing operations on wireless links cause a small *cwnd* consistently. In addition to examining the effects of hop count with background traffic, we found that the scale of a MANET may influence the quality in throughput. Fig. 5 shows the throughput on the main session in terms of hop

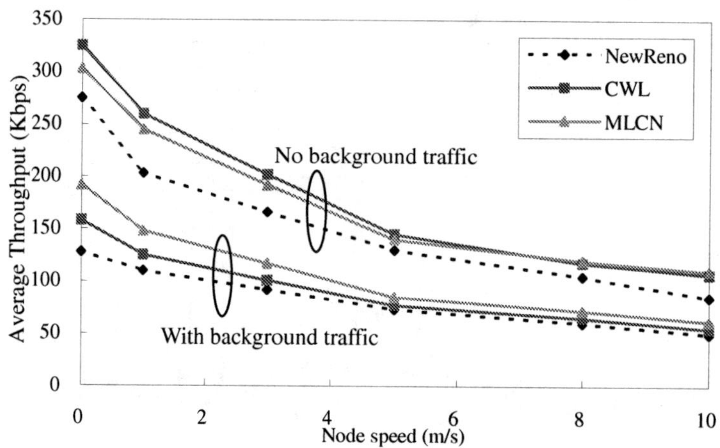

Figure 4. Throughput versus node moving speed

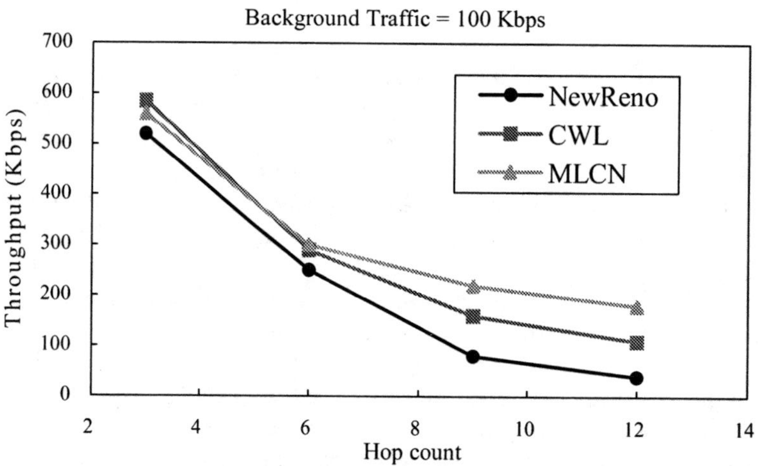

Figure 5. Throughput versus hop counts with background traffic

counts under the condition that eight background flows are generated. Consequently, high throughput is given at smaller hop counts for each mechanism. Observing the results from both figures, we can conclude that, for a dynamic and large-scale MANET, MLCN may perform outstanding in improving TCP throughput.

Next, we investigate the relationship between packet drop ratio on the observed session and moving speed of each node. The results are shown in Fig. 6 and Fig. 7. Worth noticing, both figures indicate that the packet drop ratio varies between 1% and 11% even though background traffic exists. Accordingly, this result indicates that most of packets are dropped due to contention rather than congestion. As the

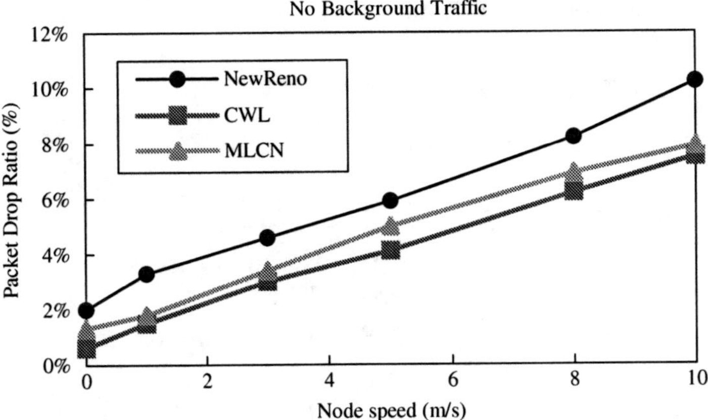

Figure 6. Packets drop ratio without background traffic

three-color strategy associated with contention notification is deliberately concerned, our approach in Fig. 7 provides lower packet drop ratio than CWL by at least 9.4% and than New Reno by 14.3%.

Interesting to us is the transmission delay of one packet delivery. Fig. 8 and Fig. 9 represent transmission delay in terms of moving speed and background traffic. Owing to estimation of backlog packets, our design helps avoid packets pending in some queues and shorten the time of one packet traveling in MANET. As a result, MLCN in Fig. 8 gives much lower latency than New Reno by roughly 50%, whereas the latency in MLCN is shorter than that in New Reno by at least

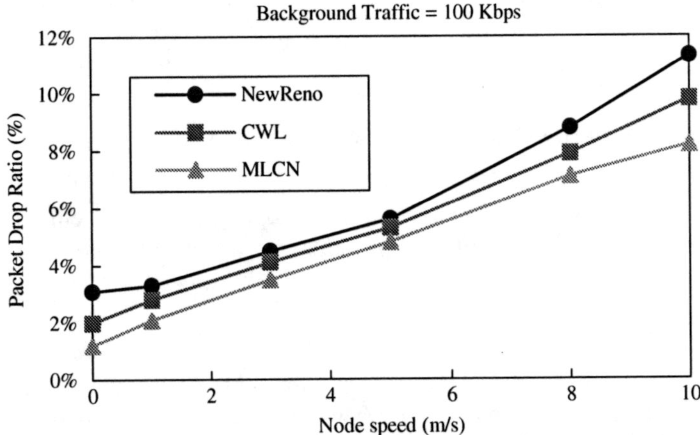

Figure 7. Packets drop ratio with background traffic

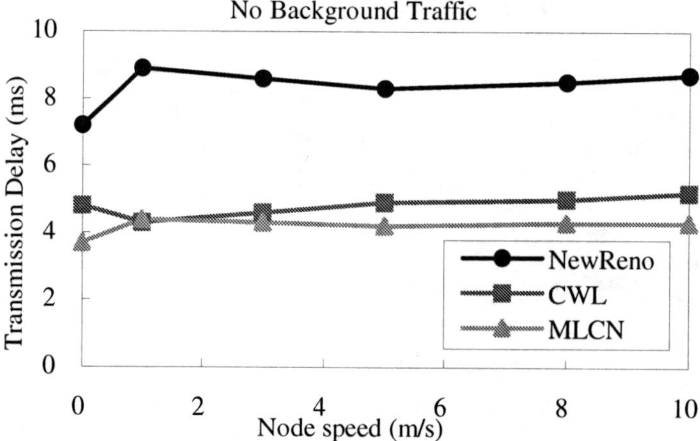

Figure 8. Average transmission delay without background traffic

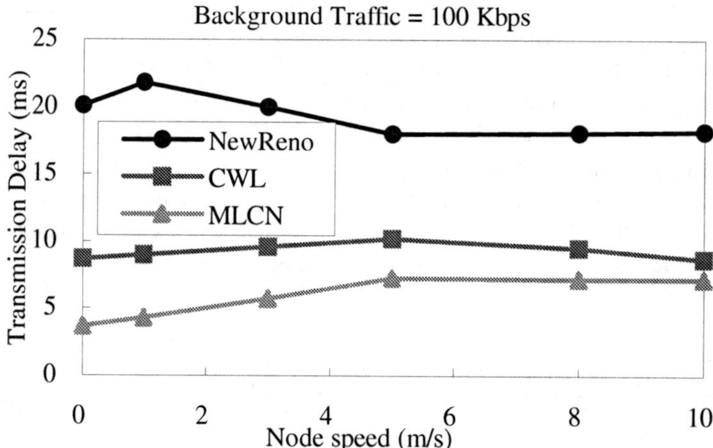

Figure 9. Average transmission delay with background traffic

60% in Fig. 9. The consequence demonstrates that adopting MLCN scheme to satisfy the requirement from time-sensitive applications is quite comparable.

5. CONCLUSIONS

To achieve high throughput and low latency in a multi-hop MANET, we proposed a cross-layer design, called MLCN in this paper. Two main designs of our approach are a MAC-layer contention notification based on the three-color marking strategy and an estimation of backlog packets at a TCP sender. We used a dynamic topology to evaluate the performance of our proposed scheme on NS-2. According to the simulation, our mechanism to control *cwnd* can offer outstanding benefits on throughput and propagation delay in comparison with TCP NewReno and CWL. In conclusion, the proposed MLCN is very suitable for a large-scale network with time-sensitive applications. Our future work in improving TCP performance is to consider a more realistic environment where wireless nodes/links may be broken and re-routing is required.

REFERENCES

1. T.L. Sheu and Y.J. Lu, "'Effective Power Control for Multimedia Streaming with QoS Constraints in Heterogeneous Networks," *IEEE Wireless Communications and Networking Conference (WCNC)*, Mar. 2007.
2. C.P. Fu and S.C. Liew, "TCP Veno: TCP Enhancement for Transmission Over Wireless Access Networks," *IEEE Journal on SAC*, vol. 21, iss. 2, pp. 216–228, Feb. 2003.
3. K. Chen, Y. Xue, S.H. Shah, and K, Nahrstedt, "Understanding Bandwidth-Delay Product in Mobile Ad Hoc Network," *Computer Communications*, vol. 27, no. 1, pp. 923–934, Mar. 2004.
4. G. Boggia, P. Camarda, L.A. Grieco, T. Mastrocristino, and G. Tesoriere, "A Cross-layer Approach to Enhance TCP Fairness in Wireless Ad-hoc Networks," *the 2nd International Symposium on Wireless Communication Systems*, pp. 498–502, Sept. 2005.
5. K.W. Kim, P. Lorenz, and M.M.O. Lee, "A New Tuning Maximum Congestion Window for Improving TCP Performance in MANET," *Proceedings of the Systems Communications (ICW'05)*, pp. 73–78, Aug. 2005.
6. Z. Fu, H. Luo, P. Zerfos, S. Lu, L.Zhang, and M. Gerla, "The Impact of Multihop Wireless Channel on TCP Performance," *IEEE Tran. on Mobile Computing*, vol. 4, no. 2, pp. 209–221, Apr. 2005.
7. M. Gunes, M. Hecker, and I. Bouazizi, "Influence of adaptive RTS/CTS retransmissions on TCP in wireless and ad-hoc networks," *the 8th IEEE International Symposium on Computers and Communication*, Sep. 2003.
8. K. Xu, M. Gerla, L. Qi, and Y. Shu, "Enhancing TCP Fairness in Ad Hoc Wireless Networks Using Neighborhood RED," *Proc. ACM MobiCom*, pp. 16–28, Sep. 2003.
9. A.K. Singh and K. Kankipati, "TCP-ADA : TCP with Adaptive Delayed Acknowledgement for Mobile Ad Hoc Networks," *IEEE WCNC 2004*, pp. 1685–1690, Mar. 2004.
10. W. Lilakiatsakun and A. Seneviratne, "TCP Performances over Wireless Link deploying Delayed ACK," *the 57th IEEE Vehicular Technology Conference (VTC)*, pp. 1715–1719, Apr. 2003.
11. D. Kim, C.K. Toh, and H.J. Jeong, "An Early Retransmission Technique to Improve TCP Performance for Mobile Ad Hoc Networks," *the 15th IEEE Personal, Indoor and Mobile Radio Communication (PIMRC)*, pp. 2695–2699, Sep. 2004.
12. K. Ramakrishnam, "The Addition of Explicit Congestion Notification to IP," *RFC-3168*, Sep. 2001.
13. The VINT Project. The UCB/LBNL/VINT/ Network Simulator – ns (version 2) Web site: http://mash.cs.Berkeley.edu/ns.

CHAPTER 14

JOINT CROSS-LAYER POWER CONTROL AND FEC DESIGN FOR TCP WESTWOOD+ IN HYBRID WIRELESS-WIRED NETWORKS

G. BOGGIA, P. CAMARDA, A. D'ALCONZO, L.A. GRIECO, AND S. MASCOLO

DEE – Politecnico di Bari, Via E. Orabona, 4 – 70125 Bari (Italy)

Abstract: In Hybrid wired-wireless networks, power control algorithms and FEC techniques have to be adopted and optimized to hide the unreliability of wireless links to TCP. To achieve the expected performance gain, a joint cross-layer optimization that takes into account also TCP congestion control is required. Recently, in literature, a theoretic framework has been proposed to address this issue when New Reno TCP is used. This work extends such a recent findings to cope also with Westwood+ TCP, a recent TCP congestion control algorithm proposed for networks with high bandwidth delay products and random losses. Numerical results have shown that, for the same propagation conditions, Westwood+ TCP requires less power and FEC codes to achieve the same goodput of New Reno TCP

1. INTRODUCTION

As wireless access networks are getting popular, even more radio links are involved in TCP end-to-end connections. Nevertheless, this medium due to factors such as interference, multipath fading, and user mobility is much more error prone than the wired one. Link failure may result in segment losses that, at the transport layer, the TCP sender erroneously interprets as signal of network congestion. Consequently, the TCP congestion control algorithm unnecessarily decreases source transmission rate underusing available bandwidth and further reducing the overall system performance [1,2]. To mitigate this problem many research efforts have been devoted to improve wireless links reliability observed by TCP [1,2,3,4]. This can be achieved both by adjusting the transmission power level and by introducing redundant data which allow errors correction (such as in Forward Error Correction protocols, FEC). Both the solutions present some drawbacks. In fact, increasing transmission power

H. Labiod and M. Badra (eds.), New Technologies, Mobility and Security, 159–171.
© 2007 *Springer.*

results in higher energy consumption (then in shorter battery lifetime), whereas redundancy introduced for error correction purposes causes waste of bandwidth resources. Therefore, the key problem is to choose the optimal values for both power and redundancy ratio for FEC purposes [3,5]. Even though the optimization of these parameters can be studied separately (e.g., in [3] and [6]), recently, a general framework has been proposed for their joint tuning [5]. The developed method, based on the maximization of a satisfaction function, which is the ratio between a function of TCP throughput and a cost function, showed its benefits in several case studies. Building on the framework proposed in [5], in this paper, we assess the impact of joint cross-layer error control and power management design on the performance of both New Reno[7] and TCP Westwood+ [8], which is an enhanced TCP congestion control algorithm based on setting the congestion window (*cwnd*) and slow start threshold (*ssthresh*) by taking into account a sender side available bandwidth estimation. In particular, classical TCP congestion control algorithms (e.g., New Reno) are based on the additive-increase/multiplicative-decrease probing paradigm, which exploits endless series of additive increasing and multiplicatively decreasing patterns of the sending rate in order to track the Internet time varying available bandwidth. Westwood+ TCP, on the other hand, has been proposed to improve the tracking capabilities of standard TCP by substituting the multiplicative decrease phase with an adaptive one, which exploits a sender side available bandwidth estimation obtained by properly conting and filtering the ACK stream. Hence, it is said that Westwood+ TCP follows an additive increase adaptive decrease paradigm [8]. It has been already shown that Westwood+ does not lose ground in presence of segment losses due to unreliable radio links, because of its adaptive rate shrinking phase, which takes into account the bandwidth used at the time of congestion [8]. Herein, we show how this performance gain can be translated in a smaller power and FEC overhead when optimal joint cross-layer power management and error correction design techniques are adopted. For that purpose, two relevant examples of physical layer have been considered: the Differential Binary Phase Shift Keying (DBPSK), which is used in IEEE 802.11, and the Gaussian Frequency shift Keying (GFSK), used in Bluetooth [9]. Numerical results show that, for the same propagation conditions and by using the same resources, Westwood+ TCP achieves higher throughput than New Reno TCP. This means that Westwood+ TCP requires less power and FEC codes overhead to achieve the same goodput of New Reno TCP. The rest of the paper is organized as follows. Section 2 summarizes the main features of Reno, New Reno, and Westwood+ TCP congestion control algorithms. In Section 3 the considered analytical framework for performance maximization by joint error correction and power management has been considered. In Section 6 results of applying such a framework to Westwood+ and New Reno TCP have been reported. Finally, in Section 5 conclusions are drawn.

2. BACKGROUND ON TCP CONGESTION CONTROL

A TCP connection is characterized by the following variables: (1) the congestion window (*cwnd*); (2) the slow start threshold (*ssthresh*); (3) the round trip time of the connection (*RTT*); (4) the minimum round trip time measured by the sender

(RTT_{min}). During the so called *slow start* phase, i.e., when $cwnd < ssthresh$, Reno TCP algorithm, for each received ACK, increases by one Maximum Segment Size (MSS) the $cwnd$; elsewhere, during the *congestion avoidance* phase, $cwnd$ is linearly increased over the time by one MSS for each $cwnd$ acknowledged segment. When 3 Duplicated Acknowledgments (DUPACKs) are received, the Reno TCP algorithm enters in the *fast recovery phase*, halves both $cwnd$ and $ssthresh$, and retransmits the segment with the lowest unacknowledged sequence number. This phase is leaved when the retransmitted segment is successfully acknowledged. NewReno differs from Reno TCP in that it does not exit from the *fast recovery* phase until all the segments within the current window are acknowledged. This feature, which is known as NewReno feature, improves the performance with respect to Reno TCP when several segments within the same window get lost [7]. In case of retransmission timeout expiration, TCP Reno halves $ssthresh$, sets $cwnd$ to one MSS, and retransmits the first segment that appears to be not acknowledged.

The key idea of TCP Westwood+ [8] is to exploit the stream of returning acknowledgment packets to estimate the bandwidth $\widehat{B}$ available for the TCP connection. When a congestion episode happens at the end of the TCP probing phase, the bandwidth estimated from the stream of ACKs corresponds to the definition of the best effort available bandwidth in a connectionless packet network. This bandwidth estimate is used to adaptively decrease the congestion window and the slow-start threshold after a timeout or 3 DUPACKs. In both cases, $ssthresh$ is set equal to $\widehat{B} \cdot RTT_{min}$, whereas, $cwnd$ is set equal to $ssthresh$ (one MSS) when 3 DUPACKs are received (a retransmission timeout expires).

3. THE ANALYTICAL FRAMEWORK

The considered scenario has been depicted in Fig. 1, where both wireless and wired links are involved in end-to-end communications. In such an environment, possible segment losses are due to either network congestion or links errors. Since usually the reliability of wired links, involved in communication, is very higher than the wireless ones, losses due to links failure can be imputed only to the wireless access network. The end system, which is transmitting on the wireless link, can reduce the segment loss probability due to link unreliability, by opportunely adjusting both the transmission power and the amount of redundancy added to data being transmitted. In fact, to face up to higher bit error probability, due to lower signal to noise ratio, the transmission power can be increased. However, it causes greater energy consumption (i.e., shorter battery life) and higher interference for neighboring communications. Lower bit error probability can be also achieved by using correction capability of coding schemes such as FEC techniques, which allow to recover possible transmission errors. Nevertheless, adding redundant information to data being transmitted, behind an optimal value, will result in bandwidth wasting and decreasing of throughput. The joint optimization analytical framework, herein considered, is based on the proposal of Galluccio et al. in [5]. It finds the optimum values of transmission power and the amount of redundancy to be added to data,

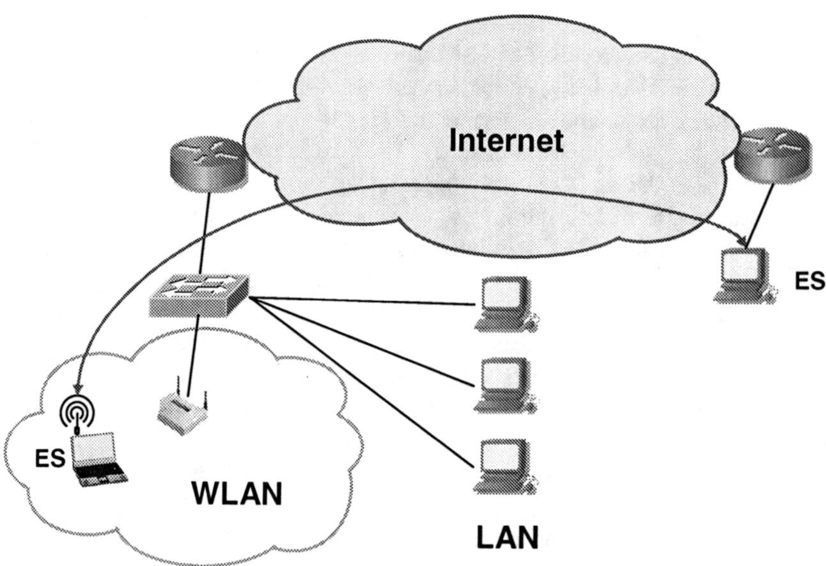

Figure 1. The considered scenario

by maximizing a *satisfaction function*, γ, defined as the ratio of an appropriate function of the TCP throughput, σ, and a cost function, c, i.e., $\gamma = \sigma/c$.

3.1. The Cost Function

The cost function, c, considers two terms: a term which accounts for energy consumption, and a term which considers the amount of bandwidth resources employed. If x is the redundancy ratio, introduced by the FEC scheme on a K bits information block, and y is the transmission power, than the considered cost function is:

$$(1) \qquad c = (1+x)\cdot(k_1 \cdot y + k_2);$$

where the constant k_1 (expressed in $[Watt^{-1}]$) accounts for the energy cost, as well as k_2 represents the bandwidth cost. Both k_1 and k_2 depend on many factors such as battery status, bandwidth availability, and user preferences. In the following, it will be assumed that $k_2 = 1$, $k_1/k_2 = 100$, that is the energy cost is higher than the resource one, since limited battery capacity for mobile users is often a critical issue.

3.2. The TCP Throughput

The considered expression for TCP throughput function σ is in the form of:

$$(2) \qquad \sigma = (b/(1+x)) \cdot f(RTT, P_{Loss});$$

where b is the bandwidth available for TCP on the wireless access network, and f is an opportune function which depends on the TCP throughput approximation formula, on the round trip time, RTT, and on the segment loss probability, P_{Loss}. In particular, the function $f(\cdot)$, that we will refer to as *normalized TCP throughput*, has the following expression:

$$f(RTT; P_{Loss}) = B(RTT; P_{Loss})/B_0; \tag{3}$$

where $B(RTT; P_{Loss})$ is the estimated TCP throughput (expressed in packets per second) when segment losses occur with probability P_{Loss}, either for congestion events or link errors, and B_0 is the TCP throughput when segment may be lost only due to congestion events, with probability P_c, i.e., it is $B_0 = B(RTT, P_c)$. It is worth to notice that it is $0 < f \leq 1$. In particular, we have $f = 1$ when $P_{Loss} = P_c$, that is segment losses are only due to congestion events, whereas it is $f << 1$ when $P_{Loss} >> P_c$, that is, wireless link unreliability gives a significant contribution to the segment loss process.

The functional expression of B depends on the adopted TCP transmission algorithm. Hence, when Westwood+ TCP is used, we have [10]:

$$B = \sqrt{\frac{1-P_{Loss}}{P_{Loss}}} \cdot \frac{1}{\sqrt{RTT[RTT-(1-P_{Loss})RTT_m]}} \tag{4}$$

where RTT_m is the minimum RTT; whereas when New Reno TCP is used we have [11]:

$$B = \frac{1}{RTT}\sqrt{\frac{3(1-P_{Loss})}{2P_{Loss}}} \tag{5}$$

Finally, let P_{block} be the error probability on a K bits block at the output of the decoder, due to wireless link failure. Assuming that block losses are i.i.d., for a segment of MSS bits, it results that $\left[1-(1-P_{Block})^{MSS/K}\right]$ is the probability of correctly receiving a segment. Since the probability that a segment is not lost due to a congestion event is $(1-P_c)$, it yields that the probability of segment losses due to either congestion event or radio link failure is given by:

$$P_{Loss} = 1-(1-P_{Block})^{\frac{MSS}{K}}(1-P_c) \tag{6}$$

Furthermore, P_{block} is related to the redundancy ratio x and the transmission power y as follows:

$$P_{block} = h(y \cdot G_{coding}) \tag{7}$$

where $h(\cdot)$ depends on the used modulation technique, and G_{coding} is the coding gain which depends on the FEC scheme and on the redundancy ratio.

3.3. Maximization of the Satisfaction Function

Since the expression of the satisfaction function depends on both the transmission power and the redundancy ratio, it yields:

$$\text{(8)} \qquad \gamma(x, y) = \sigma(x, y)/c(x, y)$$

To evaluate the optimal values of x and y which maximize $\gamma(x, y)$ it is necessary to null its partial derivatives. Nevertheless, in order to simplify such a calculation, it is possible first to find a relationship between the values of x and y which maximize the satisfaction function: therefore, by substituting this relationship in $\gamma(x, y)$, the maximum of a one variable function can be obtained numerically [5]. From eqs. (1),(2) and (8), the relationship between the optimal values of x and y is given by:

$$\text{(9)} \qquad y = \psi(x) = \frac{2 \cdot k_2}{k_1} \cdot \frac{1 + K \cdot x}{K - 3 - 2 \cdot K \cdot x}, \; x \in \left[0, \frac{1}{2} - \frac{3}{2 \cdot K}\right].$$

It is worth noting that eq. (9) does not depend on the throughput approximation formula and on the modulation technique. Substituting eq. (9) in (8), we obtain $\tilde{\gamma}(x) = \gamma(\psi(x))$, that numerically maximized gives the optimal amount of redundancy, x_{opt}. Finally, by replacing x_{opt} in (9), the optimal transmission power, y_{opt}, can be obtained as well.

4. NUMERICAL RESULTS

The afore described analytical framework has been applied considering different network congestion conditions, that is, P_c ranging from $10^{-5} \div 10^{-1}$, and RTT typical value varying from $100 \div 250$ ms. In order to asses results dependencies on the used numerical modulation technique, two relevant examples have been considered for physical layer, i.e.: (1) Differential Binary Phase Shift Keying (DBPSK), which is used in IEEE 802.11; (2) Gaussian Frequency shift Keying (GFSK), used in Bluetooth [9]. Similar results have been obtained with the two modulation schemes. Thus, due to lack of space, in the sequel, only those concerning the 802.11 case will be reported. The dependence of function $\sigma(x, y)$ on the modulation technique relies on the relationship between P_{block} and bit error probability, P_e. In fact, assuming that K is the block size, it yields:

$$\text{(10)} \qquad P_{Block}(y, G_{coding}) = 1 - \left[1 - P_e(y, G_{coding})\right]^K$$

where the functional expression of P_e depends on the modulation technique. In particular, when DBPSK is considered, it results [12]:

$$\text{(11)} \qquad P_e = 0.5 \cdot \exp\left(-A/(N_0 \cdot \Delta F) \cdot y \cdot G_{coding}\right)$$

where A is the attenuation which accounts for free-space losses, atmospheric attenuation, and multipath fading as well, N_0 is the noise power spectral density (assumed to be constant and equal to $1.379 \cdot 10^{-}20$ W/Hz), ΔF is the considered bandwidth (that is, 22 MHz for a single 802.11b channel), $G_{coding} = (Kx+1)/(x+1)$, is the FEC coding gain for Reed-Solomon encoding [13][1]. In the following we assume that the block size is $K = 12$ bits. By eqs. (11) and (10), P_{Loss} in eq. (6) can be calculated as function of the considered modulation technique, of the propagation condition, and of communication settings.

In order to show the advantages of joint FEC and power management, Figure 2 shows the maximum values of the satisfaction function, $\gamma_w(x_{opt}, y_{opt})$, for different attenuation levels, when TCP Westwood+ has been used, the network congestion probability is $P_c = 10^{-3}$, RTT_{min} =250 ms, and average $RTT = 280$ ms. In the same figure, for sake of comparison, other three curves are depicted. The one referred to as $\gamma_w(\overline{x}, \overline{y})$ has been obtained by setting both x and y to the optimal values for $A = 95$ dB, i.e., $\overline{x}, \overline{y}$. The other two curves, $\gamma_w(\overline{x}, y)$ and $\gamma_w(x, \overline{y})$, have been obtained by fixing redundancy ratio and transmission power, respectively, to the optimal values for $A = 95$ dB, and by adjusting the unfixed parameter. All the curves show a non-increasing behavior, due to higher costs needed to attain satisfactory throughput in worst communication condition. However, it can been noticed that satisfaction values obtained with joint optimization are always the highest. Very similar results have been obtained for TCP New Reno. In Figs. 3, 4, and 5, we study the effects of joint error correction and transmission power in terms of normalized throughput, f, and

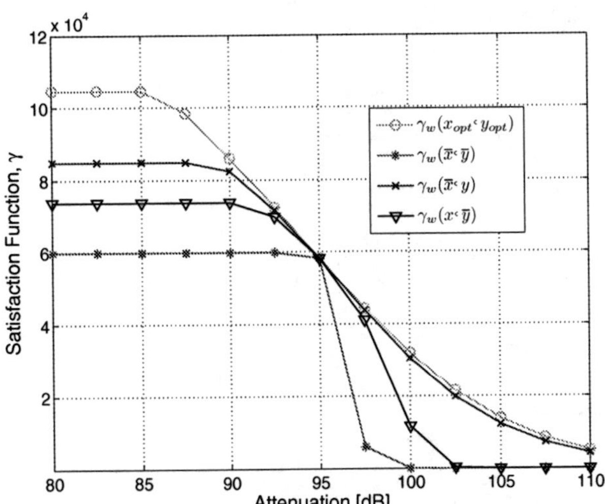

Figure 2. Maximum of satisfaction function, γ_w, for different attenuation levels

[1] Similar expressions hold for different coding schemes.

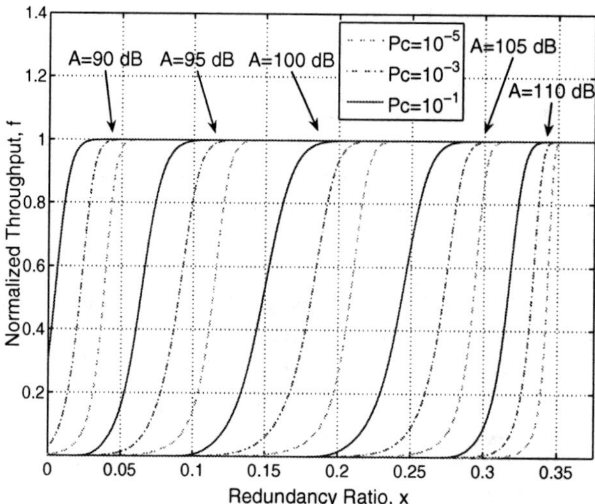

Figure 3. Normalized throughput, f

satisfaction function, γ, for Westwood+ TCP. More specifically, in Fig. 3, we show the normalized throughput f vs. the redundancy ratio x for different attenuation values, A, and congestion probability, P_c. It can be noticed that, for a given P_c, increasing x, f tends to 1, that is P_{Loss} goes to 0 and the throughput approaches the maximum for the considered congestion scenario, i.e., B_0.

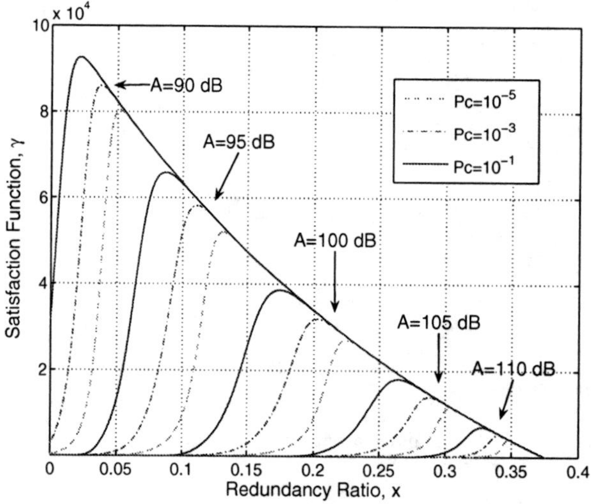

Figure 4. Satisfaction function, $\hat{\gamma}(x)$, vs. the redundancy ratio x

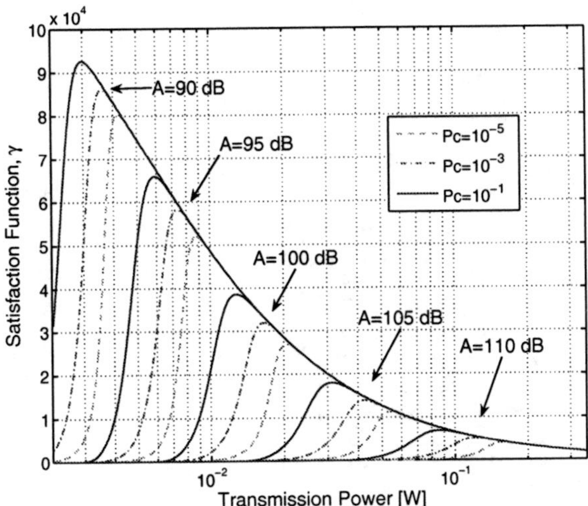

Figure 5. Satisfaction function, $\hat{\gamma}(\psi^{-1}(x))$, vs. transmission power y

Moreover, as intuition suggests, considering higher attenuation levels, the amount of redundancy to be added to improve the performance increases as well. Nevertheless, a quite counterintuitive result come out from Figure 3. That is, to obtain the same values of the normalized throughput, f, much more redundancy need to be added when the congestion probability decreases. This behavior can be explained looking at eq. (3). In fact, as the congestion probability decreases, the throughput achieved when losses are due only to congestion events, B_0, is higher. Hence, much more redundancy and transmission power are needed to reach a higher throughput B_0. From Figure 4 the dependence of satisfaction function, $\tilde{\gamma}(x)$, on the redundancy ratio can be investigated for different attenuation levels and network congestion probabilities ranging from $10^{-5} \div 10^{-1}$. It can be noticed that $\tilde{\gamma}(x)$ increases with the amount of redundancy until it gets the maximum value for $x = x_{opt}$. For a given congestion probability, P_c, as the attenuation increases (that is, either the distance increases or the propagation conditions get worst), the optimal amount of redundancy to be introduced significantly increases, but the maximum value of the satisfaction function decreases because of the increased transmission cost. Moreover, as a consequence of the behavior described in Figure 3, for the same attenuation conditions, the optimal redundancy ratio to be introduced decreases when the network congestion probability increases.

The relationship between transmission power y and the satisfaction function, $\hat{\gamma}(\psi^{-1}(x))$, has been investigated in Figure 5 for different attenuation and congestion probability values. It reveals the same dependence afore discussed for the redundancy ratio, both on A and P_c. However, in this case, the transmission power varies up to one order of magnitude with the attenuation level.

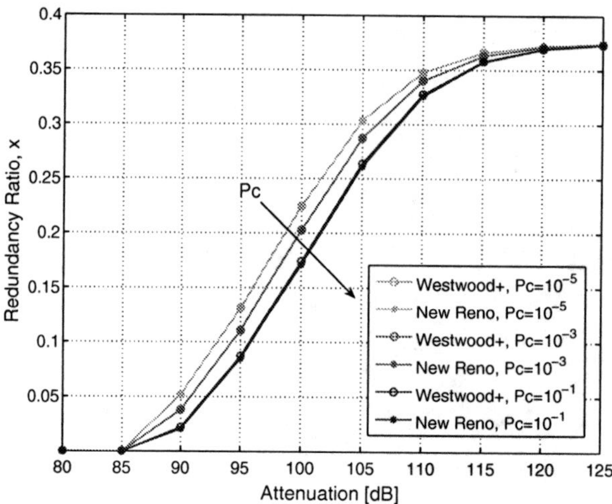

Figure 6. Optimal redundancy ratio vs. attenuation level

The optimal redundancy ratio, x_{opt}, is reported in Figure 6, as function of the attenuation and the congestion probability, for both TCP Westwood+ and New Reno. We can see that x_{opt} increases with the attenuation level and decreases when the congestion probability increases. Such an increase is more pronounced for $A = 90 \div 115$ dB, whereas it gets less important for higher and lower attenuation levels. Finally, the optimal redundancy ratio values, for the two TCP flavors under study, are quite undistinguishable. By analyzing Figure 7 where the couples of

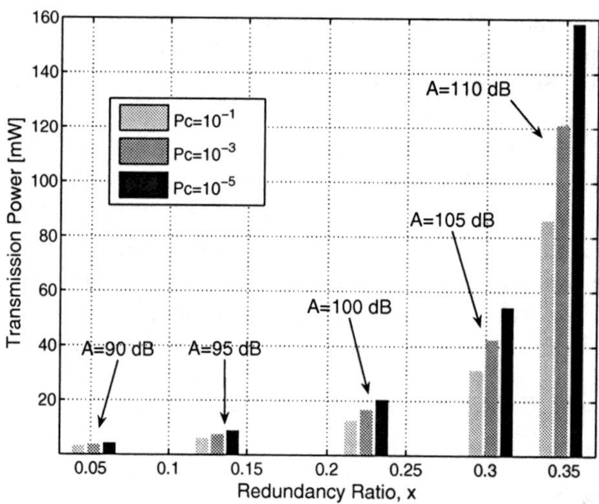

Figure 7. Optimal redundancy ratio and transmission power; TCP Westwood+

values (x_{opt}, y_{opt}) are reported, only a slight variation of the optimal redundancy ratio with P_c, for all the attenuation levels, can be noticed.

On the contrary, the optimal transmission power variations with P_c are relevant especially for severe attenuation conditions. Anyway, for the same propagation conditions, when P_c increases, both x_{opt} and y_{opt} decrease. This can be explained considering that the optimization target is the satisfaction function rather than the TCP throughput. In fact, as the wired network gets congested it becomes useless to spend both power and bandwidth to improve the wireless link reliability, since the transmitted segments will be likely dropped in the wired network. Moreover, limiting the information overhead due to FEC reduces the required bandwidth; hence, it should help in resolving the congestion event or at least it will not make it worst. On the contrary, when the congestion probability decreases, both transmission power and the redundancy ratio should be increased in order to improve the wireless link reliability and to profit of the uncongested wired network.

By applying the considered optimization framework both to New Reno and Westwood+ TCP, we obtained the similar values for the couples (x_{opt}, y_{opt}), in the same congestion and attenuation conditions. However, by substituting the optimal parameters in eqs. (4) and (5), differences in the achieved throughput can be assessed. Let $\hat{B}_w$ and $\hat{B}_r$ be Westwood+ and New Reno throughputs, respectively, when joint optimization has been performed. From Figure 8, where the throughput ratio $\hat{B}_w/\hat{B}_r$ has been reported for several P_c and A, it is evident that Westwood+ always outperforms New Reno, regardless of the attenuation level. Even though, such an advantage tends to decrease when high congested scenarios are considered.

Finally, since both $\hat{B}_w$ and $\hat{B}_r$ depend on RTT, in Figure 9 the sensitivity on its changes has been explored. Curves in Figure 9 have been drawn considering $P_c = 10^{-2}$ and RTT values varying in $100 \div 250$ ms range. For Westwood+ TCP, which

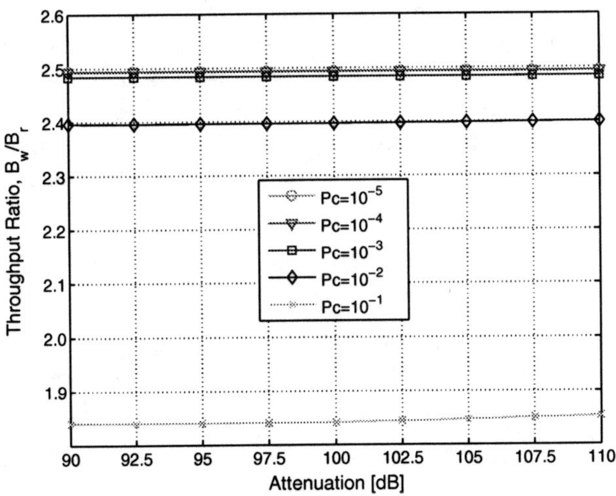

Figure 8. Throughput ratio for different congestion probability values, P_c

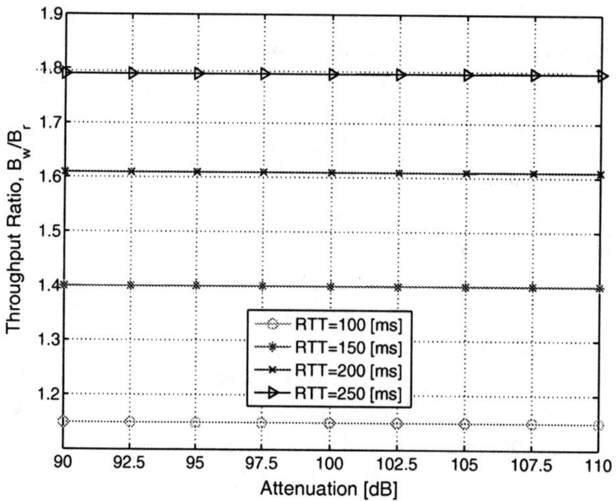

Figure 9. Throughput ratio for different RTT values

throughput expression (4) depends also on the minimum round trip time, $RTT_{min} = (RTT - 50)$ ms has been considered. Also in this case, it can be noticed that Westwood+ outperforms New Reno. Moreover, its advantage gets larger for increasing RTTs. This behavior was expected since it is well known from literature that Westwood+ TCP performance gain, with respect to New Reno, grows with the bandwidth delay product of the network [8].

5. CONCLUSIONS

In this paper, we applied a joint forward error correction and transmission power management framework to a hybrid wired-wireless network with New Reno or Westwood+ TCP connections. Results show that better performances are achieved when compared with the ones obtained by using only error or power management. Moreover for several network scenarios, Westwood+ achieves higher throughput than New Reno, by spending the same transmission power and bandwidth resources for error control.

REFERENCES

1. Balakrishnan, H., Padmanabhan, V.N., andR. H. Katz, S.S.: A comparison of mechanisms for improving TCP performance over wireless links. IEEE/ACM Transactions on Networking **5**(6) (December 1997) 756–769
2. Chaskar, H.M., Lakshman, T.V., Madhow, U.: TCP over wireless with link level error control: Analysis and design methodology. IEEE/ACM Transactions on Networking **7**(5) (October 1999) 605–615
3. Barakat, C., Altman, E.: Bandwidth trade-off between tcp and link-level fec. Computer Networks **39**(2) (June 2002) 133–150

4. Vacirca, F., Vendictis, A.D., Baiocchi, A.: Optimal Design of Hybrid FEC/ARQ Schemes for TCP over Wireless Links with Rayleigh Fading. IEEE Trans. on Mobile Computing **5**(4) (April 2006) 289–302
5. Galluccio, L., Morabito, G., Palazzo, S.: An analytical study of a tradeoff between transmission power and fec for tcp optimization in wireless networks. In: Proc. INFOCOM 2003, Twenty-Second Annual Joint Conference of the IEEE Computer and Communications Societies. IEEE. Volume 3., Phoenix (30 March-3 April 2003) 1765–1773
6. Zorzi, M., Rossi, M., Mazzini, G.: Throughput and energy performance of tcp on a wideband cdma air interface. Wireless Communications and Mobile Computing **2**(1) (February 2002) 71–84
7. Floyd, S., Henderson, T., Gurtov, A.: NewReno modification to TCP's fast recovery. IETF RFC 3782 (April 2004)
8. Grieco, L.A., Mascolo, S.: Performance evaluation and comparison of Westwood+, New Reno and Vegas TCP congestion control. ACM Computer Communication Review **34**(2) (April 2004) 25–38
9. Walke, B.H., Mangold, S., Berlemann, L.: IEEE 802 Wireless Systems. John Wiley and Sons (2006)
10. Grieco, L.A., Mascolo, S.: Mathematical analysis of westwood+ tcp congestion control. IEE Proceedings Control Theory and Applications **152**(1) (Jan 2005) 35–42
11. Padhye, J., Firoiu, V., Towsley, D., Krusoe, J.: Modeling TCP throughput: A simple model and its empirical validation. Proceedings of the ACM SIGCOMM '98 conference on Applications, technologies, architectures, and protocols for computer communication (1998) 303–314
12. Proakis, J.: Communication System Engineering. Prentice Hall International Editions (1994)
13. Reed, I.S., Solomon, G.: Polynomial codes over certain finite fields. Journal of the Society for Industrial and Applied Mathematics **8**(4) (June 1960) 300–304

CHAPTER 15

JOINT CROSS-LAYER AND DYNAMIC SUBCARRIER ALLOCATION DESIGN FOR MULTICAST OFDM NETWORKS

H.NAJAFI, H. TAHERI AND H.R. AMINDAVAR, S.H. HOSSEINI

Amirkabir University of Technology (AUT), Department of Electrical Engineering,
h_najafi@cic.aut.ac.ir, htaheri@aut.ac.ir, hamidami@aut.ac.ir, h_hosseini@algocom.net

Abstract: In multicast orthogonal frequency division multiplexing (OFDM) systems the varieties in link conditions of users decreases the throughput of users. If we assign the subcarriers which are in the best channel condition among all the subcarriers for multicast users in the base station, the network throughput can be increased by performing subcarrier allocation. In this paper, we propose a joint cross-layer and adaptive subcarrier allocation algorithm for multicast OFDM systems. This algorithm takes into consideration the physical channel condition in both physical layer and MAC layer. In MAC layer we use packet scheduling algorithm. Simulation results show that with increasing the number of users or bandwidth demanded, this joint algorithm enhances the system performance and increases throughput of multicast users besides decreasing packet loss in the channel

Keywords: Channel state information, cross-layer design, multicast, OFDM, subcarrier allocation

1. INTRODUCTION

Research on cross-layer design has recently attracted significant interest. It is concerned with adapting of information among various layers as specified in the open system interconnection (OSI) protocol layers. There are a number of papers in the literature addressing specific cross-layer issues.

Multicasting delivers data to a group of users through a single transmission, which is particularly useful for high-data-rate multimedia services due to its ability to save the network resources. Since the bandwidth allocated to each user is different in heterogeneous networks, the data rate of a multicast stream is limited by the data rate of the least capable user; otherwise it is not delivered to several users.

H. Labiod and M. Badra (eds.), New Technologies, Mobility and Security, 173–182.
© 2007 *Springer.*

Orthogonal frequency division multiplexing (OFDM) is a very promising multiple access technique to efficiently utilize limited RF bandwidth and transmit power in wideband transmission over multipath fading channels [7]. When a wideband spectrum is shared by multiple users in multiuser OFDM-based systems, different users may experience different fading conditions at all subcarriers. Each user is assigned a subset of all subcarriers by some allocation algorithm. Thus, multiuser diversity can be achieved by adaptively adjusting subcarrier, bit and power allocation depending on channel status among users at different locations [8].

In wireless system, the spectrum is scarce and the channel varies according to users due to Rayleigh fading; therefore, multicasting in wireless networks should be spectrally efficient and be able to cope with the channel variations. The channel variations among users complicates adaptive modulation because the modulation should be adjusted to serve the user who experiences the worst channel condition; thus, it is usually adapted to the worst link condition [5], or the heterogeneity in link condition is often ignored during the design of adaptive modulation for multicast transmission [6]. To cope with the channel variation between users without adaptation, the non-uniform phase-shift-keying (PSK) is used in [3] where the base layer data is encoded to constellation points that are far apart in distance from each other than the higher layer data are encoded to. In [4], an adaptive modulation for multicast data is proposed assuming that the same modulation is used for all the subcarriers in an OFDM symbol. In [2], a dynamic subcarrier/bit allocation method was proposed for multicast OFDM system in a way that maximizes the total data rate of all users.

In this paper we propose a cross-layer algorithm to improve the performance of multicast OFDM systems. The algorithm assigns subcarriers to the multicast service in physical layer and use the information of users channels and add an intentional delay to avoid packet loss in a bad channel condition.

The remainder of this paper is organized as follows. In Section 2, we describe a total overview of the system, transmission media and data type. Section 3 formulates the dynamic subcarrier allocation algorithm and we propose the method of subcarrier assignment to the multicast service. In Section 4 we present the cross-layer design of considered algorithm. Numerical results and simulations are in Section 5. Finally the conclusion is in Section 6.

2. SYSTEM ARCHITECTURE

2.1. Multicast Transmission

Multicast services have positive and negative sides for both wired and wireless communications. In wired communications where the link is relatively stable, the channel fluctuation across users is small; it, therefore, simplifies the adaptive modulation for multicast data. However, a different wired link should be connected between the source and different destinations although the source delivers the same multicast data, i.e., as the number of multicast users increase, more wired links are required.

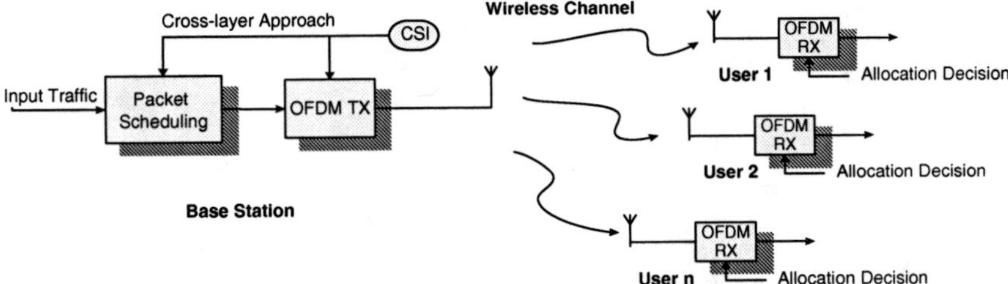

Figure 1. The cross-layer structure for mobile multicast OFDM system, base station and multicast users

On the other hand, in wireless communications, multicast data can be delivered to many users only through a *single* transmission without increasing any wired connections. It is an attractive merit of multicasting in wireless channels. However, the most difficulty of multicast transmission in wireless communications is the difference of link conditions of users.

The adaptation of the proposed cross-layer design approach is based on sharing the channel state information from physical layer with the multicast packet scheduler at the MAC layer. We investigate a downlink OFDM system with N subcarriers and M users of a multicast group (Fig. 1). Within the queue, packets are served in a FIFO (First in first out) order. Across the queues, packets are served according to certain service disciplines determined by the scheduler. Output bits from the scheduler are modulated. The resultant symbols are transmitted via the downlink channel on different subcarriers using the same amounts of transmit power. The allocation of subcarriers are determined by a dynamic allocation algorithm. Assume that the subcarrier allocation decision is sent to the mobiles via a dedicated control channel. Fig. 2 and Fig. 3 show multicast OFDM transmitter and receive supporting "n" users.

2.2. System Overview

Assume a multicast service with a Base Station (BS) and many Mobile Stations (MS). The input traffic would be voice, data, video and....

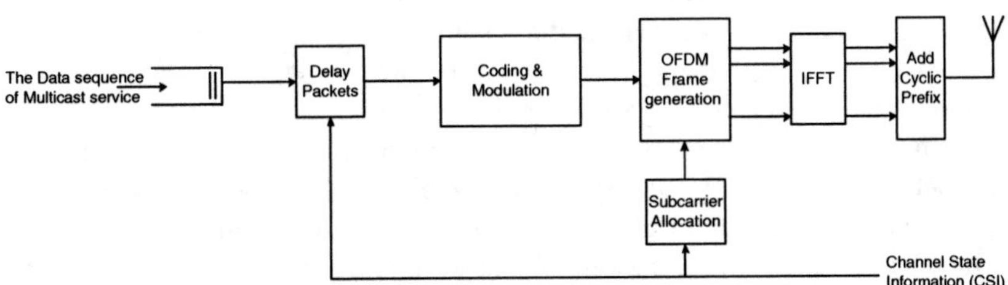

Figure 2. Transmitter structure at the base station

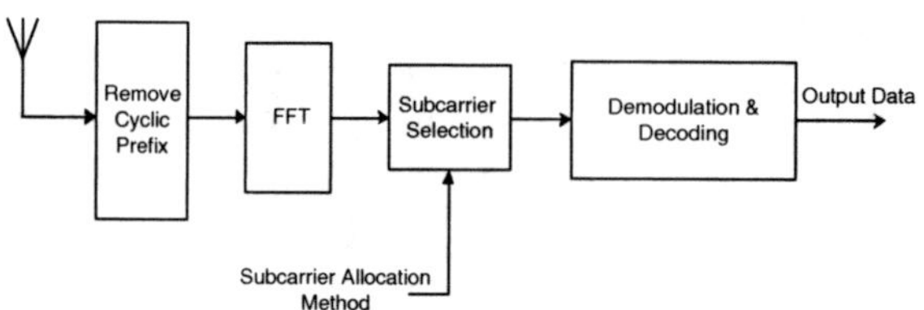

Figure 3. Receiver structure of each user

The BS gets the CSI(Channel State Information) from return channel via MS to BS. Each MS sends its channel state information to BS. The channel between BS and each MS is a frequency selective Rayleigh slow fading channel with Jakes model for Doppler shift.

The CSI which is acknowledged from each MS, is used in physical layer of BS for resource allocating in OFDM system. Assume that the subcarrier allocation decision is sent to the mobiles via a dedicated control channel.

At the receiver end, a mobile is then able to extract its data from the assigned subcarriers. Also, the CSI is used in packet scheduling for setting the delay of each packet based on the CSI of each user channel.

2.3. Traffic Model

Assume that packets arrive at the BS according to Poisson distribution. The average packet arrival rate of each user is λ packets per time unit. If the system is able to successfully transmit packets at a rate of μ packets per time unit, then it is stable only when

$$(1) \qquad \frac{\lambda}{\mu} \leq 1$$

In this paper, we concentrate on data services that are delay-insensitive.

2.4. Channel Model

We model the wireless channels as frequency-selective Rayleigh fading channel. However, we assume that each subcarrier experiences the flat-fading. Moreover, the channel is assumed to be quasi-static so that the channel gains are constant over a burst duration T, but vary from one burst to another. The burst duration T is assumed to be large enough to comprise hundreds of symbol-intervals. In particular, the channel gain, remains constant during the transmission of each frame, but, channel gains during the transmissions of different frames are assumed to be independent and identically distributed. Thus, within each burst duration, we denote the channel coefficients between the base station and the ith user at the jth subcarrier by $h_{i,j}$. We model the channel coefficients as i.i.d. complex-Gaussian random variables with zero-mean and variance of σ^2 per dimension.

3. DYNAMIC SUBCARRIER ALLOCATION ALGORITHM

The physical layer resource allocation involves the allocation of subcarriers to the users that have pending traffic in the current frame. The transmission power for all the users is the same. We assume that there is at least one multicast system that is served through the Base Station. According to the state of CSI signal, the base station knows the subcarriers which experience the worst condition due to the channel. Assume that the multicast service Bandwidth is S subchannel and has the first priority to other services. So the algorithm allocates the best S subcarriers to the multicast service and informs the receivers about them. On the other hand, each packet is delayed a specified time according to the CSI signal; it is described briefly as the Packet Delay parameter in the next section.

One of the resources in the physical layer is OFDM subcarriers. The BS assigns some of subcarriers with good CSI for a multicast service according to the results of all users channel state information. The number of subcarriers is a function of bandwidth which is required for a defined service such as data, voice, video, and so on.

The selection of subcarriers is done as a function of CSI of all MSs in the multicast group.

We define the CSI of each user that is obtained from uplink as follow:

(2) $$CSI_i = \left[H_1^i + jH_1'^i,\ H_2^i + jH_2'^i,\ \ldots,\ H_N^i + jH_N'^i\right]$$

Where $H(f) = F\,(h(t))$

The length of CSI vector is equal to the number of OFDM subcarriers since it should indicate the channel condition in all subcarriers.

$h(t)$ is the impulse response of Raleigh fading channel.

We show the OFDM subcarriers in SUB subset:

(3) $$SUB_T = [f_1, f_2, \ldots, f_N]$$

And also the subcarriers that are used for multicast service in the base station are:

(4) $$SUB_t = [f_{k_1}, f_{k_2}, \ldots, f_{k_n}]$$

$$k_i \in \{1, 2, \ldots, N\}, \forall i$$

'n' is the number of subcarriers which are used in multicast service and $1 \leq n \leq N$.

So: $SUB_t \subset SUB_T$

Now we can define the function of subcarrier selection according to the CSI vector of each user. The Γ function operates on all of CSI_i and output the best subset of subcarriers among OFDM subcarriers for multicast service.

(5) $$CSI = \Gamma\left(\{CSI_i\}_{i=1}^{M}\right)$$

M is the total number of users in the multicast group.

Therefore we show the final result of subcarriers in the following way:

$$CSI = [C_1, C_2, \ldots, C_N]$$

4. CROSS-LAYER DESIGN

The performance of the DSA algorithm decreases when the Bandwidth of Multicast service or the number of required subcarriers is comparable to the total number of OFDM subcarriers in the BS. On the other hand if the number of users increases the DSA algorithm can not find the best subcarriers of all users in a group and the chance of good subcarrier selection is decreased. Beside of these two, the condition of BS would not allow to select the best existing subcarriers of OFDM and assign some fixed subcarriers for multicast service. So we should use another technique to increase the performance of multicast service in all of the users of the group.

If the condition of channel is not so good that every user of multicast group in the base station domain can receive the transmitted signal, the data isn't sent to group and delayed up to a suitable condition. The delay is constraints to a defined threshold. Therefore the data is hold up to a determined time.

As mentioned in the previous section, the system throughput will be improved by using subcarrier allocation. This is achieved from the information of physical layer; CSI.

Another parameter that could be variable regarding to CSI is in the upper layer, MAC layer. In the situation that the channel is in a bad condition, and there is not enough subcarriers with suitable channel condition, the BS can't transmit the data to MS, but it can wait for a specific time and then start the transmission. This is done with packet delay in the MAC layer which is a function of maximum Doppler in the channel.

(6) $\quad$ Packet Delay = K(Max. Doppler)

Where K is a scalar obtained from CSI. This algorithm can improve throughput of the system for multicast services that do not use ARQ algorithms.

Since the BS doesn't transmit the signal in the bad situation of channel, another parameter which is improved in this algorithm is BS power consumption.

5. NUMERICAL RESULTS

The performance of the proposed algorithm is investigated in this section. An OFDM system with 64 subcarriers and a symbol duration of *Ts=62.5ns* is considered. The buffers for every session are of infinite lengths. Assume that a rate ½ convolutional code is applied, and QPSK modulation is used on every subcarrier.

In Fig. 4, the BER of two users with and without use of DSA algorithm is compared. User1 has a frequency selective multipath fading channel with 4 fading

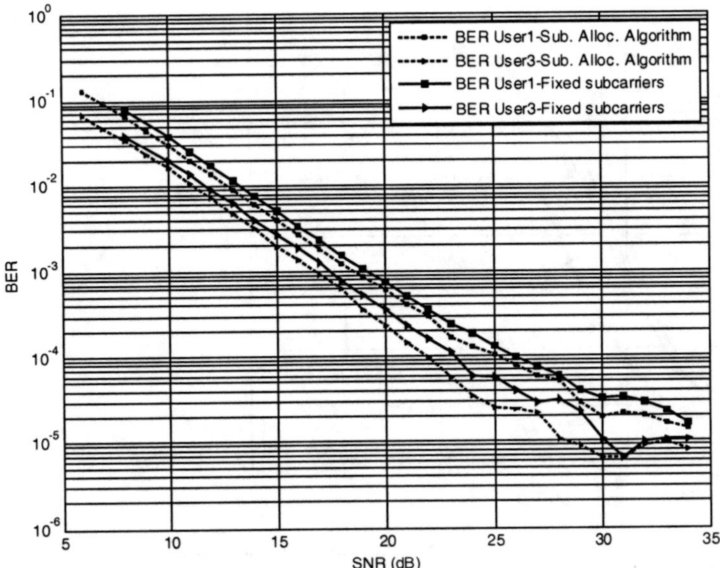

Figure 4. Bit error rate versus SNR for two users before subcarrier allocation and after use of DSA algorithm

path and user 2 has a frequency selective multipath fading channel with 2 fading path. The Doppler shift in both of them is 40Hz. According to the simulation results in Fig. 4, the SNR of user1 and user2 is decreased about 0.5dB in the same BER.

In Fig. 5, the system throughput, which is defined to be the total number of successfully transmitted packets in a time unit, is investigated for the packet-level multiplexing systems with and without Dynamic Subcarrier Allocation (DAS) and Delayed algorithm. Assume that frame has 20 OFDM symbols, then it can be seen that in the low SNR region, the throughput is mainly determined by the transmission power. In the high SNR region where the system is able to transmit more packets than those arrived, the system throughput is decided by the amount of input traffic. The figure shows that using DSA and Delayed algorithms, improve throughput of the system.

In Fig. 6, it is demonstrated that system throughput decreases when the number of the users of multicast group increases, because the chance of finding suitable subcarriers that are common in all of the users is decreased. Therefore the performance of the DSA algorithm is decreased and the other algorithm that is Cross-Layer packet scheduling, improves throughput instead. Finally, in both curves the throughput becomes stable for large number of users.

Since the high bit rate services require more bandwidth and therefore more subcarriers in the base station, the throughput is decreased faster. It means that because the number of required subcarriers is large, the number of subcarriers that achieved

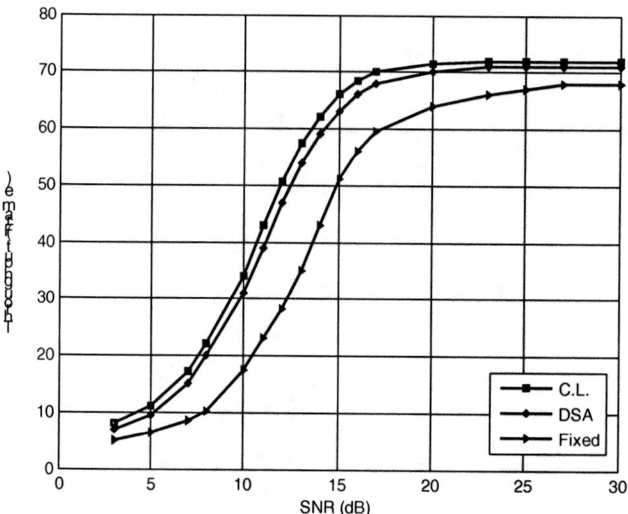

Figure 5. System throughput of multicast group with no dynamic subcarrier allocation and after use of DSA and Delayed DSA algorithm

from the DSA algorithm is not so various. When we use Delayed algorithm, the packets keep up to a suitable channel condition and therefore the packet loss is decreased and the throughput of system is improved. This improvement is more obvious in high bandwidth services. This is shown in Fig. 7. The simulation is done for 8 multicast users.

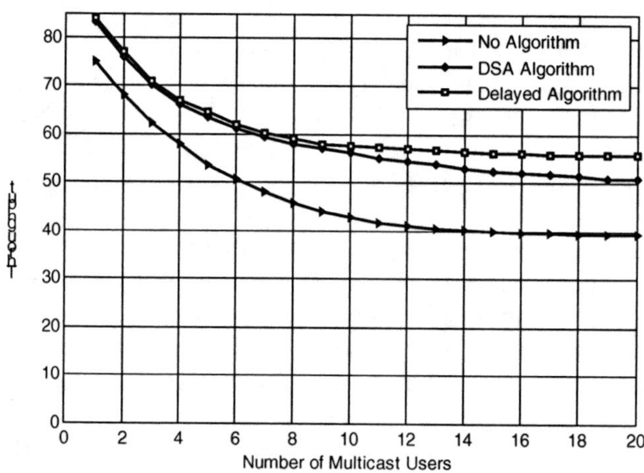

Figure 6. System throughput for different number of users

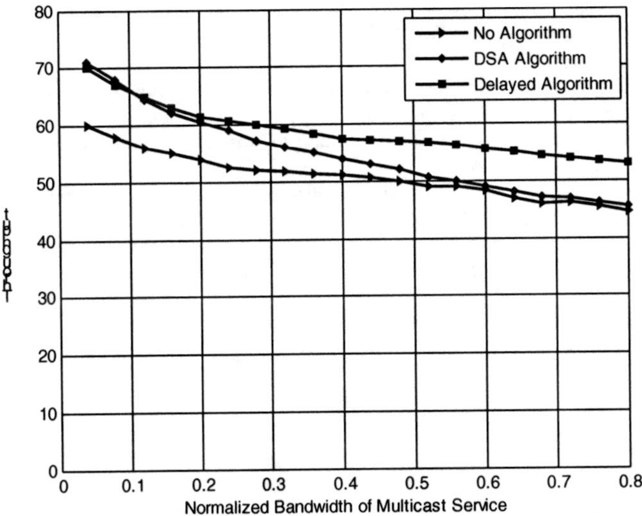

Figure 7. System throughput for various bandwidth requierment of multicast service

6. CONCLUSION

With the rapid growth of broadband wireless networks, there is a need to develop new technologies to achieve high resource utilization for packet-based services. In this paper, we assume a wireless multicast network with increasing bandwidth of demanded services and growth in the number of users of the multicast group. Therefore a cross-layer adaptive resource allocation algorithm is proposed for multicast OFDM systems. The algorithm is composed of two parts: packet scheduling and adaptive subcarrier allocation. The first part which is done between physical layer and MAC layer is Delayed Algorithm. The second part is dynamic subcarrier allocation (DSA) which is in physical layer. Meanwhile, the adaptive subcarrier allocation algorithm is to choose the best common subcarriers for multicast service and to provide guaranteed transmission efficiency in the physical layer and the Delayed Algorithm is to avoid transmission of data when the channel condition is not suitable. These two algorithms are to provide guaranteed performance and decreasing packet loss from the data link layer's point of view in a multicast system with no ARQ algorithm. Our simulation results show that the proposed algorithm is able to provide tremendous performance improvement in throughput and BER for various types of services.

REFERENCES

1. R.A. Berry and E.M. Yeh, "Cross-Layer Wireless Resource Allocation," *IEEE signal processing magazine*, pp. 59–68, Sep. 2004.
2. C. Suh and C.-S. Hwang, "Dynamic subchannel and bit allocation for multicast OFDM systems," *Proc. of IEEE PIMRC*, 2004.

3. M. B. Pursley and J. M. Shea, "Multimedia multicast wireless communications with phase-shift-key modulation and convolutional coding," *IEEE J. Select. Areas Comm.*, vol. 17, pp. 1999–2010, Nov. 1999.
4. C.-S. Hwang and Y. Kim, "An adaptive modulation method for multicast communications of hierarchical data in wireless networks," *Proc. Of IEEE ICC.*, pp. 896–900, 2002.
5. S. Yuk and D. Cho, "Parity-based reliable multicast method for wireless LAN environments," *Proc. of IEEE VTC.*, pp. 1217–1221, 1999.
6. P. Ge and P. K. McKinley, "Experimental evaluation of error control for video multicast over wireless LANs," *Proc. IEEE ICDCA*, pp. 301–306, 2001.
7. Y. W. Cheong, R. S. Cheng, K. B. Lataief, and R. D. Murch, "Multiuser OFDM with Adaptive Subcarrier, Bit, and Power Allocation," *IEEE J. Selected Areas Comm.*, vol. 17, pp. 1747–1758, October 1999.
8. U.F. Chen, J.W. Chen and C.P. Li, "A Fast Suboptimal subcarrier, Bit and Power Allocation Algorithm for Multiuser OFDM-based Systems," IEEE Comm. Soc., pp. 3212–3216, 2004.

CHAPTER 16

HANDOVER OPTIMIZATION FOR VEHICLE NEMO NETWORKS

HAI LIN AND HOUDA LABIOD

GET – ENST; LTCI-UMR 5141 CNRS; INFRES Department, 46 rue Barrault – 75634 Paris Cedex 13 – France Tel: +33 (0)1.45.81.74.36/1.45.81.78.97 Fax: +33 (0)1.45.81.31.19 hlin@enst.fr, labiod@enst.fr

Abstract: Network mobility (NEMO) provides continuous connectivity to the Internet to a mobile network when moving from one access router to another. Recently, NEMO attracts vast interest for vehicle mobile networks, where users on a vehicle are connected to a local network that attaches to the Internet via a mobile router and a wireless link. In this architecture, mobile router disruptions may have an immediate impact on a potentially large number of connections. We suggest that multiple mobile routers are installed in this kind of mobile networks. In this paper, we propose a solution to improve handover performance for multiple mobile routers based vehicle mobile network configurations. We derive a mathematical formulation for the seamless handover failure probability as a function of four key parameters: vehicle speed, vehicle length, the distance covered by mobile network (vehicle) in the overlapping area of two adjacent access routers, and the mobile router's channels scan frequency. Thanks to the obtained formulation, we have the possibility to choose the appropriate network configuration (position of APs, distance between MRs...) to achieve efficient seamless handover aiming at minimizing deployment cost

Keywords: NEMO, vehicle mobile network, mathematical analysis, handover, multiple mobile routers

1. INTRODUCTION

The need to keep connected to the Internet anywhere and anytime is highly required in recent years. As part of that trend, there have been an increasing number of both commercial systems and research projects aiming to provide broadband Internet services to public transport passengers, by deploying high-speed local-area networks (LANs) on-board public transport vehicles. Recently, Internet Engineering Task Force

H. Labiod and M. Badra (eds.), New Technologies, Mobility and Security, 183–194.
© 2007 *Springer.*

(IETF) network mobility (NEMO) proposition attracts vast interest to the people who are working on providing pervasive Internet connectivity to transport vehicles.

NEMO basic support protocol [1] is proposed as an extension to Mobile IPv6 [2]. By means of NEMO support protocol, nodes in a mobile network can keep continuous Internet connectivity anytime while the mobile network changes its point of attachment. But the loss of packets and the latency involved in handover, when the mobile network moves from its previous Access Router (AR) to its new AR, degrade the performance of the on-going session. Some Mobile IP handover improvement techniques, like Hierarchical Mobile IPv6 (HMIPv6)[3] and Fast Handover for MIPv6 (FMIPv6)[4], can be applied for NEMO to enhance its handover performance. However, previous work [5] shows that even the combined use of both these techniques cannot provide a packet lossless handover environment at IP layer.

The weakest part in NEMO architecture when a mobile network connects to Internet is the Mobile Router (MR). Hence, we propose an architecture with multiple MRs which offers multiple connections to Internet for mobile networks (specially for vehicle mobile networks). Multiple MRs located in a mobile network form a multihomed NEMO configuration [6], so the use of multiple MRs endows the mobile network with reliability, load balancing and redundancy. In this paper, our work focuses on this specific multihomed NEMO architecture to achieve handover optimization for network mobility in the case of vehicle transport application, and we use our approach based on an ICE (Intelligent Control Entity) architecture, presented in [7], to achieve this goal.

We present in detail a mathematical study to evaluate the performance of our proposition. In this formulation, we focus on the seamless handover failure probability taking into account main parameters, such as mobile network (vehicle) speed, vehicle length, distance covered by mobile network in the overlapping area of two adjacent ARs, and channels' scan frequency of MR.

The rest of this paper is organized as follows. Section 2 presents some related work on NEMO handover improvement. In Section 3, our new proposition for vehicle NEMO handover improvement is described addressing the introduction of the new entity ICE. We give a mathematical analysis to identify the critical parameters that have an important impact on handover execution performance in Section 4. Finally, Section 5 concludes this paper and highlights the future work.

2. BACKGROUND ON NEMO HANDOVER

Some previous works use two L2 interfaces to achieve a NEMO seamless handover. [8] equips MR with two interfaces, one is used for data communication, and the second is used for scanning a new AP which can provide better connectivity. Once a new better AP is discovered, the second interface performs handover; meanwhile the first interface is always used for data communication. Upon a finished handover, the second takes the role of data communication, and the first starts scanning new APs. This mechanism is called Made Before Break (MBB) handover. In [9], a multihomed MR has two egress interfaces connected to two antennas respectively. These two antennas are separately installed in a large mobile network. The front interface is used for handover improvement and the rear interface is used for data transmission.

However, two interfaces in a mobile node will inevitably interfere with each other and increase packet loss. Furthermore, one interface must not perform handover before the other has finished it. This assumption can not always be satisfied, because the handover latency involved in these proposition includes BU latency which can be rather high if the mobile network is far from its HA. In this context, we propose the ICE based architecture to improve NEMO handover in the case of multiple MRs located in mobile network. This proposition achieves seamless handover without changing any existing IPv6 policy.

In the literature, we find a lot of analytical studies that investigate handover performance in terms of many criteria such as handover delay, handover failure ratio. Until now, no such investigation has been carried out for NEMO configuration. Hence, we propose a mathematical model where the random placement of mobile router in a vehicle and the new AR detection time are investigated to analyze NEMO handover failure probability.

3. ICE ARCHITECTURE FOR NEMO HANDOVER OPTIMIZATION

3.1. ICE Architecture

To support handover optimization for network mobility, we define a new architecture (figure 1). In this architecture, we introduce a local Intelligent Control Entity called ICE which should be capable of controlling several ARs and the MRs attaching to these ARs. An ICE domain contains an ICE and several ARs. ICE collects information of all ARs within its domain, like AR's address, capacity, load, APs list and its attached MRs, in AR_INF table (Table 1). Once an MR attaches to an AR, its information should be sent immediately to the corresponding ICE, whose address is given to MR by extended Router Advertisement (RA) message. This information, like its Home address (HAddr), Care of address (CAddr), State, its attached AR, other MRs in the same mobile network, capacity, available time for this entry (TTL), and the MR which is transporting its traffic, is collected in the MR_INF table (Table 2).

Table 1. AR_INF table

Adress	Capacity	Load	APs	MRs

Table 2. MR_INF table

HAddr	CAddr	State	AR	oMRs		Cap	TTL	TMRs
				ID	Pref			

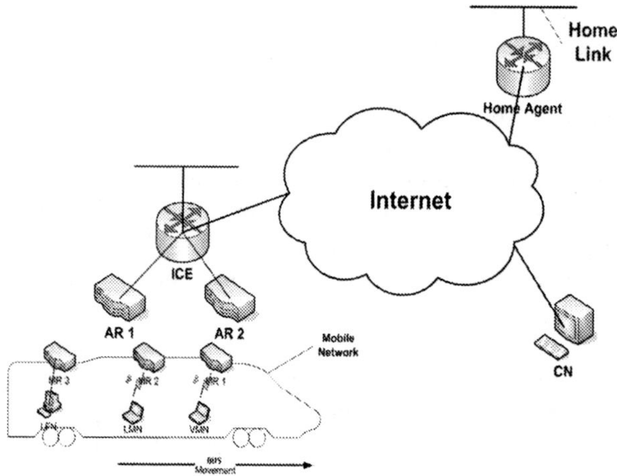

Figure 1. ICE architecture

3.2. Mobile Router States

In our proposition, according to the activity of MR, three states are defined for each MR:

- HO_ON state: MR is performing handover during a shorter period (the BU delay is removed), because during BU, the MR is IP-capable, it can transport packets. Hence, handover latency is significantly shorter than that of [8,9].
- TR_ON state: MR is transporting other MRs' traffic. At this state, MR should perform L3 handover as late as possible.
- HO_TR_OFF state: MR is neither at HO_ON state nor at TR_ON state. In opposition to TR_ON state, MR at HO_TR_OFF state should perform L3 handover as early as possible. As a consequence, two related MRs, one transporting the other's traffic, will perform handover separately.

3.3. Boundary Overlapping APs

A mobile router at HO_TR_OFF should perform network layer handover as early as possible; it does not have to wait for the signal strength of its current AP going lower than a threshold before scanning other channels. Instead, it must scan its channels frequently to find the AP belonging to new AR. But the frequent scan adds a significant additional processing, and it is ineffective to scan all the channels frequently if no handover is imminent. To resolve this issue, we define the boundary overlapping APs zone composed of APs which have the overlapping coverage zone with the APs of other ARs. When an MR at HO_TR_OFF state enters the boundary overlapping APs zone, it must scan its channels frequently to detect the AP that belongs to a new AR as early as possible; we call it boundary scan mechanism. While the normal scan mechanism (MR scans all channels to select a new AP only when the signal strength of its current AP is lower than the threshold) is always used when MR at HO_TR_OFF state stays outside of zone of boundary overlapping APs to minimize the additional processing. The information on boundary overlapping APs is stored at ICE in AR_INF table.

3.4. Handover Procedure Optimization

In this section, we describe only the handover procedure for intradomain mobility; interdomain handover is not considered.

The procedure goes through the following steps (two MRs case):

Step 1) As the vehicle moves, the MR at the front of the mobile network, MR1 for example, first predicts a handover. It sends a Handover Request (HReq) message (Figure 2) to ICE, which contains the detected L2 identifiers (such as AP MAC address).

Step 2) Upon receiving this message, ICE updates the entry corresponding to MR1 in its MR_INF table. The only parameter which needs to be updated is the state, which must be changed from HO_TR_OFF state to HO_ON state. ICE should also choose a best MR to carry the MR1's traffic. As only MR2 can be candidate in the example, it is chosen. Its state is then changed from HO_TR_OFF state to TR_ON state.

Step 3) Afterward, two messages are sent by ICE, one called Handover Response (HRep) is sent to MR1, this message contains the chosen L2 identifier and home address of the chosen MR. The other called Handover packet Transfer (HTra) is sent to AR1 which is MR1's current access router. This message instructs AR1 to set up a tunnel between itself and MR2 for the packets addressed to MR1.

Step 4) AR1 then encapsulates these packets and forwards them to MR2; the latter decapsulates these packets and sends them to Mobile Network Node (MNN).

Step 5) Once MR1 finishes handover, i.e. it generates a new valid CoA, it sends BU to its home agent as usual, it also sends Handover Finish (HFin) message to ICE, ICE changes MR1's state from HO_ON to HO_TR_OFF. Also, a message containing the changed parameters (like CoA, AR) should be sent to ICE.

Step 6) As MR1 is now IP-capable; it can receive packets, ICE sends a new HTra to AR1 to inform it, instead of tunnelling the packets to MR1's previous CoA to MR2, to tunnel these packets to MR1's new CoA, and MR2's state returns to *HO_TR_OFF state.*

In the opposite direction, MR1 should reverse tunnel the packets from the mobile network to Internet to MR2, whose address has been communicate to MR1 via the HRep packet, until it completes the BU procedure. MR2 should forward the inner packet in the tunnel to AR1. This procedure ensures that packets are not dropped due to handover and ingress filtering.

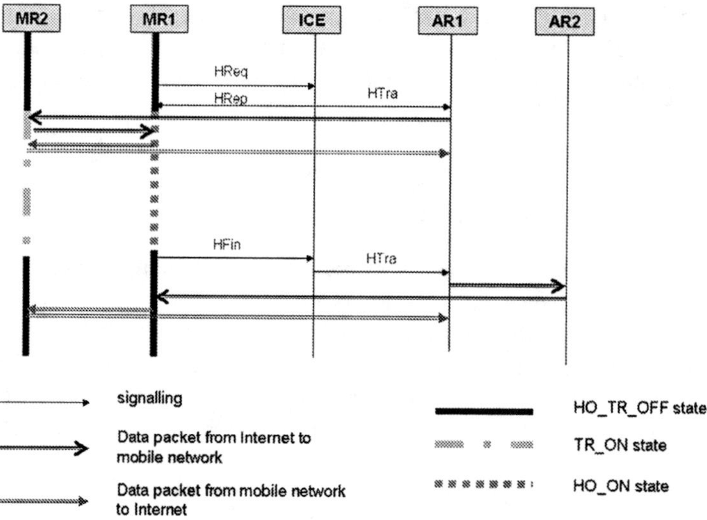

Figure 2. Handover procedure within ICE domain

4. MATHEMATICAL ANALYSIS

In this section, we derive a mathematical model that helps to evaluate the performance of our proposition in a simple vehicle case (Figure 3). In this model, we assume that MR1 and MR2 are uniformly placed in the bus whose length is supposed to d. The distance covered by bus in the overlapping area of two adjacent AR1 and AR2 is s. The bus moves from AR1's domain to AR2's domain. Two stochastic parameters considered in this case are: distance between two MRs and EPN (entry point in the new AR domain) where the MR detects its movement (begin its handover procedure) when it enters AR2's domain.

Probalility density functions (pdf) of the location of MR1 and MR2 are given, respectively, by

(1) $$f_{p_{MR1}}(x_1) = \begin{cases} \frac{1}{d}, & for\ 0 \le x_1 \le d, \\ 0, & else \end{cases}$$

and

(2) $$f_{p_{MR2}}(x_2) = \begin{cases} \frac{1}{d}, & for\ 0 \le x_2 \le d, \\ 0, & else \end{cases}$$

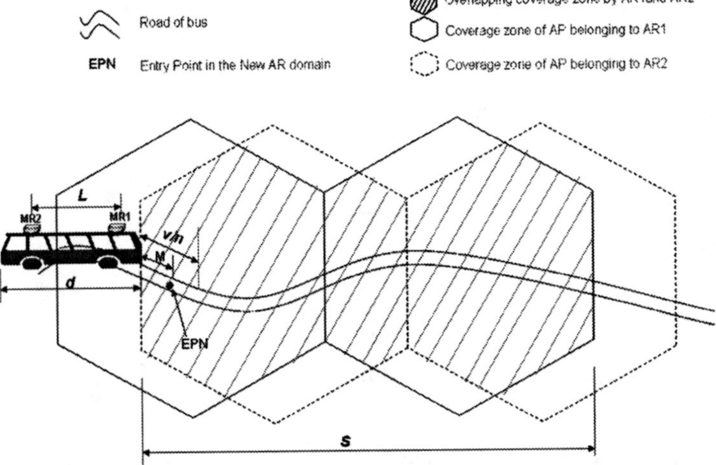

Figure 3. Scenario of vehicle mobile network

Since both MRs are randomly installed in the bus and are independent from each other, their joint pdf is:

$$(3) \qquad f_{p_{MR1}p_{MR2}}(x_1, x_2) = f_{p_{MR1}}(x_1) f_{p_{MR2}}(x_2) = \begin{cases} \dfrac{1}{d^2}, & for\ 0 \le x_1, x_2 \le d \\ 0, & else \end{cases}$$

The distance between these two MRs is given by $L = |P_{MR1} - P_{MR2}|$. The probability that this distance is smaller than l (a given value) is derived from the formula (4):

$$(4) \qquad P(L \le l) = \iint_D f_{p_{MR1}p_{MR2}}(x_1, x_2)\, dx_2\, dx_1$$

Where D is the space of location of MR1 and MR2 such that its distance $L \le l$ and $0 \le l \le d$. Obviously, $P(L \le l) = 1$ in the case where $l >$ d and $P(L \le l) = 0$ for $l < 0$. This probability can then be written as

$$P(L \le l) = \frac{1}{d^2}\left(\int_0^l \int_0^{x_1+l} dx_2 dx_1 + \int_l^{d-l} \int_{x_1-l}^{x_1+l} dx_2 dx_1 + \int_{d-l}^{d} \int_{x_1-l}^{d} dx_2 dx_1\right)$$

$$(5) \qquad = -\frac{1}{d^2} l^2 + \frac{2}{d}$$

The derivative of above function with respect to l yields by definition the desired pdf:

$$(6) \qquad f_L(l) = \begin{cases} \frac{\partial}{\partial l} P(L \le l) = -\frac{2}{d^2} l + \frac{2}{d} & for\ 0 \le l \le d \\ 0, & else \end{cases}$$

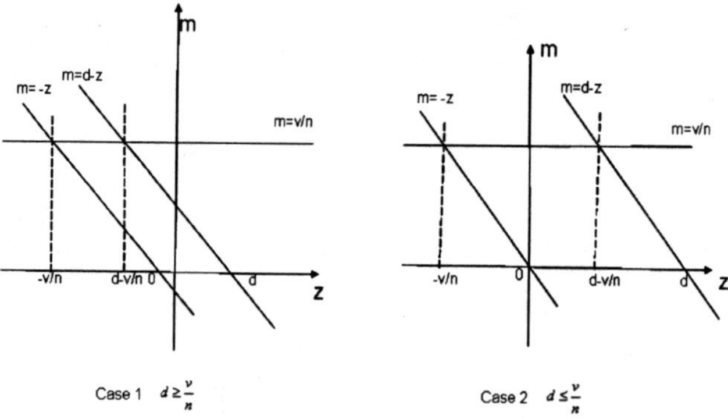

Case 1 $d \geq \frac{v}{n}$ Case 2 $d \leq \frac{v}{n}$

Figure 4. Two different cases of configuration parameters (d, v, n)

When MR enters a new AP's domain, it cannot detect the new AP exactly on the edge of its coverage zone, because of the existence of detection delay. The delay depends on the channels scan frequency. High frequency scan allows MR to detect a new AP earlier than low frequency, but high frequency also leads to more additional processing. We assume that the scan frequency when entering the boundary AP domain is equal to n times/sec; the point at which MR detects the new AP is EPN (figure 3), we note that the distance from the point, where the MR enters the new AP's domain, to the point EPN is M. As EPN is uniformly placed on the line $[0, \frac{v}{n}]$, the pdf of M is:

$$(7) \qquad f_M(m) = \begin{cases} \dfrac{1}{\frac{1}{n} \times v} = \dfrac{n}{v}, & for\ 0 \leq m \leq \dfrac{v}{n} \\ 0, & else \end{cases}$$

The amount of time MR1 will take to perform handover before MR2 also begins its handover is calculated by $T = \frac{s-M+L}{v}$ (8). We let $Z = L - M$, the pdf of Z is given by

$$(9) \qquad f_Z(z) = \int_{-\infty}^{\infty} f_M(m) f_L(z+m) dm$$

The above integral is not equal to zero when

$$(10) \qquad \begin{cases} 0 \leq m \leq \frac{v}{n} \\ 0 \leq z+m \leq d \end{cases} => \begin{cases} 0 \leq m \leq \frac{v}{n} \\ -z \leq m \leq d-z \end{cases}$$

We distinguish two different parameter configuration cases (Figure 4) to calculate the pdf of Z: case 1 where the distance d is greater or equal to $\frac{v}{n}$, case 2 where the distance d is less than $\frac{v}{n}$

Case 1: $d \geq \frac{v}{n}$

According to Figure 4, $f_Z(z)$ is given by

$$f_Z(z) = \int_{-\infty}^{\infty} \frac{n}{v}\left(-\frac{2}{d^2}(z+m)+\frac{2}{d}\right)dm$$

$$= \begin{cases} \int_{-z}^{\frac{v}{n}} \frac{n}{v}(-\frac{2}{d^2}(z+m)+\frac{2}{d})dm, & for -\frac{v}{n} \le z < 0 \\ \int_{0}^{\frac{v}{n}} \frac{n}{v}(-\frac{2}{d^2}(z+m)+\frac{2}{d})dm, & for\ 0 \le z < (d-\frac{v}{n}) \\ \int_{0}^{d-z} \frac{n}{v}(-\frac{2}{d^2}(z+m)+\frac{2}{d})dm, & for\ (d-\frac{v}{n}) \le z \le d \\ 0, & else \end{cases}$$

$$(11) \qquad = \begin{cases} \frac{1}{vnd^2}(2dvn+2dzn^2-2vnz-z^2n^2-v^2), & for -\frac{v}{n} \le z < 0 \\ \frac{1}{nd^2}(2dn-2zn-v), & for\ 0 \le z < (d-\frac{v}{n}) \\ \frac{n}{vd^2}(d-z)^2, & for\ (d-\frac{v}{n}) \le z \le d \\ 0, & else \end{cases}$$

The equation (8) can be written $T = \frac{1}{v}Z + \frac{1}{v}s$, so $z = h(t) = tv - s$, and then we can obtain the pdf of T

$$f_T(t) = f_Z(h(t))\,|h'(t)| = vf_Z(tv-s)$$

$$(12) \qquad = \begin{cases} \frac{1}{nvd^2}[2dvn+2d(tv-s)n^2-2vn(tv-s)-(tv-s)^2n^2-v^2], & for\ \frac{s}{v}-\frac{1}{n} \le t < \frac{s}{v} \\ \frac{1}{nd^2}[2dn-2(tv-s)n-v], & for\ \frac{s}{v} \le t < \frac{s}{v}+\frac{d}{v}-\frac{1}{n} \\ \frac{n}{d^2}(d-tv+s)^2, & for\ \frac{s}{v}+\frac{d}{v}-\frac{1}{n} \le t \le \frac{s}{v}+\frac{d}{v} \\ 0, & else \end{cases}$$

Using (12), the probability that MR1 fails to finish its handover before MR2 begins its handover is given by

$$(13) \qquad p_f = \begin{cases} 0, & for\ \tau < (s-\frac{v}{n})/v \\ P_T(t<\tau) = \int_{\frac{s}{v}-\frac{1}{n}}^{\tau} \frac{1}{nd^2} [2dvn+2d(tv-s)n^2-2vn(tv-s)-(tv-s)^2n^2-v^2]dt, & for\ \frac{s}{v}-\frac{1}{n} \le \tau < \frac{s}{v} \\ P_T(t<\tau) = \frac{1}{nvd^2} [-nv^2\tau^2+(2dnv-2snv-v^2)\tau-2dns+3ns^2+sv]+X, & for\ \frac{s}{v} \le \tau < \frac{s}{v}+\frac{d}{v}-\frac{1}{n} \\ P_T(t<\tau) = \frac{n}{3vd^2}\left[(\frac{v}{n})^3-(d-\tau v+s)^3\right]+X+Y, & for\ \frac{s}{v}+\frac{d}{v}-\frac{1}{n} \le \tau \le \frac{s}{v}+\frac{d}{v} \\ 1, & for\ \tau > (s+d)/v \end{cases}$$

Where, τ is the handover delay as described above (the BU delay is removed); X and Y are given as follows:

$$X = \int_{\frac{s}{v}-\frac{1}{n}}^{\frac{s}{v}} \frac{1}{nvd^2}[2dvn + 2d(tv-s)n^2 - 2vn(tv-s) - (tv-s)^2n^2 - v^2]dt$$

$$Y = \int_{\frac{s}{v}}^{\frac{s}{v}+\frac{d}{v}-\frac{1}{n}} \frac{v}{nd^2}[2dn - 2(tv-s)n - v]dt$$

Case 2: $d < \frac{v}{n}$

$$f_Z(z) = \int_{-\infty}^{\infty} \frac{n}{v}(-\frac{2}{d^2}(z+m) + \frac{2}{d})\,dm$$

$$= \begin{cases} \int_{-z}^{\frac{v}{n}} \frac{n}{v}(-\frac{2}{d^2}(z+m) + \frac{2}{d})\,dm, & for\ -\frac{v}{n} \le z < d - \frac{v}{n} \\ \int_{-z}^{d-z} \frac{n}{v}(-\frac{2}{d^2}(z+m) + \frac{2}{d})\,dm, & for\ d - \frac{v}{n} \le z < 0 \\ \int_{0}^{d-z} \frac{n}{v}(-\frac{2}{d^2}(z+m) + \frac{2}{d})\,dm, & for\ 0 \le z \le d \\ 0, & else \end{cases}$$

$$(14) \quad = \begin{cases} \frac{1}{vnd^2}(2dvn + 2dzn^2 - 2vnz - z^2n^2 - v^2), & for - \frac{v}{n} \le z < d - \frac{v}{n} \\ \frac{n}{v}, & for\ d - \frac{v}{n} \le z < 0 \\ \frac{n}{vd^2}(d-z)^2, & for\ 0 \le z \le d \\ 0, & else \end{cases}$$

As done in case 1: pdf of T can be obtained by

$$f_T(t) = f_Z(h(t))\,|h'(t)| = vf_Z((tv-s)$$

$$(15) \quad = \begin{cases} \frac{1}{nd^2}[2d(tv-s)n^2 + 2dvn - (tv-s)^2n^2 - 2(tv-s)vn - v^2], & for\ \frac{s}{v} - \frac{1}{n} \le t < \frac{s}{v} + \frac{d}{v} - \frac{1}{n} \\ n, & for\ \frac{s}{v} + \frac{d}{v} - \frac{1}{n} \le t < \frac{s}{v} \\ \frac{n}{d^2}(tv-s-d)^2, & for\ \frac{s}{v} \le t \le \frac{s}{v} + \frac{d}{v} \\ 0, & else \end{cases}$$

then

$$
(16) \qquad P_f = \begin{cases} 1, & for\ \tau > \frac{s}{v} + \frac{d}{v} \\ p_T(t<\tau) = \int_{\frac{s}{v}-\frac{1}{n}}^{\tau} \frac{1}{nd^2}[2d(tv-s)n^2 + 2dvn - (tv-s)^2n^2 - 2(tv-s)vn - v^2]dt, & \\ & for\ \frac{s}{v} - \frac{1}{n} \le \tau < \frac{s}{v} + \frac{d}{v} - \frac{1}{n} \\ p_T(t<\tau) = n(\tau - \frac{s}{v} - \frac{d}{v} + \frac{1}{n}) + X, & for\ \frac{s}{v} + \frac{d}{v} - \frac{1}{n} \le \tau < \frac{s}{v} \\ p_T(t<\tau) = \frac{n}{3vd^2}\left[(\tau v - d - s)^3 + d^3\right] + n(\tau - \frac{s}{v} - \frac{d}{v} + \frac{1}{n}) + X, & \\ & for\ \frac{s}{v} \le \tau \le \frac{s}{v} + \frac{d}{v} \\ 0, & for\ \tau < \frac{s}{v} - \frac{1}{n} \end{cases}
$$

Where, τ is the handover delay as described above (the BU delay is removed) and

$$
X = \int_{\frac{s}{v}-\frac{1}{n}}^{\frac{s}{v}+\frac{d}{v}-\frac{1}{n}} [\frac{1}{nd^2}(2d(tv-s)n^2 + 2dvn - (tv-s)^2n^2 - 2(tv-s)vn - v^2)]dt
$$

With the formulas (13) and (16), the probability P_f that MR1 fails to finish its handover before MR2 begins its handover is calculated as function of s, v, d, n, τ. Now, we can configure one or several parameters according to a given probability P_f. For example, given $d = 10$ m, $v = 10$ m/sec, $n = 2$, and $\tau = 1.5$ sec, the distance s satisfying the probability $P_f \le 0.02$ is greater than 9.51 since $d \ge \frac{v}{n}$ and case 1 is adopted.

The above analysis is made for two MRs being located in mobile network. In the case of more than two MRs, only the parameter l should be taken as the distance between two extreme MRs, other parameters remain similar.

5. CONCLUSION

In this paper, we argued that multiple mobile routers are suitable to the vehicle mobile network. We propose a solution to improve the handover performance using a new entity ICE which is responsible to manage MRs and ARs in one domain. It can choose the best MR and AR during handover execution; it can also help MR to perform handover according to different states. Also, thanks to this entity, no additional function is required for mobile router to keep its state information.

The mathematical analysis demonstrates clearly that our proposition is adapted to vehicle case where a mobile network with multiple routers is installed in bus, train or ship. In these cases, we can achieve seamless handover with no complicated deployment.

For future work, we intend to complete the investigation of the impact of the defined parameters and relationships between them. We will also make a similar mathematical analysis in interdomain handover case.

REFERENCES

1. Devarapalli, V., Wakikawa, R., Petrescu, A., and P. Thubert, "Network Mobility (NEMO) Basic Support Protocol," RFC 3963, January 2005.
2. D.Johnson, C.Perkins, and J.Arkko, "Mobility Support in IPv6," RFC 3775, June 2004.
3. H. Soliman, C. Castelluccia, K.-E. Malki, and L. Bellier, "Hierarchical mobile IPv6 mobility management," IETF, RFC 4140, Aug. 2005.
4. R. Koodli, Ed."Fast Handovers for Mobile IPv6", RFC 4068, July 2005.
5. R. Hsieh and A. Seneviratne, "Performance analysis on Hierarchical Mobile IPv6 with Fast-handover over TCP," *in proceedings of GLOBECOM*, Taipei, Taiwan, 2002.
6. C. Ng, E. Paik, T. Ernst, M. Bagnulo, "Analysis of Multihoming in Network Mobility Support," Internet-Draft, June 22, 2006.
7. Hai. LIN, Houda. Labiod, "HANDOVER OPTIMIZATION FOR HOST AND NETWORKMO-BILITY," in *Proc, WINSYS2006*, sétubal, Portugal, 2006, pp: 59–64.
8. H. Petander, E. Perera, K.C. Lan, A. Seneviratne, "Measuring and Improving the Performance of Network Mobility Management in IPv6 Network," *IEEE JOURNAL ON SELECTED AREAS IN COMMUNICATIONS*, VOL. 24, NO.9, Sep 2006 pp.1671–1681.
9. H. D. Park, D. W. Kum, Y. H. Kwon, "IP Mobility Support with a Multihomed Mobile Router," in *Proc. NETWORKING 2006, LNCS 3976*, pp. 1144–1149.

CHAPTER 17

INTRODUCING L3 NETWORK-BASED MOBILITY MANAGEMENT FOR MOBILITY-UNAWARE IP HOSTS

DAMJAN DAMIC

Siemens, IT Solutions and Services, Zagrebacka 145a, 10000 Zagreb, Croatia, damjan.damic@siemens.com

Abstract: The design of an efficient and intelligent mobility management scheme is one of the major research challenges for next-generation wireless networks. While conventional mobility solutions focus on host-based mobility, the concept of network-based mobility management has only recently emerged. Consequently, there is a limited understanding of the corresponding requirements and solution mechanisms. In this paper we identify fundamental requirements for providing the network operated mobility service for mobility unaware hosts, assuming legacy user terminals which are not implementing or using any mobile extensions or protocol stacks, while moving through visited networks. We analyze three different mobility management schemes presently being considered for diverse deployment scenarios within several standardization organizations (e.g. IETF, 3GPP, WiMAX Forum), namely the Proxy Mobile IPv4, Proxy Mobile IPv6, and the Network-based Localized Mobility Management (NetLMM). As there are currently no accepted standards for network-based mobility, this paper focuses on providing a comparative analysis using qualitative and functional merits in order to highlight potential advantages and shortcomings of each evaluated mobility scheme, and provide guidelines for feasible functional improvements

1. INTRODUCTION

Subscriber mobility is an essential functionality in wireless broadband IP networks such as public WLAN systems or WiMAX, having as such become a major topic in next generation wireless networks research. Provision of seamless communication throughout the IP-centric mobile infrastructure, and combination of diverse access technologies spanning across different domains defines the goal of numerous mobility management schemes.

This paper focuses on a new aspect of the Internet mobility problem space; the network-based approach to IP mobility management. By applying a network operated mobility scheme, subscribers can maintain IP connectivity without any

H. Labiod and M. Badra (eds.), New Technologies, Mobility and Security, 195–205.
© 2007 *Springer.*

special requirements to their end terminals while moving within a network. In this context, a legacy IPv4 or IPv6 terminal, not implementing and using any mobile extensions or protocol stacks is considered a mobility-unaware host. The aim of network-based mobility management (NBMM) is to simplify the deployment, integrate with and enhance already existing solutions, to the mutual advantages of operators and mobile users. Key benefits of NBMM are identified as: removal of mobility implementations from hosts, extended capabilities for mobility, and decreased consumption of the air-link. Additional goals include gaining optimized handoff (HO) performance and reducing associated signaling volume. Currently, there is limited understanding of the requirements and mechanisms needed to support the NBMM. The contribution of this paper is an extensive and comparative analysis of three different NBMM schemes presently considered for deployment within several standardization organizations. A number of shortcomings are identified, serving as the basis for feasible solution advancements.

The rest of the paper is organized as follows. The next section discusses requirements on NBMM. Section 3 describes three different proposals for the network-based mobility. The schemes are comprehensively compared in Section 4, with concluding remarks given in the final Section.

2. NETWORK-BASED IP MOBILITY MANAGEMENT

Mobility at the network layer is the most straightforward way of sustaining transition of user's session moving between different points of attachment. Layer 3 (L3) mobility management (MM) solves the problem of delivering packets to the topologically incorrect location, i.e., overcomes the concept with IP addresses as persistent host identifiers and locators. Independently of the underlying link-layer, L3 schemes solve the problem in the tunneling- or routing-based manner using techniques of indirection, redirection, and encapsulation. Schemes are further classified into micro/local and macro/global mobility solutions, depending on the scope of operation. The common approach to L3 MM is building on top of the widely accepted mobile IP standards, either Mobile IPv4 (MIP4) [12], or Mobile IPv6 (MIP6) [6]. The MIP6 provides basic mobility principles specifying the use of Binding Update messages (BU) to register the location, e.g., care-of address (CoA), at the Home Agent (HA) which acts as the global mobility anchor point. Subsequently, a handful of enhanced-MIP6 schemes emerged, focusing to solve known MIP6 shortcomings; reduce the registration delay and the address resolution time.

Conventional MM schemes so far relied on the mobile node (MN) as the key element in the HO execution [5]. The MN was completely aware of the ongoing HO, hence had to implement mobility support to assist in the location registration and address resolution. The NBMM schemes withdraw all mobility functions from end nodes into network elements: the base stations and routers as designated mobility agents. This way any host gains the possibility for continuous communication while moving within a domain, without needing special mobility extension at its side.

Connectivity persistence during HO is the prime requirement for NBMM, with the goal being to conceal this movement from the MN so as to deceive it being stationary. Mobility mechanisms must ensure that communication with the remote correspondent node (CN) is maintained. Incorporation of MM support entirely within the network allows operators to gain more control over the process. On the other hand, the network needs to compensate for intelligent functions which are commonly executed by the end node, among them the decision on HO initiation, utilization of radio resources, user policy enforcement, and robust signaling. The following design goals and requirements are addressed in NBMM schemes [1,5,7]:

- Support for mobility unaware hosts (preferably without modifications).
- Improvement of the handoff performance; minimize disruptions by reducing the HO latency and loss of packets.
- Optimization in resource allocation; air-link and network consumed by the mobility-related signaling and tunneling overhead.
- Transparency to higher- and lower level protocols; as not to affect the application and transport layers, also to operate agnostic of the access link-layer technologies.
- Solution envisioned for both IP versions (v4/v6) specific features.
- Preserve security aspects affected by the occurring mobility. Highlight on location privacy, integrity and authentication of the signaling flow.
- Reuse of existing standards and protocols (where applicable).
- Functional distinction between local and global mobility, i.e. not mandate global MM protocol by choosing a specific localized scheme.
- Configurable data plane between the mobility agents and anchors.

The next section discusses and compares three approaches to NBMM, currently debated on within the Internet Engineering Task Force (IETF).

3. EVALUATED NETWORK-BASED MOBILITY SCHEMES

3.1. Proxy Mobile IPv4 (PMIP4)

Despite the optimistic predictions of widespread IPv6 deployment, the majority of user terminals still operate at the IPv4 stack, and typically are not capable of executing mobility procedures. Proxy Mobile IPv4 (PMIP4) proposes extensions to the MIP4 operating within the network, enabling IPv4 devices to roam, while the dedicated network entities provide mobility support on their behalf [9]. The MM functions are handled by the Mobile Proxy Agent (MPA), a new mobility entity in the access network.

The MPA is a part of the first hop access router (AR) or the base station the MN perceives, collocated with the Foreign Agent (FA) and the DHCP function. MPA detects MN attachment to the network by the regular network access procedure, and initiates MIP4 registration with the HA on behalf of MN. Using the IP address of FA as MN's CoA, the MPA establishes a bidirectional tunnel between the HA and FA. All incoming packets from the MN are intercepted, encapsulated and sent to the HA via the tunnel. Packets coming towards the MN are decapsulated and delivered by MPA using Layer 2 (L2) forwarding according to the target MAC address. The registration procedure repeats whenever MN moves into the area of another AR; HA relocates the tunnel towards the new MPA, and terminates the previous in parallel. Throughout the entire process the MN maintains the same HoA, and is made to believe it is stationary.

An extended scenario adds additional support for IPv6 host mobility [11]. The MPA entity in the BS or AR in this case is IPv6-aware and handles the location registration and address acquisition for the legacy IPv6 nodes. MPA recognizes the attached MN, performs registration using Mobile IPv4 signaling (with IPv4 FA-CoA) to establish the IPv6-over-IPv4 tunnel with the HA. User traffic is transported through the tunnel upon the HA, decapsulated and further routed as regular IPv6 traffic. Two deployment options are proposed; MPA may incorporate the dual-stack v4/v6 router function, or just the basic IPv6 support (requires HA operating as default IPv6 router). In both cases extended MIP4 registration messages are used to acquire the IPv6 HoA, establish or relocate the MPA-HA tunnel.

3.2. Proxy Mobile IPv6 (PMIP6)

Reuse of the MIP6 protocol for NBMM is addressed by several solutions [2,3,4]. Categorized under the Proxy Mobile IPv6 (PMIP6) approach, all these schemes employ tunneling with the use of CoA, to deliver packets towards topologically incorrect HoA. The solutions are focused on localized MM, e.g., operate with a local mobility anchor within a restricted part of the network. The PMIP6 supplements the access routers with a new mobility function, the Proxy Mobile Agent (PMA) which substitutes mobility management functions missing from the MN (shown in Fig. 1).

The network access procedure and CoA-acquisition triggers the PMA at the first hop router, to issue appropriate Proxy Binding Update (PBU) and thereby initiate registration of the MN's location with the HA. The MN's mobility is anchored at a single point, so all the traffic is routed through the HA, without the possibility for direct MN-CN communication.

The PMA-HA tunnel is preestablished or dynamically built using PBU-PBA exchange. Network configuration identical to the home-link is set up at every point the MN attaches in the domain (defined through MTU, link prefix, hop limit, managed address method). The PMA has to delude the MN as not to detect the change of its physical link after the HO.

This may be done by replying with Router Advertisements (RA) or DHCP replies customized with what MN expects to receive. MN may configure HoA built from a unique IPv6 prefix, resulting with a virtual point-to-point connection with MN and AR appearing as only nodes on that IP link. MN and AR communicate through specific L2 technology, so movement between BSs in range of one AR is transparent, while the relocation to another AR causes L3 HO. When MN moves from the serving AR/PMA router (SAR) to the target AR (TAR), TAR would send

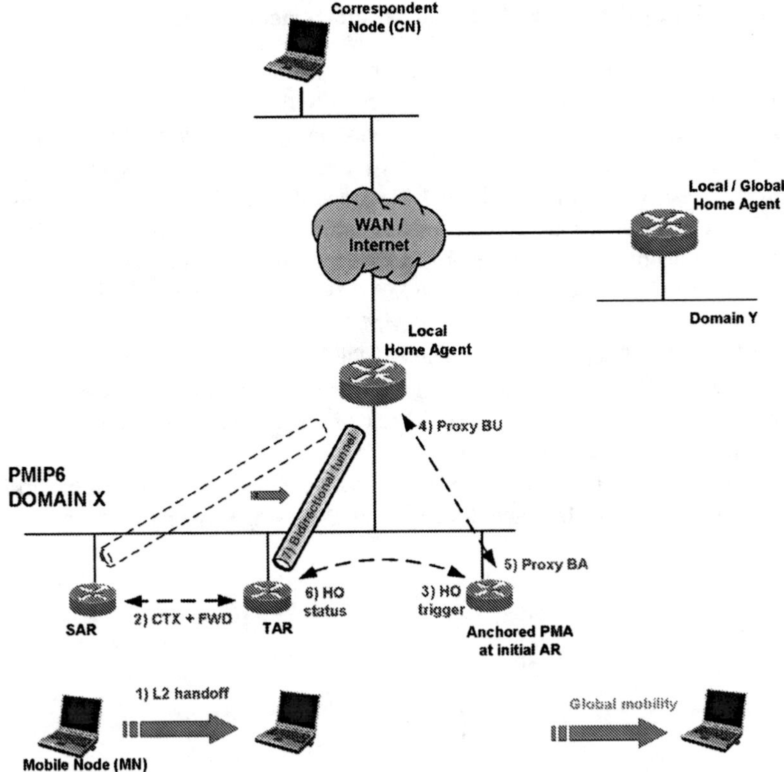

Figure 1. Proxy Mobile IPv6 architecture with the anchored PMA

the PBU at the HA, to update MN's mobility bindings and divert the packet routing towards the HA-TAR tunnel.

Proposed protocol optimizations include exchange of relevant information (context) between SAR and TAR to eliminate signaling towards external entities and shorten the HO [3]. Establishment of a temporary SAR-TAR tunnel to carry the traffic until the new tunnel is available is a way to reduce packet loss during HO (reuse of the Fast-MIP6 concept [8]). To bypass the neighbor reachability problems at the new link, [4] proposes TAR to use spurious RAs to reestablish valid default router entry at the MN. Work in [2] specifies an architecture with anchored PMA assigned for the entire session lifetime. In this case, at HO occurrence the MN changes the AR, but the PMA providing the mobility support remains the same, at the router where MN initially registered (scenario illustrated in Fig.1). Distribution of the data and control path is thereby achieved, security is strengthened, and HO is shortened (no AAA transfer between ARs). Further PMIP6 enhancements propose a solution compatible with HMIPv6 [13], by integrating PMA with the Mobility Anchor Point (MAP).

3.3. Network-based Localized Mobility Management (NetLMM)

Specification [10] elaborates on a completely new protocol design, the Network-based Localized Mobility Management protocol (NetLMM), which deals with efficient and secure MM within a confined domain managed solely by the network.

The access network architecture consists of two types of entities providing mobility support; Local Mobility Anchors (LMA), and the Mobile Access Gateways (MAG). A LMA is the router that preserves reachability to the MN by maintaining information of connectivity and associations between different entities, and mapping identities and locators of attached MNs. The MAG is the first hop AR which terminates MN's access link-layer, and provides MM by participating in the signaling on behalf of MN. Fig. 2 depicts the elements of the NetLMM architecture.

Functionally, MAG is the equivalent of the PMIP6 PMA and LMA corresponds to the Home Agent. A set of at least one LMA, with its multiple associated MAGs constitute a NetLMM domain.

The protocol defines two types of control messages; messages used for network setup and association of LMAs and MAGs, and 2 message pairs for the actual

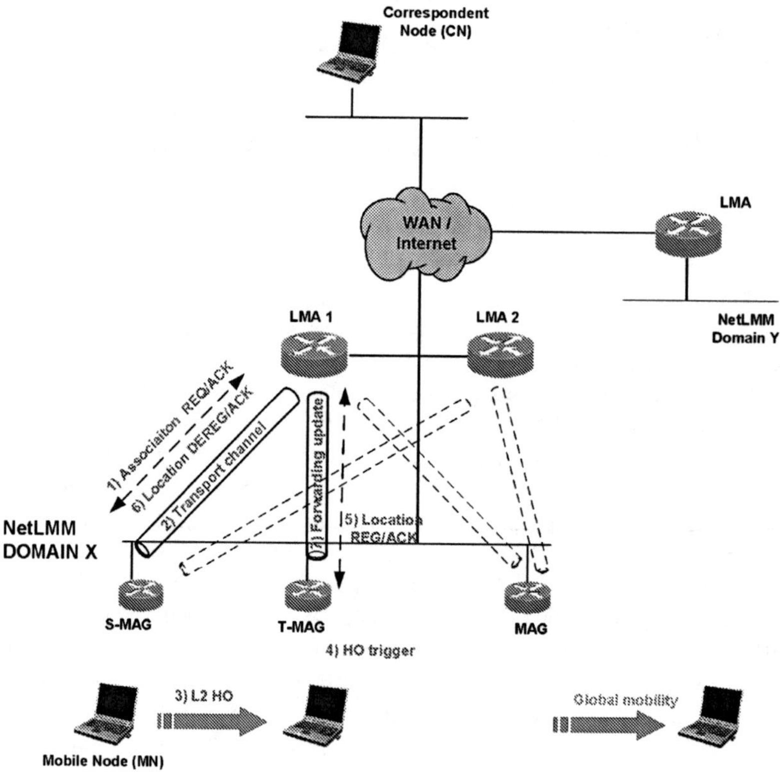

Figure 2. Handoff execution by the NetLMM protocol

mobility-related signaling. Association requests are used by MAGs to establish the control and data plane with LMAs during the startup phase, prior to any mobility handling. The transport methods, i.e., packet forwarding or encapsulation between MAG and LMA are thereby negotiated. Connectivity may be terminated by MAG or LMA through a disassociation request at any given time. MAGs may dynamically discover LMAs, and maintain association to several of them. While the association exists, heartbeat messages are exchanged to verify the connectivity status. This way the entire NetLMM domain is set up and configured beforehand, thus eliminating the related delay appearing during HO execution.

In the operational state, attachment of the MN triggers MAG to register or update MN's location and identity at the target LMA. MAG retrieves the prefix or HoA for MN configuration and supplies it in the responding RA or DHCP reply. When the HO occurs, a location registration message sent by the target MAG (T-MAG) updates MN binding at LMA, the packet forwarding is redirected towards T-MAG, and previously assigned binding is torn down when LMA sends location deregistration to the previous MAG (S-MAG). The HO process can identically repeat throughout the entire NetLMM domain.

4. COMPARATIVE ANALYSIS

This section compares the three described NBMM schemes; PMIP4, PMIP6, and NetLMM. The schemes are analyzed and matched against a list of functional and operational aspects, achieving merits for qualitative evaluation. Comprehensibly, given parameters are classified under 3 general categories of interest; features and performance, security, and deployment. We based the selection and categorization of comparison parameters by combining paradigms presented in [5,12,14].

4.1. Features and Performance

Selected criteria in the scope of performance reflects upon the conceived design goals, and include effects on the mobile node and its correspondents, achieved optimization aspect, considerations on locators and identifiers, scope of the scheme operation, implications of affected layers (e.g. link, IP and above), supported means for data forwarding, theoretical HO performance metrics, etc. Key findings are summarized below, with the highlights given in Table 1.

None of the compared schemes require modifications to the MN thereby fulfilling the essential goal (minor improvements in link detection are suggested in PMIP6 for smoother performance). With all three schemes, the mobility is anchored at a single point in the network resulting with the suboptimal routing model. Mobility signaling and tunneling overhead are effectively omitted from the air-link by all. The PMIP6 variant with temporary tunneling improves on packet loss, whereas the preconfigured forwarding channels in NetLMM reduce the HO latency. Only NetLMM accomplishes support for location-independent identifiers. The MN ID remains static across location changes, with its location defined by a MAG ID

Table 1. Network-based mobility management comparative analysis highlights

Aspect / scheme	PMIP4	PMIP6	NetLMM
MN modifications	None	optimized ND	None
Correspondent node	not affected	RO precluded	not affected
Optimization	air-link	air-link, packet loss	air-link, HO latency
Locator-identifier dependency	bound: v4 CoA–v4/6 HoA	bound: v6 CoA–v4/6 HoA	split: v4/6 addr – node ID
Operation scope	localized domain (optionally global)	localized domain (MIP6 for global)	localized domain
L2/L3 implications	p2p link, HO trig + ARP, ICMP	p2p + shared-link, HO trig + ND	P2p link, HO trig + ND
Upper-layer mob.	transparent	transparent	Conditional
Data transport - duration	tunneling (IP,GRE) - dynamic	tunneling (IP) - dynamic / lasting	conf. forwarding – preestablished
Signaling (Init / HO)	2 / 4 msgs	4–7 / 4–5 msgs	4 / 4 msgs
Per packet overhead	20–40 B	40 B	variable (40 B)
Divergence of flows	–	–	embedded support
Multicast/broadcast	supported	supported	+/–
HO efficiency	pressumed not seamless	seamless-like (not perceivable)	targets seamless HO (70 ms latency)
Security association	shared MN-HA	unique MN-HA	Certificates
Protection methods	L2 + SA	L2+IKE+SA+ESP	L2+IKE+Certs+ESP
Privacy protection	Partial	Partial	Embedded
Vulnerabilities addressed	DoS, impersonation	Replay, DoS, impersonation	Replay, DoS, MiMA impersonation
Reuse of standard	Mobile IPv4	Mobile IPv6	–
Modifications to std.	RRQ/RRP, options	PBU/PBA, options	–
Network supplement	MPA	PMA	MAG, LMA
Deployment dep.	HA, FA	HA / MAP	–
Limitations	IPv4 pool, NAT	Host-MIP6, FWs	FWs
AAA interactions	Access	MN policy, SA keys	Access
Scalability	Limited	fair (if hierarchical)	fair (load sharing)
Robustness issues	MPA, FA	PMA, HA/MAP	LMA

and the belonging address. The NetLMM ID can associate multiple interfaces and adjoined addresses, i.e. for multihoming. In PMIP4 and PMIP6, location dependent CoA is logically bound to a lasting HoA. The methods for address/locator acquisition may include DHCP, point-to-point protocol (PPP), stateless address autoconfiguration (SLAAC). Existing NBMM schemes are focused at localized MM, for micro movements between ARs of a single domain. As such, they tend to leave the choice of global mobility at will, but if one is to bypass the practical and administrative issues global MM is actually plausible.

Specification of HO triggers is left open, but commonly some L2 event is expected. By initial attachment network authentication and address request serve this purpose. The home link emulation by the mobility agents imposes them to properly adapt all affected messages bound for the MN. In PMIP4 this means responding to

ICMP, ARP and DHCP with appropriately set MAC address and addressing fields. Schemes handling IPv6 must accommodate the Neighbor Discovery (ND) protocol, and especially the advertised address prefixes. Transparency towards higher layers is achieved, although the implementation of NetLMM locator-identifier logic would impact the transport layer to some extent.

All schemes rely on the point-to-point link model to bypass the link reachability issues. Similar thing maybe done at IP layer through the use of per-MN unique IPv6 prefixes. PMIP6 may be allowed in shared link/prefix environments with some restrictions. All schemes accommodate IPv4 and IPv6 hosts, with respect to network mechanisms needed to reconcile missing support. All schemes run the data plane through the bidirectional tunnel between the mobility agent and anchor. Apart from the standard IP and GRE encapsulation supported in PMIP4 and PMIP6, NetLMM theoretically allows for configurable forwarding, negotiated by MAG and LMA at association time. Protocols with preestablished communication channels (NETLMM and PMIP6) thereby reduce the HO latency.

The useful HO metrics includes number of control messages used at initial attachment and by HO. Two messages are minimally required for location registration/update, but usually further interactions are needed. In scenarios with dynamic tunnels, previous association is terminated and resources are freed. However, more signaling does not necessarily mean an increase in latency, it depends on actual deployment. AAA interactions are always required, especially in PMIP6 to obtain user policy. Regarding the data flow over the tunnel, encapsulation and security generate overhead on every transmitted packet. In the simplest PMIP4 case, another IPv4 header of 20 bytes is sufficient, whereas for IPv6 encapsulation it is 40 bytes.

4.2. Security

The NBMM is especially affected from the security standpoint, because opening possibilities for diverse malicious attacks. A good description of expected security threats against NBMM is given in [16]. Protection of control messages is mandatory, achieved with security associations (SA) between the MN and network nodes participating in MM. The security between MN and network relies solely on L2 procedures (like EAP), being a possible target for malicious attacks. The mobility agents use AAA interactions to learn MN's keys/credentials, which they will use against each other. As MN is not involved, therefore SAs are referral and indirect.

Integrity check and authentication is simple for PMIP4 signaling; statically configured SA sharing the same MN-HA between all MPAs in the domain. For PMIP6 preconfigured keys are an option, however dynamic MN-HA keying derived for every MN is more efficient. Mutual authorization of agents is also possible using IKE exchange. Registration messages are usually protected by Encapsulating Security Payload (ESP). Protection of data plane is not required (may be IPSec-based). Privacy protection relates to hiding MN's location-identifier ties to prohibit the attacker deducing its whereabouts. Threats by external correspondents are avoided on account of traffic routed through the mobility anchor. Schemes support mapping of multiple nodes behind a single CoA, which complicates precise location resolution. NetLMM does more, using IDs without exposing location specifics. By recommended use of timestamps, strong SAs, and data plane protection, threats of message replaying, impersonation, denial of service (DoS), and man-in-the-middle attack (MIMA) are properly addressed.

4.3. Deployment

Deployment considerations are focused on practical aspects: reuse of existing standard, backward compatibility, and the extent of modifications. The NetLMM achieves noteworthy benefits on account of novel design, but the commercial widespread is in favor of Proxy MIP. Protocols in [12] and [6] need extensions to accommodate schemes defined by PMIP4 and PMIP6, respectively. Proposed modifications range from "proxy" flag in registration messages, new protocol options to carry requisite parameters, and even new message types for HO. In order to integrate with the new PMIP agents, the existing anchor implementations (HA or MAP) need improvement in message and binding handling. The NetLMM is facing even more development work to introduce its new mobility components.

The limitation of PMIP4 is lack of IPv4 address-space. If CoA or HoA addresses are assigned from a private address space, problems with address overlapping and Network Address Translation (NAT) may appear. The main shortcoming of PMIP6 is vague coexistence with host-based MIP6. In case both scenarios run simultaneously through the same infrastructure, two options seem viable; either all the hosts will run as mobility-unaware, or the network side would need to enhance logic by which it will detect and handle mobility-capable nodes. The issue with firewalls in MIP6 is generally recognized. In terms of scalability, PMIP6 scores well if inheriting the hierarchical approach. By virtue of cross-connected access domain, NetLMM is fairly scalable, whereas PMIP4 is limited as result of static SA configuration. Robustness limitations result from infrastructure breakdowns, especially if occurring at given single points of failure.

5. CONCLUSIONS AND FUTURE WORK

In this paper we have described and analyzed three schemes for NBMM. The presented qualitative evaluation has identified the gains and shortcomings of PMIP4, PMIP6 and NetLMM.

The PMIP6 and PMIP4 are aimed to simplify and ease the NBMM deployment by direct reuse of protocols and standards. This evaluation discloses this objective as not so straightforward. With the emphasis on scalability and performance limitations, we estimate PMIP4 is not capable of smooth and seamless-like HO without further improvements. For MIP6 many optimization mechanisms are already at hand, some of them already indicated as PMIP6 improvement alternatives (e.g. FMIP6 and HMIP6). If those would be mutually leveraged, the PMIP6 could run real-time services without perceivable HO interruption. The issue of simultaneous host-based and NBMM operation also needs resolution. The NETLMM aims to fulfill the goal of reaching the 70 ms HO latency (L2+L3 stages), to satisfy a theoretical seamlessness mark. Like PMIP6, the NetLMM scheme is missing on advanced optimizations, primarily the mechanisms for fast-HO, and the hierarchical architecture.

By deciding to build smart networks, hosts get exempt and operators gain more HO control. Opposing to the host-based concept, in this case the advanced MM functions must be reintroduced by the network. We suggest improvement in network discovery mechanisms, and incorporation of specific L2 mechanisms into the schemes, to enhance their overall performance. The results derived from this evaluation are envisioned as input for further quantitative evaluation in the form of simulations and performance measurements (using NS-2 simulator [15]), aiming to incorporate and verify assertions of the suggested functional improvements.

REFERENCES

1. Akyildiz IF, Xie J, and Mohanty S (2004) A survey of mobility management in next-generation all-IP-based wireless systems. IEEE Wireless Communications, Vol 11, 4:16–28
2. Bedekar A, Singh A, Kalyanasundaram S (2006) A protocol for network-based localized mobility management. (Internet-Draft <draft-singh-netlmm-protocol-00>, work in progress)
3. Chowdhury K, Singh A (2006) Network based layer 3 connectivity and mobility management for IPv6. (Internet-Draft <draft-chowdhury-netmip 6–01>, work in progress)
4. Gundavelli S *et al* (2007) Proxy mobile IPv6. (Internet-Draft <draft-sgundave-mip6-proxymip 6–01>, work in progress)
5. Henderson TR (2003) Host mobility for IP networks: a comparison. IEEE Network, Vol 17, 6:18–26
6. Johnson D, Perkins C, Arkko J (2004) Mobility support in IPv6. RFC 3775
7. Kempf J Editor (2006) Problem statement for network-based localized mobility management. (Internet-Draft <draft-ietf-netlmm-nohost-ps-05>, work in progress)
8. Koodli R Editor (2005) Fast handovers for mobile IPv6. RFC 4068
9. Leung K *et al* (2007) Mobility management using proxy mobile IPv4. (Internet-Draft <draft-leung-mip4-proxy-mode-02>, work in progress)
10. Lewkovetz H Editor (2006) The NetLMM protocol. (Internet-Draft <draft-giaretta-netlmm-dt-protocol-02>, work in progress)
11. Navali J *et al* (2006) IPv6 over network based mobile IPv4. (Internet-Draft <draft-navali-ip6-over-netmip4-01>, work in progress)
12. Perkins C Editor (2002) IP mobility support for IPv4. RFC 3344
13. Soliman H , Castelluccia C, El Malki K, Bellier L (2005) Hierarchical mobile IPv6 mobility management (HMIPv6). RFC 4140
14. Thaler D (2006) A comparison of mobility-related protocols. (IETF Internet-Draft <draft-thaler-mobility-comparison-02>, work in progress)
15. The Network Simulator ns-2. [Online: http://www.isi.edu/nsnam/ns/]
16. Vogt C, Kempf J (2006) Security threats to network-based localized mobility management. (Internet-Draft <draft-ietf-netlmm-threats-04>, in progress)

CHAPTER 18

WIMAX-BASED VERTICAL HANDOVERS FOR NEXT GENERATION NETWORKS

NADINE AKKARI[1], SAMIR TOHMÉ[2], AND MAHMOUD DOUGHAN[3]

[1]*INFRES, Ecole Nationale Supérieure des Télécommunications, Paris, France, akkari@enst.fr*
[2]*PRISM, University of Versailles St Quentin, Paris, France, Samir.Tohme@prism.uvsq.fr*
[3]*Mahmoud Doughan, Faculty of Engineering, Lebanese University, Beirut, Lebanon, mdoughan@ul.edu.lb*

Abstract: This paper describes a new architecture for mobility management in next generation networks. The proposed architecture aims to offer mobile users, roaming in next generation networks service continuity without QoS degradation. This is based on the proposed Inter-Domain Management module IDM responsible of guiding the vertical handover to WiMax network capable of offering the user the same QoS and context parameters. The mobility management is based on fast MIP handovers coupled with the Context Transfer Protocol CTP

Keywords: vertical handover; Stand-by IDM; WiMax network

1. INTRODUCTION

In next generation networks tending to integrate different access technologies for the purpose of offering the user the best possible service, the evolving WiMax, (Worldwide Interoperability for Microwave Access-802.16 standard), will play an important role in this integration. In fact, where WiFi provides high bandwidth but short coverage, and current cellular systems provide high coverage but low bandwidth, WiMax will provide both.

WiMax intended mainly for the exchange of data at home or in the office, has the potential to provide a significant improvement in cost and performance compared to existing wireless broadband access systems. Coverage will be based on large cells interconnected to provide the user with s high data rate [1].

Considering the vertical handover between UMTS and WLAN and vice versa, WiMax access technology will be considered as the "stand-by" destination network

H. Labiod and M. Badra (eds.), New Technologies, Mobility and Security, 207–216.
© 2007 *Springer.*

to where the handover will be guided instead of being dropped. This scenario will provide seamless handover with the required QoS [2].

This paper is organized as follows: Section 2 covers the WiMax specifications; section 3 describes the vertical handovers based on WiMax. In section 4, the WiMax-based vertical handovers is evaluated and a simulation study is presented in section 5. Finally, in section 6 we conclude and state some future works.

2. WIMAX 802.16E SPECIFICATIONS

Standard 802.16e will allow for the use of WiMax in situations of mobility. Travelling speeds could exceed 100 km/h, and the advantage of maintaining sessions when changing connections, or handover is possible [3].

The non-line-of-sight technology, IEEE 802.16e is based on orthogonal frequency division multiplexing (OFDM) and OFDM with multiple access (OFDMA) bringing improved levels of spectral efficiency, data throughput, and capacity compared to previous generations of radio technologies. IEEE 802.16e standards have flexible channel bandwidths between 1.5 and 20 MHz to facilitate transmission over longer ranges and to different types of subscriber platforms. The Physical characteristics of 802.16 support nomadic and mobile operation with wide area coverage or fixed/hot spot applications, OFDMA in time division duplex (TDD) and frequency division duplex (FDD) operations, and MIMO (multiple-input multiple-output) technology [3].

A broadband wireless access WiMax-enabled network includes the following key entities: The base station (BS), the mobile control point (MCP), and the Network Operations and Support Services (NOSS).

The base station provides connectivity over the radio link and manages radio link resources. It is responsible for physical layer functions (e.g., adaptive modulation and coding); radio resource management and scheduling; radio link retransmission (ARQ/HARQ); packet segmentation/reassembly; packing/unpacking; and traffic encryption and frame authentication.

The mobile control point (MCP) provides the control and mobility anchor point for a mobile station (MS) as it moves between base stations (BSs) in the access network. The MCP is responsible for device and subscriber authentication; service authorization; security key management; accounting; handover and macro diversity coordination; downlink traffic replication and distribution; and uplink traffic selection and forwarding. It is important to understand that unlike BSCs or RNCs in CDMA networks, the MCP in a WiMax network does not contain base station control functions; these functions reside in the base stations themselves.

Network Operations and Support Services (NOSS) include functions required to operate and maintain the wireless access network. These include element management; authentication, authorization, and accounting (AAA) services; and MS IP configuration services [1].

3. WIMAX-BASED VERTICAL HANDOVERS

WiMax is considered as the "stand-by" destination network in the proposed architecture when the vertical handover from WLAN to UMTS or UMTS to WLAN could not be performed (figure 1). In this case, the handover is guided to the WiMax network through the SIDM (Stand-by IDM) because of either the UMTS network or the WLAN network is loaded, or the required QoS and user profile could not be respected in the next network after the handover is performed.

In the case of WiMax-based vertical handover, SIDM, which is responsible of guiding the handover to the WiMax network, is located at the gateway between WiMax and the internet cloud as shown in figure 2.

When the vertical handover could not be performed due to a limited IDM capacity or inability to satisfy the required QoS, the handover is guided to WiMax where, due to its wider coverage and high capacity, it can accommodate higher number of vertical handovers.

Figure 2 shows the proposed architecture and the IDM modules introduced to perform handover. The IDM is divided into 3 entities. The vertical IDM (VIDM) responsible to perform the handover vertically, the horizontal IDM or HIDM responsible to perform the handover horizontally, and the stand-by IDM SIDM responsible to perform the handover to WiMax.

In what follows the handover scenario from WLAN to WiMax and from UMTS to WiMax is presented. Figures 3 and 4 show the messages exchanged based on the proposed mobility architecture. The transport of these messages is based on ICMPv6 protocol where the type and code are set to the specific type of CTP messages and the context data transfer is added in the data option field [4]. These exchanged

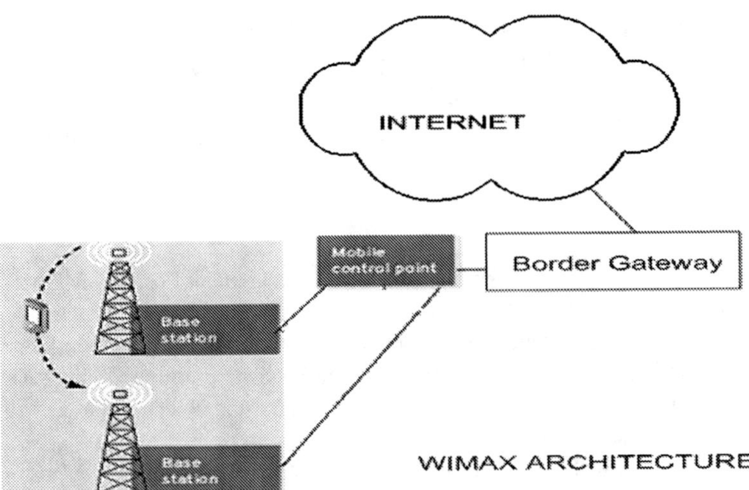

Figure 1. WiMax architecture

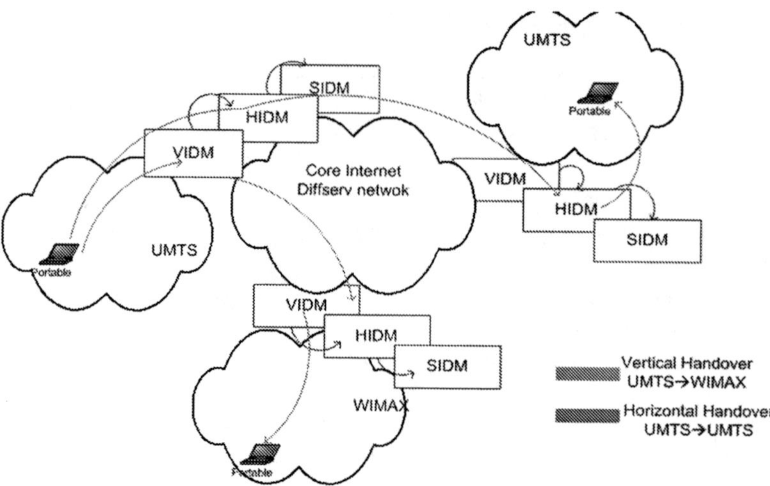

Figure 2. Integration architecture

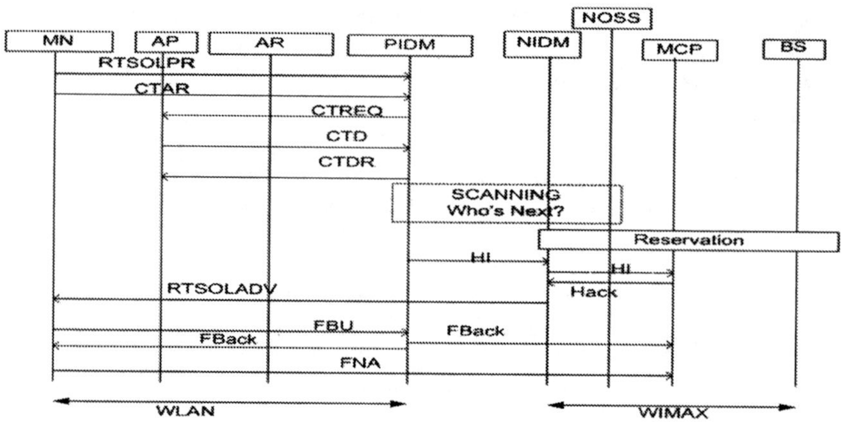

Figure 3. WLAN to WiMax

messages constitute the proposed Mobility and QoS Management Protocol MQMP. The MQMP protocol consists of the following steps:

- MN sends RTSOLPR (described in [6]) to previous acess router (PAR) to indicate HO initiation and to start the operation of acquiring a new address. The PAR will not reply directly (as in the normal FMIPv6 signaling flow [6]) but it will wait for the IDM to select the appropriate NAR.
- MN sends Context Transfer Activation Request CTAR to PAR and PIDM prior to HO. this message contains MN IP address and PIDM IP address, and the entire context to be transferred (by default). It also contains token to be used by next IDM (NIDM) for verification.

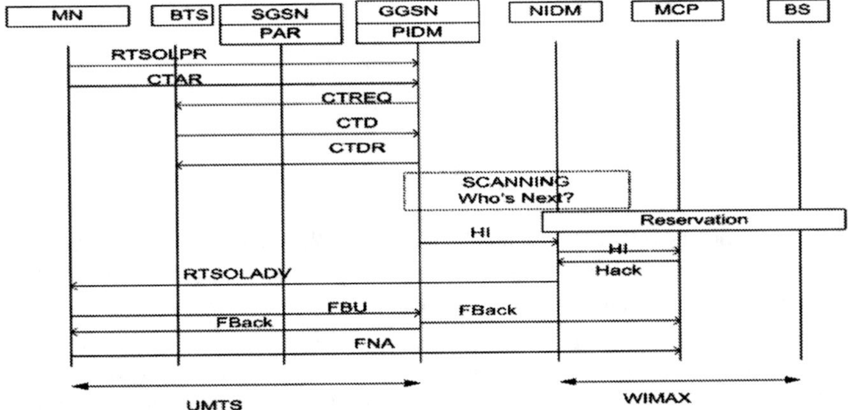

Figure 4. UMTS to WiMax

- Context Transfer Request CTREQ is sent from PIDM to PAR to update the context already available in the entry of the PIDM.
- PAR starts to send Context Transfer Data CTD to PIDM with the feature data context parameters.
- The "Who's next phase" is started. PIDM will start scanning (based on CTD) for a NIDM whose resources will satisfy the required context of MN asking for HO.
- Handover Initiation HI is sent from the PIDM to NIDM and NAR to configure the new CoA.
- All the FMIPv6 signaling flow will now proceed to perform the handover request to the selected access network.
- MN will finally send a Fast Neighbor Advertisement (FNA) to the NAR-MCP to notify him of its presence in the new access network.
- The packets destined to the old access router are no more buffered in PAR but in the PIDM which will forward these packets to the NIDM, as soon as the mobile node starts to communicate with the NAR.

The messages exchanged in the described handover scenario are similar to those exchanged in the vertical handover with the difference of the nodes involved. The new entities involved in WiMax handovers are NOSS, MCP and SIDM responsible of performing seamless handovers. In the proposed handover scenario, we assume that the node in the WiMax network are able to reserve the required resources so that the SIDM "Who's next phase" is done successfully [5].

4. WIMAX HANDOVERS EVALUATION

The proposed integration architecture with IDM support for seamless handover and QoS provisioning is based on the previously described protocol MQMP [5].

As described earlier for vertical handovers, the SIDM is now responsible to scan all the table entries to find the next IDM capable of providing the mobile with the

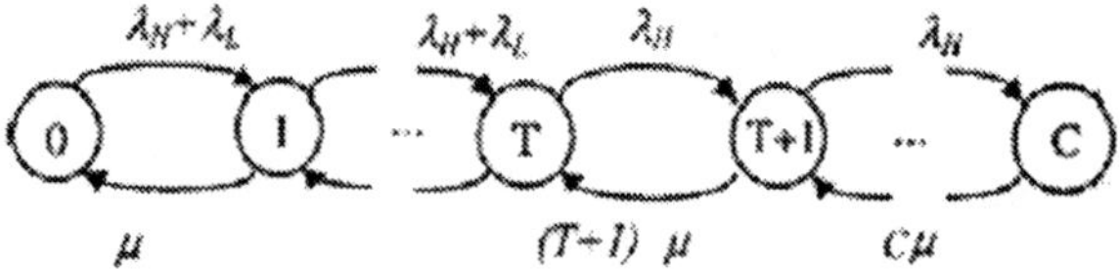

Figure 5. SIDM admission Model (UMTS, WLAN to WiMax)

required parameters ensuring a seamless handover. VIDM will either conduct the handover call to the desired destination network WLAN or UMTS, or to WiMax network if VIDM is saturated and no more calls could be accepted according to the "Who's next Policy" [5].

The handover from UMTS to WiMax and WLAN to WiMax is modeled on the following basis: Since WiMax has more bandwidth and can accommodate more calls; the following Markovien (figure 5) model is used [7]. In this model the HO calls are privileged over the local calls. The local calls are accepted until the threshold T is reached. When the number of calls is greater than T, then only the HO calls (vertical calls from UMTS or WLAN) are accepted.

To analyze the described queuing model for vertical handovers admission control [7], the following assumptions were made:

- The IDM is modeled as M/M/C/C queuing system.
- C-T interval channels are the guard band channels used only for handover calls as shown in figure 5.
- The total number of channels or the IDM capacity is $C_{WiMax} = 3^*C_{WLAN}$.
- The local traffic is a poisson process of rate λ_l and the handoff request is a poisson process of rate λ_h
- The time of stay in IDM is exponentially distributed of rate $1/\mu$.

The blocking and dropping probabilities P_B and P_D in function of the threshold T are as shown in figures 6 and 7. The choice of the threshold will highly affects the handover dropping probability as well as the blocking probability of the incoming local calls. The variation of these two probabilities with respect to the threshold is given in figures 6 and 7. The maximum allowed threshold is when T=C. As T increases, the blocking probability decreases, since the calls (local requests) are accepted until T is reached; whereas, the dropping probability is increased since the probability of dropping the handover calls is increased with no priority scheme.

UMTS to WiMax has a higher P_B due to the limited UMTS capacity. On the other hand, WLAN to WiMax has a lower P_B with the increasing IDM load, sine more bandwidth and higher IDM capacity are available.

UMTS to WiMax has higher P_D due to the higher number of handovers to a cheaper and faster network. Whereas the handover from WLAN to WiMax has lower P_D, since the users in WLAN already benefits from a low cost and high bandwidth. Figure 8 show the results.

In both cases, the user in a WiMax network will enjoy the high bandwidth and the high coverage compared to WLAN and WiMax, so the handover request from WiMax to other access technologies will not be considered.

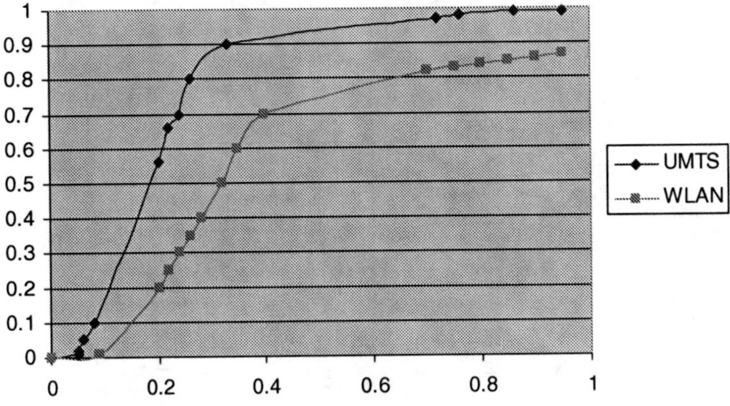

Figure 6. Blocking probability from UMTS/WLAN to WiMaX vs. T

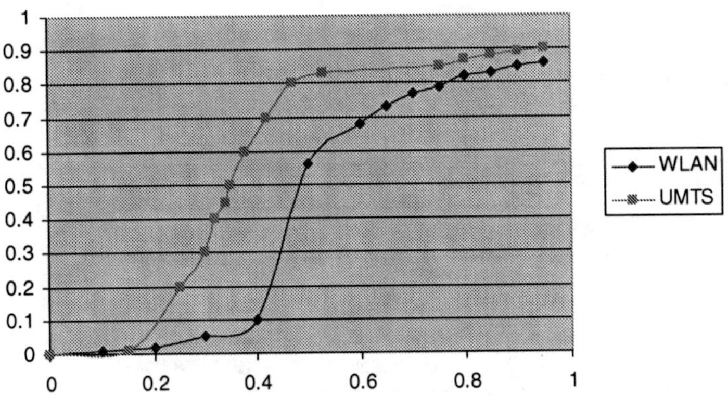

Figure 7. Dropping probability from UMTS/WLAN to WiMax vs. T

5. SIMULATION STUDY

We used the network simulator ns to simulate the analytically studied architecture. As described in table 1, the mobile is modeled as capable of receiving the best standard signal (802) which is choosing between WIFI and WiMax from one side and the UMTS network on the other side. The simulated parameters in terms of coverage, bandwidth, mobility and nodes type are shown in table 1.

The local traffic is initiated to increase the IDM load. When the mobile node will loose network coverage to start the HO process at t=80s, this traffic load is introduced on order to charge the network. Simulation time is 300s. At t=80s, the handover started. The HO represented by real time traffic represented by a CBR

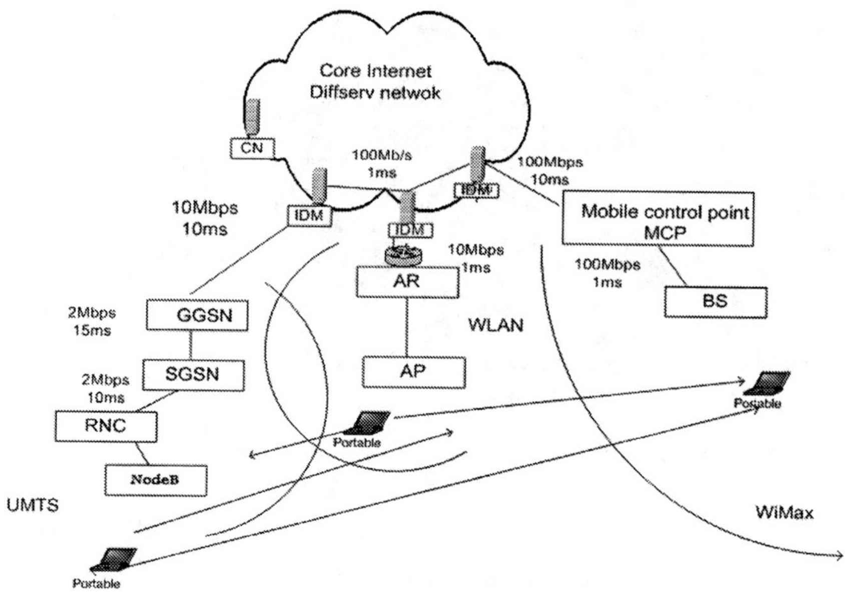

Figure 8. The simulated topology

Table 1. Simulated networks parameters

	WLAN	UMTS	WiMax
Bandwidth (Mbps)	10	2	100
Coverage (m)	300	1000	1000
Mobility	medium	High	medium
Nodes involved	AP-AR-IDM	NodeB-RNC-IDM-	BS-MCP-IDM

source of 140kbps, packet size is 150 octets. The local traffic is represented as CBR source with 130kbps.

The QoS mapping is based on mapping the WiMax service classes with UMTS and WLAN classes as shown on table 2. The WiMax service classes are: best effort, non real time with loose delay requirements, real time with variable packet size, real time with fixed packet size [3].

The simulation was decomposed into four phases as shown in table 3.

Phase1 is the initiation phase composed of FMIP initiation messages and the Context transfer messages exchanged.

Phase2 is the selection phase based on the “WHO’S NEXT” operation of the involved IDM. This phase is related to the processing time of the IDM node assumed to be 200microsec.

Phase 3 is the reservation phase in the WiMax,.

Table 2. QoS mapping (WiMax-WLAN-UMTS)

UMTS	WLAN	WIMAX
Conversational	Low latency and low jitter	Real time
Streaming	Low latency	Real time
Interactive	Low loss	Non real time
Background	Best-effort	Best-effort

Table 3. Simulated results

	WLAN	UMTS	WiMax
Phase1	100ms	500ms	20ms
Phase2	200μs	200μs	200μs
Phase3	150ms	500ms	15ms
Phase4	240ms	800ms	24ms

Table 4. HO results

	WLAN to WiMax	UMTS to WiMax
Total HO TIME (ms)	274.2ms	1024.2ms

Finally phase4 consists of the FMIPv6 operation resumed by the exchange of the 9 last messages for updates and HO termination.

The results shows that Phase1 could either happened in WLAN or UMTS, phase 2 is related to the SIDM processing time which is higher or equal to vertical IDM (VIDM) and horizontal IDM (HIDM). Phase 3 is the reservation process related to the WiMax network which is faster than UMTS and WLAN due the WiMax higher bandwidth. Phase 4 depends on whether the initial network is WLAN or UMTS as obtained. Table 4 shows the total handover delay from WLAN to WiMax and UMTS to WiMax according to the phased-simulation.

6. CONCLUSION

In this paper, we extend the Mobility and QoS Management Architecture MQMA described in [5] to include the emerging WiMax for 2 reasons. The first one is solve the problem of vertical HO from WLAN to UMTS and vice versa. In this context, WiMax mobility was considered as a stand-by destination network in order for the user requesting the vertical HO not to be blocked. The user is guided to WiMax if no resources are available in UMTS or WLAN to perform the requested HO.

The proposed MQMA architecture shows to be a global and scalable model capable of integrating different access technologies such as WLAN, UMTS and

WiMax. This integration is to provide the NGN mobile user roaming between different access technologies a seamless HO with QoS provisioning.

The results shows that if the IDM is modeled with a priority queue similar to that described above, the dropping probabilities of the handover calls, as well as the blocking probabilities of the local calls could be kept small if the threshold T is chosen to be small. In other word, if the interval from T to the total capacity C is high enough, no handover calls will be dropped.

This scheme only apply to the IDM modeling from UMTS/WLAN to WiMax since this latter could accommodate more calls, and the IDM capacity will be higher compared to that of UMTS or WLAN.

In this scenario, since the IDM is responsible to guide the HO call to the selected network based on the required QoS, QoS could be provided in the proposed MQMP framework through the IDM "who's next phase". Besides, the delay results show that, with all the MQMP processing time, the total delay is not significantly high compared to the MIP mobility protocol applied in the same context without QoS provisioning.

Currently, we are working on extending that architecture to support a client-based vertical handover scheme where, based on the client selection, the handover is guided to a cheaper network, or to a faster one according to the client preferences.

REFERENCES

1. M. Finneran, "WiMax versus Wi-Fi, a comparison of technologies, markets and business plans", June 2004"
2. E. Mykoniati, C. Charalampous, P. Georgatsos, T. Damilatis, D. Goderis, P. Trimintzios, G. Pavlou, D. Griffin, "Admission Control for Providing QoS in DiffServ IP Networks: The TEQUILA Approach", IEEE Communications Magazine, January 2003.
3. Bill Gage, Charlie Martin, Ed Sich, and Wen Tong, "WiMAX: Untethering the Internet user", Nortel Technical Journal, Issue 2, July 2005.
4. R. Koodli, C. E. Perkins, "A Contest Transfer Protocol for Seamless Mobility", draft-koodli-seamoby-ct-04.txt.
5. N. Akkari, S. Tohmé, M. Doughan. Toward a Seamless Mobility Management in Next Generation Networks. 3rd InternationalSymposium on Wireless communication Systems-ISWCS'06, Valencia, Spain.
6. R. Koodli, C. E. Perkins, "Fast handovers for Mobile IPv6", draft-koodli–04.txt.
7. L. Kleinrock, Queuing Systems, vol. I: Theory, John Wiley and Sons, 1975.

CHAPTER 19

A SYSTEM-LEVEL ANALYSIS OF POWER CONSUMPTION & OPTIMIZATIONS IN 3G MOBILE DEVICES

N. SKLAVOS[1] AND K. TOULIOU[2]

[1] *University of Patras, Patras, Greece, email: nsklavos@ieee.org*
[2] *ENST-Telecom, Paris, France, email: k.touliou@ieee.org*

Abstract: As wireless communication technology advances, more and more services and multimedia applications are supported in new generation's mobile phones. Although, the new enhanced features push the power consumption to prohibitively high values. The device size and the small battery lifetime keep limiting the available power resources. Thus, the need for efficient power management techniques arises. According to the research field of interest, power consumption issues can be approached differently and alternative power management techniques can be suggested. At a system level analysis, where CPU and memory seem to be the dominant consuming subsystems, memory hierarchy, dynamic voltage and frequency scaling technique (DVFS), are considered the most applicable solutions

Keywords: power consumption, caches, multiprocessor architecture, power management techniques, memory hierarchy, DSPs, ARM, RF, dynamic voltage, frequency scaling technique

1. INTRODUCTION

During the last decade we have witnessed an exponential growth of mobile wireless communications industry. Technology advances and consumer's demands have transformed personal communication devices, from simple voice call terminals, to rich multimedia applications platforms. New generation's mobile phones provide various services including: internet access, video teleconferencing, global positioning, high quality audio and video, as well as outstanding gaming capabilities. All these new features push the computing power requirements of mobile phones to the level of desktop computers. In contrast to wired devices however,

H. Labiod and M. Badra (eds.), New Technologies, Mobility and Security, 217–227.
© 2007 *Springer.*

the lack of a continuous power supply, poses tight limits in the overall power consumption.

The limited battery lifetime has always been bottleneck when it comes to the development of improved portable electronic products. In addition, constraints in the size and weight of mobile phones, prohibit the use of heavy and large battery packs as power sources. Although, battery technology has been improved over the years, it definitely has not kept up with the advances in other technological fields or the energy demands of wireless platforms. Apart from the short battery life span, another factor that makes power consumption a critical parameter for personal portable devices is the heat dissipation. The latter needs to remain at low levels, otherwise the system's temperature will increase, making the device too hot to be handled.

Therefore, minimizing the power consumption of wireless platforms becomes a great challenge, for the entire electronic industry, at all system levels. That's why such an intense research in many fields, including computer architecture, operating systems, computer networks, and application design has been focused on power management techniques for wireless platforms.

This paper deals with the power consumption of next generation's mobile phones from a system-level point of view. Our analysis starts with a presentation of the architecture of modern mobile phones, done in Section 2. In Section 3, a de tailed analysis is made, to show how much each one of the mobile architecture's modules contributes in the overall power consumption of the system. Section 4 explores the power management techniques used in today's portable devices. Finally, Section 5 concludes, providing also future directions.

2. SYSTEM ARCHITECTURE

The traditional architecture of a voice-centric, low end wireless phone, seems insufficient to support, the advanced multimedia applications of future mobile devices. The most popular architecture, for new generation's data-centric wireless devices will incorporate multiple processors. Each one of these will handle different tasks [6], [2]. A structure of such architecture is illustrated in the following Figure 1.

This design approach assumes the use of two basic subsystems, i.e. the cellular and the application one [2]. The former focuses on functionalities related to protocol stack and RF processing, whereas the latter implements multi-media streaming functions. Each one of these modules, and subblocks, is further analyzed in the next paragraphs.

2.1. Cellular Subsystem

The cellular subsystem is the part of the device responsible for all phone related operations of the mobile [14]. It is a digital baseband unit managing 3G modems (GSM, GPRS UMTS), network access, signal transmission and reception.

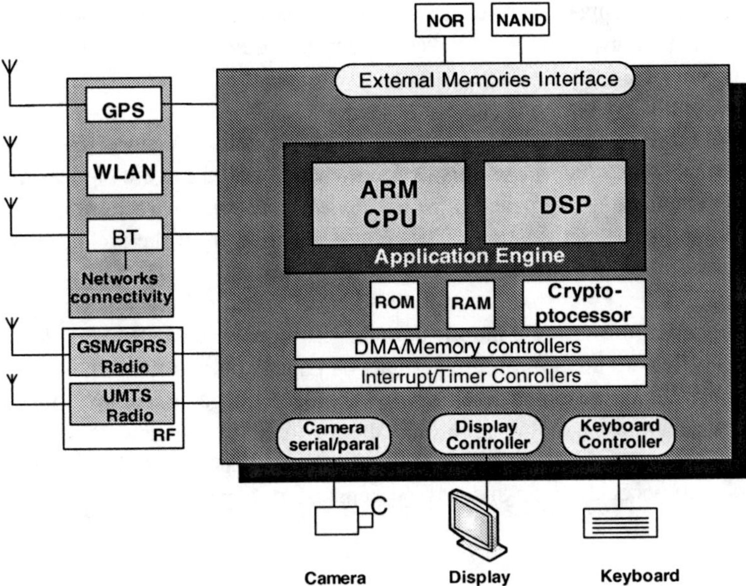

Figure 1. Structure of a multiprocessor architecture of modern mobile phones

It contains a high performance digital signal processor (DSP) for the implementation of layer-1 signal processing functions. It also includes analog chips, whose role is to handle all voice-band, mixed-signal, transit power control and radio control functions. In addition, the cellular engine is connected to the antenna and the RF parts of the mobile, e.g. receiver, transmitter, amplifiers and baseband filters.

2.2. Application Subsystem

The application subsystem deals with all the multimedia processing algorithms and the user interfaces [2]. The audio and video codecs, the image compression and processing, the communication, with all the peripherals and the connection to various networks (WLAN, Bluetooth and GPS) are tasks of this engine. The application subsystem controls all the other modules, including the cellular one, thus, it can be seen as the master control and media processing unit. In order to accomplish all these functionalities, the application engine uses two processors, i.e. a general purpose RISC processor (Reduced Instruction Set Computer) and a digital signal one, DSP. The former implements the control specific functions, whereas the latter is provided for signal processing ones.

1) RISC Processor: When comes to portable devices, the most suitable RISC processor could not be other than an ARM micro-controller, due to its low power consumption and small size. The ARM processor is connected to the boot ROM, to a SRAM, to an interrupt controller, to timers and two caches. Since, different types of external memories can be attached to wireless devices, e.g. NOR flash, SDRAM, a memory controller is needed, which is frequently supported by a direct memory access unit (DMA).

2) DSP: In contrast to the DSP of the digital baseband communication unit that deals with physical layer's algorithms this DSP core handles all the audio processing applications. Implementing from audio, voice codecs and speech recognition to noise reduction and echo cancellation, DSPs could not miss from the architecture of a rich multimedia mobile phone.

3) Video and Image Processor: Except the control specific functions and the audio processing algorithms, future mobile phones will support multi-standard video codecs (MPEG4, RealVideo etc), excellent quality image capture, on the fly image compression and 2D/3D graphics. That's why a video/ image processor unit is included in new generation's mobile phones, dealing with all the video compression and processing functions. This module consists of video encoders and decoders, a 2D/3G hardware accelerator engine, image sensors, a display controller and a camera associated module.

2.3. Hardware Cryptographic Accelerators

As wireless communication technology advances and more and more applications and services are supported by personal portable devices, the demands for security are growing. Therefore, modern mobile phones include a separate unit that handles the implementation of all cryptographic algorithms in order to ensure content protection, secure network access, terminal identity protection and transaction security.

2.4. Power Management Unit

Having mentioned already how critical is the power consumption in battery-operated devices, one can imagine the existence of a module dedicated to power management mechanisms. Controlled by the general purpose processor, this block includes software and hardware adaptive techniques in order to manage the system's power consumption and therefore maximize battery's life.

2.5. Video Peripherals

Apart the processing units we described above, mobile phones have various peripherals, which are connected with the different modules through digital serial communication links. These links are implemented and controlled once again by the master control unit, i.e. the application engine. Among the peripherals we will focus only on the display, due to its important contribution in the overall system's power consumption.

3. SUBBLOCK POWER CONSUMPTION

When studying the power consumption of today's mobile phones, there are different points of view, the subject can be analyzed from. In the architecture computer community, the energy is investigated at instruction level, whereas the attention of the network research community is attracted by power issues in wireless network protocols. Another interesting approach focuses on application-level energy consumption [5], where dissipation is measured in relation to the power loss during the execution of mobile's basic tasks, i.e. transmission and reception of the signal, email handling,mp3 song playing or web browsing.

In this paper, we explore power issues at system-level with an interest on how each one of the previously described modules, contributes in the overall energy consumption of the device. In Table 1, one can see the participation of the various semiconductor components in the total power consumption based on the results obtained in [3].

According to this Table shown above, the more power consuming subsystem is the application engine, followed by the cellular one with an important amount of energy spent for modem/RF applications. Immediately after comes the memory subsystem, in the scale of importance, that seems to contribute significantly in the consumption of power resources. Finally, the fourth-most power consuming structure of portable devices is the display, whereas a 2% of the total power is spent by the other peripherals and controllers. After obtaining a general idea on how the total energy of modern wireless devices is distributed, we can try to justify the above numbers by investigating under which circumstances each sub-block consumes power.

A) Application Engine: When comes to the application engine, the components that contribute the most in the power consumption are the CPUs of the general

Table 1. Typical energy distribution in multimedia mobile phone

Subsystem Application	Application Energy Distribution	Subsystem	Subsystem Energy Distribution
A/V and Transport	4.4%		
Video Encode	9.9%		
Audio	15.5%	*Multimedia*	39.5 %
Modem Processing Multimedia	9.8%		
Modem Processing	8.3%		
Receive	5.0%	*Modem Operation*	21.5 %
Transmit	8.2%		
Memory	19.4%	*Memory*	19.4 %
LCD Control	3.7%	*LCD*	17.6 %
LCD Driver	13.9%		
Other Peripherals / Units	2%	*Others*	2%

RISC processor, the DSP and the video/ image accelerator units, the connected caches, the interconnection buses, the register files and the remaining functional units.

A1) CPU: The power in the CPU is consumed due to the execution of the instructions and their fetching from the memories or caches [8]. Thus, the smaller the amount of code the system must fetch from the memory or caches, the less the energy consumed. For example, slower programs with better organization of fetch packets are preferred over partially filled fetched packets, which need additional fetching and thus are more power consuming. An estimation of the energy, consumed in a processor for a given task, can be obtained by multiplying for each type of instruction, the total number of the executed instructions by the base consumption of the corresponding type. In the obtained result, it is needed to be added the power consumed by the program memory accesses in order to fetch the instructions. In Table 2, one can see the consumption of different instruction types for a SPARC platform clocked at 750 MHz as obtained in [1].

Another dominant source of power consumption is the register file, meaning an array of processor registers in the CPU. Although, many mechanisms that minimize the energy dissipation in other elements have been developed, power issues related to the register file have lacked such wide research. Power loss in register file depends on the system configuration, with an emphasis on the number of integrated registers, the cache size and the existence of a branch predictor table. According to [13], having a large register set permits the temporary storage of intermediate results without the need of a main memory usage. Therefore, less memory accesses are performed and less load and store operations are needed resulting in a reduction of the power dissipation.

A2) Interconnection Buses: When implementing multiprocessor architectures, the data communication between the different contained units, becomes an important consideration. The request to the memory for read/write, the corresponding responses either from the memory or a processor cache, and the coherence control messages to the caches, are some of the messages often exchanged between the various modules. Thus, interconnection must be provisioned, in order to make possible the communication of each processor core with the memory subsystems, and between the processors themselves. The buses, the multiplexers

Table 2. Energy consumption per instruction

Instruction Type	Energy (nJ)
Load	4.814
Store	4.479
Branch	2.868
ALU (Simple)	2.846
ALU (Complex)	3.726
NOP	2.644
Program Memory Access	4.940

and the drivers, that are used to implement the communication channels, can drain a considerable amount of power in each data transfer due to their high capacitances [8].

A3) Other Functional Units: In addition, functional units, are used for the implementation of algebraic transformations, operations or architectural design techniques (e.g. pipelining, or loop unrolling). These units have also an impact in the overall power consumption. Extra registers or control units that have been added for the shake of the performance may increase the power dissipation. Thus, a trade-off between power consumption and performance must always be considered. Furthermore, the power consumed by several functional modules depends on the size and the complexity of the algorithm they perform. Energy reduction can be accomplished by dead code elimination, and by replacement of intensive energy consuming operations (multiplications) by simpler ones (shift and add operations).

Having explained how the power resources are consumed in embedded processors, it is obvious why the use of an ARM and DSP processors in modern mobile devices has not been a random choice. Both of these processors are power efficient, by optimizing code density and using parallelism/ pipelining. These techniques minimize the bus transfers and the frequency of the memory accesses. For example the ARM9 architecture used in many smart mobile devices such as Nokia phones uses [7]:

- *Two 16KB instruction and data caches for quick memory access.*
- *A 5-stage pipeline: fetch, decode, execute memory access and write register back.*
- *Advanced microprocessor bus interface to reduce the power consumption in the interconnections.*
- *32 bit data and instruction set + Thumb extension to optimize code density.*

The Thumb extension of the ARM's 32-bit architecture optimizes code density by compressing common 32-bit instructions into 16-bit operation codes [4], [7]. During the execution, the Thumb translates the codes by expanding the instructions into the expected 32-bit format in real time. This technique reduces the amount of code by 30% and therefore, the power consumption. Although, micro-processors are optimized for control specific functions, DSPs are targeted at maintaining a continuous data flow. Normally, they are used for signal processing algorithms that include operations such as modulo, multiplication, addition, fetching values from memory and incrementing address pointers. The mathematical formula that expresses these algorithms permits not only the execution of these operations in parallel, but also separates program and data spaces [9]. The latter allows a simultaneous access to program instructions and data, providing an even higher degree of parallelism, that increase the performance with a small power drain. In addition, various architectures developed such as Harvard [9] or MAC architecture make use of different design techniques, in order to reduce even further the amount of memory accesses, and the number of instructions making the DSPs the best candidates for low power devices.

B) Cellular Engine: Inside the cellular engine, the biggest energy consumer is the RF part, which deals with the communication related operations, i.e. the signal transmission and reception. One can expect that the power consumption differs depending on the phone's operation, with the device consuming more power during a call, for example, than during standby mode. This is explained by the fact that during talk time, the power amplifier consumes heavily in order to transmit the signal from the mobile terminal to the base station [14]. The works of [10] provides measurements of the power consumption during different operation modes such as standby, talking, listening, ringing and when an attempt to call is made. However, measuring the exact power consumption in the RF part of a portable phone is quite a difficult task. Several parameters like signal strength, transmitted power level and time intervals between different operating modes can affect the calculations, making the results obtained only estimates of the real consumption.

C) Memories: The memory subsystem is a critical source of power dissipation, consuming the 20% of the available resources. The power loss in memories can be either static, caused by the leakage current or dynamic, occurred in each memory access, either instruction or data access. As in the case of the instruction memory access, that was described before, the power dissipation for data access is also expressed in mW/MHz [9]. This signifies that for memories clocked only when data is needed, energy dissipation is proportional to the amount of data being transferred. Therefore, data compression techniques that minimize data and hence, the frequency of memory accesses, can reduce the memory power consumption. This stays true, however, as long as the additional computations do not add more power loss than the actual obtained power gain. As the clock frequency of the embedded processors increases and more complex applications are supported, faster memory access is required. That is the purpose of the two caches being connected to the processors, i.e. to accelerate the access to the external memories. However, caches come along with an additional power impact. Unfortunately, calculating the caches contribution in the power consumption is quite tricky. A lot of parameters need to be taken account such as the type of the cache, the used architecture (size, configuration, word line size etc.), the applied technology (line, device input/output capacitors, etc) and statistical characteristics (switching bus activity, frequency of accesses, etc) [15].

D) Display: The display and the display backlights have a significant impact on the device's power consumption. How much is this impact depends on the size of the display, since the bigger the display the more the power consumed, the manufacturer and whether it is colorful or not. Also, the architecture of the display can determine the energy loss. Displays with internal memory, for example, seem to spend less battery resources [14]. This can be explained by the fact that display content can be saved temporarily to the display's internal memory, resulting in fewer accesses to external memories and thus, to less power consumption. Finally, the display backlights accounts for important energy overheads in a battery-operated device [15]. As the brightness of the backlights increases, more energy is consumed by the related circuit, hence, a compromise between display's quality and power consumption has to be made.

4. POWER MANAGEMENT TECHNIQUES

With power consumption being so critical in wireless devices, it is no surprising why so much research has been focused on power management techniques. Having showed that the three most important contributors in the power loss is the display, the CPU and the memory, one can expect most of the employed techniques to involve these subsystems.

4.1. Energy Adaptive Displays

In order to reduce the power consumed by the display, the use of energy-adaptive display systems has been proposed [5]. According to this technique the quality of the display, meaning colors, brightness and size, is each time adjusted to the application preferences and the user's demands. Why for example, make use of the whole screen when it comes to material that can be equivalently displayed in smaller windows without loss of visual quality? This approach is further motivated by research results that showed that the window of focus, meaning the good first-order indication of the area of interest to the user – uses only about 60% of the total screen area [5]. Based on these observations, many techniques have been developed to reduce power by scaling down the quality (decrease luminescence and colors) of the non-active screen areas, while leaving the active screen area unchanged.

4.2. Low Power Memories

Although, traditional voice centric phones require a small amount of memory, integrated caches and ROM in today's phones may exceed hundred of kilobytes. Having realized how much power is consumed in memories, retaining the whole memory system active all the time results in considerable power drain. Careful studies of algorithms revealed that most of the time, fetches hit within a small cache area. Power saving techniques takes advantage of this tendency by suggesting the partition of the memory in smaller parts, called banks and then activating each time only the bank that is presently in use. In addition, advances in memory technology have resulted in the development of a new range of memories such as CellularRAM and MobileRAM designated for wireless, battery operating devices. CellularRAM is actually a pseudo SRAM that permits self-refresh and recharged operations inherent in DRAM technology, while providing lower standby and operating currents that have an impact on total power consumption [11]. Obtained experience by the development of CellularRAM led to the design of an alternative DRAM memory, named MobileRAM. The latter seems to reduce the core voltage, related to power consumption, of DRAM cells, while introducing a key new low-power feature called on-chip temperature sensor (OCTS) [11]. This feature automatically senses the temperature and chooses the most efficient memory contents refresh rate, while minimizing the standby current.

4.3. Power Management Unit

In order to reduce the overall system's power consumption, a commonly employed approach suggests the idling, stop clocking the circuits or modules that they are not used. Having grouped different functions in different clock domains, the power manager unit with the appropriate software can suspend or reduce the clock of selected units [12], [8]. Another technique called dynamic voltage and frequency scaling (DVFS), can be applied to manage the decrease of power dissipation of the application processor [15], [14], [4], [8]. This DVFS method enables the scaling of supply voltages and clock frequencies during the execution, allowing a dynamic change of the system performance. This means that the overall performance can be adjusted to the requirements and the applications demands. Lowering for example the system's performance when running applications with low needs of processing capacity can result in major power gains. DVFS technique is based on the fact that the power consumption of a device is proportional to the operating voltages and frequencies according to the following equation:

$$(1) \qquad P_d = aC_{eff}\, V^2 f$$

Taking into account, though, that minimizing the frequency signifies an increase of a task completion time, by just reducing the clock frequency we can not manage any power savings. Thus, any decrease in frequency should be accompanied by an additional decrease in the voltage, which being proportional to energy consumption can result in power dissipation reduction.

5. CONCLUSIONS & OUTLOOK

As the wireless communications are growing and the user's demands for rich multimedia portable devices with long battery life are increasing, power consumption in portable systems becomes a critical issue. Estimating the total power consumption of new generations' complex devices with multiple power consuming sources can be quite tricky. According to the research point of view, different results can be extracted and different solutions are proposed. For a system level analysis one can focus on the power consumption of the various CPUs, memories, interconnecting buses, the display and the RF part of the multi-core platform. The power related issues in wireless devices have motivated the scientific community, to find methods capable of minimizing the energy dissipation. Low power memories, energy efficient displays and power control techniques have been developed as a response to the urgent need for increased battery lifetime.

6. ACKNOWLEDGMENTS

This wok has been supported by the State Scholarships Foundation (IKY), under the Program of Scholarships for Post Doc Research.

REFERENCES

1. G. Chen, B. Kang, M. Kandemir, N. Vijaykrishnan, M. J. Irwin, and R. Chandramouli, "Studying energy tradeoffs in off-loading computation/ compilation in Java-enabled mobile devices", *IEEE Transactions on Parallel and Distributed Systems (TPDS)*, Vol 15, No. 9, pp. 795–809, 2004.
2. S. Drude, M. Atorf, L. Chivallier, K. Currie, "System architecture for a multi-media enabled mobile terminal", *Consumer Electronics, IEEE Transactions*, Vol. 51, pp. 430–437, Issue 2, May 2005.
3. F. Schirrmeister, Design for Low-Power at the Electronic System Level. ChipVision Design Systems. Available: http://www.soccentral.com/soccontent/documents/ESL_Design_for_Low_Power_ChipVision.pdf
4. L.D.Paulson, "Low-power chips for high-powered handhelds", in *Computers*, Vol. 36, Issue 1, pp. 21–23, Jan. 2003.
5. R. N. Mayo, P. Ranganathan, Energy Consumption in Mobile Devices: Why Future Systems Need Requirements-Aware Energy Scale-Down, *HP Technical Report,* HPL-2003-167, HP Laboratories Palo Alto, Aug. 2003.
6. Texas Instruments, *OMAP Platform Manual*. Available: http://www.ti.com/omap.
7. ARM, *Processor core overview*. Available: http://www.arm.com/products/CPUs/
8. P. J. M. Havinga, G. J. M. Smit, "Low Power Systems Design Techniques for Mobile Computers", Technical Report TR-CTIT-97-32 Centre for Telematics and Information Technology, University of Twente, 1997.
9. C. Moerman, E. Lambers, "Optimizing DSP: Low Power by Architecture", *Embedded Systems Technology Centre (ESTC), ASIC Service Group, Philips Semiconductors*. Available: http://www.adelantetech.com/upload/low_power.pdf.
10. E. Shih, P. Bahl, and M.J. Sinclair. "Wake on Wireless: An Event Driven Energy Saving Strategy for Battery Operated Devices", *Proc. of the Eighth Annual ACM Conference on Mobile Computing and Networking*, Altanta, Georgia, USA, September 2002.
11. O.Vargas, Infineon, "Minimum power consumption in mobile-phone memory subsystems." Available: http://pd.pennnet.com/display_article/244484/21/ARTCL/none/WIREL/Minimum-power-consumption-in-mobile-phone-memory-subsystems/
12. R. Kakerow, "Low Power Design Methodologies for Mobile Communications", in *Proc., IEEE International Conference on Computer Design*, Freiburg , Allemagne, pp. 8–13, 2002.
13. L. Wehmeyer, M. K. Jain, S. Steinke, P. Marwedel, and M. Balakrishnan, "Analysis of the Influence of Register File Size on Energy Consumption, Code Size, and Execution Time", *IEEE Trans. On Computer-Aided Design of Integrated Circuits and Systems*, Vol. 20, No 11, Nov. 2001.
14. M.P. Michael, "Energy awareness for mobile devices", *Research Seminar on Energy Awareness*, University of Helsinki, 2005
15. H. Van Antwerpen, N. Dutt, R. Gupta, S. Mohapatra, C. Pereira, N. Venkatasubramanian, R. Von Vignau, Phillips Semiconductors, "Energy-aware system design for wireless multimedia", *in Proc. Design, Automation and Test in Europe Conference and Exhibition, 2004*, Vol. 2, pp. 1124–1129, Feb.2004.

CHAPTER 20

RUNTIME SOFTWARE MODIFICATION METHOD USED ON COTS SYSTEM FOR HIGH-AVAILABILITY NETWORK SERVICE

TAKASHI IKEBE[1], YASURO KAWARASAKI[2], NAOKI UCHIDA[1], SHOICHI HIRASAWA[3], AND HIROKI HONDA[3]

[1]*NTT Network Service Systems Laboratories, Nippon Telegraph and Telephone Corporation 9–11, Midori-Cho 3-Chome Musashino-Shi, Tokyo 180-8585 Japan, ikebe.takashi@lab.ntt.co.jp, uchida.naoki@lab.ntt.co.jp*

[2] *NTT Service Integration Laboratories, Nippon Telegraph and Telephone Corporation, 9–11, Midori-Cho 3-Chome Musashino-Shi, Tokyo 180-8585 Japan, kawarasaki.yasuro@lab.ntt.co.jp*

[3] *Graduate School of Information Systems, The University of Electro-Communications 5-1, Chofugaoka 1-Chome Chofu-Shi, Tokyo 182-8585 Japan, hirasawa@is.uec.ac.jp, honda@is.uec.ac.jp*

Abstract: Generally, providing high-availability services such as a network service with COTS (commercial off-the-shelf) hardware and OS is very difficult because network services require frequent software modifications. Existing network service systems achieve high availability by using specialized systems. We present a live-patch method that enables online software modification without disrupting service on a COTS system. The live-patch method modifies user software and kernel software without rebooting by changing execution of the function to modified function. The evaluation shows the adaptability of the presented implementation as COTS systems on Linux and x86CPU SMP machines

1. INTRODUCTION

Network services are required to provide a high-availability service, especially on lifeline services such as PSTNs (Public Switched Telephone Networks). In addition, high-availability network services use specialized OSs and hardware to satisfy the required availability. The PSTN is a legacy network service; however, the requirements for service availability are the same even on NGN (Next Generation Network) services. The use of the specialized system on the NGN service is difficult because of slow service development due to lack of trained technicians and expensiveness of

H. Labiod and M. Badra (eds.), New Technologies, Mobility and Security, 229–243.
© 2007 *Springer.*

development cost. Therefore, applying a COTS (commercial off-the-shelf) system to the network service is important.

The focus of software development for providing network services is on debugging procedures to provide a high-availability network service. However, when a service requires more rich functionalities, the software becomes large and complex, resulting in many software problems. Services in the field typically have unexpected input-output, load, and function requirements. To satisfy these unexpected conditions, software modifications are needed, and frequent software modifications causes service-availability deterioration.

We present a live-patch method that enables runtime software modification of user applications and the kernel without deteriorating service availability. We also present an implementation and evaluation of a live-patch method using Linux and x86 SMP hardware in a COTS system.

2. BACKGROUND

The following two methods are used for modifying the software.

2.1. Offline Software Modification

The offline software modification method restarts software by loading a new modified load module. There are source patch and binary patch approaches to create new load modules. The source patch modifies source code of the software whenever there are problems, and new load modules are created by compiling the patched source code. The source patch approach is typically used on open-source software.

The binary patch creates a new load module by directly changing the binary code of the load module that has a problem. The binary patch approach is typically used on commercial software. Some commercial software vendors do not provide patches to users; they only provide modified load modules.

Both of the above approaches require stopping the running process or OS. Then, modified load modules are reloaded to fix the problems. These approaches are called offline software modification in this paper.

2.2. Runtime Software Modification

Runtime software modification is a method that modifies software functionalities without terminating software that is running.

Networks services need to be provided 24 hours a day, 365 days a year, achieving service availability more than 99.999% of the time. Therefore, long service disruptions due to software modification are not permitted. Other requirements of future network services include the following: the minimum interruption time for multiple modifications, applicability to all system software including the kernel without changing the application source code, and availability on COTS systems.

The minimum interruption time for network services means that modifications must be finished within the network service timeout. The timeout value differs according to service type; however, most network services require timeout values that are between 1 millisecond and less than 1 second. In this paper, the service timeout value is 100 milliseconds.

Today's network service software is complex and large; therefore, the ability to perform multiple modifications within a timeout is very important.

Systems have many software components, and services are performed by large-scale user applications. Therefore, the source code of user applications should not be changed for runtime software modifications.

Challenges in runtime software modification Unfortunately, there is no runtime software modification method that meets all of the desired requirements. Many methods have limited flexibility, for example, dynamic linking is a well-known mechanism. While systems based upon dynamic linking[1][2] can modify a running user application, they require the unmodified programs to invoke procedures through an indirection table, and statically linked procedures cannot be modified without causing performance deterioration. Some method[3] requires special process state in order to safely modify software. Some method[4][5][6] requires distributed object such as CORBA[7]. Moreover, these approaches to modify the kernel are not available today.

Some methods[8][9][10] use a virtual machine such as Java. Java Hotswap[9] enables runtime software modification of Java software. Java Hotswap is part of the Java Platform Debugger Architecture (JPDA), which is intended for effective debugging during development. A study[10] describes the improvement of Java Hotswap by expanding it for application to running systems. However, the target software is limited to Java software, and the Java virtual machine has many performance problems due to execution overhead and frequent garbage collections, which affect real-time capability.

Some methods write branch instructions[11][12]; however, in many cases, they are not applicable on kernel software. Some systems, for example, specialized OSs, only allow kernel software modification[12]. However, that approach has the limitation that the OS does not have virtual memory addressing capability because the hardware does not have multiple processors.

Our approach uses branch instruction, which is applicable to user applications and kernels on modern COTS systems with minimum interruption time of target software.

3. LIVE PATCH; RUNTIME SOFTWARE MODIFICATION ON COTS SYSTEM

In this section, we present the live-patch method, which enables runtime software modification of user applications and the kernel on multiprocessor systems. We

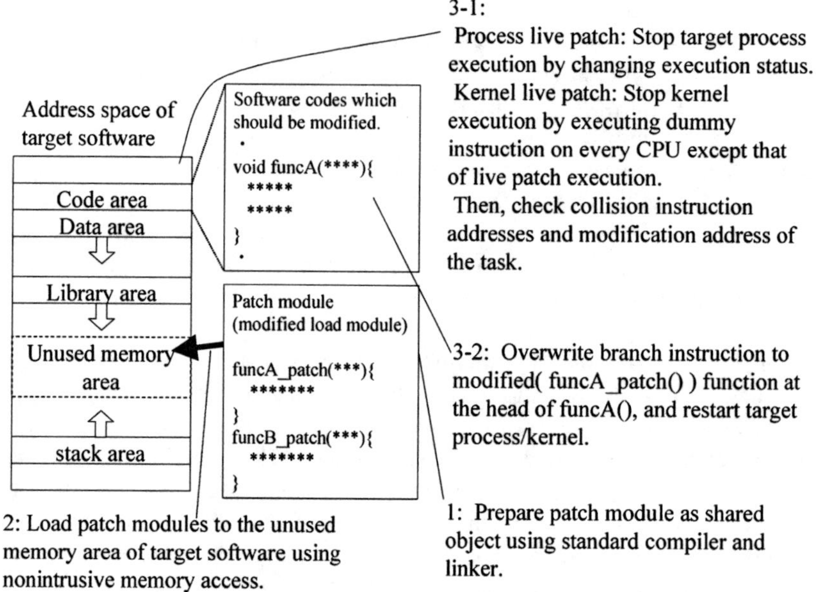
3-1:
Process live patch: Stop target process execution by changing execution status.
Kernel live patch: Stop kernel execution by executing dummy instruction on every CPU except that of live patch execution.
Then, check collision instruction addresses and modification address of the task.
Address space of target software
Software codes which should be modified.
void funcA(****){
*****
*****
}
Code area
Data area
Library area
Unused memory area
stack area
Patch module (modified load module)
funcA_patch(***){
****
}
funcB_patch(***){
****
}
3-2: Overwrite branch instruction to modified(funcA_patch()) function at the head of funcA(), and restart target process/kernel.
2: Load patch modules to the unused memory area of target software using nonintrusive memory access.
1: Prepare patch module as shared object using standard compiler and linker.

Figure 1. Method to modify software

have been working on a live-patch method[13] in the Pannus project[14], and some implementations are written in open source code.

The live-patch method achieves online software modification using the following procedures (see Figure 1).

A. Patch module preparation: Compile patch module (modified load module) from fixed source code.

B. Loading patch module: Live-patch process which performs live patch loads the patch modules that include modified routines onto the target software (process or kernel) memory area. And the live-patch process relocates the patch module.

C. Modifying execution address: The live-patch process modifies the execution address of target software. And the CPU execute modified routine whenever the original routine is executed.

For simplicity, the C language function modification case is described. The live patch for a process is called the process live patch, and the live patch for a kernel is called a kernel live patch in the following sections. The details of the above three procedures are described below.

3.1. Patch Module Preparation

The live-patch module is compiled as a relocatable, shared load module by GCC (GNU C Compiler). The object (relocatable, shared load module) of the live-patch module is widely used on a COTS OS.

3.2. Loading Patch Module

Loading patch module in process live patch The live-patch process which performs runtime modification accesses the virtual memory area of the target process through the virtual memory subsystem on the kernel.

Currently, typical COTS OSs have a virtual memory, and each process has an independent virtual address space. Therefore, there is a problem because the live-patch process cannot access the memory area of the target process directly. The virtual address space of a process is controlled by a virtual memory subsystem on the kernel, and the virtual memory subsystem controls physical memory allocation, which depends on the memory usage of a process.

Every target process could implement a specialized communication routine between live-patch processes. However, this approach is not efficient because the approach requires changing many target processes. Implementing a specialized communication routine in some target processes is difficult because of copyright issues.

Loading patch module in kernel live patch The live-patch process accesses the kernel address space through kernel memory management.

The address space of a kernel is controlled by kernel memory management that exists on the kernel. Therefore, there is a problem because the live-patch process cannot access the kernel address space directly.

3.3. Modifying Execution Address

To change the execution of an original routine to a modified routine that exists on the patch module, the following two points should be considered.

1. Method of changing execution address: The method to change the execution address of the target software whenever the unmodified routine is executed.
2. Method of writing modification: The method that safely writes an instruction that modifies the routine in a general multiprocessing system.

The above points are described in the following sections.

Method of modifying execution address The live-patch method modifies routines according to "symbols" which are information for linking and writes the modification instruction to the unmodified function/routine to change the execution address whenever the function is invoked. The symbol is data that is associated with information, such as routine location, type, and scope level in the source code. Symbols such as functions and variables[15] are created per routine and exist in load module.

The original function will be invoked from any part of the remainder of the program; therefore, modifying the callee function is much easier than modifying all caller functions.

Transition instruction The live-patch method writes the "jump" instruction to the head of the original function through memory management in the kernel.

A generic CPU has instructions to change the execution address, which are "call" and "jump" instructions. The call instruction invokes a new procedure, pushing the stack whenever executed. The jump instruction changes the execution pointer to a given address without pushing the stack. The modification is executed per symbol; therefore, a modified function needs to be able to refer to arguments of the original function that have been pushed to the stack. Whenever the new routine finishes its execution, the CPU pops the stack, and returns to the address just after the original routine invocation, as shown in figure 2. Therefore, a jump instruction is suitable for the live-patch method.

Every function is loaded to the text segment of the memory after the program has started, and the segment is marked as read-only. Segment management is also controlled by memory management in the kernel. To avoid restrictions imposed by segment access rights, the live-patch method enables privileged access through memory management in the kernel.

Another approach is changing the GOT (global offset table) and PLT (procedure linkage table) whenever all the symbols are located in the dynamic link libraries. However, there is a problem that dynamic link libraries exhibit inferior performance compared to a statically linked program. The symbols in the dynamic link libraries are resolved through GOT and PLT during dynamic linking[15].

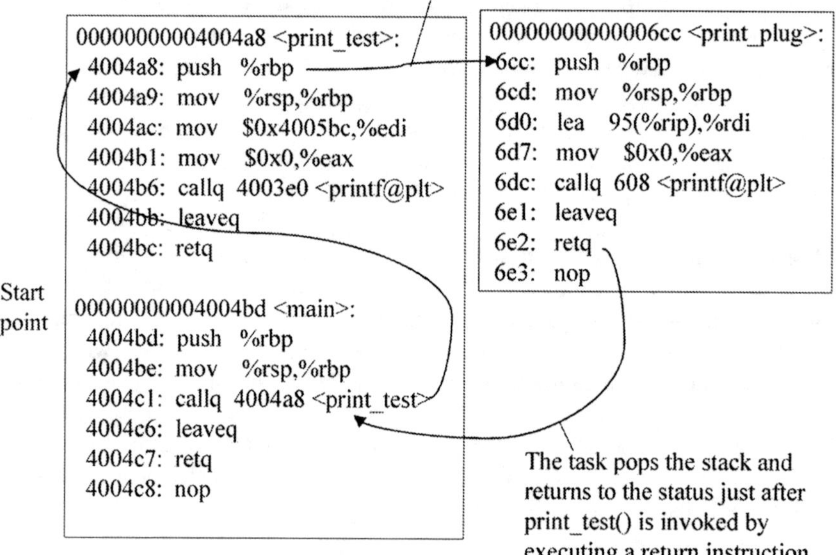

Figure 2. Change of execution path due to jump instruction overwriting

Method of writing jump instruction The live-patch method controls target software execution to safely write the jump instruction. If the live-patch process just writes the jump instruction without controlling the target software execution, the target software will crash due to incomplete instruction execution, because a multiprocessor machine runs multiple tasks simultaneously.

Control of process execution The live-patch process momentarily stops the target-process execution when writing the jump instruction in the case of a process live patch. The jump instruction size is larger than one word. Therefore, jump instruction writing cannot be performed in a single step and is required to be executed during locking the whole target process. If a jump instruction is written while the target process is not locked, the target process crashes whenever the thread of the target process executes an incomplete jump instruction at the time that the thread invokes the original symbol on another CPU.

The live-patch method changes the process execution state on the CPU scheduler that is on the kernel because the CPU scheduler controls process execution by each process's state and priority. If the process state is not ready to be run, the CPU scheduler does not assign a CPU to the process.

Then, the live-patch process checks that the stopped target process and thread execution addresses do not conflict with the address to which the jump instruction will be written. If the addresses conflict, the live-patch process restarts the target process and retries. Otherwise, the target process also crashes after resuming due to executing the wrong instruction because the jump instruction size is larger than one word. The retry count is limited to a few times to prevent a service timeout. Minimization of process stopping time is important on network service.

The live-patch method requires that the other part of an original symbol does not perform the execution at the addresses in which jump code is written to prevent the target process from crashing. The execution is typically created by the loop routine or 'ggoto'h label in C language. The live-patch method checks these executions using a source code checking tool, and judges whether the symbol is modifiable.

Control of kernel execution The live-patch process momentarily stops kernel execution while writing the jump instruction in the case of a kernel live patch. Like the process live patch, the kernel live patch controls kernel-wide execution to avoid a system crash.

The kernel live patch controls kernel executions by making all CPUs, except the CPU on which the kernel live-patch process runs, execute dummy operations such as nop (no-operation) instructions. In particular, the kernel live-patch process creates high-priority dummy operation tasks and explicitly assigns the task to each CPU except the CPU that the kernel live patch uses. After control kernel executions, the kernel live patch checks conflicts of the execution addresses of kernel threads and user processes which run in kernel mode using the same procedure as that of a process live patch.

Kernel executions are interrupts, kernel threads, and the kernel mode of user processes[16]. The kernel live patch controls the interrupts by interrupt disabling, controls the kernel threads and kernel mode of a user process with above nop method. In order to control the kernel threads and kernel mode of a user process, the process live-patch approach that changes execution status is also applicable. However, the number of kernel threads and user processes differs according to the system and load. Therefore, changes in the execution status of kernel threads and user processes in a CPU scheduler cause unpredictable delays in a network service. However, the number of CPUs in COTS hardware is limited. Therefore, the controls of CPU executions are much faster than the change of execution status in the procedure of process live patch although the approach requires that the whole system stop momentarily.

4. IMPLEMENTATION

This section describes the implementation of live patch using Linux kernel 2.6.9. The execution of the live patch follows three procedures, which are patch module preparation, loading of patch module, and changing the execution address to the patch module. The patch module preparation uses a generic compiler.

The following sections describe implementation of the second and third steps of the process and kernel live patch.

4.1. Implementation of Process Live Patch

In the implementation of the live patch, mmap3() and accesspvm() system calls on the kernel are implemented to quickly access virtual memory. The mmap3() system call enables mapping the given file to the free (unmapped) memory address of the target process without stopping the target process execution. The mmap3() system call is implemented based on existing mmap() and mmap2() system calls that map the given file to the memory area of process itself through virtual memory control. The mmap3 extend the destination memory area to any processes.

The accesspvm() system call enables a multi-word size memory read and write to used(mapped) memory addresses on any process. The accesspvm() system call also enables privileged access to the given memory address. Therefore, the live-patch process can read and write to the given address regardless of the permission given by the kernel.

Linux kernel has the ptrace() system call that also enables privileged memory access to any process for debugging purposes, and there is another live-patch implementation using the ptrace() system call based on our method[17]. However, the ptrace() system call cannot be applied to a network service. The ptrace() system call only enables one-word-size memory access and requires stopping the target process during executions. That is, when the patch module size is greater, network service availability is inferior. The effectiveness of mmap3() and the accesspvm() system call is described in the experimental results section.

Load of patch module The live-patch process accesses the target process memory address by virtual memory control that is on the kernel. The live-patch process loads the patch module to the free memory area of target memory using an mmap3() system call.

Execution transition to patch module The live-patch process momentarily stops the target process execution using a SIGSTOP signal and checks for execution address conflicts. Then, the target process writes the jump instruction using the accesspvm() system call and restarts the target using SIGCONT signal. The signal is a primitive communication method between processes. SIGSTOP is a signal that stops a process and its thread, and the SIGCONT restarts the stopped process and its thread. The SIGSTOP changes the execution status of the process to "TASK_STOPPED" and the CPU scheduler treats the process as out of scope. The SIGCONT changes the execution status "TASK_RUNNING" and the CPU scheduler treats the process as a target of CPU resource assignment.

4.2. Implementation of Kernel Live Patch

Load of patch module The live-patch process uses the init_kpannus() system call and load_patch() kernel function, which are implemented to load the patch module to the unused memory area of the kernel. The init_kpannus() system call invokes the load_patch() kernel function. The load_patch() kernel function is implemented based on the existing load_module() kernel function that loads the kernel module in the unused memory area of the kernel. The load_patch() kernel function adds administrative information such as the patch module status. The load_module() kernel function enables loading the kernel module without disrupting kernel execution, and this is applicable to a network service.

Execution transition to the patch module The live-patch process uses patch_kpannus() system calls that are implemented to safely write the jump instruction to the address of the original routine. The patch_kpannus() uses the stop_machine_run() kernel function. The stop_machine_run() is designed for a kernel-wide lock and executes a given routine on a single CPU, letting other CPUs execute a nop instruction until the end of a given routine. The nop instruction task is assigned as the highest priority to other CPUs, and CPUs execute the nop instruction task at the next CPU cycle. Normally, nop instruction tasks are executed one millisecond later on other CPUs. The patch_kpannus() gives the routine that checks for address conflicts and writes the jump instruction to the stop_machine_run().

5. EXPERIMENTAL RESULTS

The measurement of the stopping time of a target process on a live-patch implementation uses the COTS system shown in table 1.

Table 1. Evaluation environment

CPU	Pentium Xeon 2.8 GHz x2 (4 SMT with HT)
Memory	2GB
HDD	160 GB SATA-1
Kernel	2.6.9 based
Data size of function	200 bytes
Data dependability	average of 100 experiments

5.1. Results of Process Live Patch Experiment

The results graph of the experiment on target process stopping time that occurs during a process live patch are shown in figure 3. The X-axis indicates how many functions can be modified in one live-patch execution, and the Y-axis indicates the stopping time of the target process. The proposed live-patch implementation with 1 to 200 threaded target processes and live patch using a ptrace() system call with a single threaded target process are indicated in the graphs.

The result demonstrates the proposed implementation is faster than live patch using a ptrace() system call. The proposed implementation enables the modification of 5,400 functions in a single-threaded target process and the modification of 4,250 functions in a 200-threaded target process within a 100 milliseconds timeout.

However, the live patch using a ptrace() system call suddenly increases the execution time at the time when 1,000 functions have been modified, and the timeout interval is exceeded when 1,250 functions have been modified.

Detailed graphs of the stopping time of the proposed live-patch implementation are shown in figure 4, 5, and 6. The execution time required for stopping the target process and preparing for modification is shown in figure 4. The execution time required for writing a jump instruction is shown in figure 5. The execution time required for restarting the target process is shown in figure 6. The result shows the time of writing jump instructions to the target process is up to 80% of the total execution time.

5.2. Result of Kernel Live-Patch Experiment

The results graph of the experiment on kernel stopping time during kernel live patch is shown in figure 7. The number of functions that are modified in one live-patch execution is indicated by the X-axis. The stopping time of the kernel is indicated by the Y-axis on a log scale.

The kernel live-patch implementation and kexec mechanism[18] lines are shown. The kexec mechanism is a fast kernel-load mechanism that allows bypassing hardware initialization to load the modified kernel from the currently executing kernel. The measurement of kexec is performed during the initialization of the modified kernel until the system is ready.

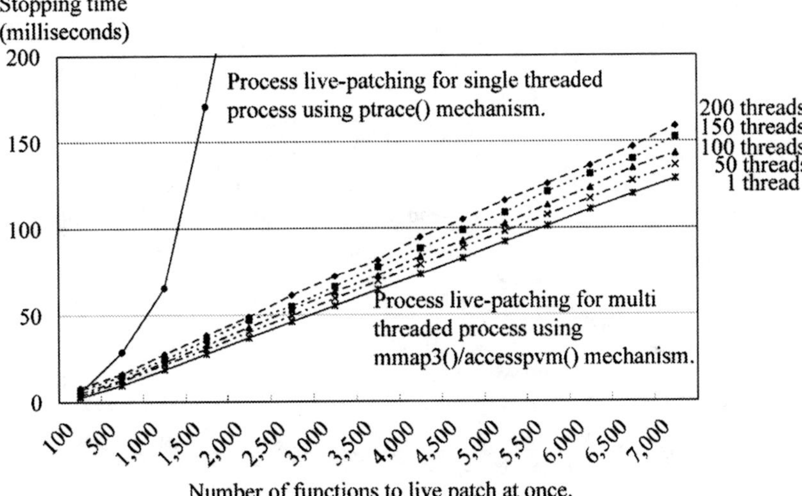

Figure 3. Target process stopping time during process live patching

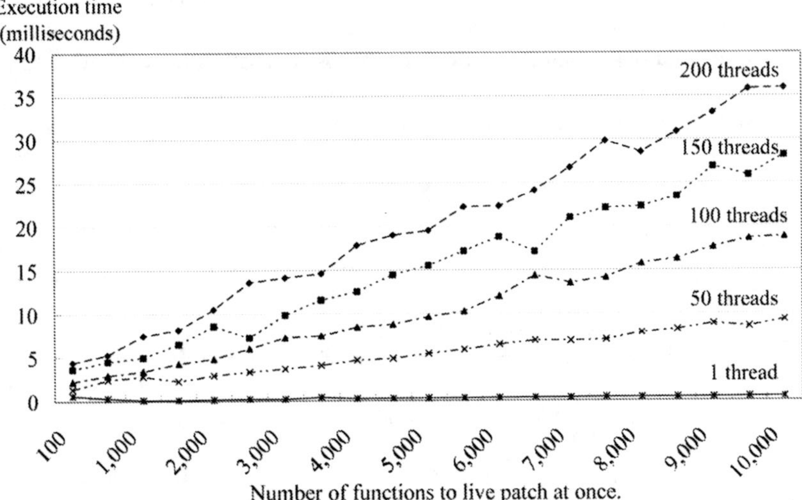

Figure 4. Execution time of stopping task execution status on CPU scheduler

6. DISCUSSION OF EXPERIMENT

Discussion of Process Live Patch A live patch using the ptrace() system call cannot be applicable to a network service, as shown in figure 3. The network service software is large, using hundreds of threads with huge functions and data.

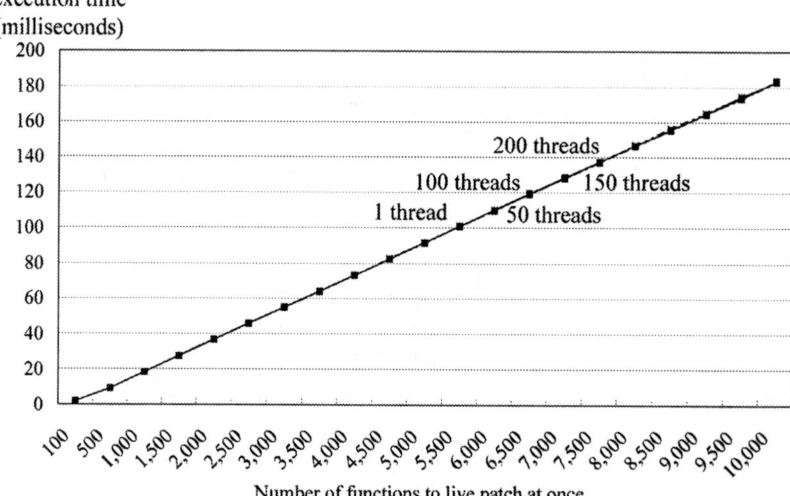

Figure 5. Execution time of overwriting jump instruction on process live patch

Modifying many functions and data of network service software at one time within 100 milliseconds timeout is very important.

As shown in figure 4, when the number of threads is greater, the execution time to stop a target process is longer. This occurs because the execution status of each thread must be changed. The result also indicates that when a large number of

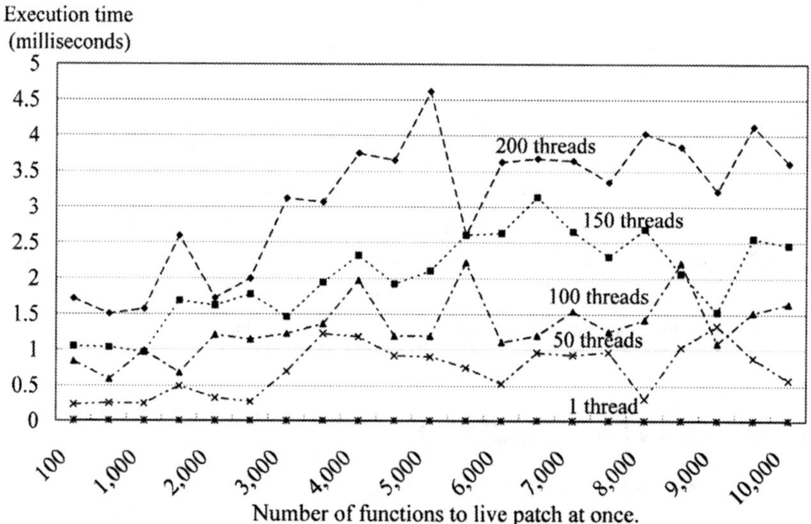

Figure 6. Execution time of restarting target process

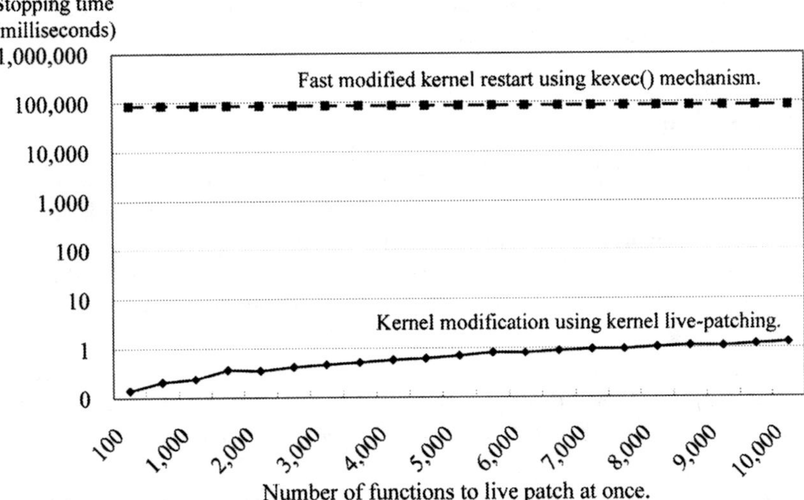

Figure 7. Kernel stopping time during kernel live patching and kexec mechanism

functions must be modified the execution time becomes longer and the probability of address conflicts increases, which causes the number of procedure retries to increase. As shown in figure 5, there is no difference in execution time due to the number of threads. However, the number of functions to modify makes the execution time longer because the increase in the number of functions causes an increase in the write times. The result also indicates that the execution of jump instruction writing takes up to 80% of the stopping time of a target process. This suggests that the improvement in jump instruction writing is important for further studies.

As shown in figure 6, when the number of threads is greater, the execution time required to restart the target process is longer. The reason is that the execution changes each thread execution status. The result demonstrates that there is no correlation between the numbers of functions and the execution time. The difference in execution times is a fudge factor. The result also indicates that the restart execution is much faster than the stop execution. This is because the restart execution does not need to check for address conflicts.

6.1. Discussion of Kernel Live Patch

As shown in figure 7, the kernel live patch is much faster than the process live patch. This is caused by the difference in target software addresses. The execution that writes the jump instruction is the same. However, the process live patch writes to the address of another process, and the kernel live patch writes to the address of the kernel. Whenever writing the jump instruction to the address of another process, an update of the information about virtual memory usage and actual memory usage is required. However, in the case of the kernel address, only the actual memory usage information update is required. That is the reason for the difference in execution times.

7. CONCLUSION

Online software modification is an essential technique for the network service systems because the service requires high availability, uninterrupted service provision, and frequent service software updates. Therefore, the existing system uses a specialized system to satisfy the requirements. However, the NGN and future network service systems also require a cost effective system by using a COTS system.

To help alleviate this problem, we presented a live-patch method that is applicable on Linux and x86 CPU SMP machines to achieve online software modification on a COTS system. The live-patch method enables online modification of user software and kernel software.

The evaluation of the presented method demonstrates that the proposed methods are much faster than other approaches. That is, the process live patch enables modification of 4,500 functions of a 200-threaded target process within 100 milliseconds. This is almost 3.6 times faster than the approach that uses ptrace() systemcall. The kernel live patch enables the modification of 10,000 functions of a kernel within 2 milliseconds. This is much faster than the approach that uses kexec mechanism. The evaluation also demonstrates that writing branch instructions to the virtual address of the target process is much slower than doing that to the address of the kernel, and the time of writing branch instructions to the target process is up to 80% of the execution time. That is, the improvement in writing branch instructions to the process makes a vast improvement in the execution time. The authors will focus on reducing the time needed to write branch instructions during the process live patch.

REFERENCES

1. J. Peterson, P. Hudak, and G. S. Ling. Principled dynamic code improvement. Technical Report, YALEU/DCS/RR-1135, Department of Computer Science, Yale University, July 1997.
2. M. Hicks and S. M. Nettles, Dynamic Software Updating, ACM Transactions on Programming Languages and Systems (TOPLAS), 27(6), November 2005.
3. D. Gupta , P. Jalote and G. Barua, A Formal Framework for On-line Software Version Change, IEEE Transactions on Software Engineering, v.22 n.2, pp.120–131, February 1996.
4. L. A. Tewksbury, L. E. Moser and P. M. Melliar-Smith, "Live Upgrade Techniques for CORBA Applications," Proceedings of the Third IFIP WG 6.1 International Working Conference on Distributed Applications and Interoperable Systems, pp.257–271, Krakow, Poland, 2001.
5. L. A. Tewksbury, L. E. Moser and P. M. Melliar-Smith, "Live Upgrades of CORBA Applications Using Object Replication," Proceedings of the IEEE International Conference on Software Maintenance, pp.488–497 , Florence, Italy, 2001.
6. L. E. Moser, P. M. Melliar-Smith and L. A. Tewksbury, "Online Upgrades Become Standard," Proceedings of the IEEE 26th International Computer Software and Applications Conference, pp.982–988 Oxford, England, 2002.
7. Object Management Group, "Online Upgrades Specification," 2002.

8. T. Ritzau and J. Andersson. Dynamic deployment of Java applications. In Java for Embedded Systems Workshop, 2000.
9. M. Dmitriev, HotSwap technology application for advanced profiling, Workshop on unanticipated Software Evolution (USE 2002), pp.14–18, June 2002.
10. T. Yamada, H. Sunaga, S. Tanaka, S. Shiraishi, and K. Koyanagi, Evaluation of Partial File Modification for Java-Based Realtime Communication Systems, IEICE TRANS. COMMUN., VOL. E88-B, NO.10, 2005.
11. O. Frieder and M. Segal. On dynamically updating a computer program: From concept to prototype, Journal of Systems and Software, vol.14, pp.111–128, 1991.
12. K. Koyanagi, H. Sunaga, T. Yamada, and H. Ikeda, Applicability Evaluation of Service Feature Enhancement Using Plug-in Modification Technique, IEICE TRANS. COMMUN., VOL. E81-B, NO. 1, 1998.
13. T. Ikebe, Y. Kawarasaki, and J. Yamanaka, Live patching method on Linux operating system for high-available network service, IEICE Technical Report, NS2005-50(2005-6), 2005.
14. NTT, "Pannus project," http://pannus.sourceforge.net
15. John R. Levine, Linkers & Loaders. Morgan Kaufmann, 1999.
16. D. P. Bovet and M. Cesati, Understanding the Linux Kernel, Third Edition, O'Reilly & Associates, Sebastopol, 2006.
17. F.Ukai "livepatch," http://ukai.jp/Software/livepatch/
18. E.Biederman, "kexec," http://www.xmission.com/ ebiederm/files/kexec

CHAPTER 21

MRC DIVERSITY RECEPTION OF OFDM WITH M-ARY MODULATION OVER FREQUENCY-SELECTIVE NAKAGAMI-*M* FADING CHANNELS

MOHAMED E. KHEDR, AHMED F. ALNAHHAL, AND ROSHDY A. ABDELRASSOUL

Arab Academy for Science & Technology and Maritime Transport, Department of Electronics and Communications, Alexandria, Egypt

Email: {Khedr, roshdy}@aast.edu, ahmed_2140@yahoo.com

Abstract: In this paper, we take a close look at the performance of MRC diversity reception of OFDM signals in multipath slowly fading Nakagami-*m* channels. We derive an exact expression for the probability density function (PDF) of a sum of Nakagami-*m* random phase vectors, which is used to derive a closed-form expression for the error-rate of OFDM signals using multichannel reception with maximal ratio combining. Consequently, the average-symbol error rate for M-ary modulation can be expressed in terms of confluent hypergeometric function, which can be easily evaluated numerically. It is observed that depending on the number of channel taps, the error-rate performance does not necessarily improve with increasing Nakagami-*m* fading parameters

Keywords: Nakagami-*m* fading channels, sum of Nakagami-*m* random phase vectors, orthogonal frequency-division multiplexing (OFDM), maximal ratio combining (MRC)

1. INTRODUCTION

Orthogonal frequency-division multiplexing (OFDM) has recently received increased attention due to its capability of supporting high-data-rate communication in frequency-selective fading environment that causes intersymbol interference (ISI). The combination of OFDM and diversity techniques has become popular in wireless communication [1]. Diversity techniques can dramatically improve the system performance over fading channel, especially in frequency-flat fading environment. In noise-limited system, it is well known MRC provides the best system performance [1]. Although, in the presence of interference, MRC is no

H. Labiod and M. Badra (eds.), New Technologies, Mobility and Security, 245–257.

© 2007 *Springer.*

longer the optimum combining scheme, its performance is comparable to that of optimum combiner, MRC is employed in many practical wireless communication system [1]. Instead of using a complicated equalizer as in the conventional single carrier system, the ISI in OFDM can be eliminated by adding a guard interval which greatly simplifies the receiver structure.

Rayleigh and Rician fading model has been widely used to represent fading environments. For example, Lu *et al.* studied the performance of OFDM M-ary differential phase keying (OFDM-MDPSK) scheme in Rician fading channel with diversity reception [2]. Propagation condition in some wireless system may not be well described by this model, e.g., microcellular system, where the fading is not as severe as Rayleigh fading [3].

The Nakagami-m [4] distribution has been employed as another useful and important model for characterizing the amplitude of the fading channel. Both theoretical and practical importance of the Nakagami-m channel has motivated intensive research into studying the performance of various communication systems operating in such channel. Previous works that studied transmission of an OFDM signal over multipath Nakagami-m channel assumed that the frequency-domain channel response samples are also Nakagami-m distributed with the same fading parameter as the time-domain channel. However, there are no experimental results presented in the literature that support this assumption. Consider an N-point fast Fourier transform (FFT) used to determine the sample frequency-domain response from the sampled time-domain response. In the case of Rayleigh fading, the faded signal samples have a joint complex Gaussian distribution. Then, the application of the FFT represents a linear transformation of jointly Gaussian random variables (RVs) and yields jointly Gaussian RVs [5]. Thus, one expects a frequency response sample to have a Rayleigh fading distribution when the time-domain signal is Rayleigh faded. However, sum of Nakagami-m random phase vectors do not, in general, have Nakagami-m distributed envelope. Therefore, it is not expected that the frequency-domain sample can be assumed to be a Nakagami-m distribution when the time-domain signals is Nakagami-m faded, except for $m = 1$, which is the special case of Rayleigh fading. Recently, Kang *et al.* claimed that the distribution of sample of the frequency-domain channel impulse response can be approximated by another Nakagami-distribution with a new fading parameter different from the time-domain fading parameter [6]. In this paper, we will use an exact mathematical analysis to show that such Nakagami-m approximations can be unreliable. The contributions in this paper are as follow: we first revisit the problem of determining the PDF of the sum of random phase vectors and recall an integral solution to the PDF of sum of Nakagami-m random phase vector. We then use this PDF expression to evaluate the exact error-rate performance of MRC diversity reception of an OFDM system in multipath Nakagami-m fading channels. We observe, through numerical example that the error-rate performance over a multipath Nakagami channel does not necessarily improve with increasing Nakagami fading parameters.

2. SYSTEM MODEL

We consider an OFDM system with N subcarriers. For each OFDM symbol, we denote the modulated data sequence as $D(0)$, $D(1)$,. . . .., D(N-1). After the inverse discrete Fourier transform (IDFT), the time-domain OFDM signal can be expressed as

$$(1) \qquad s(n) = \frac{1}{N} \sum_{k=0}^{N-1} D(k) \quad e^{j2\pi kn/N}, \quad n = 0, 1, \ldots, N-1$$

Where$j^2 = -1$. Suppose that the channel impulse response of a multipath fading channel is modeled as a finite impulse response (FIR) filter with taps $h(n)$, $n = 0, 1, \ldots, N$-1. We assume that the maximum delay of the multipath fading channel is L with $L << N$, i.e., $h(n) = 0$ for $n = L, L+1, \ldots, N$-1,and we express the frequency-domain channel impulse response as

$$(2) \qquad H(k) = \sum_{n=0}^{N-1} h(n)\, e^{-j2\pi nk/N}, \; k = 0, 1, \ldots.., \; N-1.$$

We further assume that the maximum delay is less than the length of the cyclic prefix, and perfect timing and frequency synchronization are achieved at the receiver. Therefore, ISI is not considered in our analysis.

The received signal $r(n)$, $n = 0, 1, \ldots N$-1, after analog-to-digital (A/D) sampling and removing the cyclic prefix, is input to discrete Fourier transform (DFT) processor, and the output signal becomes

$$(3) \qquad \begin{aligned} R(k) &= \sum_{n=0}^{N-1} r(n)\, e^{-j2\pi nk/N} \\ &= H(k)D(k) + N(k), \; k = 0, 1, \ldots.., N-1 \end{aligned}$$

Where $N(k)$ are independent identically distributed (i.i.d) complex Gaussian noise components with zero mean and unit variance. Equation (3) shows that each OFDM subcarrier undergoes a flat fading channel described by $H(k)$, but the distribution of $H(k)$ is only Nakagami-m if $L = 1$. Since there is no benefit from using OFDM on a flat fading channel, we consider frequency selective fading channels where the total bandwidth occupied by the OFDM signal exhibits frequency-selective fading and $L > 1$ in the model.

In this paper, we assume that the channel tap coefficients in the multipath fading channel model $h(n) = 0$, $n = 0, 1, \ldots.., L-1$, are mutually independent complex RVs, which can be written as $h(n) = |h(n)|e^{j\phi(n)}$. the amplitude $|h(n)|$ is modeled as a Nakagami-m RV with PDF

$$(4) \qquad f_{|h(n)|}(v) = \frac{2}{\Gamma(m_n)} \left(\frac{m_n}{\Omega_n}\right)^{m_n} v^{2m_n-1} e^{-(m_n/\Omega_n)v^2}, \; m_n > \tfrac{1}{2}, v > 0$$

Where $\Gamma(.)$ is the Gamma function, $\Omega = E\left[|h(n)|^2\right]$ is the expectation of $|h(n)|^2$ or the power of the nth path, and m_n is the Nakagami-m fading parameter for the nth tap. The Nakagami-m fading parameter m_n determines the severity of the channels. Tow special values of m_n are of particular interest. In the case of $m_n = 1$, the Nakagami-m fading specializes to Rayleigh fading. In the limiting case when $m_n = \infty$, the Nakagami-m fading channel approaches a static channel, and the PDF of $h(n)$ become $f_{|h(n)}(v) = \delta(v - \sqrt{\Omega_n})$, where $\delta(v)$ is the Dirac delta function. Finally, the fading phase $\phi(n)'s$ are assumed to be mutually independent, uniformly distributed over $[0.\ 2\pi)$ and independent of the fading amplitudes $|h(n)|$'s. The frequency-domain representation of the channel impulse response can also be expressed as

$$
\begin{aligned}
H(k) &= \sum_{n=0}^{L-1} |h(n)| e^{j\phi(n)} e^{-j2\pi nk/N} \\
&= \sum_{n=1}^{L-1} |h(n)| e^{j\theta(n)} \triangleq \sum_{n=0}^{L-1} (X_n + jY_n)
\end{aligned}
\tag{5}
$$

where it can be shown that the RVs $\theta(n) = \phi(n) - (2\pi k/N)(\bmod 2\pi)$ are also uniformly distribution over $[0, 2\pi)$ for different n values.

We assume that the receiver has perfect knowledge of the fading channel, that is, the phase of $H(k)$ can be perfectly estimated and compensated at the receiver. Therefore, to analyze the performance of our OFDM signal in multipath Nakagami-m channels, it is sufficient to consider only the statistical characteristics of $|H(k)| = |\sum_{n=0}^{L-1} (X_n + jY_n)|$, which is the modulus of a sum of L complex random vectors $X_n + jY_n$, $n = 0, 1, \ldots., L-1$.

3. PDF OF SUM OF NAKAGAMI RVS, & ITS APPROXIMATION

In this section, we present an integral expression for the PDF of the amplitude of a sum of Nakagami-m random phase vectors using the characteristic function (CF) approach. The joint CF of the RVsX_n *and* Y_n, which were defined in (5), is

$$
\Phi_{X_n,Y_n} = E_{X_n,Y_n}\left[e^{j\omega_1 X_n} + e^{j\omega_2 Y_n}\right] \tag{6}
$$

From [7] and [3], it can be shown that

$$
\Phi_{X_n,Y_n}(\omega_1, \omega_2) = {}_1F_1\left(m_n; 1; -\frac{\Omega_n}{4\Omega_n} R^2\right) \triangleq \Phi_{X_n,Y_n}(R) \tag{7}
$$

Where $R = \sqrt{\omega_1^2 + \omega_1^2}$ and ${}_1F_1(-;-;-)$ is the confluent hypergeometric function [8]. Now, we can write $H(k) = X + jY$, where $X = \sum_{n=0}^{L-1} X_n$ and

$Y = \sum_{n=0}^{L-1} Y_n$. Because the channel tap coefficients are independent, the joint CF of X and Y becomes

$$\Phi_{x,y}(\omega_1, \omega_2) = \prod_{n=0}^{L-1} \Phi_{X_n,.Y_n}(\omega_1, \omega_2) \tag{8}$$
$$= \prod_{n=0}^{L-1} {}_1F_1\left(m_n; 1; -\frac{\Omega_n}{4m_n}R^2\right) \triangleq \Phi_{X_n,Y_n}(R)$$

Using the inversion theorem [5], one can show that the PDF of the amplitude $|H(k)|$ is given by

$$f_{|H(k)|} = r \int_0^{\infty} \Phi_{X,Y}(R)\, J_0(Rr)\, R\, dR \tag{9}$$

Where Jo(-) is the zero-order Bessel function of first kind. Substitution of (8) into (9) yields

$$f_{|H(k)|} = r \int_0^{\infty} \prod_{n=0}^{L-1} {}_1F_1\left(m_n; 1; -\frac{\Omega_n}{4m_n}R^2\right) J_0(Rr)\, R\, dR \tag{10}$$

In [9] Abdi *et al.* derive a more general result that is valid for sum of an arbitrary number of random vectors with arbitrary statistics. Nakagami [4] stated that the PDF of $|H(k)|$ can be approximated by another Nakagami-m distribution with fading parameter m^* and Ω^*, where $\Omega^* = \sum_{n=0}^{L-1} \Omega_n$, and

$$m^* = \frac{\left(\sum_{n=0}^{L-1} \Omega_n\right)^2}{\sum_{n=0}^{L-1} \frac{\Omega_n^2}{m_m} + \sum_{n=0}^{L-1}\sum_{m=0}^{L-1} \Omega_n \Omega_m} \tag{11}$$

In [6], Kang *et al.* verified the goodness of the approximation in (11) by comparing cumulative distribution functions (CDFs for the Nakagami-m approximation with simulated data. The authors went on and used the Nakagami-m approximation to study the error-rate performance of OFDM signals in multi path Nakagami-m channels. The Nakagami-m approximation to the PDF of $|H(k)|$ can be inaccurate.

In fig. 1, by comparing the PDF of the Nakagami-m approximation and the exact analytical PDF derived in (10), we note that the Nakagami-m approximation can be poor, which contradicts the previous study in [6]. Furthermore, we will show that OFDM error-rate estimation based on the Nakagami-m approximation can be unreliable.

From (11) we can have the same Nakagami-m fading parameter m^* from many different Nakagami-m fading parameters $m_o, \ldots, m_{L-1}$. For example with $L = 2$

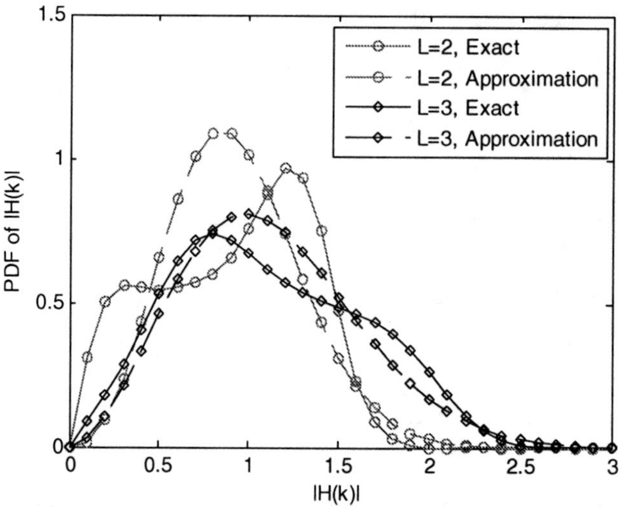

Figure 1. Exact and app PDF for $|H(k)|$ with $m_o = \ldots = m_{L-1} = 8$, $\Omega_o = \ldots = \Omega_{L-1} = 1$ and the number of channel taps $L = 2$ and 3

when $m_o = m^1 = 8$, $\Omega_o = \Omega_1 = 1$ we has the same PDF as $m_o = 5$, $m_1 = 20$, and $\Omega_o = \Omega_1 = 1$ for the exact PDF and with Nakagami-m fading parameter $m^* = 16/9 \sim 1.7778$ for the approximate PDF. Fig. 1. Show the exact and the approximation PDF when $m_o = 5$, $m_1 = 20$ and $\Omega_o = \Omega_1 = 1$ which is the same when $m_o = m_1 = 8$ and $\Omega_o = \Omega_1 = 1$. In (10) when L becomes large the shape of PDF will go to Rayleigh distribution (central limit theorem) as we will show it in section 4.

4. ERROR-RATE PERFORMANCE USING DIVERSITY RECEPTION

In this section, we study the error-rate performance of OFDM signals in multipath Nakagami-m fading channels using M independent diversity branch with MRC. The approach we use here is the MGF technique described in [10]. The conditional symbol error-rate (SER) for many coherent binary and M-ary modulation scheme can be expressed as $BQ(Ar)$ where $Q(x) = 1/\sqrt{2\pi}\int_0^\infty e^{-t^2/2}dt$, r is the fading amplitude which is $|H(k)|$ in our system model. The particular values of A and B depend on the considered modulation scheme as summarized in Table 1. Where γ is the SNR per symbol. The conditional error-rate for MRC is modified to

$$B\,Q\left(A\sqrt{\sum_{i=0}^{M-1} r_i^2}\right) \tag{12}$$

Table 1. Parameters A and B for several signaling constelations

Modulation Scheme	A	B	M
PBSK	$\sqrt{2\gamma}$	1	–
PFSK	$\sqrt{\gamma}$	1	–
QPSK, 4QAM	$\sqrt{\gamma}$	2	–
M-PSK	$\sqrt{2\gamma}\sin(\pi/M)$	2	>4
M-FSK	1	$(M-1)$	≥ 2
M-QAM	$\sqrt{3\gamma/(M-1)}$	$4(\sqrt{M}-1)/\sqrt{M}$	≥ 4
M-DPSK	$\sqrt{4\gamma}\sin(\pi/2M)$	2	≥ 2

Using the alternative representation of the Q-function [10]

$$Q(Ar) = \frac{1}{\pi}\int_0^{\pi/2} e^{-A^2r^2/2\sin\phi}d\phi \tag{13}$$

So, (12) is given by

$$B\,Q\left(A\sqrt{\sum_{i=0}^{M-1} r_i^2}\right) = \frac{B}{\pi}\int_0^{\pi/2} e^{-A^2\sum\limits_{i=0}^{M-1} r^2/2\sin^2\phi}d\phi \tag{14}$$

where $r_i = |H_i(k)|$ is the fading amplitude in the ith branch. Averaging (14) with respect to the joint PDF of the fading amplitudes, we obtain the average error-rate, denoted by $P_{e,MRC}$, as

$$\begin{aligned} P_{e,MRC} &= B\int_0^{\infty}\ldots\int_0^{\infty} Q\left(A\sqrt{\sum_{i=0}^{M-1} r_i^2}\right) f_{|H(k)|}(r_1)\ldots f_{|H(k)|}(r_i)\ldots dr_1\ldots dr_M \\ &= \frac{B}{\pi}\int_0^{\pi/2} e^{-A^2\sum r_i^2/2\sin^2\phi}d\phi\left[\prod_{i=0}^{M-1}\int_0^{\infty} r_i\int_0^{\infty}\Phi_{X,Y}(R)J_0(Rr_i)R\,dr_i\right] \end{aligned} \tag{15}$$

$$\begin{aligned} P_{e,MRC} &= \frac{B}{\pi}\int_0^{\pi/2}\left[\prod_{i=o}^{M-1}\int_0^{\infty}\frac{\sin^2\phi}{2A^2}e^{-R\sin^2\phi/2A^2}\,\Phi_{X,Y}^{(i)}(R)dR\right]d\phi \\ &= \frac{B}{\pi}\int_0^{\pi/2}\prod_{i=0}^{M-1}\mu_i(\phi)d\phi \end{aligned} \tag{16}$$

where

$$\mu_i(\phi) = \int_0^{\infty}\frac{\sin^2\phi}{2A^2}e^{-R\sin^2\phi/2A^2}\Phi_{X,Y}^{(i)}(R)dR \tag{17}$$

which is the MGF of r_i^2. Using the integral identity: [11]

$$p^{-v}\Gamma(v)F_A^n\left(v; m_1,\ldots,m_n; c_1,\ldots,c_n; \frac{\alpha_1}{p},\ldots,\frac{\alpha_n}{p}\right) = \int_0^{\infty} e^{-pt}t^{v-1}{}_1F_1(m_1;c_1;\alpha_1 t)\ldots{}_1F_1(m_n;c_n;\alpha_n t)\,dt \tag{18}$$

Where $F_A^{(L)}(\alpha;\beta_1,\ldots,\beta_n;\gamma_1,\ldots,\gamma_n;z_1,\ldots,z_n)$ is the Lauricella function of n variables [8], we now can simplify (17) to

$$\mu_i(\phi) = F_A^{(L)}\left(1; m_o,\ldots m_{L-1}; 1,\ldots,1; -\frac{\Omega_o A^2}{2m_o\sin^2\phi},\ldots,-\frac{\Omega_{L-1}A^2}{2m_{L-1}\sin^2\phi}\right) \tag{19}$$

For M-ary modulation we can express the error-rate as

$$P_{e,MRC} = \frac{B}{\pi}\int_0^{\pi/2}\prod_{i=0}^{M-1}F_A^{(L)}\left(1; m_o,\ldots m_{L-1}; 1,\ldots,1; -\frac{\Omega_o A^2}{2m_o\sin^2\phi},\ldots,-\frac{\Omega_{L-1}A^2}{2m_{L-1}\sin^2\phi}\right) \tag{20}$$

This can be evaluated numerically as illustrated in the appendix.

5. NUMERICAL RESULTS

Fig. 1 shows the precise PDF obtained from (10) and its approximation obtained from [6], which indicate that the Nakagami-m approximation can be a poor approximation as in the case of ($L = 2, 3$). We also observes for $L = 2$ the shape of the PDF have inverse properties of that when $L = 3$ where place of the two local maxima is changed, this have effect on the BER. Fig. 2 plots the exact and approximate PDF for L = 5 and 15 which indicate that when L is increased the shape goes to a Rayleigh PDF.

Fig. 3 gives the exact and approximate BERs of BPSK-OFDM for single channel reception, which a special case from our closed form in (20) "$M = 1$". We note that the approximate BERs can be a poor approximation, also the exact BERs performance does not necessarily improve with increasing Nakagami-m fading parameter. Fig. 4 indicates that the same behavior holds for a system using dual-branch reception. We observe in Fig. 4 that with two-branch MRC, the slopes of the error-rate curves become large and significant gains are obtained.

Fig. 5 gives the error-rate performance of BPSK-modulated OFDM signal with "L = 2 &3". Contrary to the two-tap case, we observe that the BERs in fact decrease with increase m when the number of taps is three.

Fig. 6 shows the error-rate performance of a BPSK-modulated OFDM signal transmitted over a multi-path Nakagami-m channel with different number of MRC channel reception, where the number of taps "L" per channel equal to 2.

Finally Figs. 7, 8 measure SERs performance of M-ary PSK and QAM modulation of ODFM signals respectively, where $L = 2$ and $m = 6$. This result was built on the conditional probability illustrated in Table 1 [10].

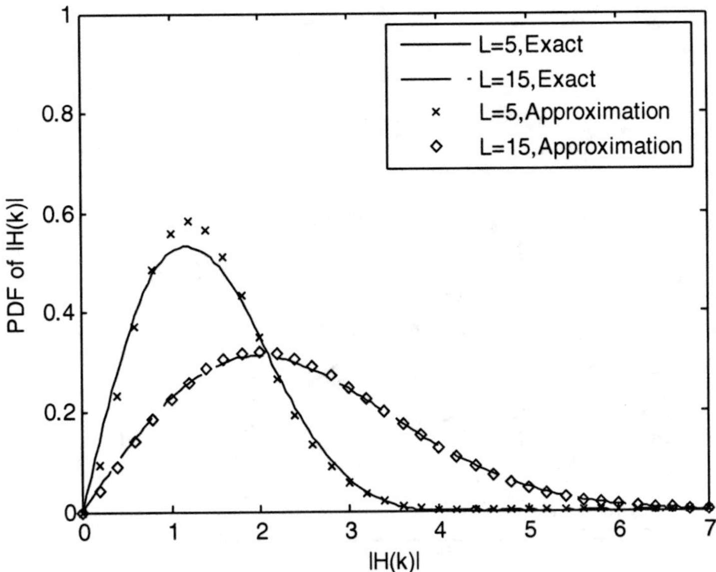

Figure 2. Exact and approximate PDF of $|H(k)|$ with $m_o = \ldots m_4 = 6$ and $m_o = \ldots = m_{15} = 6$ and $\Omega o = \ldots = \Omega L\text{-}1 = 1$

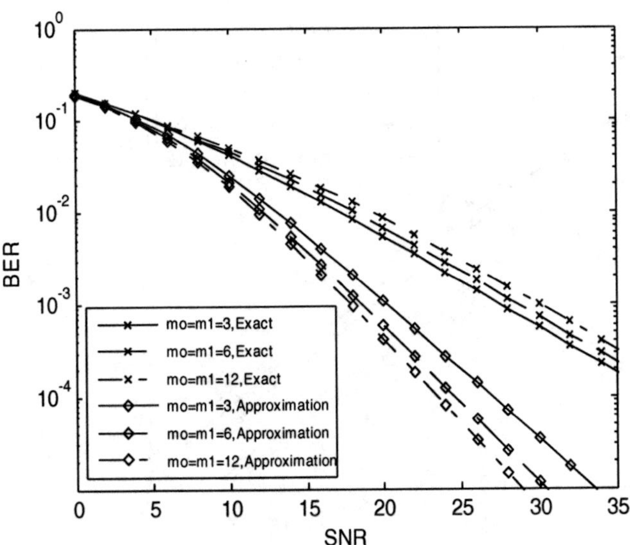

Figure 3. Exact and approximate BER of OFDM-BPSK for $m_o = m_1 = 3, 6$ and 12 and $\Omega_o = \Omega_1 = 1$

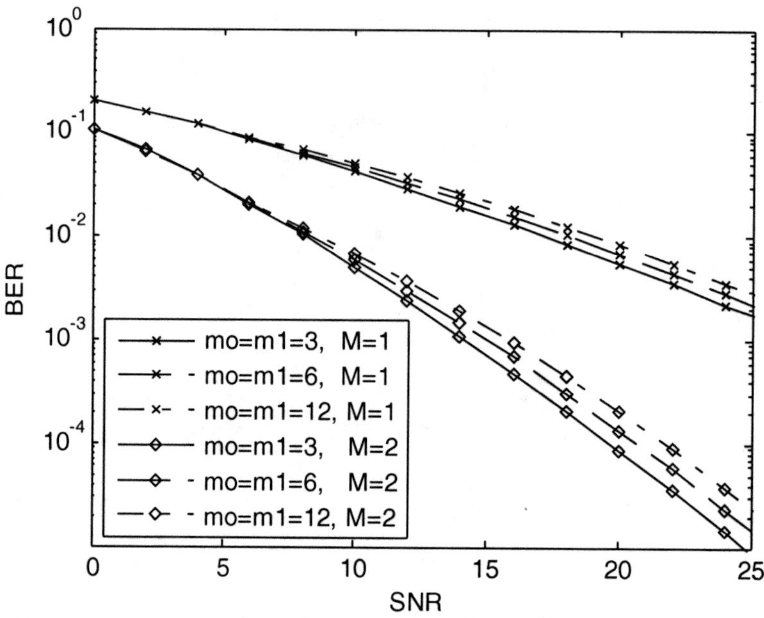

Figure 4. BERs of OFDM-BSK for $m_o = m_1 = 3, 6, 12$ for single-channel and dual-branch MRC reception

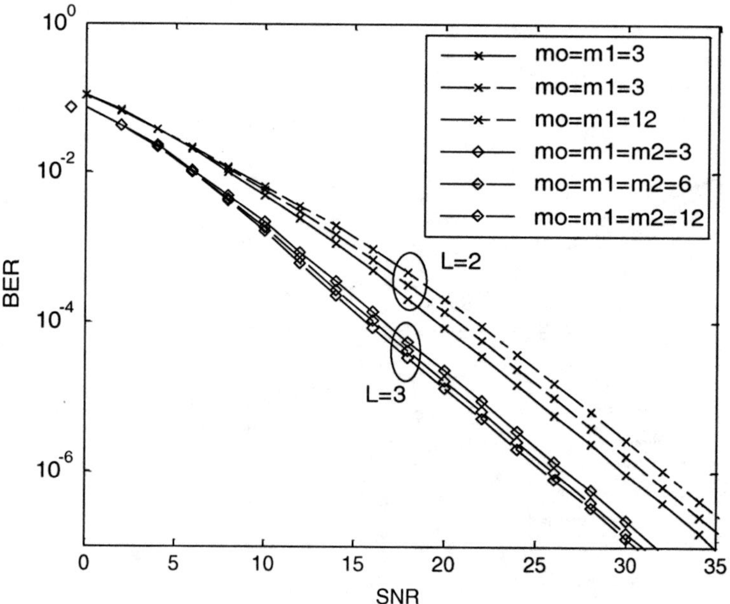

Figure 5. BERs of OFDM-BPSK for L = 2 and 3 taps when dual-branch MRC reception

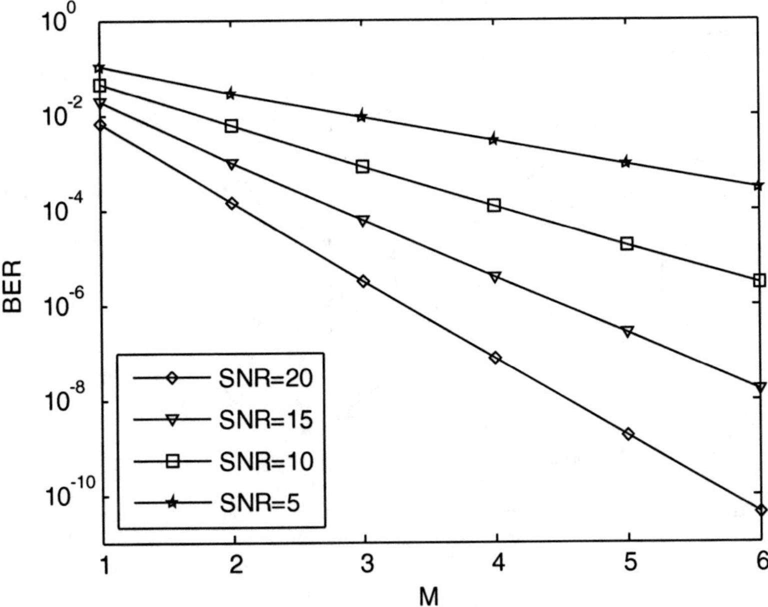

Figure 6. BER at different SNR versus the number of MRC channel reception where the number of taps per channel $L = 2$ with $m_o = m_1 = 6$ and $\Omega_o = \Omega_1 = 1$

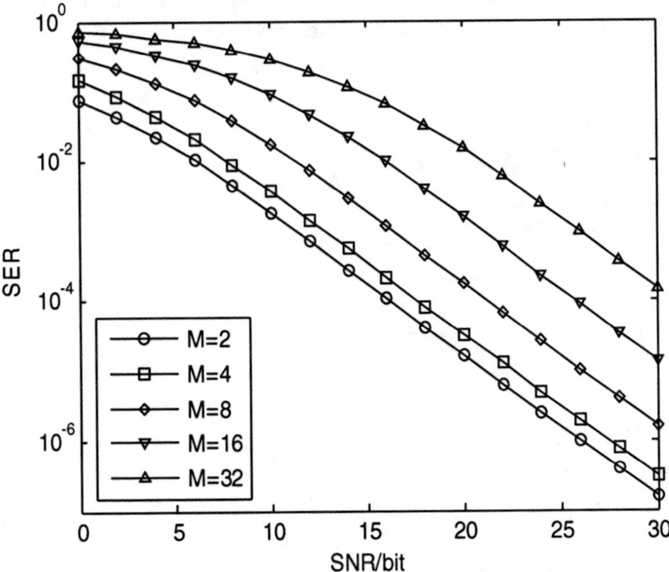

Figure 7. BER at different SNR versus the number of MRC channel reception where the number of taps per channel $L = 3$ with $m_o = m_1 = m_2 = 6$ and $\Omega_o = \Omega_1 = 1$

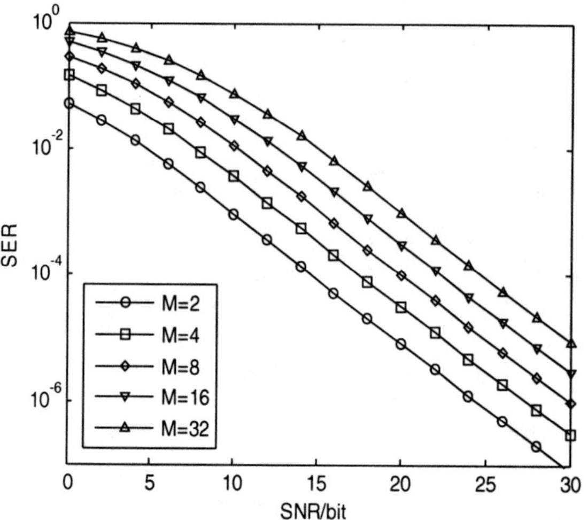

Figure 8. The BERs of MQAM OFDM for $m_o = m_1 = m_2 = 6$ for dual-branch MCR reception

6. CONCLUSION

In this paper, the performance of an OFDM system was analyzed over mulipath Nakagami-*m* channels. We obtained the exact formula for the PDF of sum of Nakagami-*m* random phase vectors, which has been used to derive an exact error-rate expression for coherently M-ary modulated OFDM signals transmitted over multipath Nakagami-*m* fading channels. Our future research will consider the

APPENDEX

NUMERICAL EVALUATION TO (20)

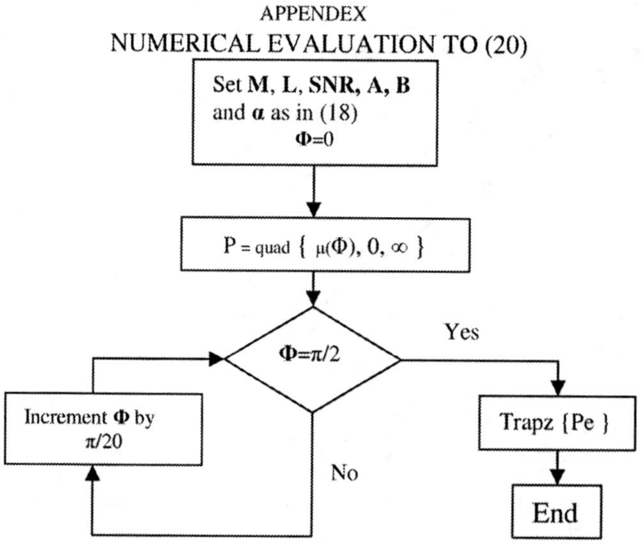

performance degradation of OFDM system over frequency-selective time-variant Nakagami-*m* channels.

REFERENCES

1. Aalo *et al.*, "Another look at the performance of MCR of schemes in Nakagami-*m* fading channels with arbitrary parameters," *IEEE Trnas. Comm.* vol. 53, no. 12, pp. 2002–205, Dec. 2005.
2. Lu *et al* "BER performance of OFDM-MDPSK system in frequency-selective Rician fading with diversity reception," *IEEE Trans. Veh. Technol.*, vol. 49, no. 7, pp. 1216–1225, Jul. 2000.
3. Beaulieu *et al.* "Precise error-rate analysis of bandwidth-efficient BPSK in Nakagami-*m* fading and cochannel interference," *IEEE Trans. Commun.*, vol. 52, no. 1, pp. 149–158, Jan. 2004.
4. M. Nakagami, "The *m*-distribution, a general formula of intensity distribution of rapid fading," in *Statistical Methods in Radio Wave Propagation*, W. G. Hoffman, Ed. Oxford, U.K.: Pergamon, 1960.
5. A. Papoulis, *Probability, Random Variables, and Sto-chastic Processes*, 3rd. New York: McGraw-Hill, 1991.
6. Kang *et al.* "Nakagami-*m* fading modeling in the frequency domain for OFDM system analysis," *IEEE Commun. Lett.*, vol. 7, no. 10, pp. 484–486, Oct. 2003.
7. ALNahhal *et al.* "Performance analysis of OFDM with M-ary modulation in sum of frequency-selective Arbitrary Nakagami-*m* fading channels," 24th national radio science conference, Egypt, March 2007.
8. I. S. Gradshteyn and I. M. Ryzhil, *Table of Integrals, Series, and products,* 5th *ed.* New York: Academy. 1994.
9. Abdi *et al.* "On the PDF of the sum of random vectors," *IEEE Trans. Commun.*, vol. 48, no. 1, pp. 7–12, Jan. 2000.
10. A. Goldsmith, *wireless communication*, copyright by Cambridge University Press, 2005.
11. Du *et al.* "Accurate error-rate performance analysis of OFDM on frequency-selective Nakagami-*m* fading channels," *IEEE Trans. Commun.*, vol. 54, no 2, pp. 319–328, Feb.2006.

CHAPTER 22

PERFORMANCE AND FLEXIBILITY OF OPEN SOURCE ROUTING SOFTWARE[1]

V. ERAMO, M. LISTANTI, A. CIANFRANI, AND E. CIPOLLONE

University of Roma "La Sapienza", INFOCOM Dept

Abstract: Routing protocols are a critical component in IP networks. Beside dedicated hardware, a great interest on routing systems based on open software is raising among Internet Service Providers. Many open source implementations of this protocol have been developed, among which Quagga and Xorp are the most used in PC-based router. In this paper we evaluate the OSPF performance of Quagga and Xorp routing software according to the test methodologies defined within the Internet Engineering Task Force. Moreover we describe a set of changes made on Quagga code in order to optimize some processes, whose algorithms were not efficient. In order to show the flexibility of an Open Source Routing Software, we have also implemented in Quagga an incremental algorithm for the evaluation of the shortest path. The realized implementation allows the shortest path computation time to be reduced of about the 97%

Keywords: Open Source Code, Quagga and Xorp Routing Software, Open Shortest Path First, Dijkstra's algorithm, Binary Heap, Incremental Algorithm

1. INTRODUCTION

The diffusion of open software implementing routing protocols, together with the big computing power of normal PCs, have been raising a big interest towards the possibility of developing a complete routing system based on open source software and standard low-cost hardware [1,2]. What still lacks is to verify if and when actually the performance of these PC-based-routers can be compared to that of commercial systems. We have developed a set of tests to analyze the performance of a router running the OSPF protocol [3], according to the IETF

[1]This work has been supported by MIUR (Ministero dell'Università e della Ricerca) throught the BORA-BORA project

H. Labiod and M. Badra (eds.), New Technologies, Mobility and Security, 259–270.
© 2007 *Springer.*

specifications [4]. In particular we have taken Black Box measures [5] of the time needed to perform the Shortest Path First (*SPF*) computation on a Personal Computer (PC) equipped with operating system Linux and Quagga [6] or Xorp (eXtensible Open Router Platform) [7] routing software. The evaluation of the *SPF* computation time raises the evidence of a lack of optimization in the Dijsktra's algorithm implemented in Quagga. A deep analysis of the code evidenced that the data structure used to implement the Candidate List during the *SPF* calculation was not optimized. So we have modified the code implementing a binary heap data structure [8]. In our previous work [9] a similar result has been obtained evaluating the *switching time*, a performance index of data plane. In order to show the flexibility of an open source software, we have implemented in Quagga an incremental algorithm.

The organization of the paper is as follows. The test-bed for the *SPF* time evaluation is illustrated in Section 2. The OSPF performance of Quagga and Xorp is shown and compared to the one of the Cisco 2801 router in Section 3. The optimization of Quagga is described in Section 4. The performance of the incremental algorithm implemented in Quagga is evaluated in Section 5. Our main conclusions are discussed in Section 6.

2. TEST-BED FOR THE EVALUATION OF THE SPF COMPUTATION TIME

The test aims at determining how long it takes for a Device Under Test (DUT) to complete the *SPF* computation. The DUT can be either a market router or a PC based router. The test configuration used is reported in Fig. 1. The network topology is made up of two real routers (a testing PC and the DUT) and a variable number of fictitious routers and networks, so that the DUT will have to find the shortest path to all the vertexes of the emulated network, a vertex being either a network or a router. The testing PC is running a C++ software allowing: i) any network topology to be generated; ii) Link State Advertisements (LSA) describing the network topology to be generated and sent to the DUT.

The IETF has been defining the measure methodology for the *SPF* computation [4]. Next we describe the procedure allowing the *SPF* to be computed. To understand the test methodology proposed, we remember that OSPF routers use

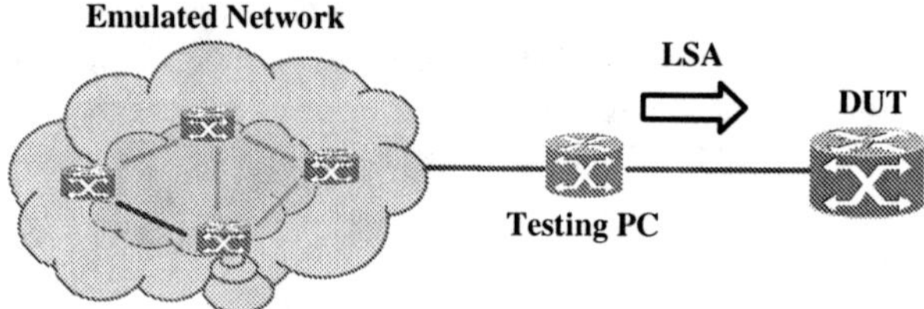

Figure 1. Test-bed for the evaluation of the *SPF* computation time

to schedule the instant in which the *SPF* computation starts to avoid to perform the calculation too many times when receiving Update LSAs [3]. So, when an Update LSA arrives, notifying for example a cost variation in an emulated network link, the *SPF* computation start time is scheduled with a fixed delay, a timer is set and the *SPF* calculation starts only when the timer expires. Moreover, another timer enforces a lag between two consecutive *SPF* computations. In particular the following two timers are defined in [3]:

- *spf_delay*: time between receiving an Update LSA and starting the *SPF* computation;
- *spf_hold_time*: time between two consecutive *SPF* computations.

The *SPF* computation time measurement consists of two different steps, with different settings of these two timers. First as shown in Fig. 2 we set both the timers to 0, so forcing the DUT to immediately start the *SPF* computation when it receives an Update LSA. In Fig. 2 we denote with *RTT* the Round Trip Time and further we assume that the propagation time is the same for the two directions Testing PC-DUT and DUT-Testing PC.

The first step of the test consists in loading the emulated network into the DUT and in sending an Update LSA at time t_{send_u}followed after a little delay by a Duplicate LSA at time t_{send_d}. The DUT processes the Update LSA in the time interval T_{u_LSA} and starts to execute the *SPF* algorithm. Once begun, the *SPF* process cannot be interrupted, and goes on till its end. Then the DUT processes the Duplicate LSA in the time interval T_{d_LSA} and sends back immediately, according to the OSPF protocol rules, its Acknowledge LSA. Thus we can use the Acknowledge LSA of the Duplicate LSA to understand when the *SPF* computation ends. In particular in this first test we measure the time $T_{totalSPF}$, which represents the time difference between the sending of the Update LSA at time t_{send_u} and the receiving of the Acknowledge LSA of the Duplicate LSA at timet_{ack_d}. As shown in Fig. 2, the $T_{totalSPF}$ time can be expressed as follows:

$$(1) \qquad T_{totalSPF} = RTT + T_{u\_LSA} + T_{SPF} + T_{d\_LSA} = T_{overhead} + T_{SPF}$$

wherein

- *RTT* is the Round Trip Time between the Testing PC and the DUT;
- T_{u_LSA} is the Update LSA processing time;

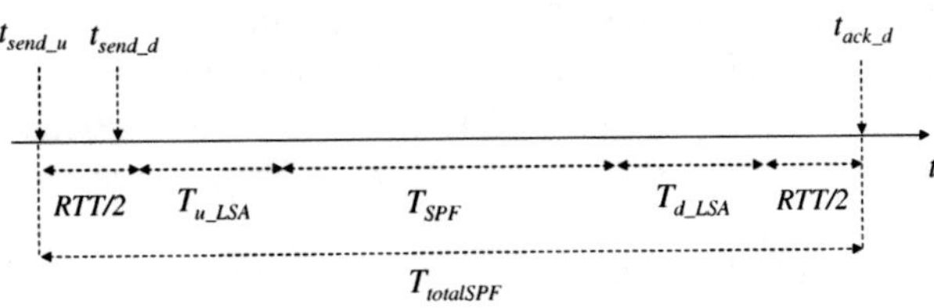

Figure 2. Time measure $T_{totalSPF}$

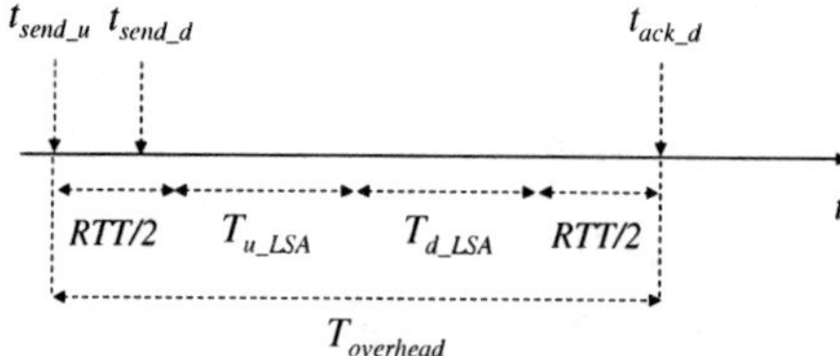

Figure 3. Time measure $T_{overhead}$

- T_{d_LSA} is the Duplicate LSA processing time;
- T_{SPF} is the *SPF* computation time;
- $T_{overhead} \equiv RTT + T_{u\_LSA} + T_{d\_LSA}$.

Hence in the performed measure we are able to evaluate $T_{totalSPF}$ but we are interested in evaluating T_{SPF}. In order to make this we evaluate $T_{overhead}$ and subtract it from $T_{totalSPF}$ obtaining T_{SPF}, that is:

$$T_{SPF} = T_{totalSPF} - T_{overhead} \quad (2)$$

We estimate $T_{overhead}$ with a second test where we set both the *SPF spf_delay* and *spf_hold_time* timers to high values (60 s).

The DUT receives the Update LSA and schedules the *SPF* computation start time but does not execute it because the timers are high. The DUT processes the Duplicate LSA and sends back the Acknowledge LSA of the Duplicate LSA as illustrated in Fig. 3. The time difference between the sending of the Update LSA and the receiving of the Duplicate LSA Acknowledge is exactly $T_{overhead}$.

3. OSPF PERFORMANCE OF QUAGGA AND XORP OPEN SOURCE ROUTING SOFTWARE

All performed tests are based on fully meshed network topologies, with each router connected to each other through a different transit network. Fig. 4 shows an example of fully meshed topology with 4 routers. It is important to remark that in representing the emulated network as a directed weighted graph [3], each router and each transit network becomes a vertex of the graph, and each network-router link becomes an edge. Each edge is labelled with a cost representing the interface cost of the link connecting a router to a network [3]. In the following the cost of all the edges will be chosen to be equal. When the Update LSA is sent as mentioned in Section 2, an interface cost is varied so that the *SPF* computation procedure is primed.

The *SPF* computation complexity will depend on the number of vertexes and edges in the graph. Now let us denote with *N*, *M* the number of vertexes and the number of edges of the graph.

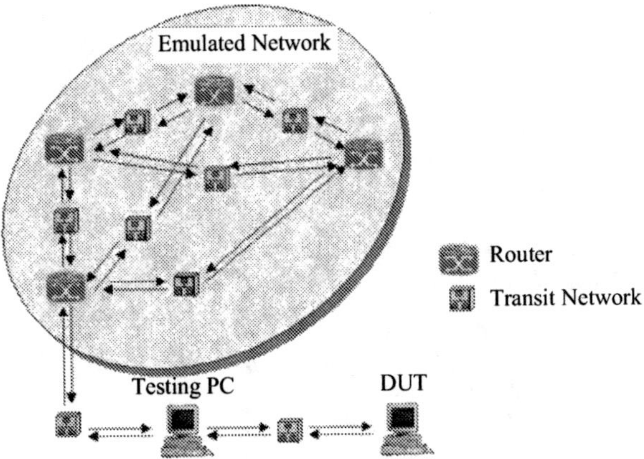

Figure 4. The emulated network topology considered in the test-bed is fully meshed

If we consider an emulated network topology composed by R routers, we have that:

$$N = \frac{R(R-1)}{2} + R \qquad (1)$$

$$M = 2R(R-1) + 2 \qquad (2)$$

The number of edges M is proportional to the number of vertexes N, in particular from (1) and (2) we can assume that $M = O(N)$.

Experimental values taken on a PC based router and on a Cisco 2801 access router are reported in Fig. 5. We report the *SPF* computation time as a function of the number of network topology vertexes. The PC used is equipped with a 2.4Ghz processor, a 512Mbyte memory and Quagga 0.98 or Xorp 1.2 routing software. Notice as the experimental values taken on a Cisco 2801 router perfectly agree with the trend foreseen by the Dijsktra's algorithm. In fact as shown in Fig. 5 we notice as the experimental measure curve of the Cisco 2801 fits the curve $0.005 \cdot N \cdot \log N$ very well. On the contrary the results obtained on a router based on the PC hardware and equipped with Quagga or Xorp routing software are quite different. Of the two Open Source Routing Software, Xorp performs much better than Quagga. For example when the number of vertexes is 5000, the *SPF* time in Xorp is about 6 s. On the contrary the *SPF* time in Quagga increases up to 16 s as the number of vertexes reaches 5000 and measured values fit on the $6.06 \cdot 10^{-4} \cdot N^2$ interpolating curve, as shown in Fig. 5.

On the basis of these results we retain that some changes are needed inside the Quagga 0.98 code, to obtain performances comparable to commercial routers. Next section is dedicated to Quagga optimization.

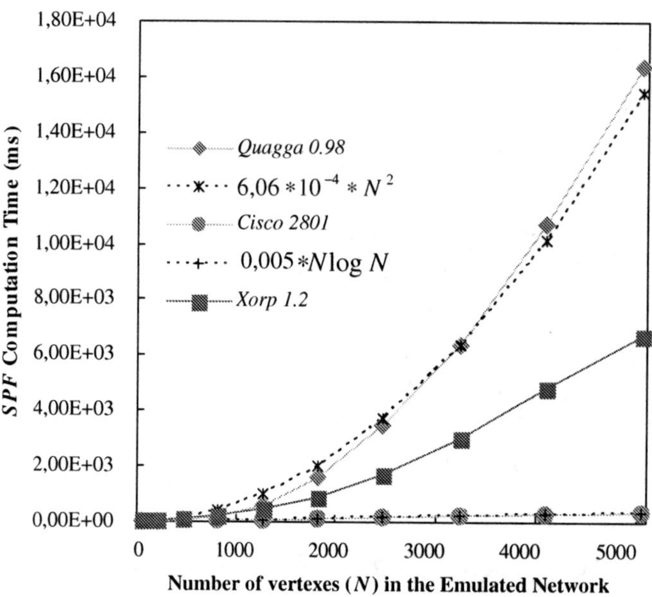

Figure 5. The *SPF* Computation Time in a Cisco 2801 router and a PC based router

4. OPTIMIZATION OF THE *SPF* COMPUTATION TIME IN QUAGGA

The *SPF* computation is based on the Dijkstra's algorithm, as described in [3]. The algorithm examines the directed weighted graph already described in Section 2, in order to find the shortest paths from a root vertex to each other vertex in the graph. All these paths give raise to a Spanning Tree of the graph. In Quagga 0.98 the directed graph is itself represented by the LSA set, stored in the LSA database. During the iterations of the algorithm all the vertexes must be extracted, one by one, from the graph and inserted into the Spanning Tree. Moreover Quagga 0.98 also uses the Candidate List, a structure that contains all the reachable vertexes that have not yet been inserted but can be reached from vertexes already inserted into Spanning Tree. The Candidate List is used as a step in the middle during the migration of the vertexes from the graph to the Spanning Tree. Each of the reached vertexes is extracted from the graph, inserted into the Candidate List and provided with a key that represents the total cost needed to reach it starting from the root and crossing the minimum cost path composed by only the vertexes that have already been inserted into the Spanning Tree. According to the Dijkstra's algorithm a vertex will be extracted from the Candidate List and inserted into the Spanning Tree only when it becomes the node with the lowest key in the Candidate List. The algorithm finishes when all of the vertexes have been inserted in the Spanning Tree and that occurs when the Candidate List becomes empty. During this procedure the Candidate List is the most stressed structure. Its management is the key point of the resulting global performances, and it is performed by four different functions: the *Extract-Min*, the *Insert*, the *Decrease-Key* and the *Lookup* functions.

In Quagga 0.98, the Candidate List is implemented with a linked list, whose elements are stored in key increasing order. It is possible to prove that in this case the Dijkstra's algorithm complexity is $O(N^2+NM)$.

Next we propose an optimization of the *SPF* algorithm. In Section 4.1 a Binary Heap data structure implementing the Candidate List will be proposed and its complexity will be evaluated. Because a Binary Heap data structure does not support the Lookup function efficiently, in Section 4.2 we will illustrate how this operation can be eliminated by modifying the LSA database data structure. Finally the results concerning the modified Quagga will be evaluated in Section 4.3.

4.1. A Binary Heap to Implement the Candidate List

We have modified the Quagga original version with a *patch* available in [10]. In the new Quagga version we have chosen the binary heap data structure to replace the sorted list used in the original version.

A binary heap is a complete and balanced binary tree with a local sorting [8]. Leaves are always inserted starting from the left, and a new level is actually created only when the previous one is complete. Thus the heap depth is always less than *logN*, where *N* is the number of nodes. Each node of the heap has a key, and the whole heap is locally ordered on these keys, so that each node has a key lower than the ones of both its children. This particular sorting ensures that the node with the minimum key is the root of the heap. The management of the tree is based on two internal functions: the sift-up and the sift-down functions. The sift-up function brings up a node with a low key toward the root node. The sift-down function, on the other side, pushes down a node with an high key toward the leaves of the heap. These two procedures are exploited in order to perform the three main functions supported by the heap structure: the *Insert,* the *Extract-Min* and the *Decrease-Key* functions. The *Insert* function takes the new node and inserts it at the end of the heap, i.e. makes it a leaf. Then it executes a sift-up procedure on the inserted node, and brings it up to its correct position. Notice that because the *Insert* function needs at most *log(N)* sift-up operations, equal to the maximum depth of the heap, its cost is *O(logN)*. The *Extract-Min* function removes from the heap the node with the lowest key. This node is always at the root of the heap, and its extraction is almost costless. After this extraction we have two sub-tree, that are themselves ordered heap, and we need to fuse them into one single heap. To achieve this result, the *Extract-Min* function takes the last leaf of the heap and puts it at the root position, then executes the sift-down procedure on it and pushes it down to its correct position. Notice that because the *Extract-Min* function needs the execution of at most *logN* sift-down operations, its complexity is *O(logN)*. Finally the *Deacrese-Key* function changes the key of a particular node to a lower value. Once the key value have been decreased, it executes the sift-up procedure on the node, and takes it to its new position. Notice that to realize the *Deacrese-Key* function, a number of sift-up operations at most equal to the maximum depth of the heap is performed. For these reasons the *Deacrese-Key* function cost is *O(logN)*. The *Insert* function and the *Extract-Min* function will be performed exactly *N* times. The number of times in which the *Decrease-Key* function is performed depends on the values of the costs of the links. However it will be always less than the number of edges, so we can assume that the *Decrease-Key* function will be performed $O(M)$ times. So the amortized costs are *O(N·logN)* for the *Insert* function, *O(N·logN)* for the *Extract-Min* function and *O(MlogN)* for the *Decrease-Key* function.

Finally, because in Section 4.2 we will show that by modifying the LSA database data structure the Lookup function is no more needed, the total cost of the new implementation of the Dijkstra's algorithm in modified Quagga becomes as expected *O((M+N)logN)*. In particular when $M = O(N)$ the amortized cost reduces to *O(N·logN)*.

4.2. The New Lookup Operation in the Candidate List

Unfortunately the changes made to implement the Candidate List rise a new problem: the binary heap need to scan one by one all the nodes to perform the *Lookup* function, as the structure is only locally ordered, thus obtaining again a $O(N)$ cost. We have modified the LSA database data structure so that the Lookup function becomes no longer needed at all. In particular for each LSA, stored in the database, we have added an information denoting if or not the LSA is in the Candidate List. In positive case the information also denotes the position in the Candidate List where the vertex associated to the LSA is stored. That allows a vertex associated to an LSA to be immediately accessed during the execution of the Dijkstra's algorithm. Further, because the sift-up and the sift-down operations may change the position of a vertex in the Candidate List, a pointer to the information of the associated LSA is added for each vertex.

4.3. Numerical Results for Modified Quagga 0.98

The test evaluating the *SPF* time on the modified Quagga 0.98 version produced experimental results that perfectly reflect the *NlogN* trend. The measured values, varying the number of vertexes in the graph, are presented in Fig. 6, and compared with the same measure taken on the Cisco 2801, on the original Quagga 0.98 version and on Xorp 1.2.

It is important to note that the *SPF* time on the modified Quagga version is always less than the time needed on the original version, proving that the optimization have been successfully done. The obtained results are also lower that the ones in Xorp and only slightly worse that in Cisco 2801. Results on Xorp can be explained after a code analysis of the software: the weak point is LSA database, implemented in a sub-optimal way.

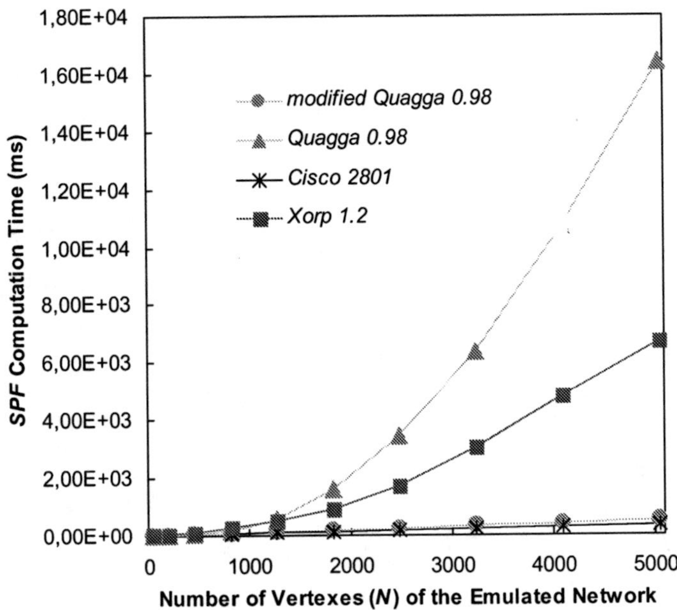

Figure 6. Performance comparison among Cisco 2801, Quagga 0.98, modified Quagga 0.98 and Xorp 1.2

5. AN INCREMENTAL ALGORITHM IN QUAGGA

When a topological change occurs in a network, the computation of the SPF may consume a considerable amount of CPU time. In literature have been proposed incremental algorithms [11–14] that make use of the SPF previously computed and update it according to the network topology changes. We have implemented in Quagga an incremental algorithm inherited from the one proposed in [12]. In particular it has been modified to take into account multipath routes to a destination node and it has been optimized for the most frequent topological changes of Internet, link failure or restoration. The incremental algorithm implemented in Quagga has been compared to the one of Cisco 2801. The comparison has been carried out by emulating on the DUT real network topologies measured within the Rocketfuel project [15]. In particular we have considered the topology of Verio, an USA Internet Service Provider whose network is composed by 893 routers and 4150 links. The *SPF* time in an incremental algorithm depends on link position and on type of change occurring, so we have chosen to measure this time when the failure and the restoration of each link of Verio occur. The *SPF* time is reported in Figs. 7, 8 in the case of failure and restoration respectively as a function of the link interested. In the figures we also report the time that the DUT takes to run the Dijkstra's algorithm, that is independent of the link position. Notice as in Quagga the *SPF* time is lower when the incremental algorithm is enabled. In particular the

time to run the Dijkstra's algorithm is 53 ms; while the average time is 1.5 ms in the incremental version. That leads to save about 97% in processing time. In the measure reported in Figs. 7, 8 it is possible to observe some peaks, for example in the case of restoration of the links 6 and 7 the *SPF* time is 27,8 ms. and 18 ms. respectively. The peaks concern links that in the shortest path tree are near the root node and are connected to a node having many descendents, so the incremental algorithm has to perform more operations.

Observing Figs. 7, 8 you can notice that though Cisco 2801 takes a time lower than modified Quagga to run Dijkstra's algorithm, in particular 34 ms. against 52 ms., the incremental algorithm implemented in Quagga is more efficient than the one implemented in Cisco 2801. In fact the average *SPF* time is 9,5 ms. in Cisco 2801 and 1.5 ms in Quagga. The incremental algorithm of Cisco 2801 allows only a saving of about 72% in processing time with respect to Dijkstra's algorithm. The measure performed on Cisco 2801 shows that there are some links whose failure causes an *SPF* time higher than the one obtained with Dijkstra's algorithm. For example in Fig. 7, the failure of link 1206 causes an *SPF* time equal to 48 ms. against the 34 ms of Dijkstra.

6. CONCLUSIONS

The aim of our work was to evaluate the OSPF performance of Quagga 0.98 and Xorp 1.2 Open Source Routing Software. A test-bed has been realized and *SPF* computation time has been measured.

The experimental results for the *SPF* time obtained on a PC running Quagga have not been good. An optimization of the Dijkstra's algorithm has been made to the Quagga code by implementing the Candidate List with a Binary Heap and by modifying the data structure of the LSA database. The measure of the *SPF* time on the modified version of Quagga are slightly worse than the one of the Cisco 2801 router.

In order to show the flexibility of an open source routing software, we have implemented in Quagga an incremental algorithm for the shortest path computation. The realized implementation performs well and allows *SPF* time lower than in Cisco 2801 to be obtained. Performance and flexibility of an open source routing software raise great expectations about the possibility of making a competitive router from a standard PC.

REFERENCES

1. M Deval, H Khosravi, R Muralidhar, S Ahmed, S Bakshi, R Yavatkar, "Distibuted Control Plane Architecture for Networks Elements", Intel Technology Journal, Vol. 7, Issue 4, November 2003.
2. A Bianco, J M Finocchietto, G Galante, M Mellia and F Neri, "Open Source PC-Based Software Routers: A Viable Approach to High Performance Packet Switching", in Proc of the third International Workshop, QoS-IP 2005, Catania, Italy February 2005.
3. J Moy, "OSPF Version 2" , Request for Comments 2328, April 1998.
4. V Manral, R White, A Shaikh, "Benchmarking Basic OSPF Single Router Control Plane Convergence", RFC 4061, April 005.

5. A Shaikh and A Greenberg, "Experience in Black-box OSPF Measurements," in Proc ACM SIGCOMM Internet Measurement Workshop (IMW) 2001, pp. 113–125, November 2001.
6. GNU "Quagga." [Online]. Available http://www.quagga.net
7. "Xorp." [Online]. Available http://www.xorp.org
8. A V Goldberg and R E Tarjan, "Expected performance of Dijkstra's Shortest Path algorithm", Technical Report 96–062, NEC Research Institute, Princeton, NJ, June 1996.
9. V Eramo, M Listanti, A Cianfrani, "Switching time measurement and optimization issues in Gnu Quagga routing software", in IEEE Globecom 2005 , St. Louis, November 2005.
10. Routing Software *Patch* GNU Quagga 0.98, http://net. infocom.uniroma1.it/projects/progetti_dip/zebra/index.htm
11. J McQuillan, I Richer, and E Rosen, "The new routing algorithm for the ARPANET", IEEE Trans. on Communication., vol. 28, pp. 711–719, 1980.
12. P Narvaez, K-Y Siu, and H-Y Tzeng, "New dynamic algorithms for shortest path tree computation", IEEE Transaction on Networking, vol. 8, pp. 734–746, 2000.
13. G Ramalingam and T Reps, "An incremental algorithm for a generalization of the shortest-path problem," *Journal of Algorithms*, vol. 21, pp. 267–305, 1996.
14. D Frigioni, A Marchetti-Spaccamela, and U Nanni, "Fully dynamic algorithms for maintaining shortest paths trees", *Journal of Algorithms*, vol. 34, pp. 351–381, 2000.
15. "Rocketfuel Project." [Online]. Available http://www.cs.washington.edu/research/networking/rocketfuel/

CHAPTER 23

EMERGING WIRELESS COMMUNICATION TECHNOLOGIES[1]

GHAÏS EL ZEIN AND ALI KHALEGHI

Member, IEEE

Abstract: This paper describes some latest development in the area of wireless communication technologies. At first, we give an introduction on the current wireless communication systems. Then, we discuss the characterization and the modeling of the propagation channel which are particularly critical in the design of the recent technologies as MIMO, UWB and time reversal. Thereafter, these techniques are presented and some results concerning the possible integration in future wireless systems are discussed

1. INTRODUCTION

Today, third generation (3G) mobile communications systems allow the integration of new services such as multimedia, packet switching and wideband radio access. In the same time, Wireless Local Area Networks (WLAN) equipments are penetrating the market. These new systems make possible the connectivity with IP networks. In this context, emerging technologies such as Multiple-Input Multiple-Output (MIMO) and Ultra-WideBand (UWB) are recognized as good solutions in the development of the forthcoming generation of broadband wireless networks. These techniques are very attractive for digital communications to increase data rates and/or to improve system performance.

The purpose of this invited paper is to highlight different aspects concerning these new technologies. In section 2 a brief summary of current wireless technologies is

[1]This work was supported in part by the CPER PALMYRE project with the financial support of Région Bretagne and the MIRTEC project with the financial support of ANR.
The authors are with the National Institute of Electronics and Telecommunications, IETR – UMR CNRS 6164 – INSA, 35043 Rennes Cedex, France (phone: (+33) (0)2 23 23 86 04; fax: (+33) (0)2 23 23 84 39; e-mail: ghais.el-zein@insa-rennes.fr, ali.khaleghi@insa-rennes.fr)

H. Labiod and M. Badra (eds.), New Technologies, Mobility and Security, 271–279.
© 2007 *Springer*.

given. Section 3 deals with the characterization and the modeling of the propagation channel, which are particularly important in the design of these new systems. Sections 4, 5 and 6 introduce respectively the principles and the features of MIMO, UWB and time reversal (TR) techniques. Some research results concerning the possible integration of these new technologies in the future wireless systems are also presented.

2. CURRENT WIRELESS TECHNOLOGIES

After the great success of 2G cellular service and the tremendous growth of the Internet, multimedia is now penetrating the mass market. Since, wireless access to the worldwide wired-line infrastructure is becoming an essential feature of modern communication networks. The first realizations of these capabilities are the 3G mobile systems. In the same way, WLAN equipments are supported by the two most prominent standards IEEE 802.11/WiFi and HIPERLAN, and allow connectivity in buildings for portable computers. However, current wireless access networks show limits in terms of data rate and quality of service (QoS). For several years, efforts have been made to improve the design of the existing systems. Fig. 1 shows the limits in the data rate of the actual communication systems that depends on the system mobility. The trend of the recent communication is to increase the data rate and to deliver better services for mobile systems (4G and 5G). In fact, the trend to increase data rates will most probably continue to reach 100 Mbps considering a moderate mobility, and up to 1 Gbps for a reduced mobility. In this context, MIMO and UWB technologies appear as new concepts to fulfill those specifications. In addition, these systems can be combined with TR technique to improve their performance.

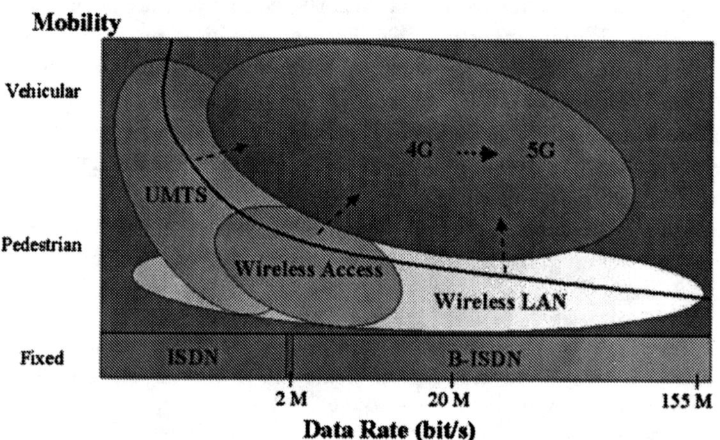

Figure 1. Current wireless technologies and trends

3. WIRELESS CHANNEL CHARACTERIZATION AND MODELING

The radio propagation of electromagnetic waves from a transmitter to a receiver is characterized by the presence of multipath due to various phenomena such as reflection, refraction, scattering and diffraction. The study of these propagation phenomena appears as an important task when developing a wireless system [1], [2].

For broadband systems, the analysis of both path loss and impulse response is required. The analysis is usually made in the time domain, which allows to measure the coherence bandwidth, the coherence time, the respective delay spread, and Doppler spread values. Also, coherence distance, correlation distance, and wave direction spread are used to highlight the link between propagation and system in the space domain. Therefore, an accurate description of the spatial and temporal properties of the channel is required for the design of broadband/multi-antenna systems, and also for the choice of the network topology. In this context, the characterization and the space-time modeling of the channel appear essential. Several methods of classification of the models are proposed in the literature [3]–[9]. Deterministic models are based on a fine description of a specific environment, when the stochastic models aim to describe the channel parameters by random laws. In practice, different sounding techniques can be used in order to characterize the propagation channel.

4. MIMO WIRELESS COMMUNICATION SYSTEMS

The MIMO (Multiple-Input Multiple-Output) principle can be defined simply. Since time and frequency domains processing are pushed to their limits, the space domain can be exploited. The main idea is to transmit multiple streams of data on multiple antennas at the same frequency. Usually, multiple receiving antennas are considered to improve the system performance (Fig. 2).

In an ideal case, it can be shown that the channel capacity grows linearly with the number of transmitting (Tx) and receiving (Rx) antennas [10]. This technique can be viewed as a generalization of space diversity and smart antennas [11]. It supposes a channel rich in multiple paths in order to exploit independent transmission channels between the Tx and Rx antennas. This transmitting and receiving structure can be modeled using a matrix representation of the channel.

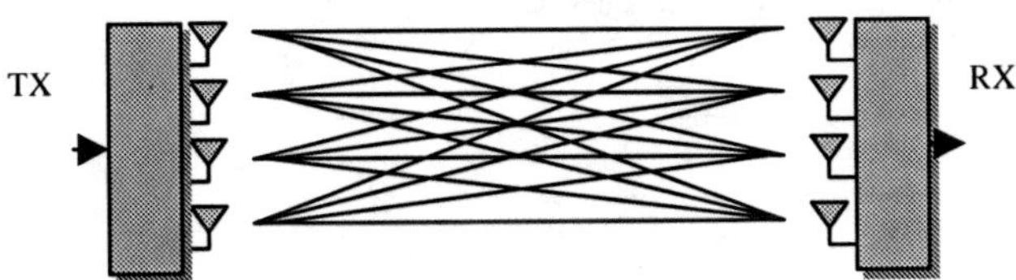

Figure 2. MIMO wireless transmission system

Information theoretic considerations allow to highlight that two fundamental mechanisms are at stake in the process of transferring information: diversity (reception of multiple de-correlated copies of the same transmitted information, by combining, allow to combat channel fading), and multiplexing (reception of multiple independent symbols of information to increase the channel capacity) [12]. The structure of the propagation channel places a tradeoff between the two types of gain.

The really large improvement in link reliability and/or data rates, and predicted by information theory relies on a fine knowledge of the propagation phenomena. Such knowledge makes it possible to choose the most appropriate coding/modulation scheme for a given environment. Transmitting and receiving antenna arrays have to be carefully designed to maximize the channel rank, *i.e.* the number of eigenmodes available for communication. In this case, correlation and dispersion measurements of channel parameters play a central role.

For our measurements, a wideband channel sounder developed in our laboratory has been used to characterize the double directional channel [13]. It operates at 2.2 GHz. It uses a periodic transmit signal based on the direct sequence spread spectrum technique. This sounder offers a temporal resolution of about 11.9 ns. In order to estimate the propagation parameters of channel multipath components, we used the Unitary ESPRIT algorithm [14]. A dense urban environment was chosen for the first campaigns of the sounder test.

The statistical analysis permits us to extract the second order statistical properties of the propagation channel and to provide important parameters for system design such as the coherence bandwidth, the coherence distances at the transmission and reception sites. Table 1 presents the average (m) and dispersion (σ) of the coherence bandwidth (in MHz) for two correlation levels (50% and 90%) in different measurement sites. Table 2 presents the coherence distances (in λ) obtained at the transmission and the reception sides.

These various coherence parameters have a profound implication in the design of MIMO communication systems. In fact, the two main performance indicators relevant to power-limited and band-limited applications, *i.e.* probability of error and data rates, both depend on the second-order behavior of the communication channel. As an example, we can observe that antenna spacing, within the transmitting and

Table 1. Coherence bandwidth results

Site	Coherence Bandwidth (MHz)			
	50%		90%	
	m	σ	m	σ
Victoire	37.7	28.3	13.3	18.4
Station	32.8	27	9	13.9
JMF	24.7	18.6	4.6	6.9

Table 2. Coherence distance results

Site	Coh. Dist. Rx (λ)				Coh. Dist. Tx (λ)			
	50%		90%		50%		90%	
	m	σ	m	σ	m	σ	m	σ
Victoire	9.73	6.03	1.8	1.8	0.23	0.10	0.06	0.03
Station	5.7	4.3	0.7	0.9	0.22	0.12	0.05	0.03
JMF	6.34	5.21	1	1.6	0.29	0.12	0.07	0.03

receiving arrays, must be larger than the local correlation distance to get sufficient transmit and receive decorrelation. Likewise, the transmit bandwidth must be larger than the channel coherence bandwidth to gain frequency diversity.

The first campaigns show that for a UMTS system with a 5 MHz bandwidth (Table 1), the channel offers little frequency paths diversity (in less than 10% of the cases) [15]. Moreover, space diversity has been obtained at the Rx base station by separating the antennas at distances close to 11 λ (1.5 m at 2.2 GHz) (Table 2). Also, the diversity at the Tx mobile position has been evaluated by spacing out the antennas at the distance 0.36 λ (4.9 cm at 2.2 GHz).

5. UWB WIRELESS COMMUNICATION SYSTEMS

Another solution under consideration is the UWB technology, which relies on transmission of series of very short pulses (< 1 ns). This technology appears as an alternative air interface for the deployment of WLAN and WPAN (Wireless Personal Area Networks) that link portable and fixed equipments [16]. It promises benefits such as high location accuracy, robustness to multipath propagation, high-data rate and low-power wireless communications. The major principle of UWB communication systems consists in the transmission of impulse radio signals, as defined by the Federal Communications Commission (FCC), whose fractional bandwidth (3 dB bandwidth divided by the center frequency) is greater than 25%, or has a bandwidth larger than 500 MHz. Thus, the FCC has permitted UWB devices to operate using a spectrum ([3.1; 10.6] GHz) occupied by existing radio services, as long as emission levels meet the proposed spectral mask.

The applications of UWB systems aim indoor home and professional environments, providing short range high-speed communications, precision location and tracking, wall and ground penetrating radar and medical imaging. In this context, the understanding of physical phenomena involved in the propagation of an impulse signal appears necessary for the modeling of the propagation channel associated with a specific environment [17]. Due to the frequency selectivity and the different time delays of multipath arrivals, the wide-band nature of the radio signal further complicates the channel modeling [18].

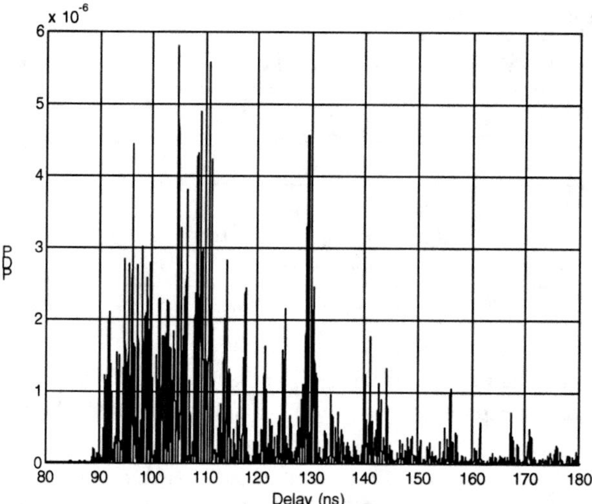

Figure 3. Instantaneous PDP of indoor UWB channel

Fig. 3 shows a sample of the power delay profile (PDP) of a NLOS (non-line-of-sight) UWB channel that is measured in an indoor environment. Due to the large number of taps at the receiver, the detection of the UWB signal is the most complicated part of the system. Traditional rake receivers can be used for signal detection. However, by considering IEEE 802.15 UWB channel models, the number of rake fingers that should be considered is in the order of 22-123 for LOS and NLOS channels. Therefore, the complexity of the receiver system is increased. To resolve this problem, the time reversal technique is proposed that moves the system complexity to the transmitter part of the communication link, which is ideal for some applications. Extremely simple non-coherent receivers can be used for low-cost and low-power sensors

6. TIME–REVERSAL WIRELESS COMMUNICATIONS

Time reversal is already known in acoustical imaging, electromagnetic imaging, underwater acoustic communication and radar domain [19–21]. Recently, some industries invested for the development of time-reversal for wireless communication. In this technique, the channel response between transmitter and receiver is measured then the time reversed version of the channel impulse response is used as a pre-filter for data communication. Mathematically, it can be shown that the convolution product of the time reversed waveform of the channel and the channel response gives the signal auto-correlation i.e.

$$(1) \qquad s(r, \tau) = h(r_0, -\tau) * h(r, \tau)$$

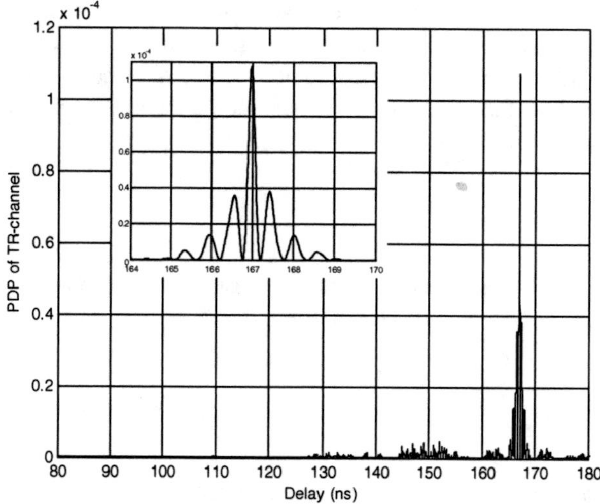

Figure 4. PDP of equivalent TR channel response

where * denotes the convolution operation. If the environment is rich in multipath, the TR will focalize spatially the received signal energy and compressed it temporally. Fig. 4 shows the measured PDP of the equivalent TR channel given in Fig. 3.

The channel response is compressed in time. Thus, the complex task of estimating a large number of taps at the receiver is greatly reduced. This implies low cost receivers with a much less need for equalization. Furthermore, due to the considerable focusing gain, better signal to noise ratio or equally higher data rate can be achieved. Spatial focusing reduces the co-channel interference in multi-cell systems.

Furthermore, the TR technique can be combined with UWB and MIMO systems to increase their performance. In UWB, it would be possible to increase the communication range by respecting FCC transmitter spectral mask limit. In MIMO, the communication can be conducted between transmitter and the intended receiver without disturbing the other receivers or users. This means that the channel rank will be maximized.

7. CONCLUSION

This paper focuses on the novel technologies in wireless communication MIMO, UWB and time-reversal. Aspects of wave propagation have been addressed. A fine knowledge of the propagation phenomena makes it possible to choose the most appropriate coding/modulation scheme for a given environment. In fact, considering different practical situations, the extracted spatio-temporal channel parameters can be used to highlight the connection between propagation and communication system. Thus, a great measurement number will be necessary to obtain significant statistical results that can give realistic MIMO and/or UWB channel models. Furthermore, the time reversal (TR) technique appears as an attractive solution that moves the system complexity to the transmitter part of the communication link, which is ideal for some applications. The TR technique can be combined with UWB and MIMO systems to increase their performance. Moreover, simple non-coherent receivers can be used for low-cost and low-power wireless communication systems.

8. ACKNOWLEDGMENT

The authors would like to thank Hanna Farhat, Ronan Cosquer, Julien Guillet, Florence Sagnard, Ijaz Haider Naqvi and Guy Grunfelder for their technical contributions.

REFERENCES

1. F. Molisch, *Wireless Communications*, John Wiley & Sons Ltd, England, 2005.
2. P. A. Bello, "Characterization of randomly time-variant linear channels," *IEEE Transaction on Communications*, pp. 360–393, December 1963.
3. K. Yu, and B. Ottersten, "Models for MIMO propagation channels, a review," *Special Issue on "Adaptive Antennas and MIMO Systems", Wiley Journal on Wireless Communications and Mobile Computing*, vol. 2, no. 7, pp. 653–666, November 2002.
4. M. A. Jensen, and J. W. Wallace, "A review of antennas and propagation for MIMO wireless communications," *IEEE Transactions on Antennas and Propagation*, vol. AP-52, n° 11, pp. 2810–2824, November 2004.
5. L. Schumacher, L. T. Berger, and J. Ramiro-Moreno, "Recent advances in propagation characterisation and multiple antenna processing in the 3GPP framework," in *Proceedings of XXVIth URSI General Assembly*, Maastricht, The Netherlands, August 02.
6. S. Salous, "Multiple input multiple output systems: capacity and channel measurements," in *Proceedings of the Seventh World Multiconference on Systemics, Cybernetics and Informatics (SCI2003)*, Florida, USA, pp. 1–5, 27–30 July 2003.
7. M. Steinbauer, A. F. Molisch, and E. Bonek, "The Double-Directional Radio Channel," *IEEE Antennas & Propagation Magazine*, vol. 43, no. 4, pp. 51–63, 2001.
8. J. Guillet, "Caractérisation et modélisation spatio-temporelles du canal de propagation radioélectrique dans le contexte MIMO," Ph.D. Thesis, INSA Rennes, July 2004.
9. G. El Zein, R. Cosquer, J. Guillet, H. Farhat, and F. Sagnard, "Characterization and Modeling of the MIMO Propagation Channel: An Overview," in *Proc. of European Microwave Week EUMW 2005*, Paris, October 3–7, 2005.
10. G. J. Foschini, and M. J. Gans, "On limits of wireless communications in a fading environment when using multiple antennas," *IEEE Wireless Personal Communications*, vol. 6, no. 3, pp. 311–335, March 1998.
11. D. Gesbert, M. Shafi, D. Shiu, P. J. Smith, and A. Naguib, "From theory to practice: An overview of MIMO space-time coded wireless systems," *IEEE Journal on Selected Areas in Communications*, vol. 21, no. 3, pp. 281–301, April 2003.
12. P. Guguen and G. El Zein, *Les Techniques Multi-Antennes pour les Réseaux sans Fil*, Edition Hermes, Paris, 2004.
13. R. Cosquer, "Conception d'un sondeur de canal MIMO. Caractérisation du canal de propagation d'un point de vue directionnel et doublement directionnel," *Ph.D. Thesis*, INSA Rennes, October 2004.
14. M. Haardt, "Efficient one-, two and multidimensional high resolution array signal processing," *Ph.D. Thesis*, TU-Munchen, 1996.

15. H. Farhat, R. Cosquer, J. Guillet, and G. El Zein, "Mobile radio channels characterization based on wideband SIMO 1 × 8 measurements at 2.2 GHz," in *Proceedings of the 5th International Conference on ITS Telecommunications (ITST '05)*, Brest, June 27–29, 2005.
16. D. Porcino, and W. Hirt, "Ultra-Wideband Radio Technology: Potential and Challenges Ahead," *IEEE Commun. Mag.*, pp. 66–74, Jul. 2003.
17. F. Tchoffo-Talom, B. Uguen, and E. Plouhinec, "A site-specific tool for UWB channel modeling," in *Proc. IEEE IUWBST Conf.*, Kyoto, Japan, May 18–21, 2004.
18. F. Sagnard, and G. El Zein, "In-Situ Characterization of Building Materials for Propagation Modelling: Frequency and Time Responses," *IEEE Trans. on Antennas and Propagation*, vol. 53, No. 10, Oct. 2005, pp. 3166–3173.
19. A. Derode, P. Roux, and M. Fink, "Robost acoustic time reversal with high order multiple scattering," *Phys. Rev. Letters*, vol. 75, pp. 4206–4209, 1995.
20. P. Kyritsi, G. Papanicolau, P. Eggers, and A. Oprea, "MISO Time Reversal and Delay-Spread Compression for Fwa Channels at 5 GHz," *IEEE Antennas and Wireless Propagation Letters*, vol. 3, pp. 96–99, 2004.
21. P. Kyritsi, G. Papanicolau, P. Eggers, and A. Oprea, "Time Reversal Techniques for Wireless Communications," *Proceedings of the 60th Vehicular Technology Conference*, 2004.

CHAPTER 24

UPGRADE OF A COMPOSITE-STAR OPTICAL NETWORK

STEFANO SECCI[1] AND BRUNILDE SANSÒ[2]

[1]*Dip. Elettronica e Informazione, Politecnico di Milano, and Dép. Informatique et Réseaux, ENST Paris, Address: ENST/INFRES, 37/39, rue Dareau, 75014 Paris, France, Email: secci@enst.fr – Tel: +33 (0)1 4581 8399*

[2]*GERAD and Dép. Génie électrique, Ecole Polytechnique de Montréal, Address: C.P.6079 succ Centre-Ville, Montréal, P.Q. CANADA H3C 3A7, Email: brunilde.sanso@polymtl.ca – Tel: +1 (514) 3406053*

Abstract: In this manuscript we tackle the optimal upgrade of an innovative optical transport network architecture called the Petaweb, which has a particular composite-star infrastructure that allows two-hop communications between edge nodes. Prior studies of the same authors have tackled the design and dimensioning problem for the Petaweb assuming TDM/WDM equipment and adopting a dedicated path protection strategy. A quasi-regular topology, more efficient than the regular, has also been proposed to minimize the quantity of fiber to install while preserving the regularity of the architecture. Exploiting the same network model, we propose an upgrade procedure for the extension of an existing optimized network, having one of the two possible topologies, under traffic increase and edge node addition

1. INTRODUCTION

With new and enhanced IP services that consume large amounts of bandwidth, the demand for optical transport services increases day by day. There is an important need for a smooth and cost-effective way to upgrade WDM networks. Such an upgrade is not an easy task given the network structure. For instance, in WDM mesh networks the use of the idle capacity by multi-hop lightpaths is complicated by the need of re-dimensioning core nodes, resizing transport links and reconfiguring a large number of optical switches.

In [1] a novel optical architecture called the Petaweb was proposed for the next generation transport infrastructure. This network is formed by edge nodes connected through core nodes as shown in Fig. 1. Every edge node is connected to every core node. Note that the core nodes are not connected to each other, forming a backbone network where all the nodes are disconnected. Given its topology, the

H. Labiod and M. Badra (eds.), New Technologies, Mobility and Security, 281–296.
© 2007 *Springer.*

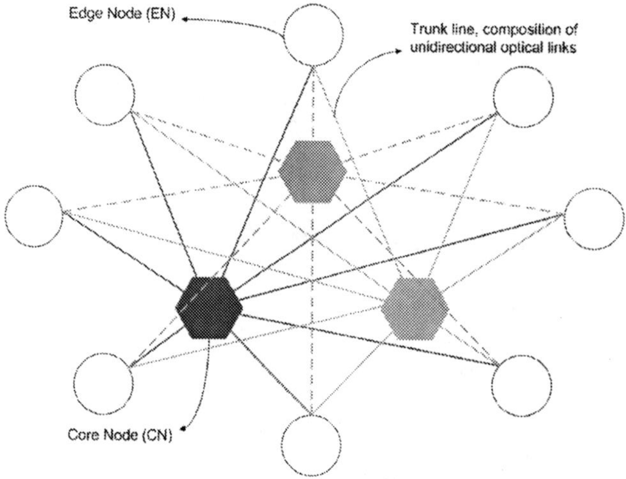

Figure 1. The Petaweb composite-star architecture

Petaweb allows for two-hop optical lightpaths between two edge nodes through a single core node.

In this architecture the classical pitfalls of the WDM network upgrades where existing network capacity may not be available because of structural bottlenecks can now be avoided. In fact, the Petaweb offers few easily-manageable and independently-configured core nodes and all its components are modular and can be extended without reconfiguring the existing equipment [2]. Moreover, given the regularity of the structure, an upgrade will not jeopardize the management of an optimized network as idle capacity can be easily allocated without compromising network management.

In previous work, we have dealt with the design and dimensioning of the Petaweb structure. In [3] the design problem was defined and an efficient resolution approach was presented. In [4] the TDM/WDM features of the Petaweb were investigated and a new design optimization was proposed. In that paper, a quasi-regular topology for the Petaweb was also introduced. Although the quasi-regular topology has a lower cost when compared to the regular one, it has the disadvantage that in the event of failures, some minor edge nodes could get disconnected. Then, the reliability issues were dealt with in [5].

Now, once the Petaweb network is designed, the question remains on how to upgrade the structure taking into account the architectural constraints. The object of this paper is precisely to tackle this issue and to present an effective formulation and resolution approach. The paper is divided as follows. In Sect. 2 the Petaweb architecture is briefly discussed. For the sake of completeness, the network model and the design problem illustrated in [4] and [5] are also presented. In Sect. 3 other network expansion problems tackled in the literature are briefly reviewed. The problem of upgrading a Petaweb architecture is presented in Sect. 4 where an Integer Linear Programming (ILP) formulation is proposed. Sect. 5 shows the results for two cases: the case for which there is only a traffic increase, and the case for which there is traffic increase and edge node addition. Sect. 6 is devoted to conclusions and suggestions for further work.

2. THE PETAWEB NETWORK MODEL

In the Petaweb, an *edge node* (EN) is an electronic node that requests bandwidth to the transport network. The connection between N edge nodes and a core node is shown in Fig. 2. Every edge node is connected to a core node through one *optical link*, composed of one or more optical fibers. We suppose unidirectional optical fibers so that an edge node has one optical link incoming from, and one optical link outgoing to, every core node. Every fiber has several optical channels and we assume that all fibers of the network carry the same number of channels.

A *core node* (CN) is a set of arrays of parallel space switches, also called switching planes. The number of switching planes s_r identifies the type r of a CN (indicated by CN-r). Note that the optical link connecting an EN to a CN-r has s_r unidirectional optical fibers, one for every switching plane. In this work, we assume three types of core nodes, with one, two and four switching planes, that is, $s_1 = 1$, $s_2 = 2$ and $s_3 = 4$. All the incoming WDM fibers are demultiplexed into their different lambda-channels, each of which is connected to the space switch of the respective array. conversion so that the Petaweb can be considered an all-optical network. Each space switch handles channels of the same wavelength; those referred to the same EN are then multiplexed into the optical link going back to that EN. Such a parallel-planes structure increases the reliability of the CNs because a hypothetical failure in a switching plane would affect only the connections on

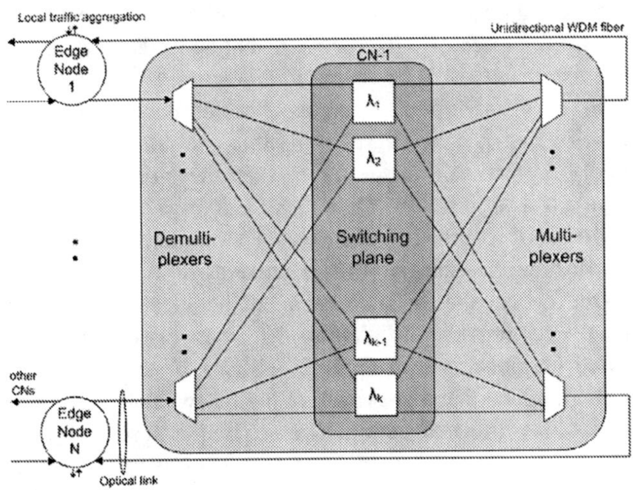

Figure 2. Parallel-planes optical core node in the Petaweb

that plane. In [2] Blouin proposed the use of TDM in the Petaweb to produce sub-channels within a wavelength channel. To integrate Time Division Multiplexing into the Petaweb, the switching cores functionalities must be specified. In [4] we proposed the replacement of the switching plane described with the compatible all-optical TDM Wavelength Space Routers of Huang [6], which multiplexes in a time-slot basis remaining in the optical domain and without any buffering operation; the behavior of such a node architecture has been recently evaluated in [7]. The EN locations define the set of potential *switching sites*. Note that several CNs can be installed in the same site. Therefore, the physical connection between an EN and a switching site can be composed of several links, given that there may be several CNs present at the site. From now on we call this physical connection an *optical trunk line*.

In [4] and [5], two possible topologies were studied. The regular and the quasi-regular topology. In the regular topology every switching plane is connected with every edge node, and no wavelength conversion is required at core nodes. The quasi-regular topology is built removing from the regular architecture the equipment that will remain unused in the dimensioned network, and require wavelength converters.

We will refer in the following to a time-slotted lightpath with the term *ts-lightpath* (as suggested in [6]) or with the acronym TLP; it is the data channel of a time-slot in a wavelength. Let us now indicate by Z_h the transport capacity of a TLP of class h and let C_{ch} be the capacity of a lambda-channel set to 10 Gbps. We also assume that there are $W = 16$ wavelengths per fiber. Then, we have $Z_2 = 10$ Gb/s, $Z_3 = 160$ Gb/s and $Z_1 = 0.625$ Gb/s [4]. Note that these bit-rate classes were chosen so that a perfect correspondence with the bit-rates of SDH and OTN interfaces is obtained [8].

To optimize the Petaweb design, a total network cost must be minimized. In our model such a cost is composed of three elements: the cost of the core node, the cost of the fiber and an additional cost to take into account the propagation delay. The *cost of the core nodes* is composed of a fixed cost f_r that depends on the type r of core and that is defined so that $f_r > f_{r-1} > \cdots > f_1$. The number of switching planes is such that $s_r = 2s_{r-1}$. An active port has a cost P scaled for higher types. Let us indicate by M the set of edge node sites; $|M|$ is thus the number of edge nodes of the network. Let γ be the scale factor for P, then the global cost of a core node of type r is $K_r = f_r + 2|M|Ws_rP\gamma^{(s_r-1)}$, with $K_r < 2K_{r-1}$. The *fiber cost* is indicated as F and is in unit of length. It is the cost of a reference fiber type, which is then scaled by a discrete function $\phi(W)$ that depends on the number of wavelengths. Let us indicate by Δ_{ij} the distance between the sites i and j; the installation of a CN-r on site i requires the installation of s_r fibers per direction for every edge node, which yields a global cost of $F_{i,r} = 2\,\phi(W)\,F\,s_r\sum_j \Delta_{ij}$.

Since in [1] the authors highlighted that a drawback of the Petaweb network may be a larger propagation delay for some connections, we decided to account it as a virtual cost of the network cost to minimize. Indeed, with proper design the traffic weighted propagation delay may be smaller than that of conventional networks. The *propagation delay cost*, indicated by β, is proportional to the distance traveled and to the lightpath bit-rate. This is an interesting addition to the classical equipment cost functions to guarantee that the solution is such that the connections between edge nodes that have the largest exchange of traffic experience as low a propagation time as possible.

The Petaweb design must respect the physical characteristics of network components. Capacity constraints concern edge nodes and optical links. The capacities can be allocated and increased only through discrete quantities: the link capacity can be increased by a multiple of the capacity of W lambda-channels at a time; the capacity of an EN depends on the number of optical fibers connected to it. Furthermore, to control the delay in buffering operations, all the TLPs of a Connection Request (CR) must be transported on the same optical trunk line, all the time-slots associated to a TLP must be transported on the same optical link, and all the TLPs of a CR must be transported contiguously in the time and in the frequency domains.

The design problem consists in finding the best composite-star physical topology for the given set of TLPs and in assigning to the TLPs their communications medium (wavelengths and time-slots). Hence it is jointly an optimal dimensioning and a resource assignment problem. It is divided into two sub-problems: Route and Fiber Allocation (RFA) problem, which treats the allocation of the resources guaranteeing an efficient routing, and the Wavelength and Time-slot Assignment (WTA) problem, which concerns the assignment of the allocated resources. The RFA problem gives rise to an ILP formulation that is solved with CPLEX, or with a specialized heuristic [3]. For the WTA problem a straightforward algorithm was devised in [4] and [5]. It assigns time-slots, wavelengths and fibers to TLPs starting form the solution of the resource allocation: each TLP-1 has one time-slot assigned, each TLP-2 one wavelength and each TLP-3 one fiber. The TLPs related to the same connection request have assigned contiguous time-slots and wavelengths, when possible. An example of the WTA solution is given in Section 5.4. A variant of the Petaweb design was recently presented in [5] where a Dedicated Path Protection (DPP) strategy was added to the network model proposed in [4] to tackle reliability issues. The idea is that for every working TLP (wTLP) a protection link-disjoint TLP (pTLP) is allocated [9]. Thus, in case of one trunk line failure all the wTLPs are recovered from the allocated pTLPs without an excessive signaling interruptions. In the 1+1 DPP case there would not be a signaling phase. In case of 1:1 DPP it makes sense to enable a shorter path for w-LPs, and a longer path to pTLPs to be used in case of failure along the working one. For this reason the optimization problem should give priority to wTLPs in the contention for short paths. The DPP strategy requires an additional constraint to allow the protection mechanism: every pTLP must be multiplexed on trunk lines different from those of the corresponding wTLP; in the Petaweb architecture this means that a pTLP must be switched to a different network site than that of its wTLP. Note that this constraint guarantees even the node protection since if a core node or part of it fails, all the affected paths can be restored by the receivers.

In this paper it is assumed that the Petaweb networks to be upgraded were first optimized using the DPP policy. To illustrate the differences in the architecture, we refer the reader to the following figures. Fig. 3a shows a 10-node quasi-regular

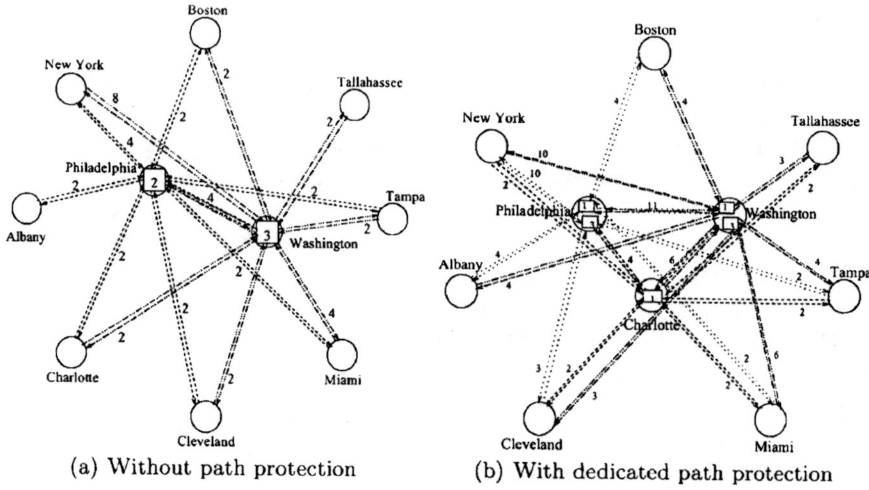

Figure 3. Optimized 10-node Petaweb. The numbers on the links represent the number of fibers per line

topology optimized without path protection [4] and Fig. 3b[1] shows the case with DPP [5]: one can notice that in the first case many edge nodes are connected to the network through only one trunk line whereas this does not happen in the second case when, instead, there is a larger quantity of fibers to install. Therefore, the path protection method produced a survivable quasi-regular Petaweb architecture.

3. REVIEWING UPDATE AND UPGRADE PROBLEMS

An increase in traffic volume imposes changes in the network configuration; there are two ways to face such an increase: by updating or by upgrading the network.

Updating a transport network means configuring new circuits to further exploit the available equipment and resources; in that case, a reconfiguration of the network virtual topology may be useful to free more resources via re-optimization and hence postponing network upgrades [10]. For example, in [11] the authors tackled the problem of accommodating an expansion of the original traffic matrix for a pre-optimized WDM mesh network with the restriction that no more physical equipment should be added to the existing infrastructure, and that only the existing idle capacity could be exploited without touching active lightpaths.

Upgrading a network means resizing its infrastructure and, optionally, reconfiguring its routes. An upgrade may require removal and/or addition of new equipment to satisfy a set of new end-to-end requests. In [12] the upgrade design problem for WDM mesh networks is solved through a methodology that exploits the idle

[1] The trunk lines connecting two switching sites are not link between core nodes, but links between edge node and core nodes, and vice-versa

capacity of an optimized network adding more resources if the idle ones are not enough. They do not consider reconfiguring the original connections.

In this article, we focus on the upgrade of the Petaweb architecture without reconfiguration; in an edge-controlled transport network, such as the Petaweb, the reconfiguration of the lightpaths routing would imply high data flows interruption for a significant gap of time. Moreover, the re-optimization of active lightpaths becomes no more an essential operation for this composite-star architecture because all the idle capacity is directly exploitable, differently than with meshed WDM networks. Another feature not explicitly taken into account in our model is the equipment removal. Even though that might be considered in some networks [13], it is not a real option in nationwide optical transport networks. In any case, this is a feature that can be easily incorporated into the model that we present in the next Section.

4. PETAWEB UPGRADE

The proposed upgrade model for the Petaweb not only considers the addition of new core node and fiber equipment, but also the exploitation of the idle capacity that is present in the initial architecture. In fact, in [5] we found that optimized survivable Petaweb networks still present a significant amount of idle capacity, which remains then available to accommodate subsequent bandwidth requests.

When the traffic volumes or the number of the connection requests between the existing edge nodes increase, they need new TLPs for which a route through a core node has to be decided. Thus, some switching sites that had few or low bit-rate Connection Requests, may now be updated with more core nodes, optical fibers and links. New switching sites may also be opened. In this work, we also consider that new edge nodes can be added to the network, which could imply that the opening of new switching sites is even more likely.

As previously stated, equipment removal is not considered in our update model. Therefore, the *upgrade cost* only includes the cost of the *new equipments* (i.e. fibers, core nodes and ports) and a propagation delay cost of the *new TLPs*. This latest term is added to make sure that the new TLPs are not routed on paths that will produce too long propagation delays.

The logic of the update model is to keep the initial Petaweb topology, whether it is regular or quasi-regular. Moreover, it is assumed that the existing optimized network was designed with a survivable strategy (DPP model) that is kept after the upgrade. Since we assume that existing core nodes and fibers cannot be removed, and that the number of new TLPs are likely to be fewer than the existing ones, the complexity of the upgrade problem is reasonably lower than that of the initial planning problem [4].

The logic of the upgrade model is to use the same type of objectives and constraints of [5] but forcing the design to keep the existing equipment, and altering the available capacity on each single link so that the media already in use is not considered to routie the new traffic.

Table 1. Notations

M	set of sites
T	set of pairs of sites $(M \times M)$, $p \in T$ is a Connection Request
O_j	a subset of T with a fixed origin site j
D_k	a subset of T with a fixed destination site k
V	set of types of core nodes
E_r	number of CN-r specimens that can be enabled in a site
(i, r, e)	triple representing a CN specimen, $i \in M, r \in V, 1 \leq e \leq E_r$
C_j	capacity, in Gb/s, of the edge node in site j, $j \in M$
H	set of TLPs classes
L_h	maximal number of TLP-h specimens for a CR, $h \in H$
(p, h, l)	triple representing a TLP specimen, $p \in T, h \in H, 1 \leq l \leq L_h$
d_{ip}	distance traveled going from the origin j to the destination k of the CR p passing by the site i: $d_{ip} = \Delta_{ij} + \Delta_{ik}$
y_{ire}	indicates if the eth CN-r specimen is enabled in the site i
x^{ire}_{phl}	indicates if lth TLP-h of CRp exists and is switched by the CN(i, r, e)
$(p, h, l+L_h)$	triple identifying uniquely the pTLP of the wTLP (p, h, l)
δ	weigh to give to the propagation delay cost of pTLPs,$0 \leq \delta \leq 1$
Ω_w	set of all wTLPs, $p \in T$, $h \in H$ and $0 < l \leq L_h$
Ω_p	set of all pTLPs, $L_h < l \leq 2L_h$
Ω	set of all TLPs, $0 < l \leq 2L_h$
χ	set of core nodes of the existing optimized network
Q^{ire}_j	pre-used capacity from site j to the existing CN (i, r, e)if $(i, r, e) \in \chi$
Q^k_{ire}	pre-used capacity from the existing CN (i, r, e) to site $k$if $(i, r, e) \in \chi$

The existing network is identified by all the enabled core nodes, the set of TLPs they commute, and by the number of fibers per optical link. From these, the used and the available transport capacity can be extracted and considered in the capacity constraints. Regarding the objective, it is worth noting that the optimization will be carried aiming at the minimization of the current total equipment costs. Thus, to assess the cost of the update, the cost of the equipment already installed will be subtracted.

The set of new TLPs (p, h, l) identifies the additional traffic volume, and the set M comprehends the pre-existing ENs sites and the new ones, if any. Thus, the solution is an optimized network with a regular or a quasi-regular topology, it indicates where the new TLPs must be routed and the equipment that have to be installed to satisfy the additional traffic. The mathematical formulation of the problem is the following. Additional notations are displayed in Table 1.

$$\min G(\overline{y}, \overline{x}) = \sum_{(i,r,e)} (K_r + F_{i,r})\, y_{ire}$$

$$(1) \qquad + \sum_{(i,r,e)} \sum_{(p,h,l)\in\Omega_w} \beta d_{ip} Z_h x^{ire}_{phl} + + \sum_{(i,r,e)} \sum_{(p,h,l)\in\Omega_p} \delta\beta d_{ip} Z_h x^{ire}_{phl}$$

$$(2) \qquad s.t. \quad y_{ire} = 1 \quad \forall (i, r, e) \in \chi$$

(3) $$\sum_{r \in V} \sum_{e=1}^{E_r} x_{phl}^{ire} + \sum_{r \in V} \sum_{e=1}^{E_r} x_{phl_p}^{ire} \leq 1 \ \forall i \in M, \forall (p, h, l) \in \Omega_w, l_p = l + L_h$$

(4) $$\sum_{(i,r,e)} x_{phl}^{ire} = 1 \quad \forall (p, h, l) \in \Omega$$

(5) $$\sum_{(i,r,e)} C_{ch} W s_r y_{ire} \leq C_j \quad \forall j \in M$$

(6) $$\sum_{(p \in O_j, h, l) \in \Omega} Z_h x_{phl}^{ire} \leq \left(C_{ch} W s_r - Q_j^{ire}\right) y_{ire} \quad \forall j \in M, \forall (i, r, e)$$

(7) $$\sum_{(p \in D_k, h, l) \in \Omega} Z_h x_{phl}^{ire} \leq \left(C_{ch} W s_r - Q_{ire}^{k}\right) y_{ire} \quad \forall k \in M, \forall (i, r, e)$$

(8) $$x_{phl}^{ire} \in \{0, 1\}, \quad y_{ire} \in \{0, 1\}$$

The objective (1) includes the cost of switches and fiber plus two cost terms to account for propagation delay: one for the pTLPs and one for the wTLPs. Note that the two terms are ponderated differently to avoid that the pTLP and its corresponding wTLP contend for the same shortest path. (2) enforces the enabling of the existing core nodes; (3) is the protection constraint; (4) ensures that a TLP must be switched only by one CN; (5) enforces EN capacity constraint; (6) and (7) enforce the capacity constraints on the idle capacity for the optical links going from every core node and every edge node, and vice versa, subtracting the already occupied transport capacity; (8) defines the binary domain of the variables.

As it was already mentioned, the upgrade cost is obtained subtracting from the final objective value the equivalent cost of the pre-existing network. Also mentioned was the fact that the upgrade aims at a regular topology. Then if the initial topology was quasi-regular and the planner intends the update to keep a quasi-regular structure, the quasi-regular topology can be extracted from the regular one.

To extract the quasi-regular topology one proceeds taking into account every optical link in the optimized regular network, looking for how much of its fibers would be used by the TLPs routed there, and disabling those fibers that would not be used at all. So, a whole optical link may be disabled in the quasi-regular topology, and, also, a whole trunk line may be disabled [4] (e.g. see Fig. 3). Moreover, even the ports associated to the disabled fibers are not considered in the quasi-regular architecture. Hence the cost reduction concerns the cost of unused fibers and ports. Note that the TLPs remain associated to the same core node than in the regular topology and that the routes are not affected by the disabling of fibers and ports.

5. UPGRADE RESULTS

The initial network status is defined by the 10-node networks dimensioned in [5]. We consider two scenarios: simple traffic increase and traffic increase with edge node additions. Moreover, we consider two types of traffic matrixes: **A** matrixes contain industrial traffic data, with many zero values; **B** matrixes are dense and are obtained from the well known gravity model, used for example in [14]. An element of a traffic matrix is a CR of an origin-destination pair, which is accommodated in the physical topology using one or more TLPs. The choice of parameters is: $E_1 = 1$, $E_2 = 1$, $E_3 = 4$, $\gamma = 0.95$, $P/F = 150$, $\beta/F = 0.1$ $[Km\,Gb/s]^{-1}$, $f_1/F = 20$, $f_2/F = 50$, $f_3/F = 100$, $C_j = 2000$ Gb/s, $L_1 = L_2 = 12$, $L_3 = 20$, $\delta = 0.9$. The CPLEX MIPGAP was set to 0.1%. We employed $\phi(W) = W$ considering that the cost of a fiber is proportional to the number of wavelengths. Other functions can also be considered. The simulations ran on a CPU AMD Opteron 64bit 2.4Ghz, 1MB cache, 16GB RAM.

5.1. Traffic Increase

In this study case the traffic of every existing Connection Request is increased by 200%. The left side of Table 2 reports the upgrade results obtained solving the formulation (1)–(8) for the 10A and the 10B pre-planned network with regular and quasi-regular topologies.

The results show that the network utilization μ_R, defined as the ratio between the used and the available capacities, increases for both topologies as illustrated in Fig. 4 (data extracted from [5]). Such an increase is more important for the regular topology. This is due to the equipment already installed that allow the new TLPs to be routed more efficiently than with the quasi-regular topology. This behavior seems to be confirmed by the average path length (weighted on the traffic unit), indicated by ν in Table 2; it is slightly bigger with quasi-regular topologies. This

Table 2. Upgrade solutions

	traffic increase				traffic increase and nodes addition			
	regular topology		quasi-reg. top.		regular topology		quasi-reg. top.	
Model	10*A*	10*B*	10*A*	10*B*	10*A*	10*B*	10*A*	10*B*
cost	856983	975431	1304076	717565	4817336	3471536	2984283	1670231
				upgrade cost distribution				
fiber	35.7%	74.9%	52.3%	69.6%	73.1%	32.2%	60.2%	75.0%
CN	5.7%	10.7%	6.3%	11.2%	8.8%	2.7%	8.9%	9.7%
delay	58.6%	14.4%	41.4%	19.2%	18.1%	65.1%	30.9%	15.3%
				global cost distribution				
fiber	70.9%	80.9%	60.8%	71.1%	75.4%	82.7%	62.6%	73.3%
CN	10.2%	11.6%	10.0%	13.4%	9.9%	10.7%	10.3%	12.1%
delay	18.9%	7.4%	29.2%	15.5%	14.7%	6.5%	27.0%	14.6%
μ_R	31.6%	22.4%	54.1%	45.6%	26.9%	17.5%	51.7%	38.3%
ν	2924	894	2953	911	1719	1151	1717	1138
time (s)	1.3	23.9	1.2	36.2	1269	128	1178	86

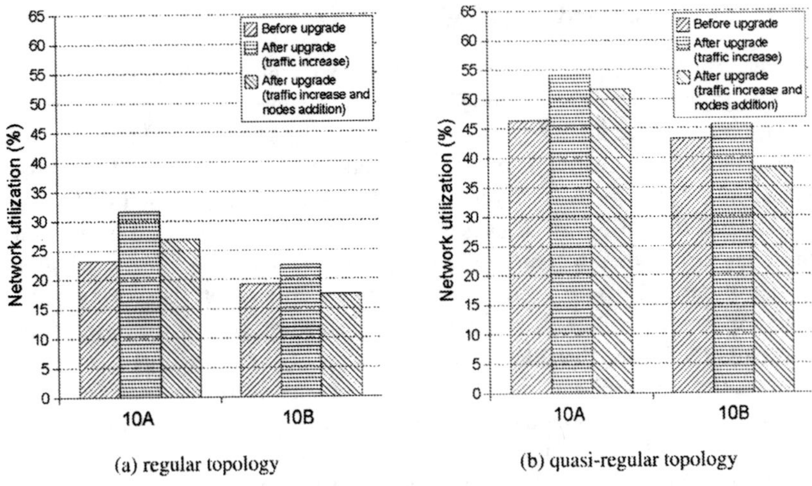

(a) regular topology

(b) quasi-regular topology

Figure 4. Network utilization before and after the 10A upgrade

seems to indicate that, if a network operator foresees to make regular upgrades of its network and want to route its TLPs in the most effective way, the cost to pay is the initial regularity.

For example, for the 10A model, the added traffic amount exploits mainly the idle capacity without enabling lots of new switching planes; indeed, the network utilization (μ_R) went from 23.19% and 46.39% (data extracted from the solutions in [5]) to 31.6% and 54.1%, and the weight of the fiber cost felt of 5–7 percentage points (as depicted in Fig. 5). In this case one can notice how the upgrade cost is bigger for a quasi-regular topology than for a regular topology; in the first case one has to install fibers that, instead, with a correspondent regular topology may have already been installed.

In Table 2 the cost distribution concerning only upgrade costs and the one concerning the whole network equipments (those installed before the upgrade together with those installed after) are portrayed. For comparison purposes the global cost distribution before [5] and after the upgrade for the 10A case are illustrated in Fig. 5. We can see that the most remarkable effect of the new TLPs and subsequent network upgrade is an increase of the weight of the cost due to propagation delays and, thus, a decrease of the fibers cost and of the core node cost weights. The upgrade cost fraction due to new fibers and core nodes is minor if compared to the one related to the whole network; on the contrary, the upgrade fraction due to the delay of the new TLPs is significantly bigger than the one related to the whole network. This confirms that the upgrade tends to exploit the existing resources rather than requiring new ones. And the difference is more evident for the upgrade of a quasi-regular topology, because the existing fibers are better exploited, and, even if new fibers are placed, the overall fibers cost weight still decreases.

5.2. Traffic Increase and Edge Nodes Addition

In this study case we increased of 200% the existing Connection Requests, and we added 4 ENs to the existing ones (the Connection Requests of the added ENs are extracted from the 34A and 34B matrixes used in [5]). The right side of Table 2 displays the upgrade results obtained for 10A and 10B with, respectively, regular and quasi-regular topologies. The resulting networks are composed of 14 ENs. As expected, the addition of new ENs causes a large upgrade cost because of the new trunk lines that are needed to connect the new ENs to at least two CNs (because of the DPP constraint). Fig. 5 shows how the equipment cost weights are still smaller than the correspondent value for the pre-existing network, but, in the case of the fiber cost, slightly higher than the value for the case with only traffic increase. Fig. 4 reflects that under EN addition the overall network utilization may decrease, as it happens for the 10B case; indeed, the new installed trunk lines are under-used with respect to the old ones that were better exploited.

Observing the upgrade cost in the two cases we notice that it is lower for an existing quasi-regular topology. When new edge nodes are added to the network, they can be integrated installing new equipment, mainly new optical links. And with quasi-regular topologies these new optical links are composed only of the essential number of fibers, and nothing more. We can conclude that the upgrade with ENs addition is more convenient if one adopts a quasi-regular topology; the cost gain is significant and the network operator may prefer to start with a quasi-regular topology and to upgrade it only in case of new ENs addition. Until new ENs have to be added, it may be possible to accommodate increases of traffic only exploiting the present idle capacity, without additional physical equipments, i.e. through updates (see section 3); this update method may be a subject for further work (similar to the method used in [11] for mesh networks).

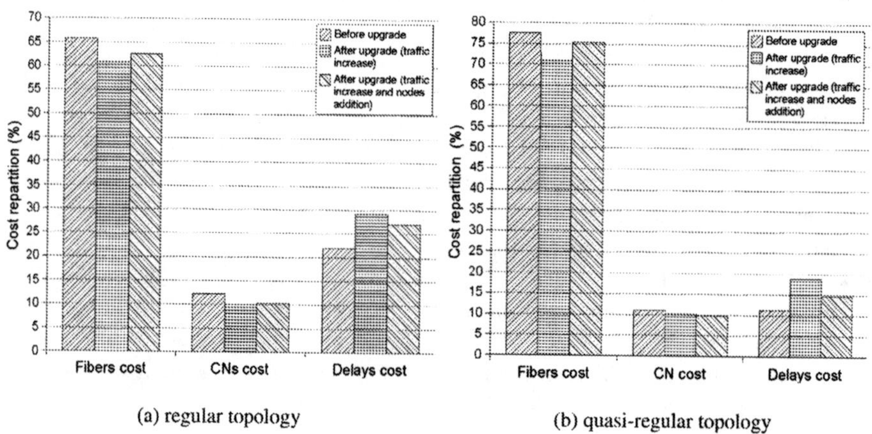

(a) regular topology (b) quasi-regular topology

Figure 5. Cost distribution before and after the 10A upgrade

5.3. Comparison with a Greedy Upgrade

In this section we comment on the results obtained applying an upgrade method based on a straightforward greedy strategy when compared with the results obtained with the method proposed in this paper. The greedy upgrade can be described as follows: when a new TLP is created, it is switched to the closest switching site with core nodes already installed. Then the two trunk lines supposed to route the TLP may be opportunely resized and new core nodes may be installed at that site. Note that the edge node capacity constraint (5) may not be respected for regular topologies. In such a case, the edge node should be replaced.

We analyze the results for the two study cases 5.1 and 5.2 The behavior of the greedy method is the same for an existing regular or quasi-regular architecture. In either case, the resulting network is suboptimal as can be seen in Table 3 where the gaps with respect to the optimal solution given by the upgrade are depicted. It can be seen that the greedy update may yield a solution costing twice as much as a solution produced by the optimized procedure. Interestingly, the worst differences are produced with the 10A matrices.

In the 10B cases the upgrade cost is not too large compared with the previous values; only one new switching plane was required. But, along with the 10A cases, we can see the worst values of fiber cost and network utilization: the route for TLPs was not carefully chosen. In terms of average path length, the greedy method gives better values than the optimal method, this can be seen by the gaps with

Table 3. Greedy upgrade solutions

	traffic increase				traffic increase and nodes addition			
	regular topology		quasi-reg. top.		regular topology		quasi-reg. top.	
Model	10*A*	10*B*	10*A*	10*B*	10*A*	10*B*	10*A*	10*B*
Cost	1874996	999035	1369553	1029584	6894998	3551366	3278257	1791541
gap	+118%	+2%	+5%	+43%	+43%	+2%	+9%	+7%
			upgrade cost distribution					
fiber	69.2%	75.6%	43.2%	66.7%	76.5%	83.9%	60.8%	69.8%
CN	10.1%	11.3%	5.8%	12.7%	9.6%	9.4%	8.7%	10.7%
delay	20.7%	13.1%	51.0%	20.6%	13.9%	6.7%	30.5%	19.5%
			global cost distribution					
fiber	75.3%	81.0%	63.7%	69.9%	77.1%	83.0%	62.8%	70.9%
CN	10.8%	11.7%	10.9%	13.7%	10.2%	10.7%	10.1%	12.5%
delay	13.8%	7.3%	25.4%	16.5%	12.7%	6.3%	27.1%	16.6%
μ_R	24%	21.5%	49.1%	47%	20.5%	16.7%	50.4%	42.6%
ν	2375	871	2375	871	1798	1074	1798	1074
gap	−23%	−2.6%	−24%	−4.5%	+4.6%	−7%	+4.7%	−5.9%

respect to the optimal solutions that are negative in almost all the instances. Clearly, plugging the new connections to the closer core nodes produces an improvement in overall path length, but this is often a more expensive choice with respect to the cost model. The upgrade cost distribution has a behavior very close to that of the global cost distribution: the greedy method can not profit efficiently of the available resources.

5.4. WTA Results

The dimensioning phase presented in the previous sections provides the equipment to be installed and the switching node assigned to each TLP. The next phase to complete the design is to apply the WTA algorithm [4] that allocates transport units (time-slots, wavelengths and fibers) to every new TLP. In this section the allocation of resources for the study case in 5 are illustrated, considering, for sake of simplicity, only the outgoing fibers of the EN in Tallahassee (the one connected by two trunk lines in Fig. 3b).

Let us concentrate on node 9. Before the update, the node had three TLP-1 for $CR_{9,8}$ and one TLP-1 for $CR_{9,10}$ opportunely protected as reported in Fig.6a from [5]. After the upgrade, the volume of the pre-existing CRs of node 9 increased and the additional traffic has to be served by nine more TLP-1, six for $CR_{9,8}$ and three for $CR_{9,10}$. Moreover, a new EN is added in the Chicago site (site 13) and the new $CR_{9,13}$ has to be served by two TLP-1.

Figure 6 illustrates the routing and the assignment of the new TLPs. As it can be noticed, the TLPs of the $CR_{9,8}$ and $CR_{9,10}$, as well as the pTLPs of the $CR_{9,13}$, occupy the free time-slots on the already installed fibers. The wTLPs of $CR_{9,13}$ are switched in a CN in site 4 and transported on the first two time-slots of the first wavelength on the only fiber connecting the EN-9 to the switching site 4. Consequently, the utilization of the pre-existing fibers increases, while the utilization of the new fiber between 9 and 4 remains reasonably low.

6. CONCLUSIONS

In this paper, a formulation and a resolution approach for the Petaweb upgrade problem were presented for the first time. The analysis shows the scalability qualities of the Petaweb architecture. On the basis of the shown results, it can be concluded that a regular topology is advisable if the network operator has a good initial budget and if frequent upgrades are foreseen; a quasi-regular topology is the best choice in case of low budget and rare upgrades, especially when the upgrade contemplates edge nodes addition.

The proposed method underlined the importance of conducting a cost-effective upgrade when compared with the common practical paradigm “plug where it is closer”, often used in the industry; indeed, such greedy upgrade provisioning method applied to the Petaweb architecture can bring a lower network utilization at significantly higher cost.

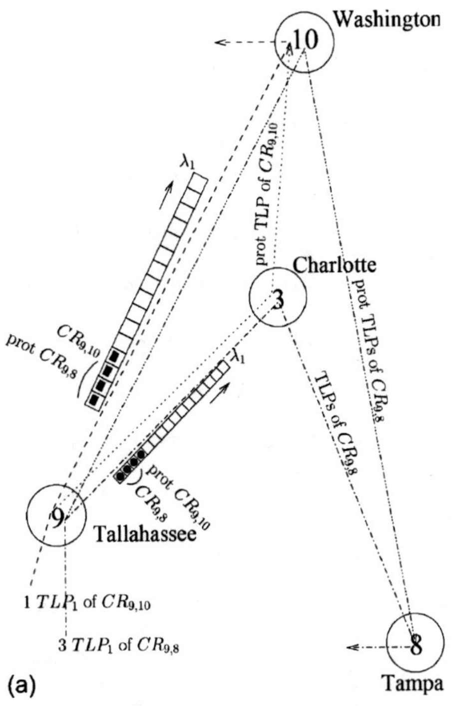

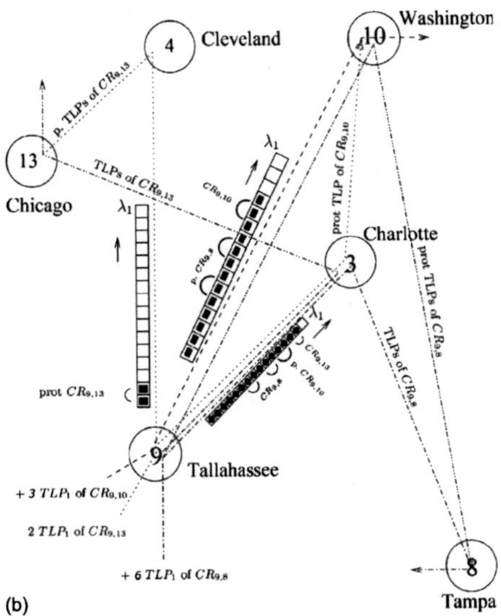

Figure 6. WTA results before (a) and after (b) the upgrade. 5.2 case (10A)

REFERENCES

1. R. Vickers, M. Beshai, "Petaweb architecture", presented at *Networks 2000: 9th Int. Tel. Network Planning Symposium*, Toronto (2000).
2. F. J. Blouin, A. W. Lee, A. J. Lee, M. Beshai, "A comparison of two optical-core networks", *J. Opt. Netw.* 1, 56–65 (2001).
3. A. Reinert, B. Sansò, S. Secci, "Design Optimization of the Petaweb Architecture", GERAD 2004–87, submitted to IEEE Transactions on Networking.
4. S. Secci, B. Sansò, "Design and dimensioning of a novel composite-star WDM network with TDM channel partitioning", in *Proc. of IEEE Broadnets'06.*
5. S. Secci, B. Sansò, "Optimization of a dedicated path protected TDM/WDM Petaweb architecture", in *Proc. of IFIP/INFORMS NAEC 2006.*
6. N. F. Huang, G. H. Liaw, C. P. Wang, "A novel all-optical transport network with time-shared wavelength channels", *J. Sel. Areas Comm.* 18, 1863–1875 (2000).
7. Y. C. Huei, P. H. Keng, "Framework for shared time-slot TDM wavelength optical WDM networks", *J. Opt. Netw.* 5, 554–567 (2006).
8. "Interfaces for the Optical Transport Networks (OTN)", G.709/Y.1331, ITU-T International Communication Union, Mar. 2003.
9. G. Maier et al., "Optical network survivability: protection techniques in the WDM layer", *Photonic Network Communications* 4, no. 3/4, 251–269 (2002).
10. Lu Shen et al., "A load-balancing shared-protection-path reconfiguration approach in WDM wavelength-routed networks", in *Proc. of OFC 2004.*
11. L. Barbato, G. Maier, A. Pattavina, "Maximum traffic scaling in WDM networks optimized for an initial static load", in *Proc. of ONDM 2003.*
12. Y. Jintae, I. Yamashita, S. Seikai, K. Kitayama, "Upgrade design of survivable wavelength-routed networks for increase of traffic loads", *Proc. of ONDM'05.*
13. S. Chamberland, B. Sansò, "Update of two-level networks with modular switches", *International Teletraffic Congress*, pp. 1009-1018 (1999).
14. S. Secci, M. Tornatore, A. Pattavina, "Optimal Design for Survivable Backbones with End-to-End and Subpath Wavebanding", *J. Opt. Netw.*, Vol.6, no. 1 (2007).

4. A GENETIC ALGORITHM-BASED APPROACH FOR COMBINING RULE SETS

Our approach relies on the idea of combining different classifiers built from some source of knowledge (common domain, for example) and adapting the resulting model(s) to a new/unseen data set (company-specific data). Combining classifiers is a difficult problem due to the size of the search space. Exhaustive and local search methods are inefficient since the problem involves a large set of classifiers that use different metrics. Genetic algorithms offer an interesting approach to this problem. The basic idea is to form a population of chromosomes (a pool of classifiers). New populations are created by combining and modifying classifiers to obtain new (ideally better) ones. In the rest of this section, we describe how we instantiate the elements of the GA (chromosome encoding, genetic operators and fitness function).

4.1. Model Coding and Fitness Function

The coding is straightforward. In our problem, a chromosome represents a rule-based classifier (rule set). Each condition and class label is encoded as a gene in the chromosome. They are listed in the order of their appearance in the rule set (Fig. 1). The objective function to optimize is the correctness of the rule set on the training data. So, the fitness of a chromosome is equal to the correctness of the classifier that the chromosome encodes.

4.2. Elitism, Crossover, and Mutation

The GA starts by copying a percentage of the chromosomes to the next population (elitism)[1] then the population is completed by crossover and mutation. Two chromosomes are selected[2] and a cut point is generated randomly in each of them. The offspring

Rule set

Rule :

NMO>1 AND NMI>22 AND SIX>1 – >class 0

Rule2:

NOD>8 AND NMI>22 – > class 1

Default class: 0

Chromosome

NMO>1 NMI>22 SIX>1 C0 NOD>8 NMI>22 C1 C0

gene1 gene4

Figure 1. A rule set and its chromosome representation: Each rule and condition in the rule set is represented in one gene in the chromosome

[1] This percentage is a parameter input to the algorithm.

[2] Fitter chromosomes get a higher probability of being selected.

Parent1

NMO>1 NMI>22 SIX >1 C0 NOD >8 NMI >22 C1 C0

Parent2

DIT>2 NMI>22 C1 NOD>8 C0 C1

Offspring 1

NMO>1 NMI>22 C1 NOD>8 C0 C1

Offspring 2

DIT>2 NMI>22 SIX>1 C0 NOD>8 NMI>22 C1 C0

Figure 2. Crossover where cutpoints (arrows) fall within a rule in chromosomes

are formed by exchanging the tails (sections after the cutpoints) of the parents. A cutpoint can fall within a rule or on a rule boundary (between two rules or between a rule and the default classification label) in a chromosome. In order to insure the validity of the resulting chromosomes, we apply the following "tuning" to our crossover operator: If cut point i_1 falls within a rule in parent p_1, cut point i_2 has to fall within a rule in parent p_2. If i_1 falls on a rule boundary in p_1, i_2 has to fall on a rule boundary in p_2. Moreover, the operator is implemented in such a way that cut points are not allowed to fall on chromosome boundaries (Fig. 2 and 3). The offspring chromosomes are mutated before being copied into the next population. Each one is scanned and each gene is changed with a certain probability. If the gene represents a condition ($NOC \leq 5$, for example), mutation consists of changing the value to which the attribute is compared to a value chosen randomly from the set of cutpoints of the attribute[3] ($NOC \leq 2$, for example). We chose cutpoints because these are more likely to change the set of cases that the rule classifies [14]. If the gene encodes a classification label, mutation consists of changing this value to one taken randomly from the domain of the classification labels (changing 0 to 1 for example).

Parent1

NMO>1 NMI>22 SIX >1 C0 NOD >8 NMI >22 C1 C0

Parent2

DIT>2 NMI>22 C1 NOD>8 C0 C1

Offspring 1

NMO>1 NMI>22 SIX>1 C0 NOD>8 C0 C1

Offspring 2

DIT>2 NMI>22 C1 NOD>8 NMI>22 C1 C0

Figure 3. Crossover where cutpoints (arrows) fall on a rule boundary in chromosomes

[3] A cutpoint of an attribute is the median of the two values of the attribute where the classification label changes [11].

4.3. Trimming

The operators described above can result in rule sets that contain *redundancy* and *inconsistency*. Both can occur within a rule or among rules. Redundancy inside a rule occurs when the rule has two conditions c_i and c_j where c_i implies c_j (example, c_i is $NOC \leq 4$ and c_j is $NOC \leq 6$). The offspring are scanned and $\forall i, j$, if c_i implies c_j, c_j is dropped. Redundancy can also occur among rules when two or more rules in the rule set have the same conditions (the order is not important). This is eliminated by keeping one of the rules only. Inconsistency, on the other hand, is allowed within a rule as long as evolution is ongoing. A rule is inconsistent when it has two conditions c_i and c_j that cannot be true at the same time (example, c_i is $NOC \leq 3$ and c_j is $NOC > 5$). Inconsistency is allowed because several experiments have shown that some conditions, although "fatal" in some combination, can have good impact when combined with other conditions. Inconsistency among rules is retained. Contradicting rules are allowed to remain in the rule set as one of them only will contribute to the classification of the rule set.

5. EXPERIMENTATION

The software quality that we experimented with is the stability of OO components (classes in an OO software system). We say that a component is *stable* when its public interface remains unchanged between two consecutive versions of the class. Otherwise, we say that the class is *unstable*. Stability of a class is affected by the structure of the class and the stress induced by the implementation of new requirements (these requirements are typically added methods) between two versions [15]. For this reason, we chose structural metrics that belong to one of the four categories: cohesion, coupling, inheritance, and size complexity as well as the stress metric. We validate our approach on the data sets described in [15] and [16].

5.1. Data Collection

STAB1[16]- Nineteen structural metrics (Table 1) were extracted from the eleven software systems[4] shown in Table 2 using the ACCESS tool and the Discover environment©[5]. Detailed description of these metrics can be found in [2], [17], [18], [3], and [19]. Fifteen subsets (of size 2, 3, and 4) were created by combining these metrics in all possible ways. The combination was based on the relationship desired among the different quality characteristics[6]. The 15 subsets of metrics were used with the 11 software systems to create $15X11 = 165$ data sets. C4.5 was

[4] JDK versions are available at http://java.sun.com/api/index.html.

[5] Available at http://www.mks.com/products/discover/developer.shtml.

[6] For example, the relationship between cohesion and coupling on one hand and stability on the other.

Table 1. STAB1-Software quality metrics used as attributes in the classifiers

Name	Description
Cohesion metrics	
LCOM	lack of cohesion methods
COH	cohesion
COM	cohesion metric
COMI	cohesion metric inverse
Coupling metrics	
OCMAIC	other class method attribute import coupling
OCMAEC	other class method attribute export coupling
CUB	number of classes used by a class
Inheritance metrics	
NOC	number of children
NOP	number of parents
DIT	depth of inheritance
MDS	message domain size
CHM	class hierarchy metric
Size complexity metrics	
NOM	number of methods
WMC	weighted methods per class
WMCLOC	LOC weighted methods per class
MCC	McCabe's complexity weighted meth. per cl.
NPPM	number of public and protected meth. in a cl.
NPA	number of public attributes
The stress metric	
STRESS	stress applied to the class

used to create 165 decision tree classifiers with these data sets. Constant classifiers (classifiers that have a single classification label) and classifiers with a training error more than 10% were eliminated and the remaining 40 were retained. These decision trees were then converted into rule sets by C4.5. Our GA uses these rule sets as experts built from common domain knowledge and combines and adapts them to the four software systems shown in Table 3. These form a data set which consists of 2920 instances and simulate the in-house data.

STAB2[15]- The previous data set is highly imbalanced. As a matter of fact, 2481 cases are stable and 439 unstable. In order to test our algorithm on a balanced data set, we used STAB2 which also involves stability. Table 6 shows the 22 software metrics used. Nine of the 11 software systems shown in Table 2 were used to build experts with C4.5 (Table 4). Fifteen subsets of metrics were created by combining 2, 3, or 4 groups in all possible ways. These subsets were used with the 9 chosen software systems to create 135 data sets. C4.5 was used to construct a decision tree from each data set. Constant classifiers and classifiers with an error rate higher than 10% were eliminated and 23 retained. The decision trees were then converted to rule sets by C4.5. Our GA uses the systems shown in Table 5 to train and test the GA (in-house data simulation).

Table 2. STAB1-Software systems used to build classifiers with C4.5

Software System	Number of versions (major)	Number of classes
Bean browser	6(4)	388–392
Ejbvoyager	8(3)	71–78
Free	9(6)	46–93
Javamapper	2(2)	18–19
Jchempaint	2(2)	84
Jedit	2(2)	464–468
Jetty	6(3)	229–285
Jigsaw	4(3)	846–958
Jlex	4(2)	20–23
Lmjs	2(2)	106
Voji	4(4)	16–39

Table 3. STAB1–Software systems used to train and test the GA

JDK version	Number of classes
jdk1.0.2	187
jdk1.1.6	583
jdk1.2.004	2337
jdk1.3.0	2737

Table 4. STAB2–Software systems used to build classifiers with C4.5

Software System	Number of versions (major)	Number of classes
Bean browser	6(4)	388–392
Ejbvoyager	8(3)	71–78
Free	9(6)	46–93
Javamapper	2(2)	18–19
Jchempaint	2(2)	84
Jigsaw	4(3)	846–958
Jlex	4(2)	20–23
Lmjs	2(2)	106
Voji	4(4)	16–39

5.2. Algorithmic Settings

We performed two sets of experiments. In the first one, we used rule sets constructed by C4.5 to seed our GA. In the second set, we used randomly generated rule sets (the size of the random rule sets ranges between 5 and 70 rules per rule set and 2 to 30 conditions per rule). This allows us to test our approach with any rule sets not

Table 5. STAB2–Software systems used to train and test the GA

Software System	Number of versions (major)	Number of classes
Jedit	2(2)	464–468
Jetty	6(3)	229–285

Table 6. STAB2-Software quality metrics used as attributes in the classifiers

Name	Description
Cohesion metrics	
LCOM	lack of cohesion methods
COH	cohesion
COM	cohesion metric
COMI	cohesion metric inverse
Coupling metrics	
OCMAIC	other class method attribute import coupling
OCMAEC	other class method attribute export coupling
CUB	number of classes used by a class
CUBF	number of classes used by a member function
Inheritance metrics	
NOC	number of children
NOP	number of parents
NON	number of nested classes
NOCONT	number of containing classes
DIT	depth of inheritance
MDS	message domain size
CHM	class hierarchy metric
Size complexity metrics	
NOM	number of methods
WMC	weighted methods per class
WMCLOC	LOC weighted methods per class
MCC	McCabe's complexity weighted meth. per cl.
DEPCC	operation access metric
NPPM	number of public and protected meth. in a cl.
NPA	number of public attributes

just ones created with a machine learning technique. In both sets of experiments, the GA evolves the rule sets and adapts them to the in-house data (tables 3 and 5 The parameters to our GA had the following values[7]: We set crossover probability to 90%, mutation rate to 10%, percentage of chromosomes copied by elitism to 10% and number of generations through which the GA iterates to 50.

[7] Several experimentations with the algorithm showed that the parameter values have very little effect on the best accuracy obtained by this GA.

5.3. Results

To accurately estimate the accuracy of the rule sets, we used 10-fold cross-validation (10 is a commonly used number). In this technique, the data set is randomly split into 10 folds. Nine of them are used to train the GA and the remaining one to test it. The process is repeated for all 10 possible combinations. Also, in order to account for the element of randomness in the GA, we repeated each experiment 30 times and averages (and standard deviations) over the 30 runs are reported in tables 7 and 8.

The tables show the correctness and J_index on both the training and the testing sets of the best rule set constructed by C4.5, the best constructed by GA, the best constructed by GA when seeded with random rule sets (GA-R), the average of randomly generated rule sets (R) and the majority classifier (Maj.- the constant classifier that always predicts the majority classification label).

Since STAB1 is unbalanced, the majority classifier has a high accuracy while it performs poorly on the minority class. This makes it very important to assess the average correctness per classification label (J_index) on this data set. Table 7 shows that the GA significantly outperforms C4.5, random guess and the majority classifier in the J_index. It also shows that GA-R outperforms the three classifiers. It is important to point out that the objective function that the GA optimizes is the correctness of the rule set encoded by a chromosome. Hence, we believe that if we include the J_index in the fitness function, the improvement will be more significant.

In the case of STAB2, the GA slightly outperforms C4.5 and significantly random guess in both the correctness and the J_index. We believe that the slight improvement is due to the noisy nature of the data. Also, improving correctness on a balanced data set is more difficult than improving it on an unbalanced data set

Table 7. Results on STAB1. Correctness and J_index (standard deviation)

	$C_{testing}$	$C_{training}$	$J_{testing}$	$J_{training}$
C4.5	85(2)	85(0.5)	50(2)	50(0.5)
GA	85.5(3)	86(1)	59(4)	60.5(3)
GA-R	85(3)	85.5(1)	55.5(6.5)	56(6)
R	29.64(30.89)	29.71(30.84)	49.79(1.08)	50.15(0.27)
Maj.	85	85	42.5	42.5

Table 8. Results on STAB2. Correctness and J_index (standard deviation)

	$C_{testing}$	$C_{training}$	$J_{testing}$	$J_{training}$
C4.5	68(6)	68(0.5)	58(4)	58(0.5)
GA	70(6)	74.5(1)	60.5(5)	65(3)
GA-R	69(6)	73(3)	60(6)	65(4)
R	28(30.27)	30(28.23)	50.11(0.21)	50.12(0.21)
Maj.	62	62	31	31

Table 9. Complexity of rule sets: rules per rule set (conditions per rule)

	STAB1	STAB2
C4.5	2–7(1–5)	2–11(1–6)
GA	1–3(1–2)	1–5(1–3)
GA-R	2–5(1–4)	2–7(1–4)

[20]. The low accuracies in general, indicate that the problem of predicting stability of software components is a difficult one.

One important quality of our GA is the low complexity of the rule sets that it constructs. These have fewer attributes and conditions per rule than those that C4.5 builds (Table 9) which makes them easier to interpret by human experts. Another interesting observation is the speed with which the GA finds the "best" rule set (within the first 10 to 15 generations when seeded with C4.5 rule sets, 35 or later when seeded with random rule sets[8]). This explains the small number of generations mentioned earlier (50).

6. CONCLUSION AND FUTURE WORK

In this paper, we proposed an adaptive approach to optimize existing software component quality estimation models. The approach starts with several rule sets and searches for a combination of rules, derived from them, that makes better rule sets with respect to a new data set. In real applications, this can be seen as taking rule sets built from common domain knowledge and finding a combination of their "expertise" that results in a rule set with a higher prediction accuracy when used on context specific data. The approach is derived from the genetic algorithm methodology. We have conducted experiments with two different data sets involving the stability of classes in an object-oriented system. One of the two stability data sets is unbalanced. The GA outperforms C4.5, random guess and the majority classifier significantly on the unbalanced data set but only slightly on the balanced one. The GA outperforms C4.5 when seeded with random rule sets. Our approach is independent of the software quality characteristic that is being estimated and hence it is interesting to see how well it performs on other characteristics (such as maintainability, reusability, etc.) especially with data containing less noise. The end result of our GA is a rule-based classifier. This provides two separate utilities: the estimation of a quality characteristic (stability in this case) and guidelines that can help software engineers to attain it. Due to the elitism operator, our technique guarantees no damage to results. In most cases, the rule set found by the GA is different from the initial one and hence, can be used as an additional guideline

[8] The difference is due to the size of the initial rule sets. Random rule sets are bigger in size than C4.5 rule sets so this provides a bigger search space.

during the development phase of a software product. In all cases, our approach results in rule sets that have a lower complexity than those constructed by C4.5. Future work involves training the GA on the J_index and comparing it to another heuristic (such as simulated annealing). Another interesting modification would be in the design of the genetic operators in a way that modifies the attributes in the conditions (not only the values) and also taking the gain factor into consideration in the objective function.

7. ACKNOWLEDGMENT

The first author's research was partially supported by a grant from CNRS-Lebanon (Conseil National de La Recherche Scientifique) and from the University Research Council at the Lebanese American University.

REFERENCES

1. G. M. Barnes and B. R. Swim, "Inheriting software metrics," *JOOP*, vol. 6, no. 7, pp. 27–34, Nov./Dec. 1993.
2. L. Briand, P. Devanbu, and W. Melo, "An investigation into coupling measures for C++," in *Proceedings of the 19th International Conference on Software Engineering*, 1997.
3. S. Chidamber and C. Kemerer, "A metrics suite for object-oriented design," in *IEEE Transactions on Software Engineering*, 1994, vol. 20, pp. 476–493.
4. J. C. Coppick and T. J. Cheatham, "Software metrics for object-oriented systems," in *CSC '92 Proceedings*, 1992, pp. 317–322.
5. Henderson-Sellers, *Object-Oriented Metrics: Measures of Complexity*. Prentice-Hall, 1996.
6. B. Henderson-Sellers, "Some metrics for object-oriented software engineering," in *TOOLS Proceedings*, 1991, pp. 131–139.
7. W. Li and S. Henry, "Maintenance metrics for the object-oriented paradigm," in *Proceedings of the First International Software Metrics Symposium*, Baltimore, Maryland, 1993, pp. 52–60.
8. W. Li and S. Henry, "Object-oriented metrics that predict maintainability," *J. Systems Software*, vol. 23, pp. 111–122, 1993.
9. M. Lorenz and J. Kidd, *Object-Oriented Software Metrics: A Practical Approach*. Prentice-Hall, 1994.
10. G. D.E. Rumelhart, Hinton, and R. Williams, "Learning representations by back-propagation errors," *Nature*, vol. 323, 1986.
11. J. R. Quinlan, *C4.5: Programs for Machine Learning*. Morgan Kaufmann, 1993.
12. J. Holland, *Adaptation in Natural and Artificial Systems: an introductory analysis with applications to biology, control, and artificial intelligence*. Ann Arbor, MI: University of Michigan Press, 1975.
13. C. Darwin, *The Origin of Species*. John Murray, 1859.
14. U. Fayyad, "On the induction of decision trees for multiple concept learning," Ph.D. dissertation, EECS Department, University of Michigan, 1991.
15. S. Bouktif, D. Azar, H. Sahraoui, B. Kégl, and D. Precup, "Improving rule set-based software quality prediction: A genetic algorithm-based approach," *Journal of Object Technology*, vol. 3, no. 4, April 2004.
16. D. Azar, S. Bouktif, B. Kégl, H. Sahraoui, and D. Precup, "Combining and adapting software quality predictive models by genetic algorithms," in *Automated Software Engineering*, 2002.
17. L. Briand, J. Wüst, J. Daly, and V. Porter, "Exploring the relationships between design measures and software quality in object-oriented systems," *Journal of Systems and Software*, vol. 51, 2000.

18. L. Briand and J. Wust, “Empirical studies of quality models in object-oriented systems.” in *Advances in Computers*, M. Zelkowitz, Ed., 2000, vol. 20.
19. H. Zuse, *A Framework of Software emeasurement.* Walter de Gruyter, 1998.
20. T. Elomaa, “In defense of c4.5: Notes on learning one-level decision trees,” in *Machine Lea-69rning: Proceedings of the 11th International Conference.* Morgan Kaufmann, 1994, p. 62.

CHAPTER 26

TRANSPORT INFORMATION COLLECTION PROTOCOL WITH CLUSTERING OF INFORMATION SOURCES

MOHAMED KARIM SBAÏ[1] AND CHADI BARAKAT[2]

[1]*Projet Planète, INRIA Sophia Antipolis, France National School of Computer Sciences (ENSI), Tunisia*
Email: Mohamed_Karim.Sbai@sophia.inria.fr
[2]*Projet Planète, INRIA Sophia Antipolis, France,*
Email: Chadi.Barakat@sophia.inria.fr

Abstract: We improve and validate TICP, our TCP-friendly reliable transport protocol to collect information from a large number of sources spread over the Internet [1]. A collector machine sends probes to information sources that reply by sending back report packets containing their information. TICP adapts the probing rate in a way to avoid implosion at the collector and network congestion. Lost packets are requested again by TICP until they are correctly received. In this work, we add to TICP a mechanism to cluster information sources in order to probe sources behind the same bottleneck together. This ensures a smooth variation of network conditions during the collection session and hence an efficient handling of congestion at the network bottlenecks. This mechanism is based upon the Global Network Positioning (GNP) Internet coordinate system. By running simulations in ns-2 over realistic network topologies, we prove that TICP with clustering of information sources has shorter collection session duration and causes less packet losses in the network than the initial version that probes sources independently of their locations

1. INTRODUCTION

Nowadays, collecting information from a large number of network entities has more and more applications. The collected data can be availability of network entities, statistics on hosts and routers, quality of reception in a multicast session, numbering of population, votes, etc. In this work, we improve the Transport Information Collection Protocol (TICP) which we proposed in [1] to collect information entirely from a large set of sources spread over the Internet. A collector machine sends probes to information sources, which send back report packets containing their information. However, some difficulties come into play when designing TICP:

H. Labiod and M. Badra (eds.), New Technologies, Mobility and Security, 311–322.
© 2007 *Springer.*

- There is a risk of network congestion due to bandwidth limitation and the large number of sources. Furthermore, all sources are not behind the same bottleneck which makes the congestion control more difficult.
- The collection traffic can be aggressive towards traffic generated by other applications. In particular, it must not penalise concurrent TCP traffic.
- The loss of probes or reports lengthens the duration of the collection session, which urges for an efficient retransmission scheme.

TICP does not only adapt the probing rate as a function of network conditions, but also tries to minimize the collection session duration by deploying an efficient retransmission strategy. Moreover, it shares network resources fairly with concurrent traffic, namely TCP traffic, by adapting its probing rate in a way similar to how TCP does.

The collector in the former version of TICP [1] probes information sources in a random order. We show in this paper that this strategy causes many problems when moving into large networks which results in longer collection sessions, higher loss ratios and out of control traffic. The reason is that only one control at the TICP collector is used to limit the traffic at the several network bottlenecks simultaneously, which is clearly suboptimal given that congestion of one network bottleneck can be hidden by the low utilization of another bottleneck and vice versa. To probe sources behind the same bottleneck together and separately from other bottlenecks, we add to TICP a mechanism to gather information sources into clusters. This mechanism is based on the modelling of the Internet by a two-dimension Euclidean space and its decomposition into clusters. We use to this end the Global Network Positioning system (GNP)[4,5] that provides Internet host coordinates. Our new mechanism makes it possible to traverse sources from the closest cluster to the collector in terms of RTT (Round Trip Time) to the farthest one, which very probably results in sources behind the same bottleneck probed together before the collector moves to neighbouring sources located behind another bottleneck. This is supposed to improve the efficiency of the congestion control and to ensure a smooth variation of its variables in TICP.

To evaluate the performances of the protocol thus obtained, we ran simulations with the NS-2 simulator [6] over realistic and complex network topologies. These simulations have shown that TICP with the new mechanism of clustering has better performances than without clustering, and that it outperforms other non adaptive data collection solutions.

The paper is organized as follows. In the second section, we describe the main functionalities of TICP. We show in section 3 that the former version of TICP has many problems and that we need to cluster sources. In the fourth section, we explain our approach of clustering. The section 5 discusses simulations results and the last section concludes the paper.

2. TRANSPORT INFORMATION COLLECTION PROTOCOL

TICP [1] is a reliable transport information collection protocol implementing diverse functionalities. We focus here on those related to error recovery and network congestion control.

2.1. Error Recovery

The TICP collector has a list of all information sources. Every source is distinguished by an identifier that can be for example its IP address. Sources whose reports are lost are probed again and a required to retransmit them until they are correctly received by the collector. To make the retransmission of reports in TICP efficient, the collection session is made as a succession of rounds. In the first round, the collector sends request (probe) packets to all sources following their ranking in the list. In a second round, the collector sends requests to sources whose reports were not received in the previous round. The collector continues in rounds until it receives all reports. This behaviour in rounds is meant to wait for transitory network congestion to disappear from one round to another and to absorb the excessive delay that some reports may experience.

2.2. Congestion Control

To control the rate of requests and reports across the network, TICP is based on a report-clocked window based congestion control similar to the TCP one [2]. The collector maintains one variable *cwnd* indicating the congestion window size in number of requests/reports. New requests are transmitted only when the number of expected reports *pipe* is less than *cwnd*. TICP adapts *cwnd* to the observed loss rate of reports. It proposes two algorithms to do so: Slow start and Congestion Avoidance.

2.2.1. Slow Start

The collector starts a collection session by setting *cwnd* to *RS* (protocol parameter) and sending *RS* request packets. After some time, reports start to arrive. Some of these reports come on time, others are delayed. A timely report indicates that the network is not congested and that the collector can continue increasing its congestion window: $cwnd = cwnd + 1$. This yields a doubling of the probing rate for every window size of probes. The window continues growing in this way until the network becomes congested. At this point, the collector divides its congestion window by two and enters the congestion avoidance phase. The protocol comes back to slow start whenever a severe congestion appears (to be defined later).

2.2.2. Congestion Avoidance

The congestion avoidance phase represents the steady state of TICP. During this phase, the collector increases slowly *cwnd* in order to probe the network for more capacity. We aim to a linear increase of the congestion window by *RS* probes every window size of probes. Thus, upon each timely report, the congestion window is increased by: $cwnd = cwnd + RS/cwnd$. When congestion is detected, *cwnd* is divided by two and a new congestion avoidance phase is started.

2.3. Congestion Detection Mechanism

TICP implements a congestion detection mechanism to compute report loss rates and to decide whether a report is on time, delayed or lost. This mechanism is based upon a timer TO scheduled at the beginning of the session and rescheduled again every time it expires.

2.3.1. Round-Trip Time Estimator

TICP sets the timer of the mechanism to an estimate of RTT (Round TripTime), using the samples of RTT seen so far. The value of the timer is computed using estimates of the average RTT and of its variance. Let *srtt* and *rttvar* be the estimates of the average and the mean deviation of the RTT. Let *rtt* be the measured round-trip time when a report arrives. The collector updates the estimates and the timer *TO* in the following way:

$$rttvar = 3/4.rttvar + 1/4.|srtt - rtt|$$

$$srtt = 7/8.srtt + 1/8.rtt$$

$$TO = srtt + 4.rttvar$$

This dynamics and the coefficients it involves are inspired from TCP. TCP maintains an estimate of RTT per couple of source and destination, whereas TICP maintains only one estimate of RTT between the collector and all sources. This estimate is adapted when moving from one source to another.

2.3.2. Detecting Network Congestion

TICP computes the report loss ratio during a time window equal to *TO*. When the timer is scheduled, the collector saves in the variable *torecv* the number of reports to be received before the expiration of the timer. Let *recv* be the number of timely reports received between the scheduling of the timer and its expiration. The collector considers then that *torecv – recv* reports were lost in the network. Consequently, it estimates the loss ratio to *1 – (recv/torecv)*. The network is considered congested if the loss ratio exceeds the Congestion Threshold (*CT*) and severely congested if the loss ratio exceeds a higher threshold *SCT* > *CT* called the Severe Congestion Threshold. *CT* and *SCT* are two parameters of the protocol. TICP sets them as follows:

$$CT = min(0.1, RS/cwnd)$$

$$SCT = max(0.9, cwnd - RS/cwnd)$$

Based on these values, network congestion for TICP means that more than RS reports were lost in a window size of probes, while severe congestion means that less than RS timely reports were received.

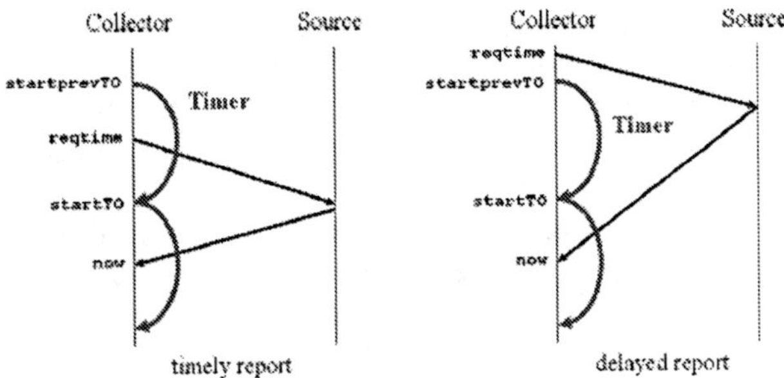

Figure 1. The two types of reports

2.3.3. Delayed and Timely Reports

A timely report is a report received before its deadline. The deadline of a report is given by the timer. A report not received before its deadline is assumed to be lost. If it arrives later than the deadline, it is considered to be delayed.

Figure 1 explains how the deadline of a report is set. Let *startTO* be the scheduling time of the timer. Let *startprevTO* be the previous scheduling time of the timer. When a report is received, the collector extracts from its header the timestamp *reqtime* indicating the time by which the corresponding probe has been sent. The report is received on time if and only if *startprevTO* < *reqtime*. The report is a delayed one in the opposite case.

3. NEED FOR CLUSTERING OF INFORMATION SOURCES

In this section, we present the drawbacks of the former version of TICP that motivated our present work. As we described earlier, the collector has a complete list of sources' identifiers. A collection session is a succession of rounds. In a given round, the collector begins by probing the source at the top of the list, then the following one and so on until the end of the list. The ranking of sources in this list has been so far done randomly and independently of any topology information. In reality, sources are more or less far from the collector. The random ordering results in variable non correlated RTTs during the collection session. Since the estimate of RTT at a given instant depends on its previously measured values, which in the case of random ordering are unrelated, this estimate seldom gives a good idea on the RTT of the next pair probe/report. This causes several problems. First, an overvaluation of RTT results in a delay in the detection of network congestion; the collector waits more than necessary for already lost reports. This delay means a waste of time and an aggravation of network congestion since the probing rate will not be reduced on time. On the other hand, an undervaluation of RTT can cause errors in the computation of report loss rate since the timer expires prematurely. Thus, some reports are declared lost while they are not. In this case, we reduce unnecessarily the size of the congestion window (*cwnd*) and hence, we increase the collection session duration.

Furthermore with random ranking of sources, packets generated can circulate everywhere in the network. At a given moment, this traffic can participate in the congestion of many bottlenecks. Since it is difficult to adapt congestion window size to network conditions on all paths from sources to collector, the Internet is considered by the original version of TICP as a single bottleneck. This version of TICP does not ensure fairness with concurrent traffic and its mechanism of congestion control is not efficient in case of large networks.

All the drawbacks described above are due to the random ordering by which information sources are probed. It is then important to cluster sources so that those close to each other are probed simultaneously. Also it is important to rank clusters from the nearest to the most distant from the collector so that to ensure that the network conditions vary smoothly and hence TICP congestion control can track them efficiently. The contribution of the present work is the addition to TICP of such a clustering and ranking mechanism together with its validation with extensive simulations.

A cluster is a group of sources located in the same neighbourhood. Our idea is that the more sources are close to each other the more their reports meet the same network conditions on their paths to the collector and the more probable they are located behind the same bottleneck. In this case, the loss of reports indicates that the common bottleneck is congested; hence the collector can handle this congestion efficiently by decreasing the probing rate. The collector probes clusters from the nearest to the farthest. This ensures a smooth variation of the congestion control parameters of TICP, for instance the rate of sending probes and the estimate of RTT. This again results in an efficient network congestion control.

4. CLUSTERING OF INFORMATION SOURCES

In this section, we describe our approach to cluster information sources. For this, we use the Global Network Positioning (GNP) system to model the Internet by a 2-dimensional Euclidean space [4]. A host is represented by a point in this space. The mathematical distance function gives an approximate value of the RTT between any 2 hosts. To ensure this, a small set of hosts called landmarks distributed across the Internet first compute their own coordinates in this geometric space. These coordinates are then disseminated to any ordinary host willing to compute its own coordinates relative to the coordinates of the landmarks [5].

The collector and information sources participate in GNP as ordinary hosts. At the end of the GNP operations, each source has a couple of coordinates $H(x_H, y_H)$ and the collector has also its own coordinates $C(x_C, y_C)$.

We define a cluster as being a set of information sources whose representing GNP points are located in a square area. The side of the square is denoted a, which is a parameter of the protocol. The central cluster is the square whose centre is the point representing the collector $C(x_C, y_C)$.

A cluster is completely defined by a couple of coordinates (X, Y) being integer values. These coordinates are those of the centre of the corresponding square relative

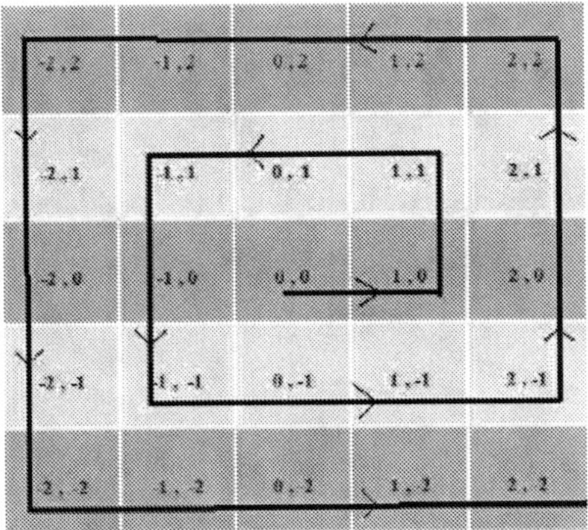

Figure 2. Order of probing information sources

to the collector coordinates and normalised by a. An information source whose coordinates equal to $H(x_H, y_H)$ belongs to the cluster (X, Y) given by:

$$X = round((x_H - x_C)/a)$$
$$Y = round((y_H - y_C)/a)$$

In order to probe information sources from the nearest to the farthest, the collector begins with the central cluster and then follows a spiral trajectory. Figure 2 gives an idea on this trajectory. One can with a simple algorithm find the coordinates of the next cluster during the collection knowing the coordinates of the current cluster.

5. SIMULATION RESULTS

In this section, we discuss the results of our simulations. We have run these simulations in ns-2 [6] in order to evaluate the performance of TICP with and without clustering of information sources. That is why we have implemented GNP and TICP in ns-2.

We generate realistic network topologies for simulations using GT-ITM (Georgia Tech-Internet Topology Modelling) [7,8]. We choose to work on transit-stub (TS) topologies which give the ability to model the complexity and the hierarchical structure of the real Internet. TS topologies model networks using a 2-level hierarchy of routing domains with transit domains interconnecting lower level stub domains. To these TS topologies, we assign latencies of 35 ms for intra-transit domain links, 10 ms for stub-transit links and 5 ms for intra-stub domain links. Figure 3 gives an example of an TS topology. Table 1 shows the parameters of the TS topologies used in ours simulations. In each simulation, we choose randomly 500 sources of information and a collector among the nodes that compose each TS topology. The parameters of TICP are set as in Table 2.

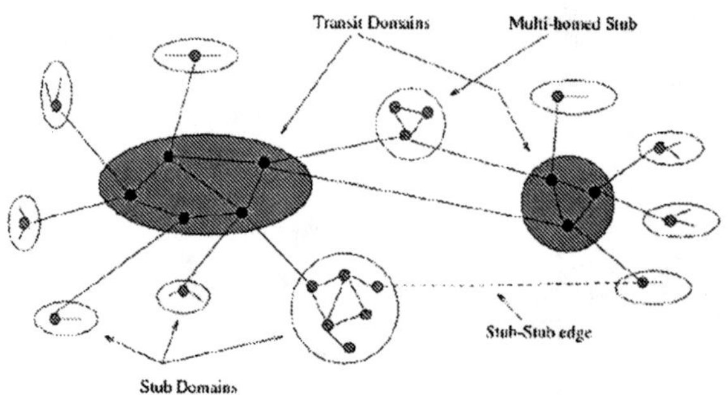

Figure 3. Transit-Stub topologies

Table 1. Transit-stub model parameters

Parameter	Signification	Scenario
T	Number of transit domains	5
N_t	Average Number of nodes/transit domain	7
K	Number of stub domains/transit node	8
Ns	Average number of nodes/stub domain	7

Table 2. TICP parameters

Parameter	Value
RS	10
probe size	100 b
report size	1500 b
a	50 ms

5.1. Network Congestion

We compare between the both versions of TICP with and without clustering of information sources. The comparison criteria are network congestion and duration of the collection session. Figure 4 illustrates an example of the evolution of the congestion window size (*cwnd*) as a function of time for TICP without clustering. First, we notice the saw tooth behaviour of TICP which adapts the window size to network conditions using the information on the loss ratio. But, we also notice that at time 17s, there was a reset of cwnd to RS following a severe network congestion (*loss rate* > *SCT*). With random probing, TICP is unable to adapt the probing rate to the available bandwidth in several bottlenecks simultaneously. Figure 5 plots the same result but this time for TICP with clustering. It is clear that in this case TICP remains in the congestion avoidance phase and that the severe congestion does not appear. This illustrates that TICP with clustering adapts the

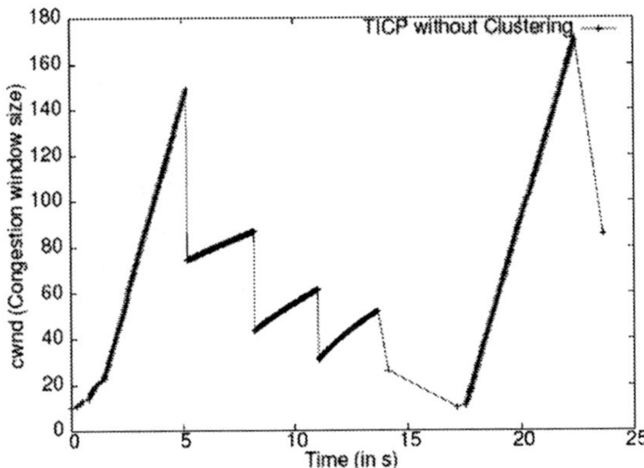

Figure 4. *Cwnd* as a function of time for TICP without clustering

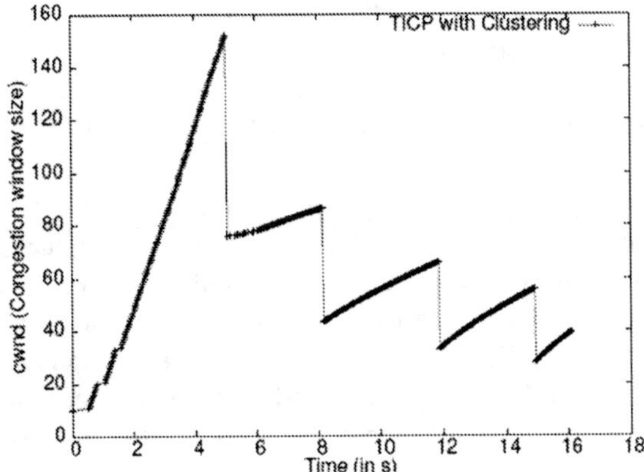

Figure 5. *Cwnd* as a function of time for TICP with clustering

window size to its right value without overwhelming the network. One can see TICP with clustering as treating bottlenecks one by one rather than at once. Note in the figures the decreasing trend in the window size, which is the result of the probing of sources from the closest to the collector to the farthest from it. TICP with clustering finishes the collection earlier because there was no congestion. The next paragraph studies the collection session duration.

5.2. Collection Session Duration

We continue the comparison between the two versions of TICP. This time we concentrate on collection session duration. Figure 6 shows this duration for several simulations of TICP without clustering. In each simulation, the order of sources in the list of the collector is different, that is why we obtain each time different

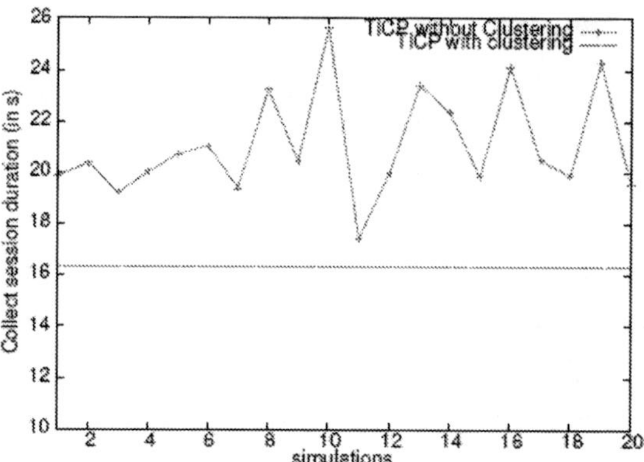

Figure 6. Collection session duration for different ordering of sources

collection session duration. For TICP with clustering, the result is the same since the topology does not change. TICP with clustering finds the good order of information sources and has the shortest collection session duration. We save on average 30% of the collection session duration by moving from TICP without clustering to TICP with clustering.

To evaluate the optimality of TICP more generally, we implement in ns-2 an information collection protocol having a constant congestion window size. For each window size, we run 10 simulations and we record the minimum of the collection session duration over them. Figure 7 presents the evolution of this duration as a function of *cwnd*. The curve has a parabolic shape: for small congestion window sizes, collection session duration is long because we have a low probing rate. For large window sizes, the network is congested which lengthens the collection session duration. The role of TICP is to find dichotomicly the good congestion window size that minimizes the collection session duration. We notice in Figure 7 how

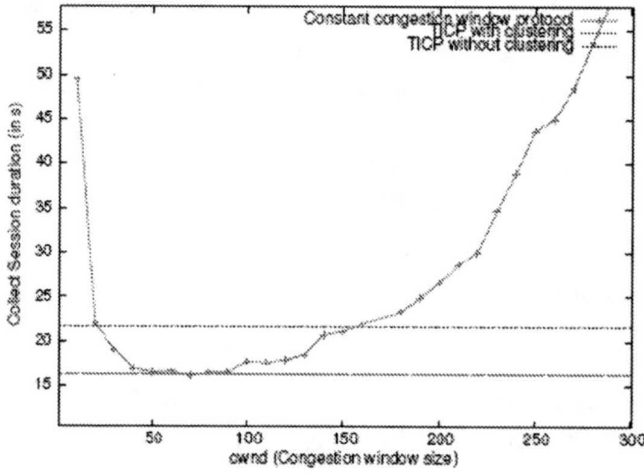

Figure 7. Optimality of the protocol

TICP with clustering manages to reach the optimum unlike TICP without clustering which yields longer durations.

5.3. Impact of Cluster Size

We vary the cluster size and we study its impact on collection session duration. Taking a very large a is equivalent to TICP without clustering since sources will be probed independently of their locations within the large cluster. Taking a very small results in clusters empty or with few number of sources which is not efficient since there will be no clustering of sources behind common bottlenecks. There should be some average a that provides the best performance. Figure 8 validates this intuition where we can see that over the network topologies we considered, a value of a around 50ms is optimal. Each point in the curve of Figure 8 is the average over 5 simulations run on different network realizations satisfying the characteristics in Table I. The number of sources is taken equal to 500.

Figure 9 studies how the number of sources impacts the choice of optimal a. We can clearly see that the optimal cluster size decreases when the number of sources increases. Compared to the value used above, the optimal a is equal to 85ms for 300 sources and to 45ms for 700 sources. Indeed, for small number of sources, one needs to increase a to group more sources behind the same bottleneck together. At the opposite, for more sources, one needs to decrease a so that the collector can better probe them depending on their locations. But, if we continue increasing the number of sources, the optimal a will stabilize and become equal to some minimum value depending on the topology. One can safely use this value for applications collecting data from a very large number of sources. We suggest that in reality, one calculates this value by running multiple collections when TICP is used for the first time, then adapts it as a function of the measured session duration to account for any change in the underlying network topology.

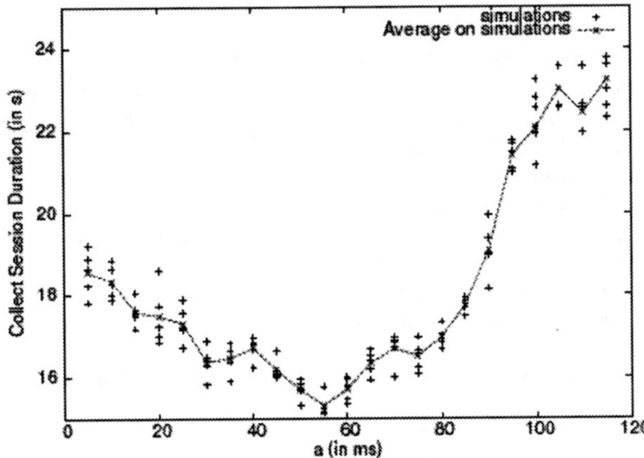

Figure 8. Impact of a on collection session duration

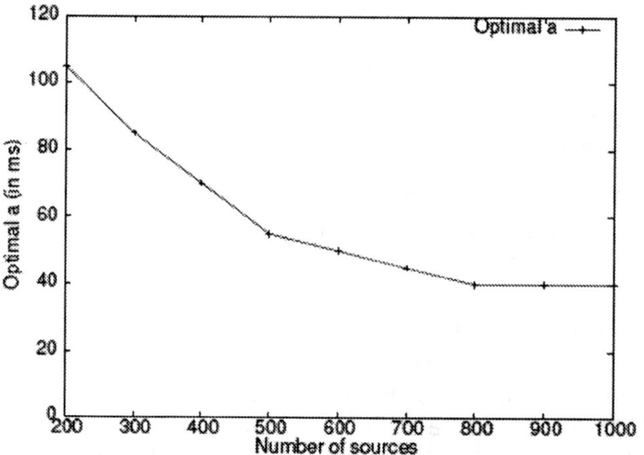

Figure 9. Optimal a as a function of the number of sources

6. CONCLUSIONS AND PERSPECTIVES

TICP is a transport protocol to collect information from a large number of network entities. It aims to control the congestion of the network and to minimize the collection session duration. To ensure a smooth variation of the congestion control parameters, we have added to TICP in this work a mechanism to cluster information sources. The simulation results show that this mechanism ameliorates the performances of TICP. In fact, it reduces loss rate and yields shorter collection session durations. However, the work on TICP is still not yet achieved. Our current research focuses on the implementation of the protocol and on its extension to account for sources of large amounts of data. In this new context, a report will be composed of several packets instead of one packet as it is now.

REFERENCES

1. Chadi Barakat, Mohamed Malli, Noamichi Nonaka, "TICP: Transport Information Collection Protocol", in Annals of Telecommunications, vol.61, no. 1–2, pp.167–192, January–February, 2006.
2. M. Allman, V. Paxson, W. Stevens, "TCP Congestion Control", RFC 2581, April.1999.
3. V. Paxson, M. Allman, "Computing TCP's Retransmission Timer", Internet Draft, April 2000.
4. T.S Eugene Ng and Hui Zhang, "Predicting Internet Network Distance with Coordinates-Based Approaches", INFOCOM'02, New York, NY, June 2002.
5. T. S. Eugene Ng and Hui Zhang, "Towards Global Network Positioning", Extended Abstract, ACM SIGCOMM Internet Measurement Workshop, San Francisco, CA, November 2001.
6. The Network Simulator ns-2, http://www.isi.edu/nsnam/ns/
7. Ellen W. Zegura, Ken Calvert and S. Bhattacharjee, "How to Model an Internetwork". Proceedings of IEEE Infocom'96, San Francisco, CA.
8. Ken Calvert, Matt Doar and Ellen W. Zegura, "Modelling Internet Topology", IEEE Communications Magazine, June 1997.

CHAPTER 27

VIRTUAL AUTHENTICATION RING FOR SECURING NETWORK OPERATIONS

NA LI AND DAVID LEE

Department of Computer Science and Engineering, The Ohio State University

lina,lee@cse.ohio-state.edu

Abstract: Securing network operations in a distributed environment is essential for today's communications yet is complex. Different than client /server architecture a distributed environment may contain peer to peer, overlay or arbitrary distributed network protocols without a centralized server for network control, and authentication has to be conducted in a distributed way to deal with malicious nodes in the network. In this paper we study authentication in a distributed environment for detecting malicious nodes when they launch attacks or disrupt applications. We propose a virtual authentication ring architecture and present a token ring authentication algorithm for detecting malicious nodes

Keywords: Distributed network; Authentication ring; DHT; Token ring algorithm

1. INTRODUCTION

In a distributed environment network operations can be disrupted by malicious nodes, and it is essential to detect and eject them for securing network operations and services.

Malicious nodes can disrupt network operations in a variety of ways. In structured peer-to-peer (P2P) and overlay networks such as CAN [9], Chord [12], Pastry [1] and Tapestry [14], a malicious node can forward a false lookup request or erroneous routing update, corrupt information stored in the system by repeatedly joining and leaving the network, and launch an attack on a specific data item by ID mapping [11]. It can even exploit millions of concurrently interactive peers as an engine for DDoS attacks against a targeted host [8]. All of the above malicious behaviors are closely related to distributed applications, such as lookup service and file sharing. However, it is usually not preferred to restrict nodes from joining and leaving the network in order to use their resources and computation power.

H. Labiod and M. Badra (eds.), New Technologies, Mobility and Security, 323–335.

© 2007 *Springer.*

Consequently, we need an authentication procedure to detect and eject the malicious nodes in a distributed environment when they launch attacks or disrupt network applications.[1]

Authentication can be conducted in either centralized or distributed way. Much work has been done on centralized authentication. Kerberos [7] is one of the popular centralized authentication protocols for client/server applications. However, for distributed applications, which are deployed in P2P, overlay, or an arbitrarily distributed network system, authentication becomes much more challenging. In addition to the lack of centralized control and management and other difficulties in a distributed environment, we emphasis two issues as follows:

(1) *Error propagation.* In distributed applications it is usually hard to trace back to the original source of malicious behaviors. Healthy nodes may have been cheated and act as accessories of malicious behaviors. For instance, for the resource lookup service in a P2P network [12], when a malicious node launches a DDoS [8] attack, it injects false information to attract traffic to a victim node by informing the whole networks that the victim has the resources they need. This false information may have been propagated to a large domain when the attack is detected and a number of healthy nodes have participated in propagating the false information. Consequently, it is difficult or even impossible to trace back and detect the original source of attack [13].

(2) *Limited Computing power and resources.* Centralized authentication is usually performed by a powerful server with abundant resources whereas distributed authentication is conducted by multiple machines, each of which usually has limited computing power and resources.

To cope with these and other difficulties in a distributed environment we need a new authentication procedure for securing network operations and services that is:

- Distributed: authentication must be conducted in a distributed manner.
- Efficient: malicious nodes should be quickly detected before error propagates yet without complicated computations.
- Robust: the architecture for the authentication has to be robust in a dynamic and distributed environment without a centralized control and management, and it can not be easily compromised.

We propose a virtual ring architecture for authentication. Different from traditional authentication approaches, which often use digital signature to identify one is actually the one it claims to be, our authentication is a ring-based network monitoring structure that aims to use distributed resources at each peer to conduct cooperative anomaly detection without any centralized authority. In this authentication ring, a node monitors its neighbors' behaviors. Upon detecting abnormal behaviors of a neighbor it sends out accusations for ejecting the malicious neighbor.

---

[1] We use the term "authentication" in a non-conventional way rather than "verification", "Byzantine agreement", "intrusion detection" or similar terms.

The detection can be done, for instance, by "sensing" any illegitimate traffic passing through. For example, when malicious nodes attempt to conduct an attack, their traffic characteristics is abnormal and thus reveals themselves. A typical scenario is as follows. Malicious nodes may start with a reconnaissance to validate connectivity, enumerate services, and check for vulnerable applications [17]. One of the most popular methods used by hackers for reconnaissance is port scanning which identifies what ports and services are open. Therefore a node can accuse its neighbor when it detects any port scanning behavior by this neighbor. In addition, many port scanners do not complete the TCP SYN/ACK sequence during a connection, and many scans use invalid packets (such as FIN or Xmas scans). All these observations enable a node to accuse its neighbors. Accusation can also be based on the packets that were sent. For example, the payload of an attack packet may contain some non-printable characters or unusual byte structure; whereas the payload of a normal packet predominantly contains ASCII characters with predefined structure, as required by the application protocol [18]. PAYL [19,20] detect anomalies by comparing byte (or n-gram) frequency characteristics of the normal and abnormal packets. If the frequency characteristic of a packet differs significantly from the normal one, the packet is deemed suspicious. In [21], Kruegel et al. have proposed six different models: length, character distribution, probabilistic regular grammar, token set, attribute presence, and attribute order for detection of http attacks. Sekar et al. [22] present an anomaly detection system based on network protocol specifications, which are defined by extended finite state automaton. Statistical features based on the state transitions are monitored to detect anomalies [18]. Thus, through observed anomaly in the packet payload passing through, one is able to detect and accuse the malicious behaviors of its neighbors if there are any.

To construct a robust authentication ring in an arbitrary network in a dynamic environment we use Distributed Hash Table (DHT) [2]. DHT has been proposed for routing [3] where one's traffic always passes through the neighbors in a ring identified by DHT table. Therefore, one's neighbors on the ring can monitor the packets passing through and anomaly can be detected.

We present an efficient distributed algorithm for detecting malicious nodes in an authentication ring. To the best of our knowledge the proposed virtual authentication ring architecture and the distributed algorithm for malicious node detection are a new approach in the published literature. Our main contributions are:

- We propose a new authentication ring architecture that can be applied to P2P, overlay, and arbitrary distributed network environment for detecting malicious nodes.
- We design a token ring algorithm that conducts authentication in a distributed way. It is efficient and the time complexity is $O(n)$ where n is the number of nodes in the virtual authentication ring.

The rest of the paper is organized as follows. The virtual authentication ring architecture is described in Section 2 and the token ring algorithm is presented in section 3 along with a time complexity analysis. The paper is concluded in Section 4 with remarks on related and future works.

2. VIRTUAL AUTHENTICATION RING ARCHITECTURE

In this section we describe the virtual authentication ring architecture. We organize nodes in a network into a virtual ring, called authentication ring. There are various ways to construct such a ring and DHT table is one of them. Authentication is then conducted on this ring architecture.

In an authentication ring, every node can accuse its neighbors as malicious nodes with misbehaviors. Based on the accusation information from all nodes in the ring, our distributed token ring algorithm analyzes and exposes the malicious nodes. Note that a node can only accuse its two immediate neighbors on the authentication ring. The accusation by a node can be based on its observations of misbehaviors of other nodes or inferences. Accusations by malicious nodes can be false.

A key question for all authentication problems is that who authenticates whom? The issue is complicated because good nodes and malicious nodes could battle to accuse each other for misbehaving in a distributed environment while there is no central "judge". When a good node detects corrupted data or mal-ware sent by a neighboring node it can accuse the sender as a malicious or compromised node. However, a good node may also be falsely accused by a malicious node. Therefore, in a distributed environment without any authentication structure each node can accuse and also be accused by all its neighbors; there is no way to authenticate in general. Consequently, we want to enforce a structure and protocol for authentication in a distributed network – that is the proposed authentication ring architecture and the token ring protocol.

The essence of our virtual ring authentication architecture is that it establishes a logical relation among nodes so that the accusations and hence authentication can be well controlled and managed. Apparently, two nodes with a long distance from each other (the distance can be physical or logical, depending on applications) usually has little information of each other and one can easily cheat the other. On the other hand, cheating can be relatively easily detected by a next hop neighbor. A good example is the overlay lookup service [10]. When a node receives a request for resource yet does not know which node has the resource it transmits the request to a neighboring node in its routing table whose logical identifier is closest to the logical identifier of the resource. A malicious node can cheat by forwarding the request to an incorrect (not closest) or non-existing node. This malicious behavior can easily be detected by its next hop logical neighbor by checking whether the request is getting "closer" to the resource identifier. However, such cheating cannot be detected by a distant node that does not have the needed routing information.

Our virtual authentication ring is constructed using Distributed Hash Table (DHT) [2] that realizes a concept of a virtual space. DHT can be used to constructs a virtual structure by reorganizing the nodes in a network. We use DHT to construct a ring of all the relevant nodes in a network and take advantage of the logical relations of the nodes on the ring for conducting mutual authentication in a distributed way. We omit the detailed construction procedure. Interested readers are referred to [2] and [3].

DHT enables us to construct a virtual ring for authentication in a distributed way without complicated computation. Furthermore, in a highly dynamic environment when nodes are joining and leaving the network the ring can be maintained dynamically, efficiently, and also in a distributed way [2].

In the sequel we assume that on a distributed network under consideration we have initialized a dynamic virtual ring so that each network node is a node on this ring. Each node in the network is assigned a distinct integer as its identifier. All the nodes on the virtual ring are ordered according to the increasing (or decreasing) size of their identifiers. A node identifier is fixed, unique and independent of its physical location (this is why it is called a virtual ring). Every node in the ring has two neighbors, a clockwise one and an anticlockwise one, which hold the closest identifiers to the node's identifier among all nodes in the ring (network). In certain distributed applications such as lookup services in P2P and overlay networks, the concept of "neighbors" embodies closest logical relationships, and neighbors have more control over a node than other distant nodes [13]. To take advantage of this virtual ring architecture and the neighbor relations for accurate and efficient authentication we require that one node can only accuse its immediate neighbors and, conversely, can only be accused by its immediate neighbors.

3. AUTHENTICATION ALGORITHM

After a virtual authentication ring is established, a token is passed around the ring from a node to its virtual neighbor in one direction (clockwise or anticlockwise) to collect and distribute information. Then a distributed authentication procedure is applied to detect malicious nodes.

In our virtual ring authentication architecture, we call a malicious node a bad node, and a healthy node a good node. Every node must evaluate its neighbors by marking on the token when it passes by. A good node must accuse a bad node but not a good node. A bad node may or may not accuse other nodes – be it good or bad. This is a very general model for authentication.

A. Information Collection and Distribution

A token is passed around the ring to collect and distribute information, that is, accusations or no accusations of each node of its neighbors. Note that the token is passed along a virtual ring in one direction by a node to its neighbor for a round.

The token contains two fields (bits) for each node to evaluate its two neighbors; it can accuse (assert TRUE) or not accuse (assert FALSE) its clockwise neighbor and anticlockwise neighbor of cheating. All fields are reset to FALSE when the authentication process starts. A node can only access to its evaluation fields but can not access to other fields in the token during the information collection phase.

Before the authentication procedure is applied, the token is passed around the ring twice: first time for collecting information and the second time for distributing information.

Phase 1 (information collection): The token is passed along the ring clockwise (or counterclockwise) to collect the accusations of all nodes. When the token completes information collection, it enters Phase 2.

Phase 2 (information distribution): In this phase, all the fields in the token are read-only to all the nodes - visible but unchangeable. The token is passed along the ring for a second round to distribute the accusation information.

In the two phases, cryptographic techniques are needed to guarantee that the token is securely passed along the ring without being viewed by an unauthorized node nor altered. We can use public key scheme to authenticate the token. Each node encodes by its own private key the content it inserts into the token. Other nodes can decrypt it by the corresponding public key and validate the content. Or we can use hash chain for authentication as done in secure routing protocols [16].

However, if one or more malicious nodes persistently scramble the whole token content or – even worse – drop the token, then there is not much one can do in a same layer. This problem is different than that in the usual token ring protocol where one has to deal with the accidental loss or scrambling of token but not purposely and persistently dropping or scrambling token by malicious nodes. One might consider transmitting token in a different layer in a secure way with a new protocol or with a cross-layer approach. It is an interesting problem yet is not a topic of this work. Without further digressing we assume that tokens can be passed along the authentication ring without being scrambled or dropped.

B. Authentication

The next step is to conduct authentication based on the information in the token and to identify the malicious nodes. Obviously, without any assumptions it is impossible to provide a valid authentication. For instance, if all the nodes are bad and none of them accuses any others, there is no way to authenticate. We make the following general and natural assumptions:

(1) There are more good nodes than bad ones in the ring, that is, the percentage of good nodes in the ring is more than 50%. If this condition does not hold, we can not tell "bad" from "good". This can be illustrated by a simple counter example. Assume that there are only two nodes in the ring – a good one and a bad one – accusing each other, and in this case, it is impossible to judge who is bad.

(2) A good node always accuses its bad neighbors but never accuses its good neighbors. This is the normal behavior of good nodes.

However, the complication of authentication comes from the mischievous behaviors of bad nodes. A bad node can have arbitrary accusing actions; it may choose to accuse or not to accuse its neighbors, no matter they are good or bad. Depending on the bad nodes' behaviors different authentication procedures are needed.

The authentication procedure is executed when the read-only token is passed around the ring the second time. Each node can read all the fields of the token – the accusations (or not) of the neighbors of each node – and use the following procedure to authenticate, that is, to distinguish the good nodes from the bad ones.

If all the bad nodes behave deterministically and in a same way, then obviously there are four cases: a bad node accuses a good node or not, and a bad node accuses a bad node or not – a total combination of mutually exclusive four cases. These cases can be easily processed as follows.

Case 1: Bad node accuses bad neighbors and does not accuse good neighbors. Since a bad node is accused by both its good and bad neighbor and a good node by neither, the accused nodes are bad and the remaining nodes are good.

Case 2: Bad node does not accuse bad neighbors and does not accuse good neighbors either. We can easily see that there is only one-way accusation: good node accuses bad node. We can first identify the accusers as good nodes and the accused as bad nodes. All these accuser-accused pairs separate the good ones from the bad ones on the ring.

Case 3: Bad node does not accuse bad neighbors and accuses good neighbors. In this case, a good node always accuses its bad neighbors and does not accuse its good neighbors, and a bad node does the opposite to its neighbors. Therefore, for each pair of nodes, they either accuse each other or not at all. We can assign a same color to two neighboring nodes which do not accuse each other (both are good or bad) and assign different colors to two neighboring nodes which accuse each other (one good one bad). We can start from an arbitrary node and color it white. We then color its neighbor white if there is no mutual accusation and black otherwise. We can continue in this way and color all the nodes either white or black. Since there are more good nodes than bad nodes, we can easily identify the white nodes as good if they out number the black ones. Otherwise, the black nodes are good.

Case 4: Bad node accuses all its neighbors no matter they are good or bad. In this case, since a bad node is always accused by its two neighbors – bad or good, a node is good when it is not accused by at least one of its neighbors. Since good nodes out number the bad ones, there is at least one pair of neighboring good nodes, which do not accuse each other, and can be identified as good nodes. Starting from these identified good nodes, we can proceed with authentication as follows. A node that is not accused by an identified good node must be good; otherwise, it would have been accused. A node that is accused by an identified good node must be bad; otherwise, it would not be accused. As for the remaining nodes, they are in a situation that each accuses all its neighbors and is also accused by both its neighbors. They are all bad but with isolated good nodes in between; there is no way to identify them.

The above four cases exhaust all possible deterministic behaviors of bad nodes. We now discuss the more complex cases that bad nodes can behave non-deterministically. For instance, some bad nodes choose to accuse its good neighbors, and some don't. This is a generalization of some of above four cases.

Case 2': Bad node does not accuse bad neighbors and may or may not accuse good neighbors. This case is a generalization of Case 2 and is harder to deal with since the behavior of bad node is non-deterministic now. However, the only difference is that now two neighboring good and bad nodes either accuse each other (as in Case 3) or there is only one-way accusation, that is, the good node accuses the bad

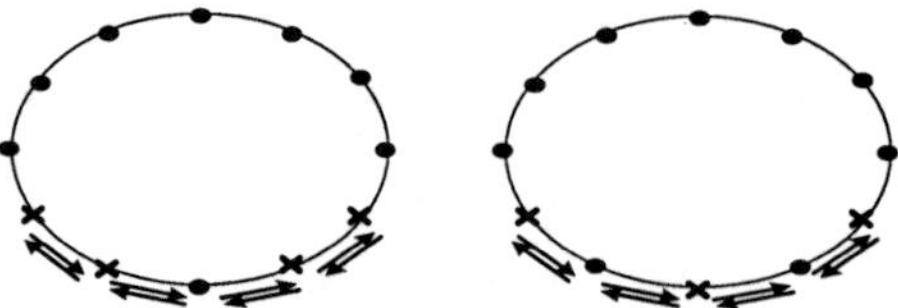

Figure 1. Accusation Scenarios

one but not vice versa. We can use the same procedure as in Case 3 except that we change color only at a pair of nodes, which either accuse each other or only one accuses the other.

Case3': Bad node may or may not accuse bad neighbors but accuses good neighbors. This case is a generalization of Case 3, and can not be solved in general. This can be shown by the example in figure 1, in which dot represents a good node and fork represents a bad one. The arrows indicate the accusation between nodes. Observe that there is no way to differentiate the two scenarios in Figure 1.

However, case 3' turns out to be solvable if we enforce isolation constraints on bad nodes, that is, there are no two contiguous bad nodes. This is a typical scenario before bad nodes start to take over the network and they are only in isolation (in the virtual authentication ring). In this case, since bad nodes accuse good nodes and vice versa, each pair of mutually accusing nodes must be a pair of good-bad nodes; it can not be a pair of good nodes since they do not accuse each other, and it can not be a pair of bad nodes since there are no contiguous bad nodes. Once we have identified the boundaries between the good and the bad by the mutual accusing pairs, a majority count can be used as in Case 1 to distinguish the good nodes from the bad ones.

Case 4': Bad node may or may not accuse bad neighbor; bad node may or may not accuse good neighbor. This case is a generalization of Case 4 and, obviously, is unsolvable in general. With the same isolation constraint as in case 3', we can solve it as follows. Since each bad node is in isolation and has two good neighbors, it must be accused by both of its neighbors. However, a good node may also be accused by both of its neighbors if they are bad. We color the nodes gray if they are accused by both neighbors. A bad node must be gray but a gray node may or may not be bad. Since there are more than 50% of good nodes, there is at lease a pair of neighboring good nodes. Since good nodes do not accuse each other, they can not be gray. Since a bad node must be gray, all the non-gray nodes are good – color them white. Now all the nodes are colored white (good) or gray (either good or bad). If a gray node is an immediate neighbor of a white node it must be bad; otherwise, it would not be accused by the white node and would have been white as well. Since bad nodes are in isolation, both its (gray) neighbor must be good – change its color to white. We can repeat the process and conclude that these consecutive gray nodes are good and bad nodes in alternation. Consequently, all the nodes are identified

C. Bad Node Behavior Classification without Learning

Our authentication algorithm is highly dependent on how the nodes behave in accusation and it seems that we need to have prior knowledge about node accusation behavior for a specific system in order to do the authentication. Indeed, we do not know beforehand the behaviors of bad nodes, which can even be non-deterministic. We might have to learn or anticipate their behaviors before applying the procedures for the four different cases as presented in the previous section. Furthermore, each case assumes that all the bad nodes behave in a same way. In reality bad nodes in a same network can behave differently and can be non-deterministic; we have discussed this issue in Case 2', 3' and 4' by assuming the non-deterministic behaviors of bad nodes but not in a most general way.

In general, even knowing the behaviors of the bad nodes, which can be non-deterministic, the problem can be still unsolvable, such as Case 3' in the previous section. However, with a natural assumption, it turns out that we can uniquely identify the bad nodes without a learning phase.

As discussed in Case 3', we have an isolation constraint on bad nodes, that is, there are no two neighboring bad nodes. This is a typical scenario before bad nodes take over the network; they are still in isolation (in the virtual authentication ring).

We now show that we can always uniquely identify the bad nodes even when they exhibit non-deterministic behaviors as follows. If there are no accusations observed (in the token) then there are no bad nodes; otherwise, a good node must accuse its bad neighbor if there are any bad ones. In the sequel we only consider the case that there is at least one accusation observed. Since bad nodes are isolated, we can classify their behavior by their actions towards their good neighbors. A bad node is called *attacking* if it accuses its good neighbor and *non-attacking* otherwise.

Proposition 1. With the malicious node isolation constraint all of them can be identified uniquely by the authentication ring protocol without any prior knowledge of their behaviors.

Sketch of proof. We discuss three cases: (i) All the bad nodes are attacking; (ii) All the bad nodes are non-attacking; and (iii) Bad nodes can be attacking or non-attacking. Obviously, Case (iii) is general case. Its proof is based on that of Case (i) and (ii).

(i) All the bad nodes are attacking. Since bad nodes are isolated, their behaviors against bad nodes do not matter. (This covers Case 1 and 4 in the previous section). In this case, a good node always accuses its bad neighbors and never accuses its good neighbors, and a bad node accuses its neighbors, which are good due to the isolation constraint. Therefore, for each pair of nodes, they either accuse each other or not at all. We can assign a white color to two neighboring nodes which do not accuse each other (both are good). The remaining nodes are not at peace - accusing both their neighbors and being accused by their neighbors.

Since there are more good (white) nodes than bad nodes, there is at least one pair of neighboring good (white) nodes. We can color the remaining nodes as follows. Starting from a white node, scan the nodes on the ring along the direction the token is being passed. When we encounter the first uncolored node, it must accuse the white node that we last scanned and also is accused by that white node; we color it black (bad). The next node must be good due to the isolation constraint; we color it white if it has not been colored yet. Starting from this white node, we continue the process until all the nodes are colored. The white nodes are good and the black ones are bad.

(ii) All the bad nodes are non-attacking. Since bad nodes are isolated, their behaviors against bad nodes do not matter. (This covers Case 2 and 3 in the previous section). In this case, two neighboring good nodes do not accuse each other, and two neighboring good/bad nodes have one-way accusation; the good accuses the bad and not vice versa. We can uniquely identify the bad nodes in a similar way as in Case (i) except that we switch colors upon encountering a pair of neighboring one-way accusation nodes, instead of a pair of mutually accusing nodes.

(iii) We have both types of bad nodes –attacking or non-attacking – on the ring but we do not know their behaviors beforehand and we want to identify all the bad nodes uniquely without learning or anticipating their behaviors. From the study of Case (i) and (ii), the procedure should be clear. Again since there are more good nodes than bad nodes, we can determine some of the good nodes and color them white if they are at "peace" – they do not accuse any neighbors nor are accused by any neighbors. We then scan the ring along the direction the token is being passed and color the remaining nodes as follows. We switch the colors upon encountering a pair of neighboring nodes, of which at least one is accusing the other. As a matter of fact, if there are mutual accusations then we have an attacking bad node, and if there is only one way accusation then we have a non-attacking bad node.

In summary, with the bad node isolation constraint, we can uniquely identify all the bad nodes without learning their behaviors – be it attacking or non-attacking.

D. Analysis of Time Complexity

We now analyze the time complexity of the authentication algorithm. Assume that there are n nodes on the ring (in the network). We pass the token all over the ring twice for collecting and distributing the accusation information with a cost $O(2n)$

Upon receiving the token the second time, each node – good node rather – detects malicious nodes. For the most complicated case, a node examines the token three times before making a decision: the first round to find the boundaries of good-bad nodes by their accusation status; the second round to color all the nodes; and the third round to finally identify all the nodes. Each run takes time $O(n)$.

Proposition 2. It takes time $O(n)$ to authenticate nodes in a network using the virtual authentication ring where n is the number of nodes in the network.

Thus our token ring authentication algorithm is an efficient one with linear time complexity.

4. CONCLUSION

Virtual authentication ring is a new authentication architecture that can be used to detect and eject malicious nodes efficiently in a dynamic, distributed and untrustworthy environment. In the ring, every node is responsible for authenticating its neighbors. With the token ring algorithm that collects and distributes the authentication information, all nodes can figure out malicious nodes in the network.

Our authentication ring is similar to that in [3], however, for a completely different application. Furthermore, for our authentication application, the ring construction with DHT is simpler; instead of having a set of neighbors, every node in the ring only has two neighbors – a clockwise one and an anticlockwise one.

There are other possible ways to expose malicious or faulty nodes in a distributed environment. For instance, Douceur and Howell [6] uses Byzantine distributed algorithm to isolate faulty nodes. However, its communication cost is high comparing with our simple ring-based authentication algorithm that takes linear time to run. Fireflies[15] provides a Byzantine tolerant solution to monitor and accuse stopping failure by multiple rings which has different goal from our work that is aiming at taking the advantage of ring structure to design a general authentication protocol as building block for monitoring and detecting malicious behaviors endangering network operations Another interesting approach is to use reputation system [5]. Some of the successful and well known reputation systems on the web are controlled and managed by a central entity. An example is the eBay feedback system. eBay enables buyers and sellers to rate other buyers and sellers, and presents that reputation information to everyone. A problem with this system is that one loses control even if there are only two malicious nodes in the network; give each other very high rank to foul the whole network. Our virtual ring authentication, however, can avoid this problem; even if two bad nodes try to conceal each other's malicious behavior, they can not evade our authentication process for: (1) They may not be neighbors in the ring and can not help each other at all; and (2) Even if they are neighbors, their bad influence can not extend across and beyond good nodes on the ring, since they will be identified by the token ring authentication algorithm. The merit of our approach is that it makes an authentication decision in a distributed way but is based on global information collected in a token that is being passed around in a robust ring structure.

The authentication process with a virtual ring is a new approach and there are problems, which remain to be resolved. We have discussed briefly the case that a malicious node can have non-deterministic behaviors and presented results on authentication with sufficient conditions. The necessary and sufficient conditions in the most general case are yet to be derived for the validity of authentication along a virtual ring. Instead of establishing an authentication ring for a whole network, we may want establish a local ring of a sub-network for a local authentication. The difficulty here is not how to construct and use the ring, since this is essentially the same as a ring for a whole network, it is on how to construct a ring that is not too long but is sufficient for the needed authentication. In a dynamic environment DHT can help to construct the authentication ring on-line. However, it remains a challenge when nodes are joining and leaving the network; how to synchronize the ring construction and maintenance and the information collection and distribution with the token to avoid inconsistency in a highly dynamic and distributed environment.

ACKNOWLEDGMENT

We are deeply indebted to the reviewers for their insightful and constructive comments and suggestions.

This work has been supported in part by NSF awards CNS-0403342 and CNS-0548403 and by DoD award N41756-06-C-5541.

REFERENCES

1. Rowstron A. and Druschel. P. Pastry: Scalable, distributed object location and routing for largescale peer-to-peer systems. In Proc. IFIP/ACM Middleware 2001, Heidelberg, Germany, 2001.
2. Baruch Awerbuch and Christian Scheideler. Towards a scalable and robust DHT. In SPAA'06: Proceedings of the eighteenth annual ACM symposium on Parallelism in algorithms and architectures, 2006, pages 318.327,
3. Matthew Caesar et al. Virtual ring routing: network routing inspired by DHTs. In SIGCOMM'06: Proceedings of the 2006 conference on Applications, technologies, architectures, and protocols for computer communications, 2006, pages 351.362.
4. Miguel Castro et al. Secure routing for structured peer-to-peer overlay networks. SIGOPS Oper. Syst. Rev., 36(SI):299.314, 2002.
5. Prashant Dewan and Partha Dasgupta. Pride: peer-to-peer reputation infrastructure for decentralized environments. In WWW Alt. '04: Proceedings of the 13th international World Wide Web conference on Alternate track papers & posters, 2004, pages 480.481.
6. John R. Douceur and Jon Howell. Byzantine Fault Isolation in the Farsite Distributed File System. In IPTPS. '06: Proceedings of the 5th International Workshop on Peer-to-Peer Systems, Santa Barbara,CA,USA, February 2006.
7. J. Kohl and C. Neuman. The kerberos network authentication service, 1993.
8. Naoum Naoumov and Keith Ross. Exploiting p2p systems for ddos attacks. In InfoScale'06: Proceedings of the 1st international conference on Scalable information systems, 2006, page 47.
9. Sylvia Ratnasamy et al. A scalable content-addressable network. In SIGCOMM'01: Proceedings of the 2001 conference on Applications, technologies, architectures, and protocols for computer communications, 2001, pages 161.172.
10. Emil Sit and Robert Morris. Security Considerations for Peer-to-Peer Distributed Hash Tables. In IPTPS'01: Revised Papers from the First International Workshop on Peer-to-Peer Systems, 2002, pages 261.269, London, UK.
11. Mudhakar Srivatsa and Ling Liu. Vulnerabilities and Security Threats in Structured Overlay Networks: A Quantitative Analysis. In ACSAC'04: Proceedings of the 20th Annual Computer Security Applications Conference (ACSAC'04), pages 252.261, Washington, DC, USA, 2004.
12. Ion Stoica et al . Chord: A Scalable Peer to peer Lookup Service for Internet Applications. In Proceedings of ACM SIGCOMM'01, UC San Diego, CA, USA, August 2001.
13. Dan S. Wallach. A Survey of Peer-to-Peer Security Issues.
14. B.Y. Zhao, J.D. Kubiatowicz, and A.D. Joseph. Tapestry: An infrastructure for fault-resilient wide-area location and routing. Technical Report. UMI Order Number: CSD-01-1141, University of California at Berkeley. 2001.
15. HÃvard Johansen et al. Fireflies: Scalable Support for Intrusion-Tolerant Network Overlays. Eurosys 2006. Leuven, Belgium. April 2006.
16. Yih-Chun Hu et al. Efficient Security Mechanisms for Routing Protocols, In Proceedings of the Tenth Annual Network and Distributed System Security Symposium (NDSS 2003), pp. 57.73, ISOC, San Diego, CA, February 2003.
17. Richard BEejtlich, The TAO of Network Security Monitoring, ISBN:0-321-24677-2
18. Prahlad Fogla and Wenke Lee, Evading Network Anomaly Detection Systems: Formal Reasoning and Practical Techniques, CCS'06: Proceedings of the 13th ACM conference on Computer and communications security, pages 59.68, Alexandria, Virginia, USA

19. K. Wang and S. Stolfo. Anomalous payload-based network intrusion detection. In Recent Advances in Intrusion Detection (RAID), 2004.
20. K. Wang and S. Stolfo. Anomalous payload-based worm detection and signature generation. In Recent Advances in Intrusion Detection (RAID), 2005.
21. C. Kruegel and G. Vigna. Anomaly detection of web-based attacks. In Proceedings of the ACM Conference on Computer and Communication Security (ACM CCS), pages 251–261, 2003.
22. R. Sekar et al. Specification-based anomaly detection: A new approach for detecting network intrusions. In Proceedings of the ACM conference on Computer and communications security (ACM CCS), 2002.

CHAPTER 28

FEDERATED DYNAMIC AUTHENTICATION AND AUTHORIZATION IN DAIDALOS

ZHIKUI CHEN

Networks and Communication Systems, Computer Center, Universität Stuttgart, Germany, Allmandring 30, 70550 Stuttgart, Germany, Tel: +49-711-68565871, Email: zhikui.chen@rus.uni-stuttgart.de

Abstract: this paper describes a dynamic authentication (AuthN) and authorization (AuthZ) (DAA) scheme based upon a virtual identity concept, as defined in the EU IST integration project Daidalos, in order to protect users' privacy and the integrity of their personal information. For multiple inter-domains, the federation concept is introduced, which states the trust relationship among different domains at different levels. A common framework to coordinate AuthN, AuthZ and users' personal information across different domains is established. The AuthN and AuthZ processes are clearly separated and implemented via SSO (Single Sign On). The Diameter protocol is used to exchange SAML assertions and AuthZ policy statements across domains and different AAA (AuthN, AuthZ and Accounting) solutions to realize service grouping management. A bootstrapping approach is used to ensure security of users' personal information

Keywords: Federation, authentication, authorization, privacy, security, bootstrapping

1. INTRODUCTION

Before a user can access services in a foreign administrative domain, the user has to first register with his home domain via the visited administrative domain. In a user registration process, the A4C (AAA, Auditing and Charging) infrastructure must be able to identify and authenticate the user, *i.e.*, determining which SLA, account, and profile are involved. It is not only the provider's needs that must be met in this process, but the user also has to authenticate with the provider to whom he is registering. Therefore, mutual AuthN is required in the registration process. Furthermore, AuthN in general has several security requirements. These include protection against replay attacks, confidentiality and resistance against man-in-the-middle attacks.

H. Labiod and M. Badra (eds.), New Technologies, Mobility and Security, 337–348.
© 2007 *Springer.*

AuthZ defines the process of verifying an object's permission to perform a particular action or not. Two different mechanism classes exist for this: (1) AuthN-based schemes require, as a precondition, object AuthN, which is utilized to check, via Access Control Lists (ACL), whether this identified object is allowed to perform the requested action(s). (2) Credential-based schemes, which apply credentials with trustworthy information provided by the algorithm performing the AuthZ process. AuthZ depends upon service specific attributes e.g. QoS service class and user specific attributes such as name, age, etc. Depending on the type of AuthZ, different credentials are submitted by the terminal client. Credential management is part of the AuthZ services as managed by A4C including creating and modifying credentials.

When a handover decision is made by a mobile terminal according to the received signal strength, the mobile terminal provides some credentials, which will be transferred to the new inter-domain access router (AR) using a handover context transfer protocol. The new AR delivers these credentials to the new inter-domain A4C server which then forwards them to the home A4C server using the Diameter protocol. The home A4C checks the credentials and sends back the results to the new inter-domain A4C and then the new AR. If all credentials are successfully verified, the service will be continued, otherwise the service will be denied and re-AuthN and re-AuthZ are necessary.

Based on the SSO principle, this paper introduces a DAA scenario within federated domains, which is bound to a user and his administrative identity management [7,9]. The state of being authenticated or registered must be bound to a certain lifetime. The length of this lifetime must be configurable and can depend on various factors including billing options. Simultaneously, in order to protect a user's privacy and to secure communication data, a new identity, the Virtual Identity, VID, has been introduced in the EU IST integration project Daidalos [4,7]. Correspondingly, a VID identifier in terms of VIDID, has been designed.

In the access network, the proposed AuthN and AuthZ consists of two phrases: a) in the access network phrase, i.e., from terminal to network access server (NAS), PANA with EAP is used to implement the functionality; b) in the second phrase, i.e., from NAS to home A4C server, the Diameter protocol is applied to deliver the user's AuthN and AuthZ AVPs including SAML assertions. When handover occurs, the VID along with its credentials will be transferred to the new domain. The VIDID and ID-Token will be verified by the home A4C via the foreign A4C. All ID-Tokens will be automatically authenticated by the home AAA server [8,10].

The rest of this paper is structured as follows. Section 2 summarizes the current related standard works. Two key concepts and a scenario of AuthN and AuthZ are described in section 3. Some privacy and security considerations are stated in section 4. Section 5 describes mobility related issues in relation to DAA. The last section 6 summarizes this paper.

2. RELATED STANDARD WORKS

2.1. 3GPP

3GPP has defined the Generic AuthN Architecture (GAA), which offers a mechanism to provide a shared secret plus certificates to two communicating entities for mobile applications. GAA is based on GSM and UMTS AuthN and key agreement protocols [2].

The 3GPP AuthN infrastructure, including the 3GPP AuthN Centre (AuC), the USIM (Universal Subscriber Identity Module) or the ISIM (IMS Subscriber Identity Module) and the 3GPP AKA (AuthN and Key Agreement) protocol running between them, is a very valuable asset to 3GPP operators. It has been recognized that this infrastructure could be leveraged to enable application functions in the network and on the user side to establish shared keys. Therefore, 3GPP can provide 'application security bootstrapping' to authenticate the subscriber by defining a Generic Bootstrapping Architecture (GBA) based on the AKA protocol [3]. The subscriber certificates support services whose provision assists mobile operators, as well as services that mobile operators provide. In GBA, a generic Bootstrapping Server Function (BSF) and the UE (User Equipment) shall mutually authenticate using the AKA protocol and agree on session keys that are afterwards applied between UE and a Network Application Function (NAF). The BSF shall restrict the applicability of the key material to a specific NAF by using the key derivation procedure. GBA bootstrapping procedures shall be modified so that in addition to GBA related information, Liberty related information (e.g. AuthN assertions or artifacts) is also carried over the Ub (an interface between the UE and the BSF) reference point [3].

2.2. TISPAN NGN

TISPAN (Telecommunications and Internet converged Services and Protocols for Advanced Networking) defines the Network Attachment Subsystem (NASS) to maintain information about IP-connectivity associated with user equipment connect to TISPAN networks in NGN (Next Generation Networks). This provides registration at access level and initialization of UE for accessing the TISPAN NGN services. The NASS provides network level identification and AuthN, manages the IP address space of the access network and authenticates access sessions. The NASS also announces the contact point of the TISPAN NGN Service/Applications Subsystems to the UE. Network attachment through NASS is based on implicit or explicit user identity and AuthN credentials stored in the NASS [1].

Access network level registration involves access AuthN which are AuthN and AuthZ procedures between the UE and the NASS to control access to the access network. Two AuthN types are considered for access networks: implicit AuthN and explicit AuthN.

Explicit AuthN is an AuthN procedure that is explicitly conducted between the UE and the NASS using a signalling procedure. Implicit AuthN does not require the NASS to explicitly conduct an AuthN procedure directly with the UE. However, the NASS performs the implicit AuthN based on identification of the L2 connection that the UE is connected to. It is a matter of operator policy whether implicit AuthN or explicit AuthN is applied.

There shall be mutual AuthN between the UE and the NASS during access network level registration. Both implicit and explicit AuthN may be used independently as the network level access AuthN mechanism, notwithstanding the fact that implicit AuthN may be a consequence of explicit AuthN (e.g., the implicit line AuthN used together with an explicit method such as PPP in xDSL access). AuthN between users/subscribers and application/service providers shall be explicit or implicit (based on trust/security assertions).

2.3. ITU NGN

NGN shall support selective AuthZ of attribute information (e.g., identity lifetime) by an attribute provider and allow separate identification, AuthN and AuthZ of users and terminal equipment. NGN shall also support a dynamic binding of user identity and terminal equipment (identity) and allow the association of a user identity to support multiple terminal equipment (identities) for certain services. A service provider may allow a user to access a service from multiple terminals in parallel using the same public and private user identity [6].

The considerations of these standards works are incomplete and are unable to protect users' privacy and personal information or to establish reliable trust relationships. The proposed VID concept with its related scenario copes better with these goals and details will be given in the following sections.

3. DAA SCENARIO IN DAIDALOS

In a Daidalos infrastructure domain, a funtional module, the ID broker, is designed to manage VIDs. A VID is a table to describe the authorized service in terms of EPP (Entity Part Profile), consisting of a service EPP and a context EPP. When a VID is created by the ID broker, a corresponding credential is produced. At the same time, a VID wallet is also proposed as a complementary service to the VID framework to provide a means to store and organize the user's avatar in a Daidalos scenario [5].

In all cases, either the user inputs his VIDID with its credential, or selects them from a VID wallet perhaps stored in a device, such as a USB stick, to be verified by the A4C and ID broker. In Daidalos, a bootstrapping process is used to provide access to the network, to register a VID/user in a network through AuthN and AuthZ and to provide access to network resources in the name of the VID user. The bootstrapping process is a sequence of events exchanged between the mobile device and the network and the associated state information maintained in the mobile device and network, which takes place after the terminal is powered-up. Within federated domains, a user can access the services according to the EPPs listed in the VID on the ID broker, providing VIDID verification was successful.

3.1. VID

Using a VID, a user can split his/her overall trace into smaller traces left in the system by using different virtual identities. A virtual identity can be considered as a pseudonym together with the data, which can be linked to this pseudonym, e.g., a pseudonym together with the location which may be seen in the context of this pseudonym. In Daidalos, a user can choose the virtual identity he/she wants to use to authenticate and register for services [5].

When the user signs a contract, an identity under which the contract and the respective profiles and rights are defined is issued. This identity, called the Registration Identity (RegID), holds the information necessary for charging its owner, and can be seen as the system representation of the contract signatory. The VIDs are always related to a RegID and can share all or none of the RegID's attributes. As such, they are privacy-enabled and possibly anonymous representations of the RegID. The following key assumptions for identities are defined [12]: 1) Each operator assigns one RegID to each customer; 2) Services are accessed only with VIDs; 3) Only operator's A4C subsystems are allowed to map the VID to the RegID; 4) Each RegID should be associated with a key pair, such as RSA, issued by the operator. This can be used for signing the credentials; 5) Each VID can be associated with a key pair issued KDC; 6) SSO over multiple operator domains will require either a globally defined name space guaranteeing ID uniqueness, or 'identity mappings' (e.g., mapping/federation of VIDs) between operators.

Privacy protection affords the user some configuration burden. Firstly, different VIDs must be configured: i.e., it must be specified what information may be revealed under which context – including, e.g., setting the access rights for accessing user information. Secondly, it must be decided when the aggregated information is too sensitive and the VID must no longer be used. Thirdly, the user must decide for each required service, which VID is to be used.

On the other hand, the security and privacy subsystem provides an enabling service for the pervasive service platform that allows users to automate the above-mentioned VID related actions. Moreover, it comprises the respective counterparts being used by third party services. The overall functionality consists of three parts: (1) privacy policy negotiation, (2) identity management and (3) credential management and access control.

The ID broker manages VIDs including creation, deletion, activation, adding EPPs and obtaining EPPs from the EPP Holder being used to store various EPPs.

3.2. Federation

A federation is a collection of realms that have established trust relationships between themselves. The level of trust may vary, but typically includes AuthN and AuthZ. In Daidalos, Federated AuthN and AuthZ (FAA) are to support and implement the AuthN and AuthZ of dynamic service provisions between the federated (mutually trusted) domains. For federated domains, only one AuthN is needed:

SSO [11]. When an authenticated user accesses another federated domain, an AuthN credential enables one application to assert the identity of the user to another.

There are three kinds of federations defined in Daidalos: (1) Data federation, whose main purpose is to exchange data. As already stated, all federations aim to exchange data, so one can say that all federations are data federations. However, for this definition, the data being exchanged is generic and as such does not serve any specific purpose. This data can also only be provider specific in the sense that it is not directly related to a user. Some problems might arise from this in relation to data consistency across providers. (2) Identity federation, where the identity of the user is exchanged between providers within the federation, given that the user is represented in the system at more than one provider. These providers exchange handles that enable them to correlate the several representations. These handles can be anonymous or not. One can say that the user's overall representation is a merger of the local representations at the different providers. As can be seen, this federation is an instantiation of the general data federation problem. It may require additional agreements since it may expect a greater level of trust between the federations. (3) Function federation, in which the main function is provided by several providers in a concerted way. I.e., no provider can provide the function alone, but if every provider provides its share, the overall function can be delivered to the consumer. This is basically distributed computing or service composition with all the known problems from these areas such as delay, access control or common data models. Moreover, it is important to define, which information a provider wants to disclose and which information the others need for the concerted functionality (involving service discovery). This might include trade-offs and mappings and blurring. Here, the result of the function will provide the information to be exchanged. A function may use 'parameters' retrieved through a more generic data federation.

Generally, there are several federations, ranging from a fully federated situation where the federated business entities exchange almost all desired details to the other extreme where no information is exchanged at all. Between these two extreme positions, there is a wide range of flavors of federation possible, as defined by the two administrative business entities involved [9].

3.3. AuthN and AuthZ Process

The AuthN and AuthZ processes in Daidalos are clearly separated and implemented via SSO, which consists of three parts: VID login, service AuthZ and handover, which will described in next section in terms of mobility.

Figure 1 shows a VID login. When selected from Key Manager/input (from external storage, such as a USB stick), the VIDID with its credentials are sent to the local foreign domain. For the inter-domain case, the VIDID realm will be locally resolved to obtain VIDID's domain address. Then, the VIDID and its credentials will be sent to its home ID Broker. The ID Broker will fetch this VIDID's private key from the KDC (Key Deployment Center), which is a functional module to generate the key specific to the VID. It then sends the VID and key to the A4C

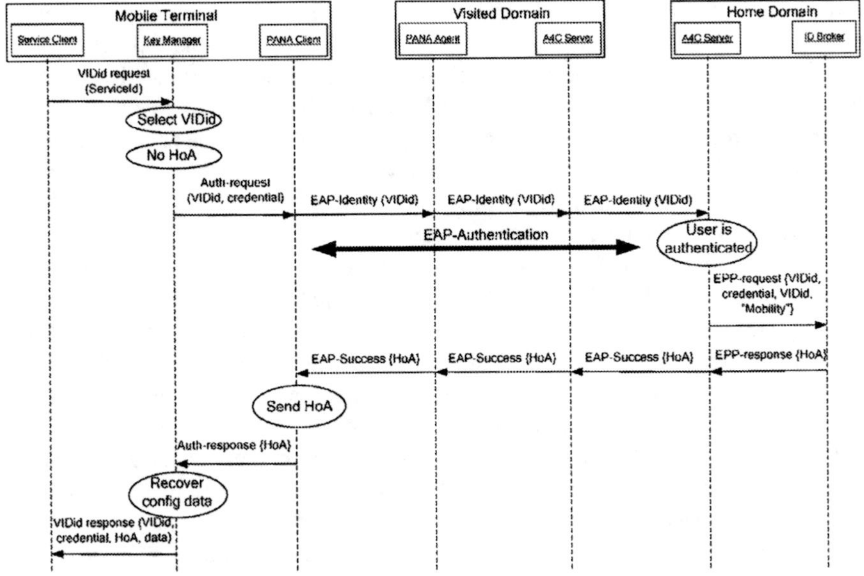

Figure 1. AuthN and AuthZ of VID

Server for AuthN. When the A4C Server authenticates the VID, it will generate an ID-Token for the VID, which includes an SAML artifact. From the Mobile Terminal (MT) to the Access Router (AR), i.e., from PaC (PANA Client) to PAA (PANA AuthN Agent), PANA and EAP are used to deliver the VIDID and its credentials. Also, from the PAA/AR to the local foreign A4C domain, Diameter and EAP are used to deliver the AuthN request. Diameter and EAP protocols are also used for communications between the A4C servers. For communications between the ID Broker, KDC and A4C Server, the AAA protocol (Diameter) is used to transfer the VID and its credentials (request AuthN phase) and the VID with its ID-Token (response AuthN phase). Figure 2 shows an example of an ID-Token.

In other words, the A4C Server requests the credentials from the VID/user by using the new EAP method (EAP-SAML request/response). After the A4C Server

Random Number	Serial Number	Artifact
Signature (by using Sender's private key)		
VID=string@realm		

Encrypted by using Receiver's public key

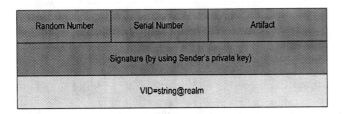

Figure 2. Content of an ID-token

verifies that the credentials are correct, it informs the AR/PAA that this user is authorized to access the network. Then the AR/PAA requests the QoS Broker to obtain QoS parameters associated with this user and whether access is possible. Note that the AR has to recover both the VID and the credentials to carry out the registration process. However, it cannot access the EAP messages because they are encrypted inside a TLS tunnel. Thus, the A4C Server sends both parameters to AR/PAA by using new Diameter AVPs: VID AVP and Credential AVP that have been added to the Diameter EAP application.

Once a VID has been AuthN and AuthZ, the EPPs, in terms of EPPID, listed in the VID can be selected. The EPPs will be fetched from the EPP Holder, and then any control policies for access will be implemented. If XACML authorizes the selected EPPs for access, the EPPs will be redirected to a service provider. The service provider will authorize the requested service if that service can be provided.

If a new service is discovered and is to be used, the privacy policy negotiation module is invoked. This negotiates anonymously with the service's respective privacy policy negotiation counterpart. The result of this policy is not only a statement about what the service guarantees to do with the personal information being revealed during the service use, but also a list of personal context information that the service needs for the requested service provision: e.g., the user's location.

Based on this negotiated result, the ID broker module selects or creates a VID in which this context information may be revealed. For this, it estimates the level of privacy invasiveness for the data being revealed in the context of this VID.

Finally, another functional module, the ID Manager, creates the respective AuthZ for the service to access the negotiated context information according to the chosen VID. This module is complemented by context management access control, which checks the AuthZ in case of a context request by, again, interfacing with the A4C subsystem of the service provisioning platform.

4. PROPOSED SCHEME IN MOBILITY

AuthN and AuthZ should continue to work properly even if users or services are temporary disconnected (because of local network limitations, or during the transfer from one network domain to another). The identities including customized services must be recognized even if the connection point to a network changes from time to time or even from a particular device to another. Also, in this case, it is necessary to allow multiple AuthZ frameworks to work independently (within administration domains) as well as together.

4.1. Terminal Mobility

Figure 3 depicts a federated heterogeneous domain communication diagram, where a handover is initiated. In the new domain, a user, before getting access to the network, has to be authenticated and registered. In the SSO, the new AuthN and AuthZ are implemented with transferred credentials from the old access router.

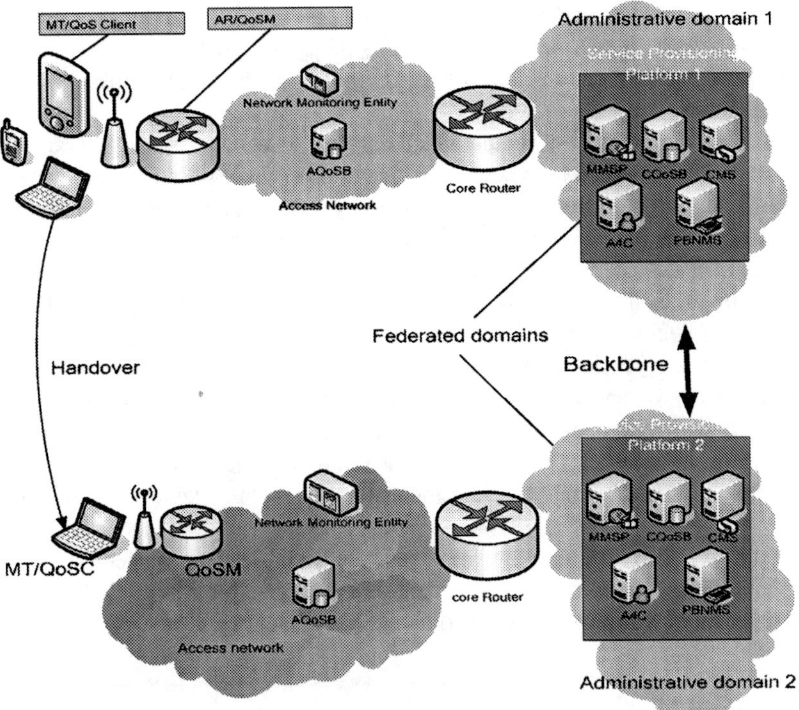

Figure 3. Federated heterogeneous domains

The credential management is part of the AuthN services which is managed by the A4C subsystem including creating and modifying credentials. When the handover decision is made by the mobile terminal according to the received signal strength, the mobile terminal provides some credentials with its VIDID, which will be transferred to the new inter-domain AR using a handover context transfer protocol. The new AR delivers the credentials (ID-Token) to the new inter-domain A4C (A4C.f_new) server which firstly checks whether the credentials (credential signature) is valid and then forwards them to the home A4C Server via the Diameter protocol. The home A4C (A4C.h) server checks the credentials and sends back the result to the new inter-domain A4C Server and then the new AR. If all credentials are successfully verified, the service will be continued, otherwise the service will be denied and re-AuthN and re-AuthZ are necessary.

In the new domain, if a user wishes to change his service preferences, the new services should be authorized. Actually, it becomes a new service.

4.2. Session Mobility

Session Mobility is the seamless media transfer of an ongoing communications session from one device to another. There are two cases: firstly, to transfer a complete session to another device (whole-session mobility, e.g. audio + video + signalling) and secondly, partial session mobility (e.g. just transfer video to another device, because the MT detects that a high definition device is available and the user accepts its use).

Session mobility involves both transfer and retrieval of an active session. Transfer means to move the session on the current device to one (whole-session mobility) or more (partial session mobility) other devices. Retrieval means to remotely transfer a complete or partial media stream of a communications session on another device back to the local device; for complete or partial mobility respectively. This may mean to return a media stream to the device on which it had originally been before it was transferred to another device. For example, after discovering a large video monitor, a user transfers the video output stream to that device using partial mobility approach. When the user leaves, the user returns the stream to the user's mobile device for continued communication. In the whole-session mobility case, a participant in an audio call on the user's IP phone may leave the user's office in the middle of the call and transfer the call to a mobile device as the user is leaving.

For session mobility, the new device should be authorized by a specific VID, which should depend on context management. For example, for a large LCD in the classroom, teachers will have the right to use it, but students will not.

4.3. Network Mobility (NEMO)

Network mobility involves Mobile Routers (MR). Security and AAA support will allow MRs to use the same infrastructure (ARs) as the Daidalos Mobile Nodes (MN). A MR arriving at a visited domain must authenticate itself within this domain before having access to the network, just like any other MN arriving at this network. For this purpose, it must use the PANA protocol and, in a more detailed way, the

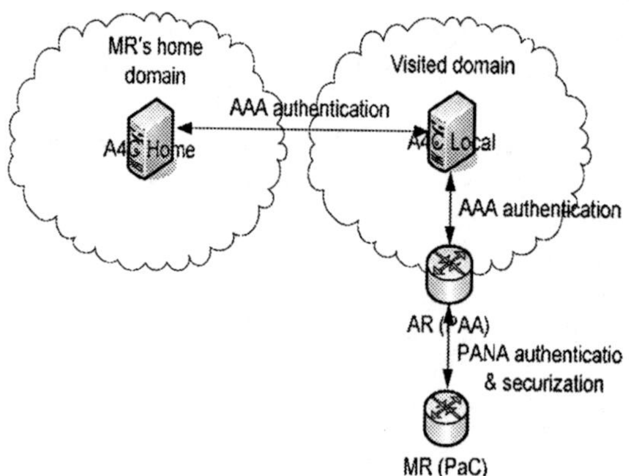

Figure 4. Mobile Router authentication in the visited domain

PANA Client (PaC) module for gaining access to the network. Because all the access networks must have a PAA, the MR is going to be able to authenticate itself using this PAA, see Figure 4.

5. SECURITY AND PRIVACY CONSIDERATIONS

Security and privacy refers to two aspects: firstly, a user arriving at a network to access the network and secondly, a user interested in gaining access to other services, which are provided by the network. The user will be interested in accessing these services confidentially, so the user is required to obtain an identity, credentials plus other information (e.g. address of the service). For satisfying the above, a bootstrapping scenario is used for which a service bootstrapping process is required whose entities are in charge of managing access to the services in a provider's domain and are able to collaborate with similar entities in other provider's domains. This collaboration could be realized using the EAP keying framework, defining a procedure to transfer cryptographic material or other user information to a local domain (for roaming scenarios) in order to improve the performance of the system. The root-key is used for providing keying material to all of the entities/services which were using the AAA-Key (or a derived key from this AAA-Key) obtained from an EAP method. In some sense, this approach for example applied to key distribution is related to the GAA/GBA model; see 2.1.

In this way, in our proposed scheme, to provide the mobile node with information to allow a user to access the services based on profiles or contracts, different VIDs ensure user privacy based on user needs. In order to allow a user several VIDs and to have control over privacy issues, the bootstrapping architecture as defined allows the user to request a different VID and associated parameters such as credentials either during the AuthN phase or indeed at any other moment in time.

6. CONCLUSION

The SSO and federation protocol defines a request and response protocol by which a user is able to authenticate to one or more service providers and federate (or link) configured identities among multiple domains. A service provider issues a request for AuthN to an identity provider (ID token or credential). The identity provider responds with a message that contains AuthN and AuthZ information, or an artifact that points to AuthN and AuthZ information.

The VID framework provides the possibility to instantiate several virtual users (even being physically just one user) all potentially using the same physical device or different physical devices. The bootstrapping of the different VIDs is strictly connected with the AAA aspects developed for network DAA. VIDs impact mobility in the sense that users can move a VID without actually moving the physical device. Traditional host mobility is therefore extended with a new concept of VID handover. The network and the terminal are therefore required to handle mobility with a different granularity depending on users' profiles and requirements.

7. ACKNOWLEDGMENT

The work presented in this paper was partially funded by the EU project IST-2004-026943 "Daidalos II" (Designing Advanced Network Interfaces for the Delivery of Location Independent, Optimized Personal Services) [4].

REFERENCES

1. ETSI TS 282004: TISPAN, NGN functional architecture: Network attachment sub-system (NASS) Version: 1.1.1, 2006–06.
2. 3GPP TS 33980, Interworking of Liberty Alliance Identity Federation Framework (ID-FF), Identity Web Services Framework (ID-WSF) and Generic Authentication Architecture (GAA), version: 7.2.0, 2006–09.
3. 3GPP TS 33220, Generic Bootstrapping Architecture (GBA), version: 7.5.0, 2006–09.
4. Daidalos IST Project: http://www.ist-daidalos.org.
5. Daidalos deliverable, D341, "Architecture and design: A4C, security and privacy framework", 2006–12.
6. Marco Carugi, Identification requirements in NGN, Identity workshop of ITU, 2006–12.
7. Zhikui Chen, "A Scenario for Identity Management in Daidalos", IEEE CNSR2007, Canada.
8. Olivereau, A.; Gomez Skarmeta, A.F.; Marin Lopez, R.; Weyl, B.; Brandao, P.; Mishra, P.; Ziemek, H.; Hauser, C., "An Advanced Authorization Framework for IP-based B3G Systems". Proceedings of the 14th IST Mobile & Wireless Communications.
9. Aguiar, R.L.; Jaehnert, J.; Gomez Skarmeta, A.F.; Hauser, C., "Identity Management in Federated Telecommunications Systems". Proceedings of the Workshop on Standards for Privacy in User-Centric Identity Management 2006, Zurich, 2006.
10. Fitzgerald, W.; Doolin, K.; Mahon, F.; Gomez Skarmeta, A.F.; Butler, S.; Schlosser, P.; Weyl, B.; Hauser, C.: "Daidalos Security Framework for Mobile Services". Proceedings of eChallanges 2005, Ljubljana, 2005.
11. Daidalos deliverable, D321, "Architecture and Design: Interdomain and federation concepts", 2006–12.
12. B. Weyl, P. Brandao, A. F. Gomez Skarmeta, R. M. Lopez, P. Mishra, C. Hauser, H. Ziemek, "Protecting Privacy. of Identities in Federated Operator Environments", IST-4th Wireless Mobile Summit 2005.

CHAPTER 29

SECURE AND FAST ROAMING IN 802.11 WLANs

HASSNAA MOUSTAFA AND GILLES BOURDON

France Telecom R&D 38–40 rue du General Leclerc - 92794 Issy Moulineaux Cedex 9, France
{ hassnaa.moustafa, gilles.bourdon } @orange-ftgroup.com

Abstract: The mass deployment of IEEE 802.11 based wireless local area networks (WLANs) and the popularity of portable devices created an urgent need to support voice and multimedia applications. However, these applications require fast handoff among access points (APs) while users are on the move for maintaining quality of connections and services continuity. In this paper, we firstly discuss the WLAN roaming problem showing the importance of minimizing re-authentication latency for real-time multimedia applications' support. We then present IEEE standards support to roaming capabilities mainly focusing on 802.11i and 802.11r standards, and we point up some vendor-specific solutions for roaming enhancements, based on 802.11i. Furthermore, we present the IETF efforts in enhancing authentication performance for seamless roaming. We end this paper by showing the advantages and shortcomings of the presented roaming support approaches, highlighting some critical factors that need to be fulfilled in WLAN deployments for achieving fast and secure roaming

Keywords: WLAN, 802.11, fast handoff, roaming, authentication, access control

1. INTRODUCTION AND BACKGROUND

IEEE 802.11 Wireless LANs (WLANs) have gained a lot of popularity due to its low cost and relatively high bandwidth capabilities. On one hand, the IEEE 802.11 standard enables the deployment of low cost WLAN services. On the other hand, the unlicensed and free spectrum used by 802.11 networks allows high speed wireless access deployment: 2.4 GHz in 802.11b/g and 5 GHz in 802.11a with theoretical throughput ranging from 11Mbps in 802.11b to 54 Mbps in 802.11g/a. In addition, 802.11n that is expected to be ratified in early 2008 is designated with the objective of high throughput, and will operate in both the 2.4 GHz and the 5 GHz spectrum permitting a throughput of more than 100 Mbps.

The high demand for wireless access together with the low deployment costs of WLANs have led to wide deployment of public WLAN hotspot services by

H. Labiod and M. Badra (eds.), New Technologies, Mobility and Security, 349–360.
© 2007 *Springer.*

many providers, including startups and telecom operators [1]. This rapid adoption makes many people believe that 802.11 will become the fourth generation cellular system (4G) or a major part of it, and creates an urgent need to support real time applications while wireless users are on the move. However, the small cell size of WLAN creates frequent handoffs during mobile nodes (MNs) roaming, resulting in delays or disruption of communications. In this paper we focus on layer-2 handoffs during MNs roaming, where we address the problem of fast and secure roaming of MNs. A major component that consumes latency during the handoff process lies in discovering the new available access point (AP) and associating to this AP. The handoff latency especially impacts real-time applications as voice over IP (VoIP), imposing excessive jitter. These applications however require fast handoffs among APs to maintain the quality of the connections as well as the continuity of services. The ITU-TG 114 recommendation gives a maximum end-to-end delay of 150 ms for VoIP applications [2], considering one-way transmission. Table 1 gives an idea on the delay impact on VoIP applications, according to the ITU-TG 114 recommendation. In [3], a study found that the observed layer-2 handoff latencies are from 60ms to 400ms (252ms on average) depending on the vendors of wireless cards and APs, and that the phase of discovery of the next AP is a dominating factor accounting for more than 90% of the overall handoff cost.

With the proliferation of WLANs, the next generation wireless technology leap into supporting real-time applications such as VoIP over WLANs seems promising. Hence, there is a hard need to resolve the handoff latency problem incurred while MNs transition from one AP to another, taking into consideration inter-domain as well as inter-operator roaming. Since most service providers cannot cost-effectively deploy as many APs as needed to achieve satisfactory wide area coverage, supporting inter-operator roaming is an appropriate strategy for enlarging their service's area. In such roaming scenarios, users may connect to the Internet via APs owned by providers that are unknown to them for whom a trust relationship may not exist. Consequently, efficient security mechanisms protecting both the users and the network are required. In this context, authentication and authorization are important in allowing only authorized users to get connections to the network and hence to access the offered services. However a considerable delay can result during the authentication process, which can impact the continuity of services. Improving authentication delay is thus a key issue for achieving seamless roaming across networks and domains.

Table 1. Delay impact on VoIP applications

Delay	Impact on application quality
< 100–150 ms	Delay not detectable
150–200 ms	Acceptable delay, light delay or hesitation noticeable
> 200–300 ms	Unacceptable delay, normal conversation impossible

Authentication, Authorization and Accounting (AAA) pose a main challenge in WLAN roaming. Since AAA should be continuously carried out for each user while moving, there is a need to avoid heavy authentication process that consumes delay and impacts the services, aiming to provide a seamless roaming. Although the web based Universal Access Method (UAM) is recommended as the best current practice for inter-provider roaming by the Wi-Fi Alliance [4], this method is known to be vulnerable to many different attacks such as impersonation of an AP, dictionary attacks and service theft by means of address spoofing [5]. The 802.11i standard [6] was developed to address these problems. It requires mutual authentication between a MN and the network, and uses MAC layer encryption between MNs and APs which prevents service's theft by means of address spoofing. On the other hand, the new IEEE 802.11r standard [7] is developed to address issues faced by real-time applications, implementing security and quality of service enhancements. This new standard attempts to minimize the APs' transition time while still providing the services offered by 802.11i and 802.11e standards. The problem of fast and secure roaming is also being studied within the IETF. The HOKEY WG [8] is treating this problem through studying efficient handover keying solutions, serving intra and inter-domain roaming. Also, the CAPWAP WG [9] is developing a standard protocol which enables an Access Controller (AC) to manage a collection of Wireless APs. The CAPWAP architecture and protocol are promising in enhancing the authentication performance, through centralizing the authentication and policy enforcement functions for a wireless network.

This paper considers the fast and secure roaming problem in WLANs, considering the related standardization activities. The rest of this paper is organized as follows, Section 2 gives an overview on 802.11i, discussing its roaming capabilities. Section 3 presents some vendor-specific solutions based on 802.11i to enhance WLAN roaming. Section 4 describes the key features of the new 802.11r standard, showing the role of this standard in enhancing roaming in WLANs. Section 5 presents some activities carried out at the IETF related to the secure roaming issue. Section 6 summarizes the paper and presents our outlook.

2. ROAMING CAPABILITIES IN 802.11I

802.11i [6] is adopted as the WLAN security standard, offering access control via mutual authentication between the MN and the network while protecting the confidentiality and integrity of the air interface between the MN and the AP. An authentication model based on 802.1X [10] is used between the MN, the AP with which it associates, and the Authentication Server (AS). Figure 1 gives a general description of 802.11i.

The MN firstly associates with an AP within its range in order to exchange authentication messages with the AS. Extensible Authentication Protocol (EAP) [11] is the end-to-end transport protocol and EAP over LAN (EAPoL) transports

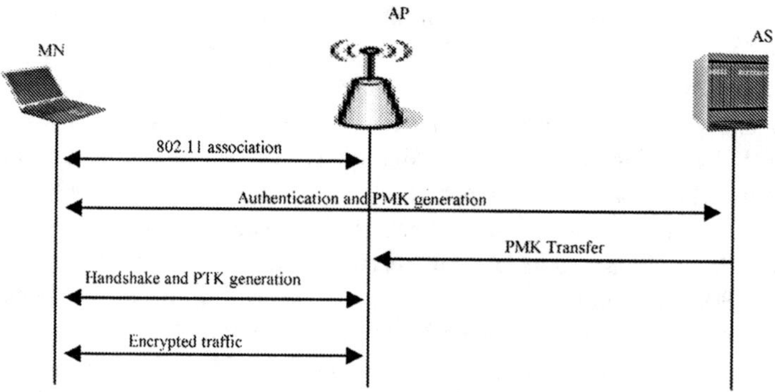

Figure 1. 802.11i authentication

EAP over 802.11 LAN and implements a port-based access control. RADIUS [12] can be used to transport EAP over IP establishing an authenticated channel between AP and AS and securely transporting the generated key from the AS to the AP. The MN and the AS generate a secret Pairwise Master Key (PMK) using a negotiated EAP-method, which is then transferred by the AS to the AP with which the MN is associated. Finally, a handshake process is carried out by the MN and the AP to generate a Pairwise Transient Key (PTK) derived from PMK and other parameters. The PTK is used for encryption and integrity protection of the subsequent 802.11 traffic as well as in the encrypted transfer of the Group Transient Key (GTK) which is used in broadcast traffic from the AP to the associated MNs.

A problem arises when the MN roams from one AP to another, it needs to perform a full 802.1X authentication that could take up to hundreds of milliseconds or even seconds to complete depending on the AS load, devices performance, and traffic conditions. Such latency affects real-time applications in a way that can impact the continuity of services. Aiming to minimize this delay, the *PMK caching* and the *Pre-authentication* are two options defined in 802.11i standard for fast roaming. WPA2 (Wi-Fi Protected Access 2) is created by the Wi-Fi Alliance providers [13] to implement and commercialize 802.11i standard considering these roaming enhancement options. PMK caching allows the storage of the PMK resulting from the first security association of a MN trying to optimize the re-authentication process. The following section gives more details on this method. Pre-authentication, also known as "*fast associate in advance*", allows the AP with which the MN is associating to communicate through the wired network with AP(s) to which the MN is expected to roam, pre-authenticating the MN to these APs. This method reduces the handoff delay, however, it adds more load on the AS and consumes bandwidth and APs storage resources. This is detrimental, especially if the MN does not roam to the new AP(s).

3. VENDOR-SPECIFIC SOLUTIONS FOR FAST ROAMING

A number of solutions are proposed by some vendors for optimizing authentication, aiming to support real-time applications during MNs roaming. These solutions are based on 802.11i authentication and are mainly employing authentication keys caching, switched architectures, and keying management and hierarchy.

Keys caching: i) Trapeze Networks developed the PMK caching mechanism [14], which is adopted in the 802.11i standard. This mechanism allows MNs' "*fast roam back*" to previously visited APs, through storing in a cache the PMK resulting from the first security association with these APs. During re-authentication with the previously visited APs, the stored PMK is presented by the MN and hence there is no need to carryout a full 802.1X authentication. Although this method succeeds to optimize the re-authentication process, the handoff delay could not be neglected in real-time applications. ii) Airespace, Atheros and Funk software developed the Proactive Key Caching (PKC) [15], in which each MN gets a PMK during the initial authentication and uses this PMK to authenticate to new APs during roaming. During roaming, the new APs validate the used PMK through soliciting the AS. PKC eliminates the authentication from scratch with each new AP during roaming and is expected to save about 60% of the 802.11i authentication process.

Switched Architecture: i) Opportunistic PMK storage is developed by Symbol technology [16] for switched wireless environment. Each WLAN switch manages the keys of all APs linked to it. The PMK resulting from the association with an AP can thus be used by all other APs linked to the switch. Tests carried out by Symbol show that using this mechanism allows the handoff between APs to be accomplished in less than 40ms. ii) Airespace, Atheros and Funk software apply the PKC mechanism in switched architectures [17]. When a MN is roaming among APs linked to the same switch, it uses the PMK obtained in the first authentication with the first AP and the switch verifies the PMK validity through checking its cache. This approach can allow seamless inter-switch roaming through the context transfer for MN's information among switches. iii) Proxim Corporation developed ORINOCO [18] as a switched WLAN system for supporting intelligent and secure roaming. ORINOCO integrates MN's pre-authentication and inter-switch tunneling for transferring authentication keys of each MN. This approach especially aims at providing seamless voice traffic in highly partitioned networks. iv) Aruba proposes a mechanism integrating pre-authentication and encryption keys re-utilisation. This allows secure and fast. roaming while minimizing the overhead of encryption key generation [19]. In this approach, PMKs and encryption keys are stored in a centralized switch, with no need to distribute and synchronize them to APs. In distributed approaches in which encryption is carried out at APs, anticipation of MNs' mobility takes place and encryption keys are transferred to the corresponding APs. Although this mechanism saves a considerable delay, no key refreshment mechanisms are considered which decreases the security level.

Keying hierarchy: Cisco proposes an optimized re-authentication mechanism [20] using the concept of WDS (Wireless Domain Service) and allowing a new key hierarchy. WDS plays the role of a centralized authentication entity, where both

APs and MNs authenticate to the AS through the WDS. Consequently, the WDS creates a shared key with each successfully authenticated entity whether an AP or a MN. This key is named BTK (Base transient Key) and is derived from the NSK (Network Session key). When an authenticated MN roams to a new AP, the AP requests the authentication key of this MN from the WDS. Based on this key, both the MN and the AP derive their encryption keys skipping the 802.1X authentication phase. This approach can assure an optimized re-authentication however in a restricted manner only based on Cisco LEAP authentication protocol.

4. IEEE 802.11R ROAMING CAPABILITIES

An IEEE Task Group (TGr) was formed to address the roaming issues within 802.11 WLAN. This TG targets a new fast roaming standard, 802.11r [7], attempting to minimize the handoff latency while supporting 802.11i security and 802.11e QoS features. This standard is expected to be ratified in June 07. In order to improve the performance of real-time applications, 802.11r ensures that most of the authentication process is carried out before the MN actually begins roaming. The 802.1X authentication is firstly carried out to generate the PMK once the MN joins the network. Keys derived from the PMK, corresponding to each authenticated MN, are generated and are distributed to all authenticated APs in the subnet. When a MN roams across a new AP, it finds its corresponding authentication key, saving the overhead of a complete authentication process. In contrast to 802.11i, the 4-way handshake and the 802.11e traffic specification (TSPEC) negotiation are completed during re-association, further reducing the latency incurred during the MN transition.

802.11r specifies a new key management system, illustrated in Figure 2, including a three level keying hierarchy and corresponding key derivation algorithms. Two levels of Key Holders (KHs) are introduced: i) level 0 KH (R0KH) storing top-level keys that are referred to as PMK-R0s, and ii) level 1 KH (R1KH) storing second-level keys that are referred to as PMK-R1s. PTK is the third level of the key hierarchy and is stored at APs. A *Mobility Domain* (MD) consists of a single R0KH, all its associated R1KHs and all APs associated with R1KHs. A common key hierarchy is accessible to all APs in the same MD. Indeed, KHs are logical entities that can be separate physical entities or can be located within APs. For example, a controller that manages a number of APs can be R0KH, with each AP serving as R1KH. Otherwise, each AP can serve as its own R0KH and R1KH. The three level keys are derived as follows: i) PMK-R0 is mutually derived by the MN and the R0KH from the last 32 octets of the MSK (Master Session Key) and is stored at both R0KH and the MN. ii) PMK-R1 is mutually derived by R0KH and the MN, and is delivered from R0KH to R1KHs within the same MD, and iii) PTK is derived by R1KH and the MN based on PMK-R1. After a MN performs an initial association with an AP, the keys are derived and R0KH distributes the PMK-R1 to R1KHs within the MD. The standard defines a secure three party protocol allowing the R0KH to distribute PMK-R1s to R1KHs. If the MN transitions to another AP

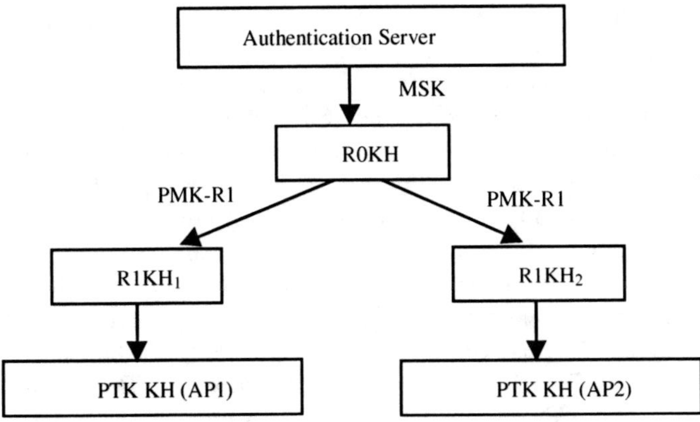

Figure 2. 802.11r key hierarchy

in the same MD, the R1KH associated with this AP should have a copy of the PMK-R1, and no 802.1X authentication is carried out.

Although 802.11r standard is being developed to solve the handoff delay in WLAN environment while assuring the security of connections as well as the QoS, some issues are still not resolved within this standard. In general cases, full EAP authentication is needed when the MN roams beyond the domain of R0-KHs, thus limiting 802.11r solution to intra-domain authentication. Another consideration is that there needs to be a key transfer protocol between R0-KH and the R1-KHs. In other words, there is either a star configuration of security associations between the KHs and a centralized entity that serves as R0-KH, or if KHs are located within the AP, there will be a full mesh of security associations between all authenticators which is undesirable. Furthermore, in 802.11r architecture, the R0-KH may actually be located closed to the edge, thereby creating vulnerability: If R0-KH is compromised; all PMK-R1s derived from the corresponding PMK-R0s will be compromised.

5. IETF ACTIVITIES FOR ROAMING PROVISION

The IETF focuses on EAP in providing solutions for authentication during network's access. However, the EAP model is not efficient in mobile networks roaming scenarios. When a MN moves from one authenticator to another, it is expected to run an EAP method irrespective of whether it has been authenticated to the network recently and has unexpired keying material. A full or even a reduced round trip EAP method execution involves several round trips between the EAP peer and the server causing handoff delay, which is not suitable for real-time applications. The HOKEY WG [8] has been initiated at the IETF 67th meeting, addressing the handover keying problem. This WG is planning to implement a generic mechanism reusing the derived EAP keying material for handover scenarios. AAA infrastructure and EAP framework will be used in the key generation and management in a mean that is agnostic to the type of technology, and is therefore applicable to both intra-technology and inter-technology handovers. The solutions of the HOKEY WG will be independent on the used EAP methods, and no new methods will be proposed. The main issues that are being addressed in this WG concern: re-authentication with the same authenticator, re-authentication with different authenticators in the same mobility domain (defined as handoff), re-authentication with different authenticators in different mobility domains (defined as roaming), and pre-authentication using unexpired keying materials previously acquired in a previous authentication.

The CAPWAP WG [9] is developing a protocol between what is termed an Access Controller (AC) and Wireless Termination Points (WTPs), enabling an AC to manage a collection of WTPs independent of layer 2 technology. The AC and WTP can be mapped to a WLAN switch and APs respectively. One of the goals of CAPWAP is to centralize the authentication and policy enforcement functions for a wireless network, where the authenticator is always implemented at the AC. The AC may also provide centralized bridging, forwarding, and encryption of user traffic, enabling reduced cost and higher efficiency. The inter-authenticator handoff addressed in the HOKEY WG also applies during handoff between ACs. Indeed, CAPWAP inter-controller handoff is a topic that still needs to be studied and the re-authentication work that is being studied within the HOKEY WG can potentially address it in an effective manner.

6. SUMMARY AND OUTLOOK

Roaming in WLAN is a critical mobility feature for real-time applications such as VoIP. It is important to maintain acceptable voice quality during the roaming process while also supporting security features. Indeed, the delay for setting up a connection with an 802.11 AP, including the authentication and the encryption key generation, ranges from 150 to 350 milliseconds. In some extreme cases, this delay may reach 800 milliseconds [15], which is not suitable for real-time applications. Consequently, 802.11i standard specifies the PMK caching and pre-authentication mechanisms to optimize the authentication delay while roaming. However, these mechanisms could not scale well and have seen limited deployment support since they are optional. Vendor-specific solutions, developed to optimize the handoff delay, are mainly based on PMK caching and WLAN switched architectures for centralization and reutilization of PMK keys. These mostly necessitate high security level during PMK storage, otherwise illegitimate MNs could connect to the network via PMK compromising. Table 2, gives a comparison of the different roaming enhancement approaches based on 802.11i.

We generally notice that the pre-authentication approach, also known as *fast associate in advance*, allows roaming enhancement while inducing extra load on the AS, unused contexts in APs, and bandwidth waste especially if the MN does not transition to the expected APs. On the other hand, the optimized re-authentication approach, where the *fast roam back* is an example, can accelerate re-associations

Table 2. Different roaming approaches based on 802.11i

	802.11i	PMK storage by Trapeze Networks	PKC	PKC in switched WLAN archi- tecture	PMK Storage by Symbol Technology	ORINOCO of Proxim	Aruba Centralized Key Storage	CISCO Fast Roamin
Standardized	✓	✓	**X**	**X**	**X**	**X**	**X**	
Based on 802.11i	✓	✓	✓	✓	**X**	**X**	**X**	**X**
Pre-authentication	✓	**X**	**X**	**X**	**X**	✓	**X**	**X**
Optimized reauthentication	✓	✓	✓	✓	✓	**X**	✓	✓
Soliciting the AS for each authentication	✓ In pre-authentication case	**X**	**X**	**X**	**X**	**X**	**X**	**X**
Re-negotiating encryption Keys	✓	✓	✓	✓	✓	✓	**X**	✓
Switched Architecture	**X**	**X**	**X**	✓	✓	✓	✓	**X**
VoIP Support	**X**	✓ According to a study carried out by *Trapeze networks*	Not tested	Not tested	✓ Tests done by *Symbol Technology*	✓	Not tested	✓

with APs but does not always consider associations with new APs. For the later to be assured, centralized storage of authentication keys should take place among APs, however this does not solve inter-domain handoff. Some standards efforts are being carried out aiming to provide solutions for fast and secure roaming. Although, 802.11 TGf proposed the Inter Access-Point Protocol (IAPP) for secure exchange of MNs' security context between APs during handoff, this work is considered as a Recommended Practice since July 03. It was also shown that the PKC mechanism allows less handoff delay compared to the expected one when using IAPP. 802.11r standard (expected to be ratified in June 07) is aiming to provide fast secure roaming. This new standard is expected to enhance multimedia applications performance in roaming scenarios. A performance study was carried out by Intel Corporation in [21], evaluating the performance of a prototype of 802.11r compared with the 802.11i standard. In this study, several sessions of a 2-way G.711u codec VoIP stream were run (64 kbit/sec and 20 ms interval). It was shown that when using 802.11r in roaming scenarios, VoIP application can achieve shorter transition time and reduced packet loss allowing an improvement in voice quality. Table 3 [21], gives a figure of the average roaming time and the packet loss using 802.11i authentication and 802.11r fast transition mechanism.

Another study was carried out by Intel Corporation in [22] using simulation. It has been shown that roaming latency and hence packet loss and voice quality are better in 802.11r authentication when compared to basic 802.11i. The experiments use a 64 kbit/sec constant bit rate for the voice calls (conforming to the ITU G.711 Law codec). 802.11r shows approximately 30 ms lesser delay compared to 802.11i, considering that both authentication approaches are not impacted by the increase in the backend delay. Also, 802.11r shows significant decrease in packet loss compared to 802.11i thanks to the very short latency at the connection process. Table 4 [22], illustrates the packet loss during roaming when using 802.11i and 802.11r, considering different backend latencies. For an end-to-end latency of not more than 150ms in VoIP applications, it is suggested that the backend latency should not go over 100ms even for 802.11r authentication.

Although, 802.11r seems promising in roaming scenarios compared to 802.11i, the fact of distributing MNs' authentication keys among APs belonging to the same SMD does not consider the overall network performance. This may cause a degradation of the overall network performance if an AP having the strongest signal strength is loaded to its full capacity. In this context, 802.11k standard (expected to be ratified in early 07) is addressing the resources management during

Table 3. Effect of roaming on performance

Authentication approach	Average roaming time	Average packet loss %
Basic 802.11i	525 ms	1.8
802.11r fast transition	42 ms	0.2

Table 4. Packet loss in 802.11i and 802.11r

Backend Latency (ms)	Number of Packets Loss	
	802.11i	802.11r
0	4	1
50	10	5
150	22	11
250	35	17

authentication through providing information to discover the best available AP. MN can thus connect to one of the underutilized APs in case of APs' loading.

The IETF activity for fast roaming provision mainly concerns the HOKEY WG. The main objective is providing optimized authentication solutions during roaming of MNs. As these solutions target intra-technology as well as inter-technology handover, some additional support should be provided from lower layers. A convergence layer might be needed enabling handover between entities of different technologies in a seamless manner. Seamless handover between different technologies is studied within the IEEE 802.21 standard; however, security is not in the scope of this work.

REFERENCES

1. Wi-FiHotSpotList.com. http://www.wi-fihotspotlist.com.
2. International Telecommunication Union, ITU-TG 114 recommendation. http://www.itu.int.
3. Mishra A, Shin M, and Arbaugh WA (April 2003) An Empirical Analysis of the IEEE 802.11 MAC Layer Handoff Process. ACM Computer Communications Review.
4. Anton B, Bullock B, and Short J (February 2003) Best Current Practice for Wireless Internet Service Provider (WISP) roaming. Wi-Fi Alliance–Wireless ISP Roaming (WISPr).
5. Wang H, Prasad R, Schoo AP, Bayarou M, Rohr K, Rohr S (2004) Security Mechanisms and Security Analysis: Hotspot WLANs and Inter-operator Roaming. ACM WMASH'04.
6. IEEE Std. 802.11i (July 2004) Amendment 6: Medium Access Control (MAC) Security Enhancements.
7. IEEE P802.11r/D3.0 (September 2006) Draft Amendment to Standard for Information Technology–elecommunications and Information Exchange Between Systems–LAN/MAN Specific Requirements.
8. IETF HOKEY WG (Handover Keying), http://www.ietf.org/html.charters/hokey-charter.html
9. IETF CAPWAP WG (Control and Provisioning of Wireless Access Points), http://www.ietf.org/html.charters/capwap-charter.html
10. IEEE Std.802.1X (June 2001) Port-based Network Access Control.
11. Aboba B, Blunk L, Vollbrecht J and Carlson J (June 2004) Extensible Authentication Protocol (EAP). RFC 3748.
12. Aboba B, Calhoun P (September 2003) RADIUS (Remote Authentication Dial In User Service) Support For Extensible Authentication protocol (EAP). RFC 3579.
13. The Wi-Fi Alliance, http://www.wi-fi.org
14. Trapeze Networks Press Releases (June 2004) Approval of landmark IEEE security standard enables WLAN vendors to add stronger encryption to product portfolios. http://www.trapezenetworks.com/news/pressreleases

15. CommsDesign NewsLetters (August 2004) Caching Technique Eases WLAN Roaming. http://www.commsdesign.com/design_center/wireless/news/
16. Technical White Paper (February 2005) Securing Enterprise Air: Understanding and Achieving Next-Generation Wireless Security with Symbol Technologies and 802.11i. Symbol Technology, http://www.symbol.com/assets/files/SecureEntAirWP.pdf
17. AirSpace Technology White Paper (2005) Authentication And Encryption In An Enterprise Wireless LAN, http://www.airespace.com/technology/technote_auth_enc_wlan.php
18. Wi-Fi Technology Forum. Proxim's Unveils ORINOCO Wireless LAN Switching System. http://www.wi-fitechnology.com/printarticle849.html
19. Aruba Technical Report (2006) Mobility in an 802.11i Enables Wireless LAN. http://www.arubanetworks.com/pdf/802.11i-mobility.pdf
20. Cisco Application Note (2004) Cisco Fast Secure Roaming. http://www.cisco.com/warp/public/cc/pd/witc/ao1200ap/prodlit/cifsr_rf.pdf
21. Bangolae S, Bell C and Qi E (2006) Performance Study of Fast BSS Transition using IEEE 802.11r. International Wireless Communication and Mobile Computing Conference (IWCMC'06).
22. Yap C (2005) Issues with real-time streaming applications roaming in QoS-based secure IEEE 802.11 WLANs. 2nd IEEE International Conference on Mobile Technology, Applications and Systems.

CHAPTER 30

WORMHOLE DETECTION METHOD BASED ON LOCATION IN WIRELESS AD-HOC NETWORKS

KYUHO LEE, HYOJIN JEON, AND DONGKYOO KIM

Graduate School of Information and Communication, University of Ajou
Woncheon-dong, Youngtong-Gu, Suwon, Republic of KOREA {im295, jinsclub, dkkim}@ajou.ac.kr

Abstract: In mobile ad-hoc networks, multi path routing is vulnerable since network topology is constantly changed and nodes join or leave the network frequently. So malicious nodes can drop or modify, fabricate packets for interruption of communication because of no guaranteeing reliability of nodes. Wormhole attack is one of the serious attacks which form a serious threat in the networks, especially against ad-hoc wireless routing protocols. Detection of this attack which uses colluding nodes is more difficult than attacks by single node. In this paper, we present a countermeasure for detection of wormhole attack, which efficiently mitigates the wormhole attack in ad-hoc networks. This method is based on location information of each node in DSR protocol. Each intermediate node appends its own ID with location information into a request packet and source node analyzes distance of between each node in a reply packet

1. INTRODUCTION

In mobile ad-hoc networks, since mobile nodes join or leave the networks frequently and network topology is constantly changed, it is difficult to manage network. So, many researchers and laboratory have studied about secure communication and management of networks in wireless ad-hoc networks. Especially routing techniques which is different with wired environment is needed for forwarding messages and providing services in wireless ad-hoc networks. In order to satisfy this requirement, many kink of routing protocol is proposed. [1–3]

However these routing protocols are focused on exact forwarding of messages, thus they are exposed many type of attacks and vulnerable. To solve above problems, some researchers have proposed secure routing protocols like CSER[4], ARAN[5], SRP[6], SEAD[7], SAODV[8], etc.

But these secure routing protocols only provide defense of attacks which execute by single malicious node. In other words, they can defense attacks that single node

H. Labiod and M. Badra (eds.), New Technologies, Mobility and Security, 361–372.
© 2007 *Springer.*

modify or drop routing packets in ad-hoc networks. Thus these secure routing protocols cannot counter attacks executing by colluding malicious nodes. The colluding attacks like wormhole attack is difficult to be detect than noncolluding attacks and damage is extensive. And the more computing environment is developed well, the more services based on ad-hoc networks increase, so secure method for defense of attack at routing is more required. Specially, colluding attack like wormhole attack which damages seriously to networks, nodes.

Thus, we present a countermeasure for detection of wormhole attack which is typical colluding attack which is based on location information of each node in DSR protocol. Each intermediate node appends its own ID with location information into RREQ packet and source node analyzes distance of between each node in RREP.

In this paper, we make the following contributions:

- We provide a primitive that prevents a node from corrupting route by wormhole.
- We describe a wormhole attack presented at early studies.
- We develop a protocol that detects a wormhole using location information of nodes in ad-hoc networks
- We analyze the efficiency and overhead of our mechanism and security analysis.

2. RELATED WORK

The wormhole attack in wireless networks was independently introduced by Dahill [9], Papadimitratos [6], Hu [10], and some mechanisms is proposed for detection and prevention of wormhole attack.

Early studies for countermeasure of wormhole used cryptographic mechanism. So it can prevent a network from external attacker who doesn't know secret key.

Hu *et al.* [10] introduced the concept of geographical and temporal packet leashes for detecting wormhole attack. In the geographical leashes, the location information and loosely synchronized clocks verify the neighbor relation. If the distance between nodes exceeds a maximum distance of 1 hop, it suspects that path as wormhole. In the temporal leashes, the packet transmission distance is calculated as the product of signal propagation time and the speed of light. It requires accuracy time synchronized clocks.

Capkun [11] proposed MAD (Mutual Authentication with Distance-Bounding) without use of clock synchronization. They show how to detect wormhole attacks using MAD. Each node sends one bit challenge to another node, then the node responds to sender instantaneously. The sender measures the round trip time of the challenge bit to estimate the distance between the nodes. But, this approach should use special hardware that can respond to a one-bit challenge without any delay.

DelPHI [12] lets the sender to check whether there are any malicious nodes sitting along its paths to the receiver. This approach uses the delay per hop count information of some disjoint paths between the sender and the receiver to check whether a path includes wormhole or not through assume that the delay per hop of a wormhole path is larger than normal's. However if attack node send the packet to another attack node using high transmission rate, value of delay per hop on wormhole path is similar to the value of normal path.

Hu, Evans *et al.* [13] used directional antennas to prevent a wormhole attack. It presented a method for secure neighbor discovery using the secret key and directionality of the antennas. It can prevent attacker node from forging as neighbor of one node but the requirement of directional antennas on all nodes may be infeasible for some deployments.

LiteWorp [14] is based on local monitoring at each Guard node. Guard node observes that whether a node forwards a packet from a previous node to next node without modifying the packet. And all nodes manage table of neighbor in 2-hop for wormhole detection. But since this mechanism use continuous monitoring at all nodes, it consumes much resources of each node.

3. WORMHOLE ATTACK

Wormhole attack used colluding malicious nodes damages to route discovery processing, in wireless ad-hoc networks. It interrupts correct routing process, so communication of some node is not available.

3.1. Wormhole Attack in DSR

Wormhole is a kind of tunnel building in networks. Actually, a distance between two nodes is more than 1 hop but using this tunnel, the distance becomes 1 hop as neighbor. Thus route between nodes is established including malicious nodes consist of wormhole, since it is quite probable that a route which has smaller hop count is selected.

Example of basic wormhole is shown in Figure 1. Assume that S wants to send a data to D. S starts route discovery sending RREQ. And each intermediate node appends its own address. At this time wormhole node X forwards RREQ to Y directly using tunnel between X and Y. Thus route between S and D will not be <S-A-X-B-C-E-Y-F-D> but <S-A-X-Y-F-D>. After establishing route, wormhole nodes X, Y are available

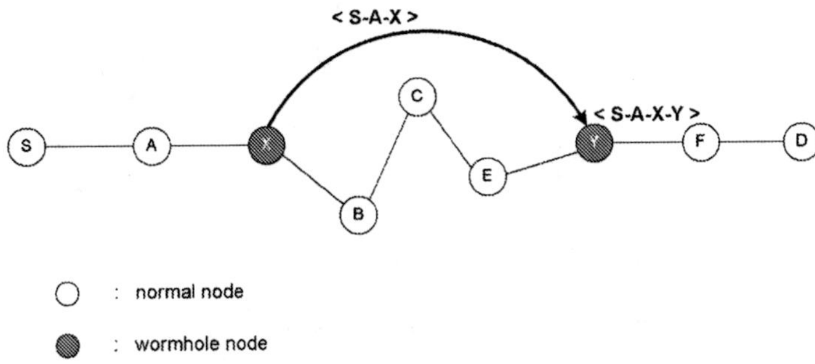

Figure 1. Wormhole attack example in DSR

drop or modify packets between S and D. If wormhole node is deployed at critical position, many communication of network is affected by wormhole.

3.2. Classification of Wormhole Attack

Wormhole is classified based on the techniques used for launching it [14].

1) Wormhole using Encapsulation

Consider Figure 2 in which nodes A and B try to discover the shortest path between them, in the presence of the two malicious nodes X and Y consisting of wormhole. If node A broadcasts a route request (RREQ), X gets RREQ and encapsulates it in a packet destined to Y through the tunnel that exists between X and Y. If Y get the packet from X, Y decapsulates the packet and obtains RREQ, and then Y rebroadcast the RREQ.

2) Wormhole using out of band channel

This wormhole is launched by having an out-of-band high-bandwidth channel between the malicious nodes. This channel can be achieved by using a long-range directional wireless link or a direct wired link. Node A is sending a RREQ to node B, nodes X and Y are malicious having an out-of-band channel between them. X forwards the RREQ to Y. and then Y rebroadcasts the packet to its neighbors. So B gets two route requests <A-X-Y-B> and <A-C-D-E-F-B>. The first route is shorter than the second.

Wormhole attack is classified based on the exposure of malicious node [15].

3) Open wormhole attack

In this type of wormhole, the attackers include themselves in the RREQ packet header following the route discovery procedure. Other nodes are aware that the malicious nodes lie on the path but they would think that the malicious nodes are direct neighbors.

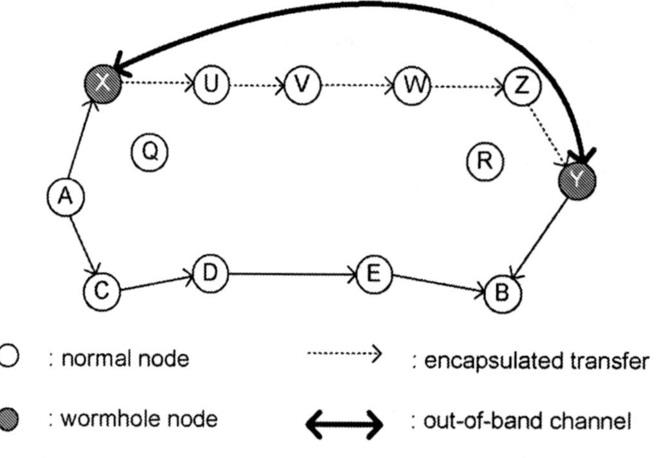

Figure 2. Techniques used for launching wormhole

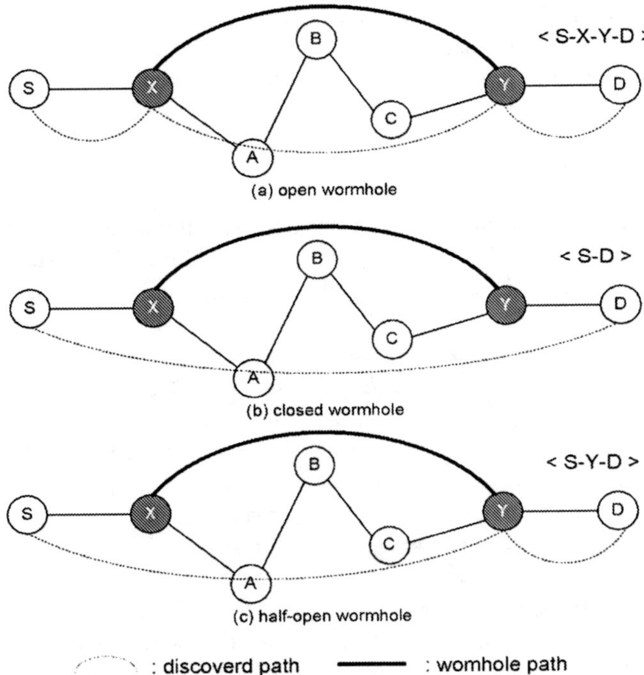

Figure 3. Three type of wormhole attacks

4) Closed wormhole attack

The attackers do not modify the content of the packet, even the packet is a DSR route discovery packet. Instead, they simply tunnel the packet from one side of wormhole to another side and it rebroadcasts the packet.

5) Half open wormhole attack

One side of wormhole doesn't modify the packet and only another side modifies the packet following the route discovery procedure.

4. DETECTION OF WORMHOLE ATTACK

In this section, we describe the method for wormhole detection based on location information of each node.

4.1. System Assumptions

Our wormhole detection mechanism assumes following conditions.

- The link are bidirectional, which means that if a node A can hear node B then B can hear A.
- All nodes have a hash function and digital signature generation · verification function.

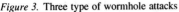

- Public key is distributed by Bootstrapping Security Association [16].
- All nodes have own location information.

Our mechanism is based on DSR (Dynamic Source Routing) protocol.

4.2. Detection Mechanism

Our mechanism for wormhole detection is based on location information of each node in DSR protocol. Each intermediate node appends its own address with location information into RREQ packet and source node analyzes distance of between each node in RREP.

When a source node S wants to discover a route to destination node D, it initiates route discovery. It constructs a RREQ (Route Request) including the address with location of node S and node list to accumulate the address and location information of intermediate nodes forwarding the request to D. S also appends its own public key and broadcasts it. Each intermediate node receiving the packet appends its address with location information to the node list and rebroadcasts the packet. If an intermediate node finds that its address is already on the node list, it discards the packet.

When the RREQ reaches its destination, node D constructs a RREP (Route Reply) packet. It extracts the accumulated path with locations in the RREQ, includes a copy of it in the RREP packet, and digitally signs the accumulated nodes list. It then appends its public key and the signature to the packet, and unicasts the packet on the reverse of this accumulated path.

Intermediate node receiving the RREP check to see if the own location is the same as the location in nodes list of RREP, if true the node forwards.

When source node S receives the RREP packet, it first verifies the RREP's validity. To verify the packet S verifies the D's signature and comparing it against the nodes list of RREP. If the signature verification passes, S checks if each distance between nodes is less than maximum distance of 1-hop allowed by networks using location information in RREP. If location information of two nodes i, j are (x_i, y_i)

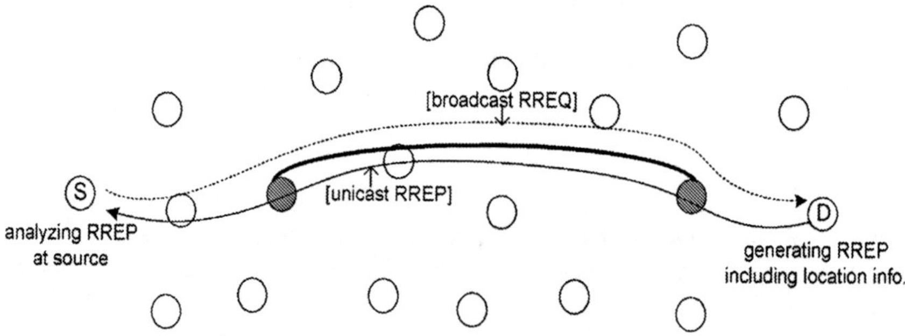

Figure 4. Concept of wormhole detection in our mechanism

$$\begin{array}{lllll} S & \rightarrow & * & : & (\text{Rq, S, D, \# } (A|P_s)) \\ A & \rightarrow & * & : & (\text{Rq, S, D, \# } (A|P_s), (A|P_A)) \\ X & \rightarrow & Y & : & (\text{Rq, S, D, \# } (A|P_s), (A|P_A, (A|P_X)) \\ Y & \rightarrow & * & : & (\text{Rq, S, D, \# } (A|P_s), (A|P_A, (A|P_X, (A|P_Y)) \\ F & \rightarrow & * & : & (\text{Rq, S, D, \# } (A|P_s), (A|P_A, (A|P_X, (A|P_Y(A|P_F)) \end{array}$$

Figure 5. RREQ message flow for the route discovery of between S and D of figure 1.
Legend: * indicates broadcast address. Rq is request tag. # indicates sequence number. A|Pi is address with location information of i

Digital signature of nodes list

D → F : (Rp, SR(D, F, Y, X, A, S), S, D, #, ($A|P_S$, $A|P_A$, $A|P_X$, $A|P_Y$, $A|P_F$, $A|P_D$)) + [($A|P_S$, $A|P_A$, $A|P_X$, $A|P_Y$, $A|P_F$, $A|P_D$)]$_{PriK\text{-}D}$ + PubK-D

F → Y : (Rp, SR(D, F, Y, X, A, S), S, D, #, ($A|P_S$, $A|P_A$, $A|P_X$, $A|P_Y$, $A|P_F$, $A|P_D$)) + [($A|P_S$, $A|P_A$, $A|P_X$, $A|P_Y$, $A|P_F$, $A|P_D$)]$_{PriK\text{-}D}$ + PubK-D

Y → X : (Rp, SR(D, F, Y, X, A, S), S, D, #, ($A|P_S$, $A|P_A$, $A|P_X$, $A|P_Y$, $A|P_F$, $A|P_D$)) + [($A|P_S$, $A|P_A$, $A|P_X$, $A|P_Y$, $A|P_F$, $A|P_D$)]$_{PriK\text{-}D}$ + PubK-D

X → A : (Rp, SR(D, F, Y, X, A, S), S, D, #, ($A|P_S$, $A|P_A$, $A|P_X$, $A|P_Y$, $A|P_F$, $A|P_D$)) + [($A|P_S$, $A|P_A$, $A|P_X$, $A|P_Y$, $A|P_F$, $A|P_D$)]$_{PriK\text{-}D}$ + PubK-D

A → S : (Rp, SR(D, F, Y, X, A, S), S, D, #, ($A|P_S$, $A|P_A$, $A|P_X$, $A|P_Y$, $A|P_F$, $A|P_D$)) + [($A|P_S$, $A|P_A$, $A|P_X$, $A|P_Y$, $A|P_F$, $A|P_D$)]$_{PriK\text{-}D}$ + PubK-D

Figure 6. RREP message flow for the route discovery of between S and D.
Legend: Rp is reply tag. # indicates sequence number. PriK-i is i's private key and PubK-i is i's public key

and (x_j, y_j), the distance of between i and j is

$$D_{i,j} = \sqrt{(x_i - x_j)^2 + (y_i - y_j)^2}$$

If any of the each distance checks fail, S discards the RREP packet and that path which has a distance more than maximum distance of 1-hop is suspected of malicious path including wormhole.

5. SECURITY ANALYSIS

We consider several possible attacks executed by wormhole on our wormhole detection protocol using the example topology of Figure 8.

Attack 1: Assume that the wormhole is open wormhole.

Node S will receive a RREP including nodes list <S-A-X-Y-F-D>. After analyzing distance between nodes at node S, node S can detect wormhole. Since the distance between X and Y is more than maximum distance of 1-hop allowed by networks.

Attack 2: Assume that the wormhole is closed wormhole.

Node S will receive a RREP including nodes list <S-A-F-D>. After analyzing distance between nodes at node S, node S becomes aware of wormhole on the route.

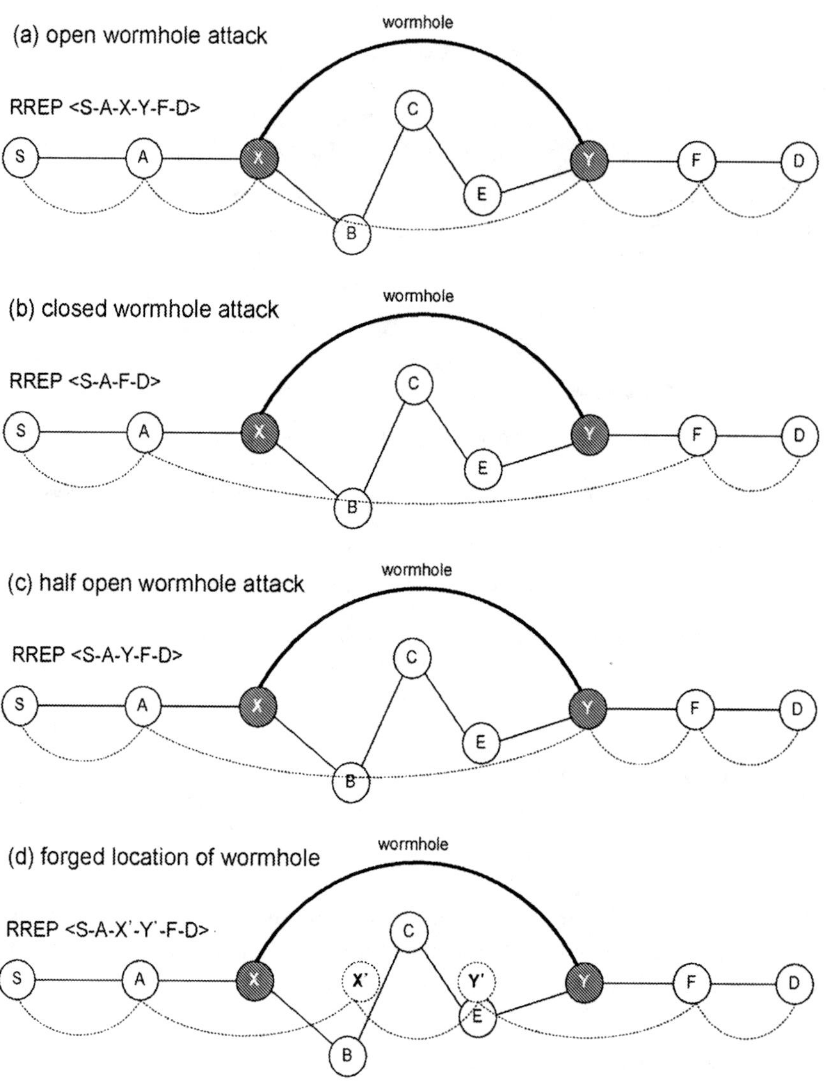

Figure 7. Security analysis using some cases

Because the distance between A and F is more than maximum distance of 1-hop allowed by networks.

Attack 3: Assume that the wormhole is half open wormhole.

Node S will receive a RREP including nodes list <S-A-Y-F-D> or <S-A-X-F-D>. After analyzing a distance between nodes at node S, node S can detect existence of wormhole. Because the distance between Y and F, or between X and F is more than maximum distance of 1-hop allowed by networks.

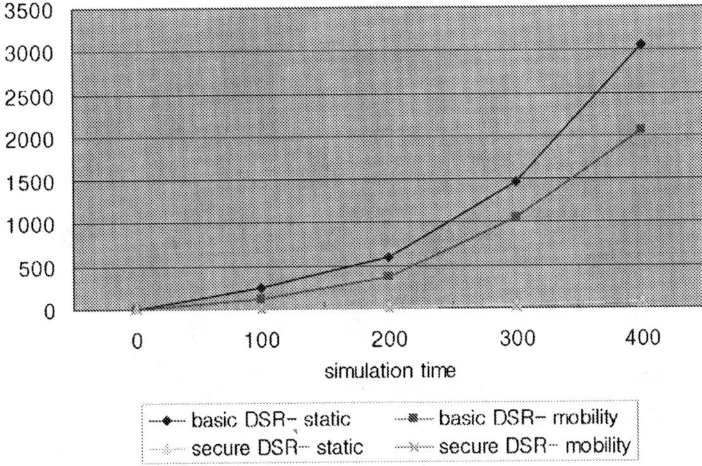

Figure 8. Cumulative number of dropped packets with and without our mechanism

Attack 4: Assume that wormhole nodes X and Y forge their own location information.

Node S will receive a RREP including nodes list <S-A-X'-Y'-F-D> in figure 8, (d). After analyzing distance between nodes at node S, although the distance between X and Y is less than maximum distance of 1-hop, the distance between A and X' or between Y' and F is more than maximum distance of 1-hop. Thus node S becomes aware of wormhole on route.

Attack 5: Assume that intermediate malicious node modify a RREP for forging a location of nodes.

Node S will receive a RREP modifying the location of nodes list. However, node S would not accept the modified RREP since a signature of D would not pass S's authentication check.

6. SIMULATION RESULT

We performed a set of simulations using the popular *ns-2* simulator to evaluate the overhead and performance of our security mechanism. We used a random way point model for our mobility scenarios in the setting described in table 1. We compare our mechanism to basic DSR.

6.1. Efficiency of Our Mechanism

In our simulation, assume that the wormhole nodes pass routing packet but drop all data packets. So when using basic DSR, dada packets through a route including wormhole is dropped by malicious wormhole nodes. However our mechanism

Table 1. Environment for the ns-2 experiments

Nodes	50
Scene	1000m × 1000m
Wireless range	250 m
Number of sources	10
Traffic rate	4 pkts/s
Traffic type	CBR(UDP)
Bandwidth	2 Mbps

avoids that a route including wormhole is established. Thus it prevents from dropping data packets by malicious wormhole nodes as shown figure 9.

6.2. Packet Overhead of Our Mechanism

Our mechanism does not support the DSR optimization [2] since it performs end-to-end signature authentication of routing packet and verification of whether a node is authorized to send a RREP packet. Therefore, an intermediate node cannot reply from its cache, since it cannot be verified whether that node had the right to send that message. Figure 10 illustrates the packet overhead of basic DSR versus DSR installed our mechanism.

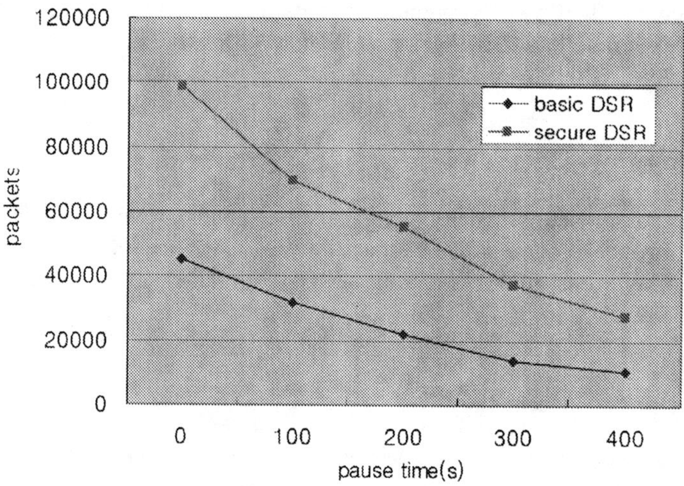

Figure 9. Packet overhead of our mechanism

6.3. Computation Cost Estimates of RSA

Our mechanism uses digital signature based on public-private key algorithms like RSA. But computational cost of RSA is much more than symmetric key algorithms like SHA-1, HMAC. For example, the time to perform RSA operation on 14Mhz Palm V is 27.8 seconds for 1024 bits RSA signature generation, and 0.758 seconds for RSA signature verification since the Palm V was not designed with security in mind[17].

However hardware crypto accelerators like ds5250 of MAXIM Corp. can perform RSA operations in less than a 6ms[18]. If so, use of digital signature is available.

7. CONCLUSION

In mobile ad-hoc networks, since mobile nodes join or leave the networks frequently and network topology is constantly changed, it is difficult to manage network.

The more computing environment is developed well, the more services based on ad-hoc networks increase, so secure method for defense of attack at routing is more required. Specially, colluding attack like wormhole attack which damages seriously to networks and nodes. However, secure routing protocols proposed previously provide defense of attacks which execute by only single malicious node.

Thus, in this paper we proposed a mechanism to detect wormhole in ad-hoc networks. We achieved this through the use of location information of each node on route selected by DSR. Every time a source node receives RREP, the source node analyzes whether each distance between nodes included in RREP is more than maximum distance of 1-hoc allowed by networks for wormhole detection.

In future work we plan to study a countermeasure after detection of wormhole and a mechanism using symmetric key for reduction of computational cost.

REFERENCES

1. C. E. Perkins and E. M. Royer, "Ad hoc On-Demand Distance Vector Routing", In Proceedings of the 2nd IEEE Workshop on Mobile Computing Systems and Applications, pp. 90–100, New Orleans, LA, February 1999.
2. D. B. Johnson, D. A. Maltz, and J. Broch, "DSR: The Dynamic Source Routing Protocol for Multi-Hop Wireless Ad Hoc Networks", in Ad Hoc Networking, edited by Charles E. Perkins, Chapter 5, pp. 139–172, Addison-Wesley, 2001.
3. Rajendra V. Boppana and Satyadeva Konduru. "An Adaptive Distance Vector Routing Algorithm for Mobile, Ad Hoc Networks.", In Proceedings of the Twentieth Annual Joint Conference of the IEEE Computer and communications Societies(INFOCOM 2001), pages 1753–1762, 2001.
4. B. Lu, U. W. Pooch, "Cooperative Security-Enforcement Routing in Mobile Ad Hoc Networks", Mobile and Wireless Communications Network, 4th International Workshop, pp. 157–161, 2002.
5. K. Sanzgiri, B. Dahill, B. N. Levine, C. Shields, E. M. B. Royer, "A Secure Routing Protocol for Ad Hoc Networks", Proceedings of the SCS Communication Networks and Distributed Systems Modeling and Simulation Conference, 2002.
6. P. Papadimitratos and Z.J. Hass, "Secure Routing for Mobile Ad hoc Networks", Proceeding of the SCS Communication Networks and Distributed Systems Modeling and Simulation Conference (CNDS 02), 2002.

7. Y. C. Hu, D. B. Johnson, A. Perrig "SEAD: Secure Efficient Distance Vector Routing For Mobile Wireless Ad Hoc Networks", Mobile Computing Systems and Applications, Proceedings 4th IEEE Workshop, pp.3–13, 2002.
8. M. C. Zapata "Secure Ad hoc On-Demand Distance Vector Routing", ACM SIGMOBILE Mobile Computing and Communications Review, Vol 6, issue 3, New work, USA, pp.106–107, 2002.
9. B. Dahill, B. N. Levine, E. Royer, and C. Shields, "A secure routing protocol for ad-hoc networks", Electrical Engineering and Computer Science, University of Michigan, Tech. Rep. UM-CS-2001-037, 2001.
10. Y. C. Hu, A. Perrig, and D. B. Johnson. "Packet Leashes: A Defense against Wormhole Attacks in Wireless Ad Hoc Networks", In Proceedings of IEEE INFOCOM 2003, pp.1976–1986, 2003.
11. S. Capkun, L. Buttyan, and J. P. Hubaux, "SECTOR: Secure Tracking of Node Encounters in Multi-hop Wireless Networks", in Proceedings of the 1st ACM workshop on Security of ad hoc and sensor networks (SASN 03), pp.21–32, 2003.
12. Hon Sun Chiu, King-Shan Lui, "DelPHI: Wormhole Detection Mechanism for Ad Hoc Wireless Networks", In Proceedings of Wireless Pervasive Computing 2006 1st International Symposium, pp.1–6, Jan 2006.
13. L. Hu and D. Evans "Using Directional Antennas to Prevent Wormhole attacks", In Proceedings of Network and Distributed System Security Symposium, pp.131–141, 2004.
14. Khalil, I. Saurabh Bagchi Shroff, N.B., "LITEWORP: a lightweight countermeasure for the wormhole attack in multihop wireless networks", In Proceedings of the International Conference on Dependable Systems and Networks,(DSN'05), pp.612–621, 2005.
15. W. wang, B. Bhargava, Y. Lu, and X. Wu, "Defending against Wormhole Attacks in Mobile Ad Hoc Networks", under review at Wiley Journal Wireless Communication and Mobile Computing 2005, pp. 15:1–21, 2005.
16. Bobba, L. Eschenauer, V.D. Gligor, and W. Arbaugh., "Bootstrapping Security Associations for Routing in Mobile Ad-Hoc Networks.", Technical Report TR 2002–44, University of Maryland, May 2002.
17. N. Modadugu, D. Boneh, and M. Kim, "Generating RSA keys on a handheld using an untrusted server", in RSA 2000, 2000.
18. MAXIM Homepage, "http://www.maxim-ic.com/appnotes.cfm/appnote_number/2033"

CHAPTER 31

AN EFFECTIVE DEFENSE AGAINST SPOOFED IP TRAFFIC

HIKMAT FARHAT

Computer Science Department, Notre Dame University, Zouk Mosbeh, Lebanon

Abstract: The problems presented by Denial of Service (DoS) attacks are aggravated by IP spoofing. In this paper we propose a new approach for IP spoofing detection and real-time prevention. The proposed method depends on the inability of attackers with spoofed source IP address to complete TCP transactions and on the concept of path signatures. Simulations based on real-world Internet topologies shows that 95% of spoofed packets are dropped by the border routers employing the proposed scheme. Using the concept of partial matching of signatures coupled with priority queueing of packets at border routers, the proposed method can be deployed in an incremental fashion with immediate benefit for ISPs who deploy the scheme. In addition, a filter aggregation technique, based on an analysis of BGP dynamics and substantiated by extensive measurements, is presented which allows the proposed scheme to be highly scalable and feasible for deployment on current generation hardware

1. INTRODUCTION

Denial of Service attacks are one of the most serious threats facing the Internet today and also one of the most difficult problems to solve. Many recent studies [1] have shown that while most DoS attacks are not publicized the threat is prevalent. Even worse, attackers often compound the problem facing legitimate servers by using IP source address spoofing. Due to the stateless nature of the Internet Protocol and the routing decision which is based on the packet's destination address, address spoofing is simple and very effective in evading detection. Even the simplest approach to defending against DDoS, installing filters at border routers, can be defeated by IP spoofing. The attacker(s) can choose randomly an IP address as the source for different packets and therefore make the scheme infeasible. Therefore, detecting and blocking packets with spoofed source address has been actively pursued in the research community.

H. Labiod and M. Badra (eds.), New Technologies, Mobility and Security, 373–383.
© 2007 *Springer.*

Most solutions to the aforementioned problem fall into two categories. Traceback [2–5] schemes are representatives of the first category where the goals is to determine the source of the attack. These methods, however, do not specify how an attack can be blocked once the source is identified. In addition such schemes are computationally infeasible when the number of attacking hosts is large. The second category of defenses attempts to restrict the address space available to the attacker(s) and thus reducing the severity of IP spoofing rather than completely eliminating it [6][7].

The Implicit Token Scheme (ITS) has been proposed recently by the author of this paper as a method to mitigate DDoS attacks [8]. The key idea in ITS is that attackers cannot complete the TCP three-way handshake if they use spoofed source addresses. In this paper, we improve the basic ITS method to allow for incremental deployment and higher scalability. Incremental deployment is supported by a priority queueing scheme for partial path matching and scalability is improved using an analysis of the Border Gateway Protocol (BGP)[9] route announcements. Incidentally, the two additions improve the method in other areas especially by obviating the need to estimate the number of hops a packet has travelled.

The rest of the paper is organized as follows. The ITS method is reviewed in Section 2. Incremental deployment and scalability improvements to the ITS are discussed in Section 3. The results of simulations are presented in Section 4. Other research on DDoS and IP spoofing are discussed in Section 5. The conclusion and future directions are presented in Section 6.

2. THE BASIC METHOD

The Implicit Token Scheme (ITS) [8] provides protection against IP spoofed traffic by having ISPs install filters at the border router as shown in figure 1. These filters are based on entries in a database that map a given IP address to the network path that a packet from that IP address takes to reach the victim. The basic ITS method requires intermediate routers to add a *deterministic* mark to the identification field of the IP header of transiting packets. The collection of marks by all intermediate routers is called the packet's *signature*. According to [10] a mark of a 2-bit hash related to the IP address of the intermediate router and its peer is optimal. Therefore each entry in the database contains the source IP address and the corresponding path signature. Such entries are used as filters at the border router by dropping any packet that does not contain the valid path signature. New entries are added to the database only when a client completes the TCP three-way handshake and therefore making absolutely sure that the added entry belongs to a legitimate IP address and not for a spoofed one. We assume that there is a "Lookup" function to retrieve a given entry based on the IP address. Once such an entry is retrieved, the signature therein is compared to the signature stored in the database.

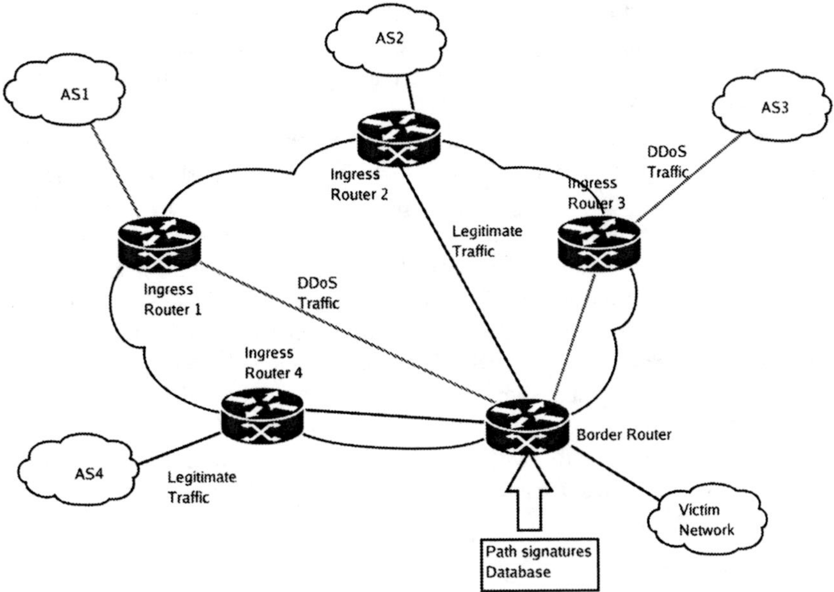

Figure 1. Defense model against DDoS attacks

2.1. Building the Database

The database contains entries, each composed of the source IP address and the corresponding path signature stored in the 16-bit IP identification field of the IP header. This field is marked by the routers along the path, from the source to the destination, where each router contributes 2 bits. Populating the database, however, is not an easy task.

The first approach that comes to mind is to add entries to the database during *peace time* (i.e. when the server is not under attack) and store the path signature for every IP address that connects to the target. There are many reasons that render this approach impractical. First, the path signature may change from the time it is recorded to the time it is used as a result of routing changes. Second, during a spoofed attack the target sees a large number of previously unseen addresses, which leads to a large number of false positives [11]. Finally, a prospective target would require about a year to collect half of the Internet address space [11].

A superior method is to add paths signatures and their corresponding IP address after the clients completes the TCP three-way handshake. This way a perimeter router can allocate a large portion of the bandwidth, say 95%, to non-spoofed packets and the remaining 5% to rest. This method has none of the above- mentioned shortcomings. First, the path signature is up to date since it is the same one that is present during TCP connection establishment. Second, the database will have the exact number of entries needed because any legitimate packet must arrive after the TCP handshake and therefore its corresponding entry will be present. Third, by delaying the addition of an entry until the TCP handshake is complete we are sure that the entry is correct because any attacker using a spoofed address will not be able to complete the TCP handshake.

While the above method protects all established connections, an attacker can still deplete the 5% of bandwidth allocated to connection establishment and denies legitimate users from establishing a new connection. To solve this problem ITS introduced a SYN cookie mechanism in the perimeter routers. A SYN cookie [12] is a special value of the Initial Segment Number (ISN) that allows a device to respond to a TCP connection request in a stateless manner. In ITS it is used to allow a perimeter router to respond to TCP requests on behalf of the target. When a perimeter router receives a TCP SYN segment having destination IP address equal to that of the target it responds with a SYN+ACK on behalf of the target without maintaining any state information. Only when the third segment of the TCP handshake is received correctly, will the perimeter router forward the packet to the target. The algorithm used for packet filtering is shown in Figure 2.

2.2. Packet Filtering

The path signature of a packet is a sequence of deterministic stamps marked by intermediate routers and stored in the 16-bit identification field of the IP header. Each router adds two bits, which represent a hash of the router IP number, to the identification field. When a packet with identification value id arrives at a router with router mark M, the value of the identification field is modified as shown below:

$$id = id \ll 2 + M$$

```
FILTER AT A BORDER ROUTER
for each packet pkt
      do
         if pkt.SYN=1
            then
                 sendCookie
                 Exit
         if pkt.SIG = Lookup(pkt.SOURCE).SIG
            then
                 forward packet
         elseif checkCookie(pkt.ACK)= TRUE
            then
                 forward pkt
                 insert pkt.SIG in D
         else
                 drop pkt
```

Figure 2. Packet filtering using exact match

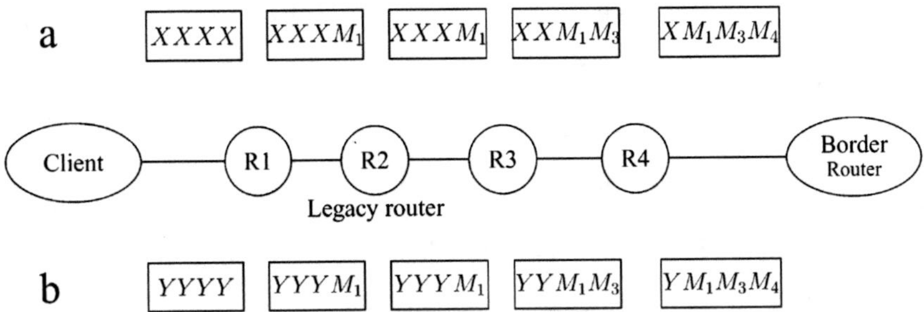

Figure 3. The effect of the presence of legacy routers on path signatures

Where "≪" is the bit-wise left-shift operator. This way router stamps are always contiguous even in the presence of legacy routers. Since the IP identification field is 16-bits there are room for 8 stamps. Figure 3 illustrates this process by using an 8-bit field for simplicity.

Since every packet sent by a client initially contains a different value in the identification field it is hard for the perimeter router to distinguish the original bits from the bits marked by intermediate routers. The solution adopted by ITS is to estimate the number of hops a packet has traveled and use it to mask out the original bits. The idea is that most OSs use a well-known set of initial TTL values and by comparing the final TTL value one could estimate the number of hops. See [8] for details. This method has two problems. First, it is not always possible to accurately determine the number of hops. More importantly, in the presence of legacy routers the number of hops that a packet has traveled is not equal to the number of router stamps, even if we could accurately determine the number of hops. Figure 3 illustrates this problem where for simplicity we have shown only 4 router marks. A client sends a packet with initial value of XXXX in the identification field and the signature is stored by the perimeter router as $XM_1M_3M_4$. At a later stage the same client sends a packet with initial value of $YYYY$ and the value received by the perimeter router is $YM_1M_3M_4$, which does not match the stored value for that particular source IP address. The key idea, developed in this paper, is that even though the signatures do not match there are three contiguous router marks that match, $M_1M_3M_4$. This fact will be used in the next section to implement priority queuing.

3. IMPROVEMENTS

3.1. Packet Filtering

In the new filtering scheme the system maintains 9 different priority levels $0, \ldots, 8$ where 0 is the lowest priority. When the border router receives a packet it executes the algorithm in Figure 4. If the SYN bit is on, a reply containing a cookie is send to the client. If the SYN bit is not on, the router looks up the source IP address in the database as shown on line 7. If it does not exist it means that

MODIFIED FILTER AT A BORDER ROUTER

```
1   for each packet pkt
2       do
3          if pkt.SYN=1
4             then
5                sendCookie
6                Exit
7          if SIG=Lookup(pkt.SOURCE).SIG
8             then
9                n ← CountMatches(SIG,pkt)
10               add pkt to queue n
11         elseif checkCookie(pkt.ACK)=TRUE
12            then
13               add pkt to queue 0
14               insert pkt.SIG in D
15         else
16               drop pkt
```

Figure 4. Modified packet filtering using partial match

the client has no active connection, therefore the router checks if the packet is a reply to a cookie (line 11). If it is, then the signature is added to the database and the packet is given priority 0 (the lowest). If the signature exists in the database (lines 9 and 10) then the number of contiguous matches between the packet signature and the one stored in the database is computed. This number of matches is used to assign a priority to the packet. As an example, consider Figure 3 again. Since the stored signature is $XM_1M_3M_5$ and the signature of the received packet is $YM_1M_3M_5$ then the number of matches is three and the packet is given priority three. In the general case the number of matches is computed as shown in Figure 5 and this is in turned used in the improved algorithm shown in Figure 4.

The simplest attack is the brute force attack where the attacker randomly chooses a source IP address for packets. In this case a random source IP address has the probability of 2^{-32} of matching the IP address of a legitimate client. Suppose that the system has 5000 active TCP connection at a given moment, then the probability of an attacker packet with random source IP of matching a legitimate IP would be about one in a million.

We can also show the effectiveness of our proposed method using a simple scenario where only the target AS deploys ITS. In this case there will be only one router stamp for every packet: the stamp of the ingress router of the target AS (refer to Figure 1). Since the router stamp is 2-bits then given a source IP the probability that the path signature also matches is $\frac{1}{4}$. Thus if the attack is highly distributed and the attacker *knows the IP address of all the active connections* only 25% of the attack packets will have the same path signature as a given IP, p_1. Of course if

```
COUNTMATCHES(sig, pkt)
1  count ← 0
2  if pkt.sig&3 = sig&3
3      then
4          count ← count + 1
5          sig ← sig ≫ 2
6          pkt.sig ← pkt.sig ≫ 2
7          goto 2
8  return count
```

Figure 5. Comparing the number of matches

more ASs deploy ITS the scheme becomes much more effective and IP spoofing can be eliminated almost completely.

3.2. Scalability

A major concern in the ITS method is scalability. Keeping track of all active flows and their path signature requires a large amount of resources. The idea is that packets from the same source network tend to follow the same path for a given destination. This is confirmed if we look into how routes are propagated using BGP. In BGP each node only selects the *best* route and propagates *that single* route to its neighbors. Now, if a group of hosts have the same *egress* point in their AS then they will enter the neighboring AS at the same *ingress* point since the *egress* router has a *single best* route to that particular destination. Therefore the path followed by hosts belonging to the same IP prefix *tend* to follow the same path, at least at the AS level. To confirm our hypothesis we performed experiments using traceroute servers [13]. Each experiment consisted of tracing the path from 10 different traceroute servers to a 50 network prefixes extracted from BGP tables in Route Views [14]. We kept only the 36 prefixes in which at least 10 clients were active. Out of the 36 prefixes only 2 had more than 2 different paths from the server to the destination network. Based on this analysis a border router can keep a single filter containing the path signature associated with an IP prefix. When a new connection, from the same prefix and having a different signature, is established, the border router updates its database. Therefore a border router needs to keep only about 150,000 filters which is the approximate number of announced prefixes [15].

4. SIMULATIONS

A series of simulations were performed using real-world topological data from Skitter [16]. We selected randomly 600 hosts: 100 were used as clients and 500 as attack sources. The IP address and path signature of clients were manually added to the border router database. The attacking sources send data at the constant rate of 10M packets/s while the clients send at the rate of 1M packets/s. The data rates were chosen such that target's link is staturated. A total of d intermediate routers marked packets from sender to victim were d was used as a parameter with $d = 1, 2, 4$. The d routers for each path were selected randomly with the condition that the border router of the target AS for particular path is among them. The metric used to measure the performance of our method is the fraction of bandwidth of the link between the border router and the target consumed by the attacking packets.

A simple priority queueing was performed where all high priority packets are processed before processing the low priority ones. For example, all priority 7 packets were forwarded before priority 6 and so on. While this is a simplistic model to adopt it serves as a good measure to gauge the effectiveness of the scheme. Two attack scenarios were considered.

In the first scenario the attacking hosts send packets with the source IP address randomly generated. Even when only the border routers of the target AS stamped packets, the attacking packets consumed less than 1% of the total bandwidth. A quick analysis will show us that this not a surprising result. The probability that a packet with random IP addresses matches the IP address of one of the 500 clients is 500×2^{-32}. Therefore, an attacking host needs about 1000 seconds to get a single attacking packet to match a "legitimate" one.

In the second scenario, one of the attacking hosts, with source IP address ip_1, was assumed to have established a normal TCP connection to the target and therefore its

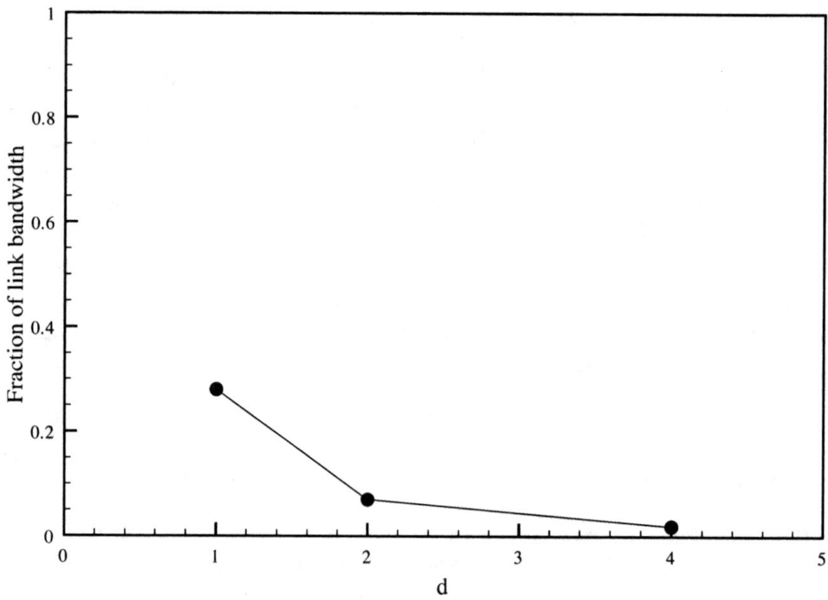

Figure 6. Fraction of link bandwidth consumed by spoofed traffic as a function of number of routers deploying ITS

source IP and path signature were added to the database. The remaining attacking hosts used ip_1 as the source IP address. The results for $d = 1, 2, 4$ are shown in Figure 6. The results for $d = 1$ are very close to our analysis while the results for $d = 2$ and $d = 4$ shown that the effectiveness of ITS in eliminating IP spoofing grows almost exponentially with d.

The simulation results have clearly shown that using ITS can largely eliminate IP spoofing even if not universally deployed. Furthermore, an ISP can immediately benefit from deploying ITS without waiting for other network providers to do so ($d = 1$).

5. RELATED WORK

There has been extensive research done on mitigating the effect of IP spoofed. Most of the work was on traceback methods[17][2][18][3][4] to help victim networks to trace the origin of the DDoS attack . These methods require routers to stamp, with a certain probability, a mark in the IP header. When enough such packets have been collected the victim starts the process of reconstructing the path(s) that the attack packets have followed. However, the cost of the reconstruction algorithm becomes prohibitive when the number of attacking hosts is large.

One of the earliest proposals for IP spoofed traffic detection is ingress filtering by Ferguson et al. [19]. Ingress filtering requires that an AS blocks any traffic leaving the AS and having source address not belonging to the AS. While effective, ISPs are reluctant to deploy such scheme for lack of incentive. A different approach to ingress filtering is the SAVE protocol proposed by Li. et al. [20]. Using how the BGP protocol works, Park and Lee [6] showed how to construct filters to block spoofed IP packets. The downside is that such filtering scheme requires the cooperation of thousands of autonomous systems. A similar approach was proposed by Duan et al. [7] by introducing the concept of the feasible upstream router and deeming any packet that does not originate from such a router a spoofed packet. The number of feasible upstream routers, however, is a large one and the method while correct blocks only a small fraction of spoofed traffic. Jin [21] proposed a scheme where packets with spoofed addresses are identified by their hop count. The idea is to build a table in "peace time" that maps the source address of clients to the number of hops they need to reach the victim. During a DoS attack each packet is compared to the corresponding entry in the table. In practice, routes change very frequently, which makes the table entries obsolete very quickly.

A statistical approach for DDoS mitigation was recently proposed by Kim et al. [22]. Their basic idea is to build traffic profile during "peace time" and drop traffic that does not conform to the profile during an attack. The method is a promising one by we believe that a smart attacker can slowly inject traffic to modify the base profile. An improvement to the basic method was given by [23].

To our knowledge, Yaar et al. [10] where the first to use a *deterministic*, rather than a probabilistic path identification as was done in traceback schemes. They made a rather substantial assumption, a foreknowledge of malicious signatures, then used such signatures to distinguish malicious from legitimate users. Even if one assumes that the malicious signatures can be clearly identified the number of malicious and legitimate users having the same signature grows as the number of attackers grows, leading to substantial *collateral damage*.

6. CONCLUSION AND FUTURE WORK

We have proposed an improved version of the Implicit Token Scheme (ITS) used to defend against spoofed IP traffic by using a deterministic path signatures along with an algorithm for assigning priority to packets according to their signatures. The scheme was shown to be effective by performing simulations using real-world data. Incremental deployment of the method was demonstrated to be feasible by using the concept of partial matching of signatures. Even if only 2 intermediate routers on the path between client and server implemented ITS the effect of IP spoofing can be reduced by 90%. BGP analysis as well as traceroute "experiments" show that the scalability of the system can be improved by aggregating path signatures belonging to the same network prefix.

We are currently performing extensive measurement on the relationship between IP prefixes and their network path. Measuring the performance of ITS and the overhead incurred by routers is an important aim that will be studied in the future.

REFERENCES

1. D. Moore, G. Voelker, and S. Savage, "Inferring internet denial of service activity," in *Proceedings of the 10th USENIX Security Symposium*, 2001.
2. S. Savage, D. Wetherall, A. Karlin, and T. Anderson, "Network support for IP traceback," *IEEE/ACM Trans. Netw.*, vol. 9, no. 3, pp. 226–237, 2001.
3. D. Song and A. Perrig, "Advanced and authenticated marking schemes for IP traceback," in *Proceedings of IEEE INFOCOMM.*, 2001.
4. M. Sung and J. Xu, "IP traceback-based intelligent packet filtering: a novel technique for defending against internet DDoS attacks," in *10th IEEE International Conference on Network Protocols*, pp. 302–311, 2002.
5. A. Yaar, A. Perrig, and D. Song, "FIT: Fast Internet traceback," in *Proceedings of IEEE INFOCOMM*, Mar. 2005.
6. K. Park and H. Lee, "On the effectiveness of route-based packet filtering for distributed DoS attack prevention in power-law internets," in *Proc. ACM SIGCOMM*, pp. 15–26, 2001.
7. Z. Duan, X. Yuan, and J. Chandrashekar, "Constructing inter-domain packet filters to control ip spoofing based on bgp updates," in *IEEE InfoCom*, 2006.
8. H. Farhat, "Protecting TCP services from denial of service attacks," in *LSAD '06: Proceedings of the 2006 SIGCOMM workshop on Large-scale attack defense*, pp. 155–160, 2006.
9. Y. Rekhter and T. Li, "A border gateway protocol." RFC 1771, 1995.
10. A. Yaar, A. Perrig, and D. Song, "PI: A path identification mechanism to defend against DDoS attacks," in *Proceedings of the IEEE Symposium on Security and Privacy.*, pp. 93–107, 2003.
11. M. Collins and M. K. Reiter, "An empirical analysis of target-resident DoS filters.," in *IEEE Symposium on Security and Privacy*, pp. 103–114, 2004.

12. D.J. Bernstein http://cr.yp.com/syncookies.html.
13. http://www.traceroute.org.
14. Route Views, *University of Oregon Route Views Project*, Available at http://www.routeviews.org/.
15. S. Uhlig and B. Quoitin, "Tweak-it: BGP-based interdomain traffic engineering for transit ASs," in *Next Generation Internet Networks, 2005*, pp. 75–82, 2005.
16. CAIDA's skitter initiative http://www.caida.org.
17. D. Dean, M. Franklin, and A. Stubblefield, "An algebraic approach to IP traceback," *ACM Trans. Inf. Syst. Secur.*, vol. 5, no. 2, pp. 119–137, 2002.
18. A. Snoeren, C. Partridge, L. A. Sanchez, C. E. Jones, F. Tchakountio, B. Schwartz, S. T. Kent, and W. T. Strayer, "Single-packet IP traceback," *IEEE/ACM Trans. Netw.*, vol. 10, no. 6, pp. 721–734, 2002.
19. P. Ferguson and D. Senie, "Network ingress filtering: Defeating denial of service attacks which employ IP source address spoofing," *RFC 2827*, 2000.
20. J. Li, J. Mirkovic, M. Wang, P. Reiher, and L. Zhang, "SAVE: source address validity enforcement protocol," in *Proceedings of IEEE INFOCOMM*, 2001.
21. C. Jin, H. Wang, and K. G. Shin, "Hop-count filtering: an effective defense against spoofed DDoS traffic," in *Proceedings of the 10th ACM conference on Computer and communications security*, pp. 30–41, 2003.
22. Y. Kim, W. C. Lau, M. C. Chuah, and H. J. Chao, "Packetscore: A statistics-based packet filtering scheme against distributed denial-of-service attacks," *IEEE Trans. Dep. Sec. Comp.*, vol. 3, no. 2, pp. 141–155, 2006.
23. P. E. Ayres, H. Sun, H. J. Chao, and W. C. Lau, "Alpi: A DDoS defense system for high-speed networks," *Selected Areas in Communications, IEEE Journal on*, vol. 24, no. 10, pp. 1864–1876, 2006.

CHAPTER 32

ON THE SECURITY OF QUANTUM NETWORKS: A PROPOSAL FRAMEWORK AND ITS CAPACITY

QUOC-CUONG LE[1], PATRICK BELLOT[1], AND AKIM DEMAILLE[2]

[1] *ENST-LTCI, Paris, France,* [2] *EPITA-LRDE, Paris, France*

Abstract: In large Quantum Key Distribution (QKD)-based networks, intermediate nodes are necessary because of the short length of QKD links. They have tendency to be used more than classical networks. A realistic assumption is that there are eavesdropping operations in these nodes without knowledge of legitimate network participants. We develop a QKD-based network framework. We present a percolation-based approach to discuss about conditions of extremely high secret key transmission. We propose also an adaptive stochastic routing algorithm that helps on protecting keys from reasonable eavesdroppers in a dense QKD network. We show that under some assumptions, one could prevent eavesdroppers from sniffing the secrets with an arbitrarily large probability

1. INTRODUCTION

The problem of transmitting a secret key from an origin to a destination on the network was considered for a long time, and currently solved in most of Internet applications by using Public Key Infrastructure (PKI). PKI relies on unproven assumptions about the computing power of eavesdroppers and the non-existence of effective algorithms for a certain mathematical hard problems. Thus, PKI cannot meet higher security level requirements. Quantum Key Distribution (QKD) technology is an alternative that provides unconditional security, but supports only point-to-point connections. Besides, QKD has some significant limits on throughput and range [1,2]. Moreover, QKD large networks are always vulnerable as some nodes may be controlled by eavesdroppers. That makes an open question: how to build large QKD-based networks capable of supporting extremely high secret key exchange between network participants?

This paper studies the model of a partially compromised QKD network in which two members want to establish a common key with almost-certainty that this key will not be eavesdropped. The contributions are (i) a model of partially compromised

H. Labiod and M. Badra (eds.), New Technologies, Mobility and Security, 385–396.
© 2007 *Springer.*

QKD networks, (ii) the use of percolation theory techniques to find where almost-certainty can be achieved, (iii) a proposal based on stochastic routing capable obtaining a given secrecy level requirement.

In 2, we introduce the context and our problem statement. In 3, we use percolation theory to show where almost-certainty can be achieved and we also present an adaptive stochastic routing algorithm. We analyze it in some attack strategies. Relations with other works is presented in 4 and we conclude in 5. The proofs of the theorems are given in Appendix.

2. A PROPOSED QUANTUM NETWORK FRAMEWORK AND THE PROBLEM STATEMENT

QKD-based networks, also called quantum networks, are the special purpose networks that aim to an extremely high level security of secret key transmissions between two arbitrary member pair Alice and Bob on the network. Nowadays, the most famous quantum network is DARPA quantum network that, with its specific characteristics of consisting of a few nodes, should be considered as a special model for quantum network. A general fully architecture of large quantum networks is still in discussion. However, we should consider two prominent foreseeable characteristics of quantum networks as follows: (1) Links between two adjacent nodes of the network are perfectly trusted and (2) Quantum networks need more intermediate nodes than classical ones to totally cover the same region.

Direct QKD links are proven perfectly secure by information theory. This is a perfect mean for secret key transmissions. There are also some realistic applications [3–6, 1, 2] Unfortunately, QKD links are only implemented for the short distances. The current records of these links are 150 km in fiber and 23 km in free space, with a typical rate of 1 Kb/s or so. In order to overcome this limitation, we think of a chain of successive terminals such that each terminal is capable establishing a direct QKD link with the previous and another one with the next terminal. This is also called QKD data relays. Note that such a relay is not quantum repeater and data appears unencrypted for all the relays on the transmission. To ensure that data is secured in the transmission, one must ensure that all the intermediate relays are not eavesdropped. The DARPA quantum network was constructed based on this idea. This implies the key assumption that one has already protected all the DARPA network's nodes. Such an assumption can be acceptable in the context of a few node network as the DARPA network, however, this is unconvincing as we must deal with more large networks.

In this paper, as we want to solve the problem of large quantum networks, we should not use the assumption of ultimately trusted nodes as in DARPA network's model. We consider a new assumption: Eve cannot eavesdrop on all the nodes, but on some nodes without leaving any trace. Such an assumption seems to be more plausible in realistic contexts. With the new assumption, the choice of a good topology for quantum network becomes more important. Eve could prefer to eavesdrop on some nodes than others. It means that the probability of being attacked for some nodes could be more than the others. Thus, if quantum network topology presents some backbone nodes as that of Internet, then certainly Eve prefers to attack these nodes than others. The concentrated architecture may be not good for quantum network problem. It is better if we could make Eve confused in choosing the attack targets. In such a situation, the best attack strategy for Eve is to choose randomly targets according to a uniform distribution. In this paper, we restrict our attention to such an attack strategy.

A fully standard reference model for quantum network is still in discussion. What the best topology should be is an open question. As mentioned above, the concentrated topology as that of Internet which features some backbone nodes may be not good because such a topology helps Eve choose her targets. If we assume that Eve only attacks and gains control on some proportion of all the nodes, then a distributed topology that makes Eve confused in choosing her target can improve the global security. The short distance covered by today's QKD-links also influences the topology network design. Although the maximum length of quantum links is about 150 km, because of rates and costs, 30 km-long quantum links are more likely to be used. As a first step in studying QKD networks, we restrict our attention in a simple square grid, a 4-connected lattice, see Fig. 1, large enough so that we can neglect nodes on the boundaries. This models roughly a large region meshed with QKD links.

2.1. Problem Statement

Eve can control any node with probability $p_a \in [0, 1]$. These nodes are called by *attacked* or *unsafe* nodes; the others are *safe*. Alice and Bob can be any node. Alice wants to send securely the key K to Bob. They do not know whether a node is safe but they know the probability $p_s = 1 - p_a$ with that a node is safe. Every message that passed over one or more unsafe nodes is considered as being eavesdropped by

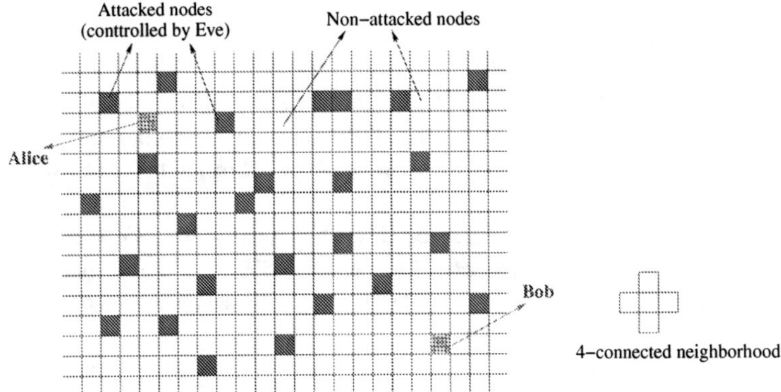

Figure 1. Two-dimensional lattice network

Eve, but there is no way to verify whether a message is eavesdropped or not. We consider to a key transmission method as follows:

1. Alice uses a stochastic routing algorithm to send to Bob N blocks (or random messages) $M_1, M_2, .., M_N$; all are the same length as the key K.
2. Alice and Bob computes $K = \sum_{i=1}^{N} \oplus M_i$ where $\oplus$ is the bit-wise XOR.

According to Information Theory [7], even if Eve intercepts all the blocks M_i but one, the key K is safe.

One can ask for the reason of using stochastic routing. As is well known, almost traditional routing algorithms, e.g. those used on the Internet, are deterministic. As they are tailored to be efficient, one can guess the path that will be taken with a high probability even though there are an almost infinite number of paths connecting two points on the network. On our key transmission method above, such routing algorithms compromise security. By contrast, stochastic routing is better: each message takes independently a random path and Eve is confused in choosing the path to be attacked.

We state our problem in this framework: if Alice uses a stochastic routing algorithm to send messages, then how many messages Alice must send so that at least one message is not intercepted? More precisely: *Given an arbitrarily small real number ϵ, find N_0 such that if Alice uses a stochastic routing algorithm to send $N \geq N_0$ messages, then we have the probability $(1 - \epsilon)$ that it exists at least one message not being intercepted by Eve?*

3. THE PROPOSED SOLUTION

A *safe path* is a path that only consists of safe nodes. Otherwise, it is said *unsafe*. Obviously, if there is no safe path from Alice to Bob then there is no solution. We first find the existence condition for solution. Percolation theory helps address this question in 3.1. Then, we present our stochastic routing algorithms and its primary results in 3.2. We use our algorithm to answer to the initial question in 3.3.

3.1. Percolation Theory

Suppose we immerse a large porous stone in a bucket of water. What is the probability that the center of the stone is wetted? In formulating such a situation, John M. Hammersley and Simon R. Broadbent, in 1957, gave birth to the percolation theory [8].

In 2 dimensions, percolation model can be described as follows. We focus on a regular graph $G = \mathbb{Z}^2$, with vertex set V and edge set E. Let the vertices be independently *open* with probability $p \in [0, 1]$. All the edges are assumed open. Consider a path π in G as a sequence $\pi = v_1, v_2, \ldots$ of adjacent vertices. A path *open* iff all the vertices v_i are open. Obviously, the central vertex of the stone is wetted iff there is a open path from it to a vertex on the boundary.

The goal of percolation theory is to describe the transition phase from non-existence to existence of a infinite wetted vertex cluster. The existence of an infinite wetted cluster is equivalent to having an unbounded open path starting from the origin. We denote by $u \leftrightarrow v$ the existence of an open path between two vertices $u, v \in V$. The *wetted cluster* or *open cluster* $C(v)$ of the vertex v is defined as $C(v) = \{u \in V : u \leftrightarrow v\}$.

The central quantity studied in percolation model is the probability that the cardinality of $C(v)$ is infinite for a vertex v, also called the *percolation probability* $\theta(p) = \{\Pr\left(|C(v)| = \infty\right)\}$.

Perhaps, the most important result of percolation theory is to well define a critical value of p, also called percolation threshold or critical probability p_c, that separates the globally disconnected and globally connected states for the unbounded lattice (see Fig. 2).It is defined by $p_c = \max\{p : \theta(p) = 0\}$.

Nowadays, there are many variant of the basic percolation model. One studied the percolation in a number of various structures and dimensions. Results were presented as a mixture of rigorous results, numerical estimates and conjectures. However, its polyvalence and efficiency in characterizing non-linear phenomena led the scientific community to use this theory to model complex systems such as biological systems, social networks and economic systems.

In this paper, we focus on the 2-dimensional site percolation as sketched above. Note that if we restrict our attention only on the existence of a safe path from Alice and Bob, then there is no difference between our framework and percolation model: safe nodes and safe paths are equivalent with open nodes and open paths, respectively. Thus, we could use some important properties that have been proven in percolation [9]:

1. The probability that a node belongs to the infinite wetted or open cluster, or percolation probability $\theta(p)$, is a non-decreasing and continuous function with respect to p, except possibly at the percolation threshold p_c, where it is at least non-decreasing and continuous from the right (see Fig. 2).
2. The number of infinite wetted or open clusters k_0 must take no other value than 0 or 1: $k_0 = 1$ if $\theta(p) > 0$, 0 otherwise.

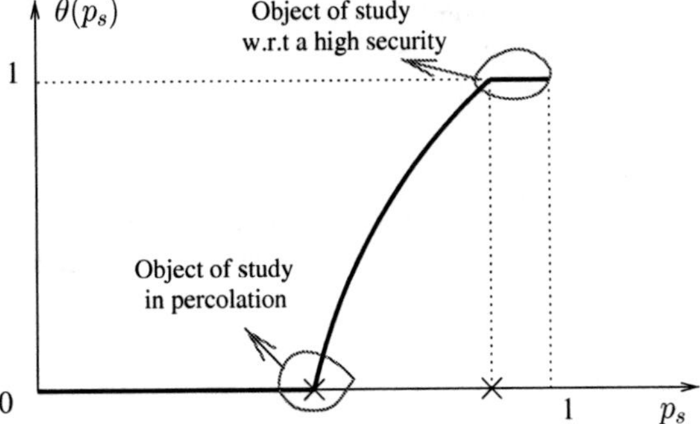

Figure 2. Two different objects of study

The fundamental goal of quantum network is to find a more higher security level. Situations that lead to a small probability of having at least one safe path should be taken out of interest. We should focus only on the values of p in the region where $\theta(p)$ is equal or almost equal to 1 (see Fig. 2). For such p, Alice and Bob belong almost certainly to the infinite safe cluster. This implies there exists almost certainly at least a safe path between them. The probability τ(Alice,Bob) that Alice and Bob can be safely connected is $\tau(\text{Alice,Bob}) = (\theta(p))^2$.

We have the following idea on the lower bound p_0 for the interesting region of $p: p_0 = \inf\{p : (\theta(p))^2 \geq (1-\epsilon)\}$. As $\epsilon \to 0$, we have $p_0 \to p_0'$ where $p_0' = \inf\{p : \theta(p) = 1\}$. The threshold $p_c \sim 0.6$ for 2-dimensional site square lattice percolation is obtained by numerical estimates. Motivated by this idea, we did simulations [10] that show some critical threshold in our quantum network problem. In this paper, we want to re-use one of them, this is $p_0 = p_0' \sim 0.91$.

3.2. Stochastic Routing Algorithms

Introduction to stochastic routing algorithms Traditional routing algorithms, such as those used on the Internet, are mostly deterministic. As they are tailored to be efficient, they are guessable. This is not good in our framework. The basic idea of stochastic routings is sending randomly a packet to one of possible paths. When a node needs to forward a packet, it randomly chooses one of its neighbors, not necessarily the most "efficient" one. This makes the emergence of a new concept called next-hop probability distribution: the next-hop choice is random, but according to the next-hop probability distribution. The main challenge in stochastic routing is how to determine the next-hop probabilities that could maximize a given specific goal. In quantum networks, the top priority is the security and the other metrics for performance evaluation of routing algorithms are less considered.

A constant-length stochastic routing algorithm. Called L-SRA(l), it is a stochastic routing algorithm that takes l as input and tries to transmit a message using random path of length l. If l is less than the distance d between Alice and Bob then L-SRA(l) returns no path. For $l \geq d$,there can be different paths $\pi_1, \ldots, \pi_m$. In such cases, for each message, L-SRA(l) will choose randomly a path π_i among $\pi_1, \ldots, \pi_m$ according to a probability distribution that holds:

1. $\forall i, 1 \leq i \leq m, 0 \leq Pr(\text{L-SRA(l) takes } \pi_i) \leq 1$
2. $\sum_{i=1}^{m} Pr(\text{L-SRA(l) takes } \pi_i) = 1$

Theorem 1. The probability that L-SRA(l) chooses successfully a safe path to send one message depends only on the safe probability p and the length l, not on the distance d between Alice and Bob:

$$(1) \qquad Pr\,(1, p, d, \text{L-SRA(l)}) = p^l$$

A proposed routing algorithm. Called K-SRA(k), it is built based on L-SRA(l). K-SRA(k) receives an input value $k \geq 2$, and considers only the paths with lengths

$d, (d+1), \ldots, (k*d-1)$. For each message, K-SRA(k) chooses randomly a value l among $d, (d+1), \ldots, (k*d-1)$ according to a uniform distribution. Once l was chosen, K-SRA(k) uses L-SRA(l) to send the message.

Theorem 2. The probability that K-SRA(k) chooses successfully a safe path to send one message depends on the safe probability p, the input parameter k and the distance d between Alice and Bob:

$$\beta = Pr\ (1, p, d, \text{K-SRA(k)}) = \frac{p^d * (1 - p^{(k-1)*d})}{(k-1)*d*(1-p)} \tag{2}$$

3.3. Some Attack Strategies of Eve

We consider 2 strategies of Eve:
1. *Dynamic attack.* Frequently, Eve re-chooses the set of attacked nodes.
2. *Static attack.* Eve chooses once for all the set of attacked nodes.

Theorem 3. If Eve does a dynamic attack, then the probability that there is at least one safe path in N routings of K-SRA(k) depends on N, the safe probability p, the input parameter k, and the distance d between Alice and Bob:

$$Pr\ (N, p, d, \text{K-SRA(k)}) = 1 - (1-\beta)^N \tag{3}$$

where β is evaluated in the formula 2.

Lemma 1. If Eve executes a dynamic attack, given ϵ and K-SRA(k), then we have the threshold N_0 responding for the initial question:

$$N_0 = \lg(\epsilon)/(1 - \lg(\beta)) \tag{4}$$

where β is evaluated in the formula 2.

Theorem 4: If Eve does a static attack, then the upper bound of the probability that there is at least one safe path in N routings of K-SRA(k) depends on N, the safe probability p, the input parameter k, and the distance d between Alice and Bob:

$$Pr\ (N, p, d, \text{K-SRA(k)}) \leq 1 - (1-\beta)^N \tag{5}$$

where β is evaluated in the formula 2. And the equality is possible when $N \leq 4$.

Lemma 2: If Eve executes a static attack, given ϵ and K-SRA(k), we have the threshold N_0 responding for the initial question:

$$N \geq \lg(\epsilon)/(1 - \lg(\beta)) \tag{6}$$

where β is evaluated in the formula 2. And the equality is possible when $N \leq 4$.

4. OTHER WORK

Percolation theory. The context of percolation theory has many similarities with our problem context. The significant method used to solve percolation problems is simulations and statistics that report the percolation probability and an approximate formula that describes the system state at the phase transition. In this paper, our ambition is not to find approximate formulas that describe the evolution of a certain process. In contrast, it is to feature a quantum network framework based percolation, and beyond it to find solutions for security problem. The rigorous thresholds, formulas featuring the region of interest for p (see Fig. 2) is one of our future works.

Stochastic routing. The main challenge in stochastic routings is how to compute the next-hop probabilities that maximize a given goal. In literature [11,12], stochastic routing can be re-formalized as an abstract game between two players: the designer of the routing algorithm and the attacker that attempts to intercept packets. It is assumed that the attacker has a finite resource, i.e. she only intercepts packets at some nodes on the network. This is a zero-sum game in which a designer seeks a strategy to minimize the cost that he has to pay for a packet being safely transmitted and the attacker wants to maximize this cost. Such a problem was studied in [12].

Previous works on stochastic routing focus on performance metrics (latency, throughput, acceptance rate, etc.), which are not of major importance to quantum networks. What matters is sensitivity to eavesdropping and security. As the main goal of our works is to investigate the possibility of achieving an extremely high security level, the object of study is also different: this is the overall state of a set N paths. One grid 4-connected topology as proposed in this paper can suit to quantum network, but also makes previous works on stochastic routing become useless.

Quantum network. The first quantum network, DARPA Quantum Network, was built to test the strength of such systems in the real-world applications. It consists of three sites (Cambridge, Harvard, and Boston University) and became fully operational in October 2003 [13]. It relies on trusted relays: one must trust the security of all the participants, and be sure that eavesdroppers cannot sniff any information on any of the nodes. In realistic contexts, nobody can be sure that he does not reveal any information for eavesdroppers. Moreover, in a larger quantum network, the assumption that one can trust all the nodes becomes unacceptable. In this paper, we studied large quantum networks in a more realistic context: nodes are not totally trustworthy, there is a probability that nodes are controlled by eavesdroppers.

5. CONCLUSIONS

We investigated the constraints of quantum networks and the ineluctable probability that some nodes are attacked. The existence condition of an extreme high security level for key transmission was analyzed using percolation-theory based methods.

CHAPTER 33

AN ADAPTIVE SECURITY FRAMEWORK FOR AD HOC NETWORKS

VINCENT TOUBIANA[1], HOUDA LABIOD[1], LAURENT RAYNAUD[2], AND YVON GOURHANT[2]

Ecole Nationale Supérieure des Télécommunication (ENST)[1], LTCI-UMR 5141 CNRS, GET/ENST/INFRES Department, 46 rue Barrault – 75634 Paris Cedex 13 – France

Email : {labiod, toubiana}@enst.fr,

France Telecom R&D [2], 38–40 rue du Général Leclerc, 92794 Issy Les Moulineaux Cedex 9 – France

Email : {laurent.raynaud, yvon.gouhrant }@orange-ft.com

Abstract: Security in Mobile Ad hoc NETworks (MANETs) is a very prolific research topic. Many original solutions have already been proposed to solve important security flaws regarding routing protocols conceived for these emerging networks. However many proposals just focus on a small part of the security problems and only few global framework have been proposed yet. In this paper we define a new framework that may be used to combine different security mechanisms and provide a global security solution. The global framework is thus very generic, allowing a large panel of security proposals to be implemented. To improve the security of the framework, a new structure, named MacroGraph, is defined applying multipath routing and trust management. Mainly, self-control, flexibility, adaptability, autonomy and distribution are the key features addressed in our architecture that fulfills application security requirements

Keywords: Ad-hoc network, Adaptive Security, Multipath, Framework, Authentication, Network Security Management, Trust

1. INTRODUCTION

Besides they are not widely deployed nowadays, Mobile Ad hoc NETworks (MANETs) have already been the subject of many proposals to improve their security. Securing MANETs turns out to be a very challenging issue as every feature of this network is source of security problems.

Although numerous solutions have been proposed concerning security in MANETs, most of them just secure a part of the security functions and rely on other

H. Labiod and M. Badra (eds.), New Technologies, Mobility and Security, 397–406.
© 2007 *Springer.*

mechanisms to secure the other ones. Consequently, only few solutions provide both a secured routing protocol and an authentication mechanism. However, there is a deep dependence between the two functions as secure routing often relies on pre-established authentication and key exchange to secure the communications. Actually, authentication requires a secret to be shared between two nodes, and nodes have to rely on a secured routing protocol to distribute this secret.

To avoid this dead lock, a majority of the secured routing protocols require a bootstrapping step where nodes safely exchange secrets which will be used later for authentication or suppose the existence of safe channel to exchange secrets. So it appears that most proposals have too strong prerequisite to be widely deployed in MANETs.

In this paper, we propose a new security framework combining different security solutions to offer a complete set of security functions. We define the six modules of the framework in a generic way, allowing direct implementation of already proposed solutions. Our solution is based on reactive routing protocol since it is proved that they perform better than other protocols in highly mobile network. We also propose original implementations of the modules to take advantage of the framework's design. Our proposed modules rely on efficient and low-cost multipath routing and trust management to enforce the security.

Observing different mechanisms, we noted that no one adapt the security requirement to the secured application. Even if such adaptation is now commonly used in Quality of Services protocols, no one exploited it in a security solution. Nevertheless, different applications have different security goals, and our framework named Adaptive Secure Multipath for Ad hoc Networks (ASMA) adapts the security level to the objective of the application and so reduces the resource consumption and delays.

This paper is organized as follows. Section 2 gives a review on the security solutions proposed for ad hoc networks including authentication and routing protocols. Section 3 provides a description of our proposed framework. Section 4 highlights the main advantages of ASMA and section 5 provides concluding remarks and highlights our future work.

2. RELATED WORK

To be able to authenticate each other, usually nodes need to share a secret which will be used for next authentication operations. Secured communication can not be established before authentication so nodes have to know a part of the secret without connecting to the authenticated node. In SUCV (Statistically Unique and Cryptographically Verifiable) 1 Casteluccia et al. propose to compute the node address from its public key. Although this process seems to be an ideal solution to the authentication problem, this merely reports the problem, because nodes still have to bind that address to an identity. Furthermore, every node may forge its own identity, thus SUCV is highly vulnerable to Sybil attacks 2. Another idea to address authentication and secure routing is to rely on trust. Regarding the authentication field, trust is the probability that a node is really what he claims to be. To provide authentication based on trust Capkun and al. rely on "Small World Theory" to propose [4] which is very similar to PGP. However in practice the probability to get the wanted certificate through a chain is very low[5]. Yi and Kravets [6] suggest enhancing this mechanism associating trust values to certificates.

Authentication is intensively used in secured routing protocols to protect key exchange between the source and the destination because these protocols often rely on powerful cryptography algorithms. ARIADNE [7] relies on symmetric cryptography to provide end-to-end security for both routing and forwarding. ARAN [8] uses asymmetric cryptography for hop-by-hop encryption, but such mechanism assumes that the source trusts every node in the route because if there is at least one untrusted node, the whole route is untrusted. Recently some proposals [9] introduce secured multipath routing. These protocols provide security through redundancy but use totally distinct paths resulting in high overload of the network and resource consumption. Z. Ye et al. [10] show that with a moderate node density, the number of available node-disjoint paths is too low to provide efficient robustness.

The existing solutions are very costly and do not address the whole problem of security vulnerabilities. What needed is a flexible, adaptable and affordable security solution, which provides greater autonomy. Therefore, it is required to review the way in which security is designed and performed in MANETs in order to identify and to alleviate its weaknesses. In this context, we propose a new approach which reveals itself as a suitable candidate to make a balance between security requirements, system flexibility and adaptability in the case if a global security context.

3. DESCRIPTION OF THE ASMA FRAMEWORK

The framework design just includes a description of the six composing modules and the communication primitives, providing more flexibility and allowing modification and plug-in addition. The ASMA framework takes place between the transport layer and the routing layer of the OSI model. ASMA can be implemented without important modifications on the other layers. It is just needed from the routing protocol to support multipath routing extension. In addition, the transport layer should inform ASMA about the applications requirement used for security adaptation. This type of extension field is widely used by the transport layer to inform about the Quality of Service requirement of an application.

3.1. Definitons

Our solution is based on a new structure name MacroGraph constituting the network as it can be perceived by the source of a communication 12. A Macrograph is illustrated in Figure 1.

ASMA relies on three main concepts: trust, risk and knowledge. The following definitions are adopted:

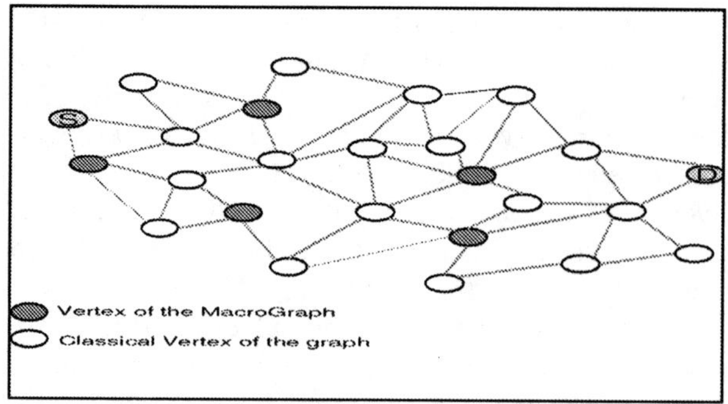

Figure 1. Example of MacroGraph

Knowledge. Here we define the "knowledge relation" as follow: *node A knows a node B if A and B have authenticated each other.*

Trust. We refer to the definition of [3] to propose the trust function: *Trust (S,I,A,C)* is the trust that node *S* has in an agent *I* to perform an action *A* in a network's context *C*. A trust should be differenced regarding action and context as an attacker could change its behavior depending on the requested action and the context. As we use multipath routing, we assume that agent *I* is either a node or a path.

Reputation, also referred as indirect trust, is often less valuable than direct observation, but it remains necessary to initialize trust about out of range nodes. The resulting trust value is a weighted average of recommendations and observations. Obviously the credit of a recommendation depends on the node which emits it.

Risk. The risk reflects the minimum security level that a node requires to perform an action. This information, included in every packet involving the action, is used by intermediate nodes to decide how to process a given packet. In the rest of this article, we say that *S* trusts *I* to perform action *A* if :$Trust(S,I,A,C) > Risk(S,A,C')$. The application context C' reflects the security requirement regarding the application (e.g.: "low security" for a streaming application). Obviously only known nodes may be trusted.

The six defined modules (Fig. 2) rely on the previous definitions and structure to fulfill the available security requirements.

3.1.1. Modules

The MG Manager (MGM): This module is the interface with the upper layer and is specific to our architecture. The MG is modeled as an oriented graph specified for a particular couple (source, destination). MG(S,D) designates the MG connecting S to D. Trust and multipath routing are the key elements of this structure and allow quantifying the security of an operation. The MG vertices are nodes trusted by the source which belong to a path linking the source to the destination. The MG vertices, $V_{MG(S,D)}$, are nodes trusted by the source which belong to a path linking the source to the destination. The edges regroup paths connecting two vertices and are

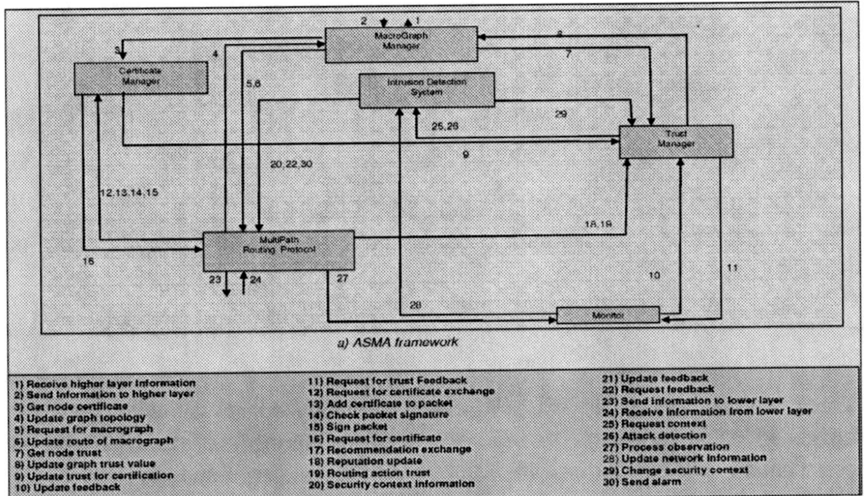

Figure 2. Global ASMA node architecture modules

noted $E(\overrightarrow{I, J})_{MG(S,D)}$, where I is the input and J the output. The weight of an edge is calculated by its input and is viewed as an indicator of trust. The MGM expects from the upper layer (arrow 2) a set of security parameters used to compute the Risk, thus ASMA adapts the security mechanisms to the application requirements. Consequently resources consumption and delays are reduced, and more nodes are allowed to participate to a communication. In the following, we briefly describe the different operations performed by the MGM.

Referring to the trust manager (arrows 7–8) and the routing protocol (arrows 4–6), MGM attempts to build a MG connecting the source S to the destination D which reaches the wanted security level. When no suitable graph to D is in cache, a Route REQuest (RREQ) is broadcasted on the network via the multicast routing protocol. The routing protocol returns to MGM the list of vertices and edges (including weights) which will allow the establishment of the macrograph. Recommendations about paths and nodes should be signed to assure the integrity of this information (arrow 14). Finally it asks recommendation from the Trust Manager about node on the paths (arrow 6).

MGM computes, maintains and updates MacroGraphs based on information provided by the other components (arrows 3–7). ASMA computes two generic types of trust: transitive trust derivation[15] of a path; parallel trust computation[15] of multiple paths. The trust associated to a communication between node S and D is computed using this recursive formula:

If the requested security level is reached, the route is reported to be established, otherwise the MGM informs the upper layer that no corresponding route is available and indicates the reachable security level.

For every vertex J in MG(S,D) from D to S:

$$\mathrm{Trust}(S->J) = \underset{J\in O(I)_{MG(S,D)}}{\mathrm{PTC}} (\mathrm{TTD}[\mathrm{Trust}(S \rightarrow I), E(I,J)])$$

Done
TTD (Transitive Trust Derivation): compute the path trust
PTC(Parallel Trust Computation): compute multipath trust $E(I,J)_{MG(S,D)}$: weight of the edge going from I to J
$O(I)_{MG(S,D)}$: set of I outgoing neighbors in MG(S;D).

Once the communication starts, the source receives feedback allowing the Trust Manager (TM) to update trust value associated to intermediate agents. Based on these accurate trust estimations, the MGM modifies the graph and may reduce the graph to prevent unnecessary resource consumption. If the trust of the macrograph is too low, the MGM consider that the route is broken and try to discover a new one using local repair optimization.

The Certificate Manager (CM): Symmetric cryptography appears as a suitable solution for authentication in MANETs since it does not consume important resources. However this approach requires storing a key for every possible destination. Thus the key exchange mechanism is the main bottleneck of symmetric cryptography based solutions. As a consequence, most of the key exchange protocol for MANETs are specified for asymmetric cryptography and rely on certification.

CM manages, issues certificates and signs certificates of known nodes, proceeds to certificates exchanges and performs the cryptographic operations: signatures of packets addressed to known nodes and verifications of the received packets. When there is a doubt regarding the authenticity of a certificate, CM requests information to the TM about the nodes which sign the certificate and then decides to accept or to reject it.

Until it is able to strongly authenticate (for example using external devices as in [16,17]) CM accredits low trust to received certificates, so a freshly connected node should solicit multiple agents to reach a suitable security level. Once a node is authenticated, the trust associated to recommendation it provided should increase.

The Trust Manager (TM): It stores all the information obtained about the trust level of paths and other nodes. It is a central part of ASMA which communicates with every module to update its data about routing (arrows 18–19), certification (arrow 9) or forwarding, and it is also queried by these modules when they have to gather information about nodes and paths. When important variations in trust are observed concerning some nodes or paths, the TM informs the MGM (arrow 8) so it is able to update the corresponding MG.

Numerous proposals have been suggested to introduce trust in ad hoc networks [15,18,19,11]. Most of these proposals can be implemented in the ASMA framework with minor modifications. Actually we just expect from the TM that its output values are defined in [0;1] to be compliant with our previous definition of trust. This requirement is common and many trust models define trust in this interval. Compliance with 15 and 11 which define trust in [−1;1] and assign negative values to malicious nodes , may be achieved by taking into account distrust.

According to the MG design, a *MG(S,D)* vertex I informs S about the achievable degree of trust regarding $\underset{E(I,J)_{MG(S,D)}}{\longrightarrow}$ for J in $O(I)_{MG(S,D)}$. If it already used a path to a node in $O(I)_{MG(S,D)}$, I probably has fresh information about them.

When I observes $E(I, J)_{MG(S,D)}$ fall below a threshold, it use local repair to discover more paths. If not enough path are available, I sends a Route ERRor (RERR) to S.

When it has no past interactions with a node, a node rely on reputation which is decisive at start, but will be less important afterwards since TM obtains direct feedback. Trust in the recommender is also updated 11 depending on the difference between recommendation and measured trust.

The Intrusion Detection System (IDS): Information is stored in different modules no correlation can be established to detect an attack. IDS gather important information from modules aggregate them, and then deduce the network context. If it appears that every node is fair the IDS could increase the global trust (arrow 29) resulting in fewer resources need allocated to security. When there is not a sufficient amount of information on the network and that an attack is suspected, IDS may request more feedback from MT (arrow 11). IDS should also collect information from the application layer (e.g. anti-virus and firewall). The detection of numerous attacks at the application layer means that the network is not well protected. Attacks may be suspected if the correlation of different parameters corresponds to an attack scheme. Monitored parameters are the trust of the nodes, the routing information (loss and retransmission rate), certificate management (a certificate request receive a lot of certificate response).

The Monitor (MT): When the routing protocol obtains feedbacks, it forwards them to the MT which analyzes and dispatches them to the appropriated module. The most common feedback is packets' acknowledgement, certificate confirmation, error message, recommendation update... Other feedback could be associated to a specific type of action. The MT receives information about the recommendation trust of other nodes from the TM. An appropriate weight is then associated to a feedback regarding the node which emit it. Received information is transmitted to the corresponding module by MT. When the network is safe, less feedback is transmitted; such adaptation also reduces the resource consumption.

When the MT is not able to access to a sufficient amount of feedback, it queries the routing protocol for more feedback (arrow 5). Solutions have already been proposed to obtain feedback about neighbors and may be implemented as part of the MT [20,21,22].

The Multipath Routing Protocol (MRP): As previously explained, the framework modifies the routing protocol to allow multipath routing which is required to the establishment of MG. Actually, multipath routing increases the security achievable through redundancy [9] and allows more feedback exchanges.

ASMA also requires that the ID of every intermediate node I in $V_{MG\,(S,D)}$ is included in the routing packets thus the trust can be estimated for every subgraph. Other packets do not have to include all the vertices and edges; but just an extension containing the MG. These requirements impose just small modifications to AODV [23] and almost no one to DSR[24] and DYMO[25]. Thus the routing protocol can be implemented by every declination of these protocols supporting multipath routing. Most of the multipath routing protocols discover node-disjoint paths; however this is a too strong requirement thus we just assume that because node in $V_{MG\,(S,D)}$ are trusted, they can belong to multiple paths.

The changes only take place in the route treatment and the structure of the RREP. Every node I in $V_{MG\,(S,D)}$ computes the trust associated to the edge by which it received the RREP and appends this computed value to the packet (arrow 6). Consequently$I$ computes trust value about $\underset{E(I,J)_{MG(S,D)}}{\longrightarrow}$ for every J in $O(I)_{MG\,(S,D)}$, thus the trust value are more accurate as I probably often interact with its neighborhood. Besides computation load is balanced between nodes in $V_{MG\,(S,D)}$. Route maintenance information is transmitted to the macrograph manager thus the macrograph is maintained up to date.

4. ASMA FEATURES

ASMA owns the following characteristics:

1) Adaptive Application-oriented approach: ASMA adapts the security requirement to the applications and to the system resources.
2) Reactive: the solution can adapt to the network environment.
3) Robust: Multipath routing protocol provides redundancy and more security compared to unipath protocol.
4) Distributed: ASMA is deployed in every node, thus nodes are totally autonomous.
5) Dynamic Trust reliant: Dynamic trust relationship can be used to increase trust and reduce resource consumption. Trust is defined for nodes and.
6) Generic and Flexible: Generic modules and their interaction are described offering compliance with a large panel of solutions.

Thanks to its advantageous features, ASMA is viewed as the first approach forming then a new category of security frameworks for MANETs. Aiming at analyzing ASMA against a comparable solution [19], we find that ASMA provides a detailed description of the node's architecture. The interactions between the modules are explicitly explained. Furthermore, ASMA is a flexible and generic solution since it can merge existing mechanisms as authentication and secured routing to enhance the overall security of the network.

5. CONCLUSION

In this paper we give the main guidelines for designing a new framework to deal with security in MANETs. This framework called ASMA combines six security mechanisms to offer a complete set of security functions. We adopt a modular approach enhanced by an appropriate trust model suitable to multipath reactive routing protocols. Since dynamic trust is supported, no bootstrapping phase is needed. ASMA combines efficiently key management, routing and forwarding operations to securely transmit data packets through the network. Our ongoing work concerns the evaluation of our mechanisms through Network simulator 2 (NS2) simulations. For future work, we intend to investigate the impact of the critical parameters of the framework on the global performance taking into account different network configurations (diverse mobility and traffic scenarios).

REFERENCES

1. G. Montenegro and C. Castelluccia. "Statistically Unique and Cryptographically Verifiable (SUCV) Identifiers and Addresses". in Proc. of the Network and Distributed System Security Symposium (NDSS'02), 2002
2. J.R. Douceur, "The Sybil Attack", in Proc. of the 1st International Workshop on Peer-to-Peer Systems, 2002
3. D. Gambetta, "Can we trust trust?", Trust, Making and Breaking Cooperative Relations. basil Blackwell (1990) p. 213–237
4. S. Capku, L. Buttyan, and J.-P Hubaux. "Small worlds in security systems: an analysis of the pgp certificate graph", in Proc. of the 2002 workshop on New security paradigms, 28–35, 2002
5. P. Zimmermann. "The official PGP user's guide". MIT Press 1995
6. S. Yi, R. Kravets, "Composite Key Management For Ad Hoc Networks", in Proc. Of the 1st Annual International Conference on Mobile and Ubiquitous Systems: Networking and Services (MobiQuitous'04), 2004
7. Y.-C. Hu, A. Perrig, and D. B. Johnson. "Ariadne : A secure on-demand routing protocol for ad hoc networks". In MOBICOM, 2002.
8. K. Sanzgiri, D. LaFlamme, B. Dahill, B. Neil Levine, C. Shields, E. M. Belding-Royer, "Authenticated Routing for Ad Hoc Networks, IEEE Journal on Selected Areas in Communications, vol. 23", no. 3, march 2005
9. R. Mavropodi, P. Kotzanikolaou, C. Douligeris, "Performance Analysis of Secure Multipath Routing Protocols for Mobile Ad Hoc Networks", in Springer WWIC 2005, p.269–278, 2005
10. Zhenqiang Ye, Srikanth V. Krishnamurthy, Satish K. Tripathi "A Framework for Reliable Routing in Mobile Ad Hoc Networks" in Proc. Of *IEEE INFOCOM 2003*
11. Y. L. Sun, Z. Han, W. Yu and K. J. Ray Liu, "A Trust Evaluation Framework in Distributed Networks: Vulnerability Analysis and Defense Against Attacks" in Proc of IEEE Infocom 2006
12. Vincent Toubiana, Houda Labiod "ASMA : Towards Secure Adaptive Multipath in MANETs", in Proc of *IFIP MWCN 2006*
13. T. Li, Z. Wan , F. Bao , K. Ren, R. H. Deng, K. Kim, "Highly reliable trust establishment scheme in ad hoc networks", Computer Networks 45 (2004) : 687699
14. T. M. Chen, V. Venkataramanan, "Dempster-Shafer Theory for Intrusion Detection in Ad Hoc Networks", *IEEE Internet Computing* Volume: 9 Issue: 6 Date: Nov.-Dec. 2005 Journals style),"
15. A. Jøsang and S. Pope. *Semantic Constraints for Trust Transitivity.* Second Asia-Pacific Conference on Conceptual Modelling (APCCM2005), Newcastle, Australia, January-February 2005
16. J. M. McCune, A. Perrig, M. K. Reiter "Seeing-Is-Believing:Using Camera Phones for Human-Veri.able Authentication ", in Proc of the 2005 IEEE Symposium on Security and Privacy (S&P'05)
17. M. Cagalj, S. Capkun, J.-P. Hubaux , "Key agreement in peer-to-peer wireless networks"; Proceedings of the IEEE Volume 94, Issue 2, Feb. 2006 Page(s):467–478
18. P. Michiardi and R. Molva, "Core: A collaborative reputation mechanism to enforce node cooperation in mobile ad hoc networks," *Communication and Multimedia Security*, September 2002.

19. S. Buchegger and J. L. Boudec, “Performance analysis of the confidant protocol,” in *Proceedings of ACM Mobihoc*, 2002.
20. K. Bradley et al.,“Detecting Disruptive Routers: A Distributed Network Monitoring Approach”, IEEE Network, vol. 12, no. 5, 1998
21. D. Djenouri, N. Ouali, A. Mahmoudi, and N. Badache, “Random Feedbacks for Selfish Nodes Detection in Mobile AdHoc Networks”, IPOM 2005
22. Sergio Marti, T.J. Giuli, Kevin Lai, and Mary Baker, “Mitigating Routing Misbehavior in Mobile Ad Hoc Networks”, in The 6th ACM International Conference on Mobile Computing and Networking, August 2000.
23. Charles E. Perkins, Elizabeth M. Belding-Royer and Samir R. Das, “Ad hoc On-Demand Distance Vector (AODV) Routing”
24. David B. Johnson, David A. Maltz, Yih-Chun Hu, Rice University, “The Dynamic Source Routing Protocol for Mobile Ad Hoc Networks”
25. I. Chakeres and C. Perkins, “Dynamic MANET On-demand (DYMO) Routing”

CHAPTER 34

LOCALIZATION WITH WITNESSES

ARUN SAHA AND MART MOLLE

Department of Computer Science and Engineering, University of California, Riverside, Riverside, CA, 92521, Email: mart@cs.ucr.edu

Abstract: Localization protocols enable an entity (called the verifier) to determine the physical location of another entity (called the prover), even if the prover maliciously advertises a false location or tries to corrupt the verifier's time measurements by time-shifting its responses. Unfortunately, the correctness of such protocols is critically dependent on the verifier's ability to make high-resolution time measurements and on the prover's ability and trustworthiness to send its response by the mandated time. To address these problems, we propose the idea of incorporating passive witnesses into the localization protocol. All witnesses monitor the same bilateral packet exchange between the prover and lead verifier and later report their respective inter-packet time measurements to the lead verifier for further processing. We show how the extra information provided by the witnesses can eliminate the threat of response-time shifting by a malicious prover. We also pose the question, how can we combine multiple localization observations to a single localization estimate? While analyzing that, we observe that the localization estimate is sensitive to the relative position of the prover among the verifiers

Keywords: Ad-hoc network, Sensor network, Localization, Time-Difference-of-Arrival

1. INTRODUCTION: A CASE FOR LOCATION AUTHENTICATION

The membership to an adhoc network or a sensor network is generally dynamic. It is interesting to explore the situation when the membership criteria to such wireless, self organizing networks is based on proximity and relative distances among the devices.

We consider a wireless network where requests for membership roles are granted if the requesting device is "sufficiently" close to the existing network, i.e. existing

[*] A. Saha is currently with Cisco Systems

H. Labiod and M. Badra (eds.), New Technologies, Mobility and Security, 407–424.
© 2007 *Springer.*

members. The semantics of "sufficient" closeness is decided by the network. A straightforward approach to address this requirement is to make the requesting device mention its position while sending the membership request. In a perfect world where all devices are going to be truthful, there is no problem. However, there might be some incentives to be part of such a wireless network which might tempt a malicious device to claim any arbitrary position of its choice.

Example 1 Some wireless sensors are spread on a environmental experimental testbed to keep the temperature, humidity in control. A rival organization who want to steal the experimental procedure might place some sensors outside the testbed yet inside the transmission range and attempt to join the sensor network.

Example 2 There is a query to an environmental sensor network, whichever sensor is closest to a particular position is asked to report the temperature. A malicious sensor who is not actually the closest to the target can claim its position such that it appears closest and subsequently responds to the query with incorrect data. This might either raise a false alarm or subvert a true alarm.

Example 3 All laptops which are inside a building are assumed to be carried by employees or their guests; they are allowed to join the network and access the Internet if requested. Here, people outside the building carrying laptops might be tempted to claim an inside position, thereby gain access to the Internet, and unauthorizedly use the bandwidth.

In these ways, there might be different kind of undue advantages to be gained if a malicious device can join a position/proximity based network when it actually is not located in a place to do that. This is the motivation to correctly determine the location of such a requesting device. There are two minor variations. The requesting device can claim a position which the existing members then verify. Or, the existing members can determine the location of their own. Whichever it is, the existing members do not trust the claimer but collaborate among themselves to decide.

2. BACKGROUND

The first notion is who is performing the localization activity. In one approach, known as self-localization, the mobile entity collect information from the neighborhood and determine its own location. One such example is Global Positioning System (GPS), where the GPS receiver device receives/collects information from the GPS satellites and determines its own location. In the other approach, the neighbors of the mobile entity, whose location is of interest, collect information, and combine them to determine its location. One example of this approach is E-911 calls from cellphones where the location of the caller is determined by the cellphone base stations.

In the last decade, a number of localization systems were proposed, mainly based on infrared, ultrasound, radio signal and ultra wide band. In our work, we refer to localization in the context of mobile ad-hoc network or sensor network, here the communication between the entities are wireless radio communication.

The localization algorithms are mainly based on (i) received signal strength, (ii) angle of signal arrival (AoA), (iii) or time of signal arrival (ToA) measurements, or their combinations. One important variation of ToA system is the Time Difference of Arrival (TDoA) system.

Another broad categorization of localization approaches is based on whether the distance between the entities are measured or not. Time based systems finally convert the time measurements to distance measurements. The localization systems which are based on the distances between the entities are known as range-based systems, e.g. [1]. Others are called range-free or range-independent, e.g. [2]. Another important characteristic is whether the localization is infrastructure based or ad-hoc. The GPS self-localization or the E-911 localization are based on infrastructure.

The entities participating in the localization can use traditional omnidirectional antenna or smart directional antenna. Smart antennas can transmit and receive energy in one direction as opposed to disseminate in all directions. However, we feel that they will defeat the simplicity of the system since we are targeting the entities in mobile ad-hoc or wireless sensor networks using standard networking protocols.

In ToA based solutions, each verifier executes a distance bounding protocol and determines an upper bound of the distance to the prover. Accurate timing measurements are required to obtain the round trip signal propagation time (and hence distance) between the verifier and the prover. The basic distance bounding is proposed by Brands and Chaum [3] where the verifier sends a single-bit challenge and the prover responds with a single-bit response "immediately after" receiving the challenge. Such challenge-responses are carried out for multiple rounds and the verifier measures the round-trip time at each round. The verifier then computes the upper-bound of the distance based on the maximum of the round-trip times. The above concept is applied and further extended by Capkun and Hubaux [4], Capkun, Buttyan and Hubaux [5], Hancke and Kuhn [6], Reid *et al.* [7] etc. These approaches have multiple implementation constraints as mentioned in [8], [9].

There are significant differences between ToA based solutions and TDoA based solutions. In ToA, the disadvantages are: (i) multiple (at least three) verifiers have to undergo separate challenge-response dialog with the prover, and (ii) the prover's delay between receiving the challenge and sending the response called response delay (note that it happens for each challenge-response, and they are independent!) affects the final localization result. However, in ToA the verifiers need not be time-synchronized, but in TDoA that is a requirement.

The model of localization in this work is the following. There is a set of entities, called verifiers, who want to localize another entity whom we call prover. The verifiers and the prover use omnidirectional radio-frequency communication. The localization system is range based without any infrastructure. Also note that, in this paper, we are mainly focusing on localization concepts ignoring the cryptographic security of the protocol messages; in reality these concepts need to be tightly coupled as in [4], [8], [9].

3. THE WIRELESS SECURE LOCALIZATION PROBLEM

The problem can be formalized as the following. The formation of a wireless network A is based on locality constraints. In most cases, it is a single hop network i.e. every member can receive any transmission by other members. The n members of the network are designated as $A_1, A_2, ...A_n$. There is new node U which claims to be in the vicinity of the network and wants to join. The existing members collaborate to determine the location of the requesting node and decide on the request.

3.1. Assumptions

The goal of the verifiers is to localize the prover using the existing standard network hardware and protocols.

3.1.1. Trust model

The existing members of the wireless network A_i $(i = 1...n)$ mutually trust each other and co-operate, they are known as verifiers. The new entity, known as the prover, is completely untrusted. The prover can be located anywhere with respect to the verifiers, and can take any amount of time in responding to messages.

3.1.2. Mobility

The wireless entities can be possibly mobile. However, we assume that during the execution of the localization protocol, the group of wireless entities are relatively in rest. For example, a group of wireless entities might actually be a fleet of cars in a highway and all of them are moving at a constant speed.

3.1.3. Co-ordinate System

We assume that there is a local co-ordinate system. The wireless entities included in the network know their location in that coordinate system. The entities may be possibly equipped with GPS receivers but that is neither necessary nor sufficient to solve the problem.

3.1.4. Transmission Range

A good number of wireless entities already included in the network must be able to receive the transmission from the requesting entity. The entities which will be receiving transmission from the requesting entity are the verifiers. The verifiers will take part in the execution of the protocol.

4. USING WITNESSES TO PREVENT DISTANCE FRAUD ATTACKS

4.1. Motivational Example of a One-Dimensional Network

Figure 1(a) shows a typical timed-echo challenge-response dialog between the verifier V and prover U using the space-time representation described in [10].

For simplicity, we will temporarily assume that the network is a one-dimensional broadcast network, such as a coaxial cable shared Ethernet segment. In addition, we summarize each message by a single arrow, representing the start-frame delimiter at the beginning of its transmission, rather than a shaded region covering its entire transmission time. In this example, V starts a timer at point (1) from which the challenge spreads in both directions away from V as time advances down the page. After propagation delay α_{VU}, the leading edge of the challenge arrives at point (2), where U can start to receive it. After formulating a suitable reply, U begins to transmit its response at point (3). After a further propagation delay α_{UV}, the leading edge of U's response arrives at point (4), where V stops its timer and continues to receive the remainder of the message. Unfortunately, even with perfect timing accuracy, V still cannot determine its distance to U by measuring the inter-packet time between points (1) and (4), τ_V, because U can vary the inter-packet time between points (2) and (3), τ_U. Indeed, we say that a malicious prover is launching a *distance fraud attack* if it secretly changes τ_U from the value expected to force V into calculating an incorrect value for $\alpha_{VU} \leq \alpha_{VX} \equiv \frac{1}{2}\tau_V$.

In Figures 1(b)-(d) we show how a passive witness W to the same challenge-response dialog between U and V can help the verifier avoid distance fraud attacks. (Recall that the physical layer is assumed to be an omni-directional broadcast channel, so the same challenge transmission that left V at point (1) also reaches W at point (5), and the same response transmission that left U at point (3) also reaches W at point (6).) First, in Fig. 1(b) we assume that witness W_1 is located on the opposite side of V from prover U. In this case, the inter-packet time measurements τ_V made by V between points (1)–(4) and τ_{W_1} made by W_1 between points (5)–(6) will be identical, which shows only that the rays from U to W_1 and from V to W_1 must be parallel, and hence U and W_1 must be on opposite sides of V. In this case the witness cannot help the verifier to localize the prover.

Next, in Fig. 1(c) we assume W_2 is located somewhere between V and U. In this case, the triplets $(1) - X - (4)$, $(5) - X - (6)$, and $(2) - X - (3)$ all form similar triangles, from which we obtain

(1) $$\tau_{W_2} \equiv \tau_V - 2\alpha_{VW_2}$$

and hence that U must be further from V than W_2 in the same direction. Finally, in Fig. 1(d) we assume that W_3 is located beyond U on the same side of V. This time, the inter-packet time measurement τ_{W_3} made by W_3 between points (5)–(6) must be exactly the same as τ_U ***whether or not U tries to spoof V's measurement of*** τ_V. Therefore, in the case of Fig. 1(d), V can easily determine the location of U by substituting all known values into Eq. (1) by solving for the unknown propagation delay:

(2) $$\alpha_{VU} = \frac{\tau_V - \tau_{W_i}}{2}$$

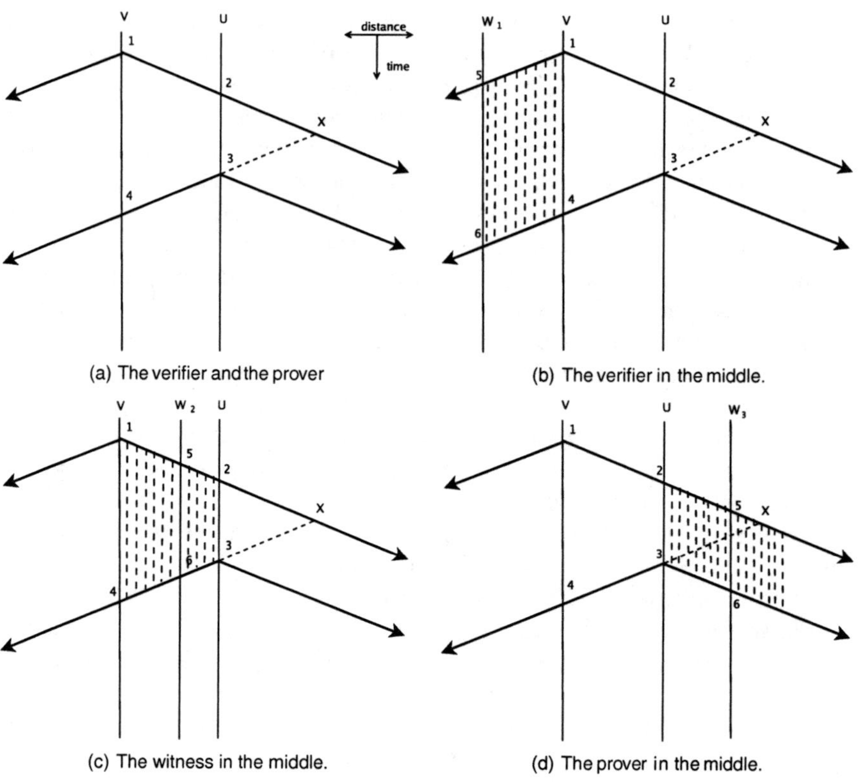

(a) The verifier and the prover

(b) The verifier in the middle.

(c) The witness in the middle.

(d) The prover in the middle.

Figure 1. Witness can help localization in one-dimension. 1(a) shows the verifier and the prover. Subsequent figures show three possible relative locations of the witness with respect to the verifier and the prover (Figures are not to scale.)

4.2. General Approach for Wireless Broadcast Networks

The simple one-dimensional example described above can be generalized to create a novel and *secure* localization technique for a group of trusted co-operating nodes, say $\{A\}$, that communicate over a wireless, planar[1] broadcast network. Our approach uses a single bilateral query-response dialog between one member of the group, V called the *lead verifier*, and an untrusted non-cooperating prover, U. One of the verifiers is (s)elected as lead-verifier. Other group members who are within transmission range of both the lead-verifier and the prover, serve as passive witnesses. The technique does not depend on distance-bounds or RTT from verifier to prover. The basic concept is similar to the Time-Difference-Of-Arrival (TDoA)

[1]The generalization to three dimensional geometry is straightforward but tedious, and will not be considered further.

techniques as summarized in [8]; however we apply it differently for localizing a possibly-fraudulent prover.

Similar to other localization protocols (e.g., [11]), there is an initial untimed setup phase in which U approaches the group for verification of its claimed location so it can participate in various location-dependent activities. The group chooses a lead-verifier, V, and a set of witnesses, and then V instructs U to prepare for the bilateral message exchange phase. This second phase consists of only two messages as shown in Figure 2: the *challenge* sent by the lead-verifier V in Figure 2(a), and a *response* sent by prover U in Figure 2(b). Notice that the witnesses are completely passive during this phase: they receive all messages while sending none of their own.

Once the prover receives the (real) challenge, it computes the response and transmits it. The inter-packet time from the instant that the prover starts receiving the challenge until the instant that it starts transmitting its response is called the **response delay** and denoted τ_U. During this second phase, all active group members monitor the packet exchange between V and U and record the interpacket-time from the instant they either start receiving (if a witness) or transmitting (if the lead verifier) the challenge, until the instant they start receiving the response. Therefore, at the end of the challenge-response dialog, the active group members report their respective inter-packet time measurements: τ_V for lead-verifier V and τ_{W_i} for the ith witness.

Let the distance and signal propagation delay between two entities X and Y be denoted by D_{XY} and α_{XY} respectively. If the signal propagation speed in the medium is v, then $\alpha_{XY} = D_{XY}/v$. Now for witness W_1 we have,

$$\begin{aligned} \tau_{W_1} &= \alpha_{VU} + \tau_U + \alpha_{UW_1} - \alpha_{VW_1} \\ &= D_{VU}/v + \tau_U + D_{UW_1}/v - D_{VW_1}/v \end{aligned} \tag{3}$$

For another witness $W_2 (W_2 \in A \setminus W_1)$, we have,

$$\tau_{W_2} = D_{VU}/v + \tau_U + D_{UW_2}/v - D_{VW_2}/v \tag{4}$$

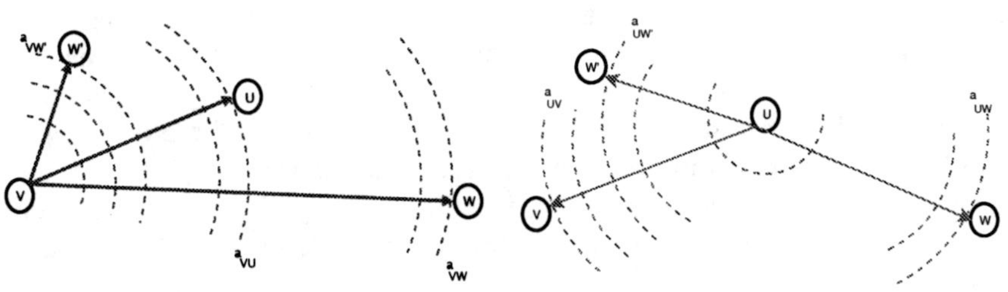

(a) The verifier transmitting the real challenge.

(b) The prover transmitting the response.

Figure 2. Different stages of the localization protocol. V is the lead-verifier, U is the prover, and W, W' are witnesses. Due to the nature of the broadcast medium, all messages are heard by the witnesses as well. Signal propagation delay from entity i to entity j is denoted by a_{ij}

Subtracting Eq. (4) from Eq. (3),

$$\tau_{W_1} - \tau_{W_2} = \frac{1}{v}\{(D_{UW_1} - D_{UW_2}) - (D_{VW_1} - D_{VW_2})\}$$

Transposing,

$$D_{UW_1} - D_{UW_2} = v(\tau_{W_1} - \tau_{W_2}) + (D_{VW_1} - D_{VW_2}) \tag{5}$$

Let us assume that the entities are located in a two dimensional plane. Suppose the positions of U, V, W_1 and W_2 be $U(x_U, y_U)$, $V(x_V, y_V)$, $W_1(x_{W_1}, y_{W_1})$ and $W_2(x_{W_2}, y_{W_2})$ respectively. The distances D_{VW_1} and D_{VW_2} are known,

$$D_{VW_1} = \sqrt{\{(x_{W_1} - x_V)^2 + (y_{W_1} - y_V)^2\}} \text{ and}$$
$$D_{VW_2} = \sqrt{\{(x_{W_2} - x_V)^2 + (y_{W_2} - y_V)^2\}}$$

Substituting similarly in Eq. (5) we obtain,

$$\begin{aligned}&\sqrt{\{(x_U - x_{W_1})^2 + (y_U - y_{W_1})^2\}} - \sqrt{\{(x_U - x_{W_2})^2 + (y_U - y_{W_2})^2\}}\\ &= v(\tau_{W_1} - \tau_{W_2})\\ &+ (\sqrt{\{(x_{W_1} - x_V)^2 + (y_{W_1} - y_V)^2\}} - \sqrt{\{(x_{W_2} - x_V)^2 + (y_{W_2} - y_V)^2\}})\end{aligned} \tag{6}$$

Since V, W_1 and W_2 are mutually-trusted and co-operating, they can exchange their location and the independently measured τ values, i.e. τ_V, τ_{W_1}, τ_{W_2} respectively. So all the terms in Eq. (6), except (x_U, y_U), are known. Hence Eq. (6) is the locus of the unknown position (x_U, y_U). In particular, the locus is one of the two arcs of a hyperbola.

We observed above that *any* two verifiers can find out the locus of the prover. Similarly, another independent locus of the prover can be formed by combining the time interval data of a different pair of verifiers. The two loci, i.e. the two equations, can be solved to find out the location of the prover. Thus any three verifiers, including or excluding the lead-verifier, can localize the prover. Other things like distance, round trip time and response delay can be easily derived.

Positions of two witnesses can be initialized during the network startup. Moreover, once a prover's position becomes known, it can possibly be used as a witness in future.

Note that the Eq. (6) does not depend on the response delay τ_U. Hence the final solution is independent of response delay at the prover. Thus this technique is resistant to Distance Fraud attack where the prover can intelligently enlarge and reduce distances to fool a set of three verifiers and spoof a different location [1], [4].

4.3. Results:

The above analysis leads us to the following results:

- Any verifier-pair can form the locus of the prover.
- Any verifier-triplet can localize the prover.
- The location found by a verifier-triplet is independent of the response delay (τ_U) at the prover.

5. ACCURATE MEASUREMENT OF THE TIME INTERVAL τ_{W_i}

There are two challenges with regard to the measurement of the Time Interval τ_{W_i}:

- The time interval measurements should be fine grained. Radio signal travels in vacuum at a speed of $c \approx 3 * 10^8 m/s$. So, the context of localization application and the tolerances are important. For example, when we target localization within a room or building, we should target error tolerance in meters, not *hundreds* of meters. Reversing the argument, if we want the distance estimation within error tolerance of meters, then the maximum error tolerance in timing measurement must be in the order of nanoseconds.
- The time interval measurements should be free from clock skew. The time intervals measured by all the verifiers should be based on a single clock as there might be clock skew among the verifiers' local clocks. Of course that single clock should meet the required high precision requirements.

In this section we describe how the above two challenges can be addressed.

5.1. Features of Wireless Communication

There is one feature in wireless communication which is distinct from the wired world. When there is a one-hop connection between two wired hosts, then whatever they transmit in the medium is known only by those two hosts and no other host.[2] On the other hand, since the wireless medium is basically broadcast in nature, any unicast message transmission can be "heard" not only by the intended receiver but also by all other entities in the sender's transmission range. In particular, if the channel is not reserved by some other neighbor, a wireless entity *has* to receive all packets in the medium at least to determine whether the packet is destined for it. This is what we mean by "heard".

5.2. Fine Grained Time Interval Measurement Cannot be done in Software

For correct localization in range-based systems, the true signal propagation time is required; the time interval measurement should exclude time spent in all other

[2]However, generally two hosts seldom have direct one-hop connection, the connection goes through some network devices e.g. repeater, switch or router. That is, direct single hop connections are seen between a host and a network device or between two network devices.

activities. If the desired time interval is measured in the application layer of the verifier/witness, then it includes the time of the message traversal through the verifier/witness protocol stack. Also, if the response is generated at the application layer of the prover then the message traversal time through the prover protocol stack is also included in the measured interval. Traversal time through the protocol stack includes time for passing the message among different layers of hardware and software. This additional delay is unpredictable and of much higher order of magnitude than the propagation delay. Similar arguments hold true if the time interval is measured in operating system software. The reader is requested to refer [8, Sec. 3.1.9] for additional details and sample figures regarding this.

5.3. Fine Grained Time Interval Measurement with Help from Hardware

Our approach is to perform the measurement in a place inside the device which is very close to the device to medium interface. Using current technology we can achieve it by moving the measurement task all the way down to bottom of the protocol stack, where we utilize the first hardware component adjacent to the medium-dependent signaling interface, the PHY.

There are two main functions performed by the Physical Coding Sublayer (PCS) inside the physical layer (PHY): (i) generation of continuous code-groups to be transmitted on the channel, and (ii) processing of the code-groups received. In addition to the code-groups corresponding to the frame data, the transceivers use some control code-groups. One such control code-group is Start Frame Delimiter (SFD), which indicates the arrival of a new frame.

We assume that the verifiers, i.e. the PHY of the verifier, are able to detect the reception time of a specific marking code-group when they receive a frame. By specific marking code-group, we mean something particular like Start Frame Delimiter (SFD) code-group of the frame.

The lead-verifier can measure the time interval τ_V as follows: it starts a timer when the SFD of challenge frame is transmitted, and stops it when the SFD of response frame is received.

Due to the inherent property of the wireless communications mentioned above, the challenge frame transmitted by the lead-verifier will be "heard" by all the other verifiers i.e. witnesses. Hence the witnesses can measure the time interval as the elapsed time between the following two events: (i) reception of the SFD-CG of the challenge frame from the lead-verifier, and (ii) reception of the SFD-CG of the response frame from the prover.

Let us now examine the solution in little more detail. Once a witness hear the dummy Challenge before the real Challenge from the lead-verifier, the witness understands that a challenge-response dialog is about to begin. The witness then keep its transceiver in ready-to-receive state until finally the response arrives from the prover.

However, the PHY is not capable of interpreting the contents of a received frame. So, the PHY by itself cannot know when the challenge frame is going to come, and correspondingly start the timer on receipt of the marked code-group. But, with instructions from the higher layers, the PHY can do so. Therefore, once the dummy Challenge is received, the higher layer instructs the PHY that now is the time that the PHY should start the timer on the receipt of the marked code-group of the next arriving frame. Once instructed, the PHY will start the timer on the receipt of the marked code-group of the next arriving frame and stop the timer on receipt of the same marked code-group in the subsequent frame.

5.4. Measuring Time Interval Using Common Clock

When an entity receives a frame, it changes state from IDLE to RECEIVING, and goes through a synchronization process. The synchronization process is responsible for determining whether the underlying receive channel is ready for operation. After bit synchronization, the receiver knows the bit transmission rate of the sender. In other words, the receiver acquires the clock rate of the sender's PHY. Once that is done, the receiver can setup a local timer with frequency equal to the sending PHY's transmitter clock.

Thus when an witness "hears" or receives the challenge frame, it sets up a local timer with the frequency of the lead-verifier's PHY transmitter. This timer is used measure the time interval τ_{W_i}. The interval is measured in terms of the number of clock ticks of that timer. One clock tick interval if that timer is equal to the symbol time of the lead-verifier's PHY. All the witnesses use this method. This way all the measured time intervals are in terms of a single clock and free from clock skew errors.

5.5. Discussion

The PHY will report the time interval to the higher layers only if it is requested, and that reporting will be much later after the receipt of the response frame. Also note that, there will always be maximum of one such time interval measurement result in the PHY. If the higher layer instruct the PHY to make another time interval measurement, then the PHY will overwrite its previous measured value. This is because the main localization algorithm is carried in the application layer, the PHY needs to make measurement once for each execution of the algorithm.

We note that due to the multipath nature of the wireless channel, frame transmissions will experience multi-path delay spread. A code-group radiated using an omnidirectional antenna, will take multiple paths (as a consequence of reflections from various objects) to arrive at the receiver. In other words, the receiver will receive multiple copies of the same signal, each of which may have a different amplitude, phase and delay. One received symbol will interfere with other copies of its own. Due to this fact, the exact reception time of a code-group is difficult to characterize. One possible approximation is to consider the first copy since the line-of-sight path will frequently be the quickest.

6. COMPUTATIONAL ISSUES

6.1. Measurement Errors

Like any other measurement, the time intervals τ_{W_i} noted by the verifiers are subject to error – the measured time intervals might be little too high or little too low. Such measurement noise will affect the locus of the prover and subsequently its location. In that case, the solution points of two different verifier-triplets will not be exactly the same. But, if the measurement errors are not large, we can expect all those solutions points to be scattered around the actual location (unknown to the verifiers) of the prover. There might be some extraneous solution points however.

6.2. An Over-Determined System

Since any three verifiers can collaborate to localize the prover, then the next natural question is the following. If there are more than three verifiers, which three of them will be chosen to localize the prover? For each arbitrary choice of a verifier-triplets, we can determine a possible position for the prover. If there are n verifiers, then the number of verifier-triplets can be formed is $N = \binom{n}{3}$. We will get one (two in some cases) solution point from each set. In total, there will be approximately N solution points, i.e. N possible locations for the prover. Such a system is often referred to as an "over-specified" or "over-determined" system, a potential drawback of using an over-determined system relates to the fact that hyperbolic localization algorithms can calculate more than one mathematically valid position [12]. The situation is like a fallacy and counter-intuitive from the statistical point of view. When we had less information it was easier to conclude, when we have more information it is difficult!

6.3. Combining Multiple Solution Points

Now, what is needed is some method to use *all* these solution points to make a *single* final estimate about the location of the prover V. The most naive choice, the mean of all the solution points as (mean of all x-coordinates, mean of all y-coordinates), is not good because of the fact that arithmetic mean is highly affected by the outliers. However, the median of the solution points might be good. (Zhang *et al.* [13] takes the median of K distance-estimates.) There, one option is to output the point (median of all x-coordinates, median of all y-coordinates) as the final estimate. Another option is to find the two-dimensional median of the points, i.e. the central-most point among all. One simple way to do that is to find the 2D-median as described below.

2D median: Construct a convex polygon with a subset of the solution points, such that all the remaining solution points which are not the vertexes of the polygon are inside the polygon. Then discard the solution points included in the polygon, and repeat the process with the remaining solution points. In this way of repeatedly peeling-off outer points, the central-most solution points (maximum of three) can be found. One of these, or their mean, can be the final estimate.

Another approach of combining the multiple solution points is the following: Imagine all the solution points obtained from different sets of verifiers as different measurements of the same signal and use them to make a final estimate. Kalman Filtering is one possible way to do that.

6.4. Kalman Filtering

Kalman filtering is an optimal, recursive, discrete data processing algorithm. It addresses the general problem of trying to estimate the state of a discrete-time controlled process that is governed by linear stochastic difference equation [14]. The algorithm predicts the state ahead, makes a measurement, then combine the prediction and measurement such that the error covariance is minimized. Again it makes prediction for next stage and so on.

6.5. Kalman Filtering to Combine Multiple Solution Points

We experimented with Kalman Filtering to estimate the prover's location. First, we obtained all the possible solution points by pairwise solving all the hyperbola equations. Then we passed the solution points one by one through the Kalman Filter. After sufficient number of steps, the estimate converges.

However, as we observed in our simulation experiments, the order in which different solution points are considered by the filtering algorithm significantly affect the final estimate. The same set of solution points processed in different order produces different final estimate. Thus there is a need to find out a way to order the different solution points such that the final estimate is as close to the actual location as possible.

We believe that the orientation of the verifier-triplet, and the location of the prover relative to that orientation is of importance. Recall from the earlier example of one dimensional network, if the witness is on the opposite side of the prover (Fig. 1(b)), then the verifier cannot localize the prover. Some solution points and their associated verifier-triplet are more significant than others. More significant solution points should be treated earlier than the less significant ones. We propose the following heuristics:

1. If the solution point lies within the triangle formed by the verifier-triplet, then that solution point is more significant than if the solution point lies out of the triangle.
2. If the verifier-triplet is almost collinear, then the solution obtained from them will be poorer than from the verifier-triplet which constitute a well-formed triangle.

3. A solution point which is closer to all locus curves is expected to be nearer to the actual location compared to a solution point which is not. To achieve that, the normal distance from each solution point to all the locus curves are found and added up. Then the solution points are ordered in decreasing order of aggregate distances to be considered by the Kalman filter

In the following section we analyze the first heuristic.

6.6. Sensitivity of Prover Location w.r.t. Verifier-Triplet

If the solution point lies outside the triangle formed by the verifier-triplet, then it is very sensitive to measurement error. In such cases, a little measurement error displaces the probable solution points by a (relatively) large amount. This is shown in the following example.

Three verifiers, $A_1(-5, 0)$, $A_2(0, 5)$ and $A_3(8, 0)$ are trying to localize a prover. In our experiment, we consider:

- Two locations of the prover V: (i) inside the triangle $\Delta A_1 A_2 A_3$ as $P_1(1, 2)$, (ii) and outside the triangle as $P_2(4, 15)$.
- Two methods of localization: (i) ToA [3] (intersection of circles) method, (ii) and TDoA (intersection of hyperbolas) method.
- Two cases of measured data: (i) with no measurement error, (ii) and with measurement error (distance error for ToA, difference of distance error in TDoA).

Table 1 connects the cases described above to the diagrams shown below. Curves obtained when there is no measurement error are shown with solid lines, the dot-dashed lines show the curves obtained with measurement error. Measurement error was injected by increasing the 'distance' (in ToA) or the 'difference-in-distance' (in TDoA) by a small amount.

The solid-line circles in Fig. 3(a) show how the three verifiers localize P1 when there is no measurement error. But when there is measurement error (here, in A_1's measurement), the three circles do not intersect at any single point, however there are two intersection points which are *very close* to the actual

Table 1. Experiment summary

	Prover P1(1, 2)		Prover P2(4, 15)	
	No Error	With Error	No Error	With Error
ToA	Fig. 3(a)	Fig. 3(a)	Fig. 4(a)	Fig. 4(c)
TDoA	Fig. 3(b)	Fig. 3(b)	Fig. 4(b)	Fig. 4(d)

[3]The solution proposed in this document (§4) uses the TDoA method. Still we consider the ToA method in this example since that is another widely used method.

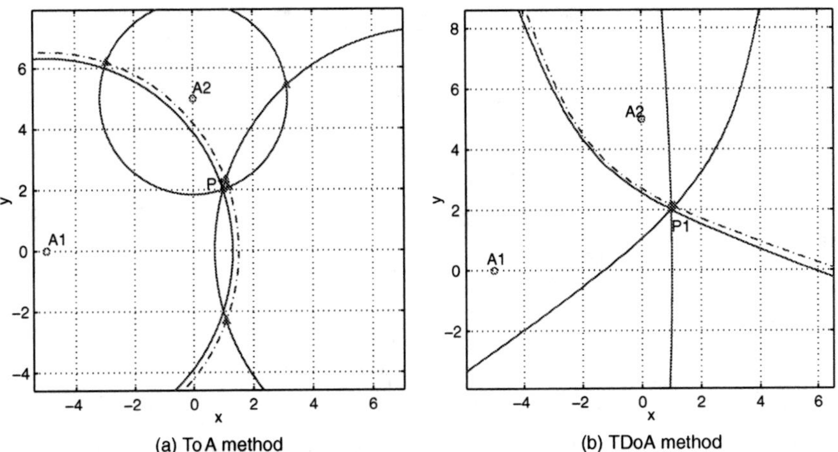

(a) ToA method

(b) TDoA method

Figure 3. Sensitivity of errors when the prover is at location (1,2) which is inside the triangle formed by the verifier-triplet

solution point. In Fig. 4(a), we see how the three verifiers localize P2 when there is no measurement error. However, Fig. 4(c) shows that when measurement errors are present, the derived intersection points are *not very close* to the actual location of P2. The following simple logic indicates that the intersection points moves more when they are outside the triangle as opposed to when they are inside the triangle. The Fig. 3 and the Fig. 4 have the same zoom level. In Fig. 3 where the prover is inside the triangle, the intersection points generated from erroneous measurement are almost superimposed on the actual location point. In Fig. 4 where the prover is outside the triangle, the intersection points generated from erroneous measurement are distinctly visible from the actual location point.

The same observations are repeated for the TDoA method as shown in Fig. 3(b), Fig. 4(b), and Fig. 4(d).

In a different experiment, the measurement noise with both positive and negative values are considered. When there is no measurement error, verifiers A_1 and A_2 generate the hyperbola H_{12} (see Fig. 5). With a small positive offset added to the difference-in-distances, they generate the hyperbola $H_{12}p$ shown with dashed line. Similarly, with a small negative offset added to the difference-in-distances, they generate the hyperbola $H_{12}m$ shown with dotted line. H_{13} and H_{23} are the hyperbolas generated from (A_1, A_3) and (A_2, A_3) pairs respectively. The three hyperbolas — H_{13}, H_{23}, and one from $H_{12}p$ and $H_{12}m$ — gives four intersection points that form a diamond shaped patch area surrounding the actual location of the prover V. If the errors injected in the experiment is the upper bound of permissible measurement error, then that patch area denotes the possible place where the prover might actually be located. The area of the patch quantifies the

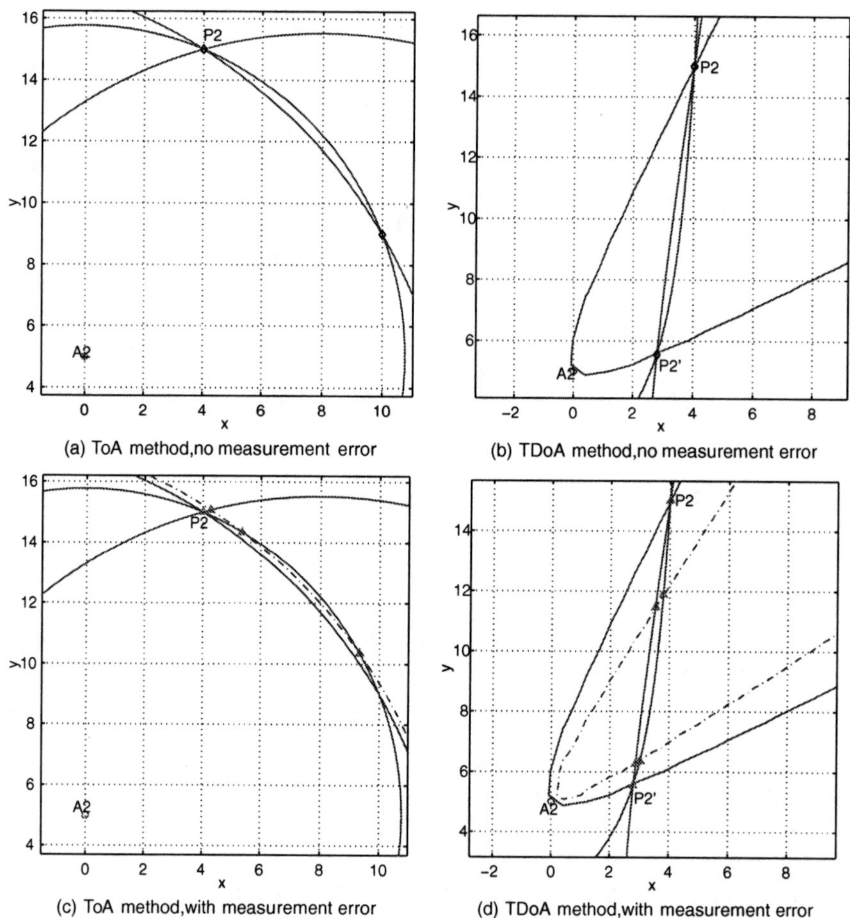

Figure 4. Sensitivity of errors when the prover is at location (4,15) which is outside the triangle formed by the verifier-triplet

uncertainty of measurement: the lesser the area, lesser is uncertainty in prover's localization. If the system of three hyperbolas yield two solution points, then there will be two patch areas; however they will surround the respective solution points (see Fig. 5(b)).

Fig. 6 shows the patch areas for different locations of the prover. The actual prover location is denoted by an asterisk inside its patch. For dual solutions, there are some patches with no asterisks inside. Note in the figure that, for prover locations closer to the verifier triangle – like (−2, 2), (−1, 2), (1, 2), (2, 2), (−2, −2), (2, −2) – the patch is almost invisible. This intuitively suggests that when the prover is actually located closer to the verifier triangle, the uncertainty in localization is lesser.

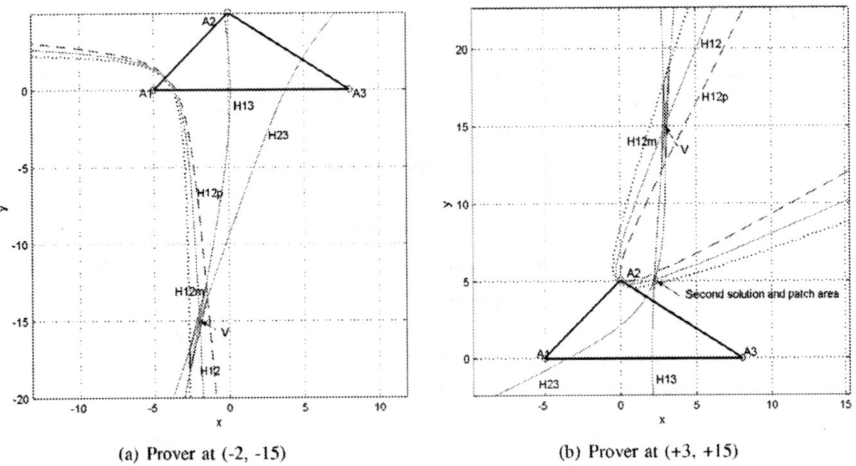

(a) Prover at (-2, -15) (b) Prover at (+3, +15)

Figure 5. The patch areas surrounding the prover location

In ToA localization, where distance-enlargement attacks are possible, there is a philosophy where a verifier-triplet accepts a prover location only if the location is inside the verifier triangle. See for example the "Point in the triangle" test in [4] or the similar "Point in a polygon" test in [13]. In TDoA localization, where distance-enlargement attacks are *not* possible, the above philosophy might still hold true.

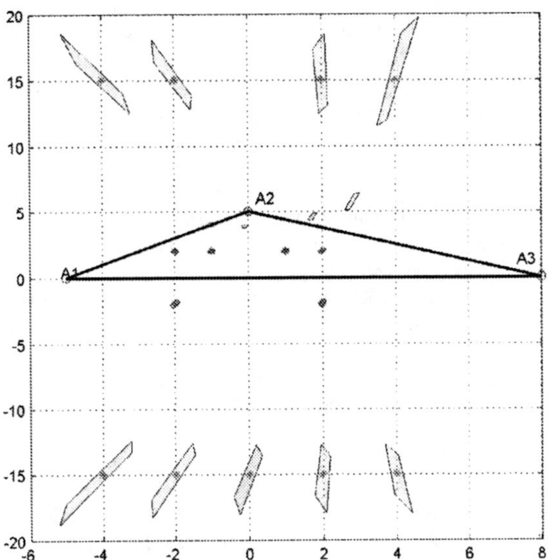

Figure 6. Patch areas surrounding different possible prover locations

7. CONCLUSIONS

This paper analyzes some problems of localization in infrastructure-less wireless networks. We propose an approach where a passive listener can help in the localization process. Unlike some other systems where all the verifiers need to synchronize to the global clock and measure the time of occurrence of one specific event, here the verifiers need to measure a time interval between two specific events. Our solution is secure in the sense that it does not depend on the time taken by the prover to compute the response of the challenge from the verifier. We propose cross-layer co-operation between localization software and network interface hardware to measure that time interval in high precision as required for localization. Further, we introduce the problem as to how multiple erroneous observations from multiple verifiers can be combined to a single less-erroneous solution.

REFERENCES

1. S. Capkun and J.-P. Hubaux, "Secure positioning in wireless networks," *IEEE J. Select. Areas Commun.*, vol. 24, no. 2, pp. 221–232, Feb. 2006.
2. L. Lazos and R. Poovendran, "Serloc: secure range-independent localization for wireless sensor networks," in *WiSe '04: Proceedings of the 2004 ACM workshop on Wireless security*. New York, NY, USA: ACM Press, 2004, pp. 21–30.
3. S. Brands and D. Chaum, "Distance-bounding protocols," in *Advances in Cryptology – EUROCRYPT ' 93*, ser. Lecture Notes in Computer Science, T. Helleseth, Ed., vol. 765, International Association for Cryptologic Research. Springer-Verlag, Berlin Germany, 1994, pp. 344–359.
4. S. Capkun and J.-P. Hubaux, "Secure positioning of wireless devices with application to sensor networks," in *IEEE INFOCOMM*, vol. 3, Miami, USA, Mar. 13–17, 2005, pp. 1917–1928.
5. S. Capkun, L. Buttyan, and J.-P. Hubaux, "SECTOR: Secure tracking of node encounters in multi-hop wireless networks," in *ACM Workshop on Security of Ad Hoc and Sensor Networks*, vol. 1, 2003.
6. G. P. Hancke and M. G. Kuhn, "An RFID distance bounding protocol," in *IEEE/CreateNet SecureComm 2005*, Athens, Greece, Sept. 5–9, 2005.
7. J. Reid, J. M. G. Nieto, T. Tang, and B. Senadji, "Detecting relay attacks with timing based protocols," 2006. [Online]. Available: *http://eprints.qut.edu.au/archive/00003264/*
8. A. Saha, "Cross Layer Techniques to Secure Peer-to-Peer Protocols for Location, Adjacency, and Identity Verification," Ph.D. dissertation, University of California, Riverside, 2006. [Online]. Available:
9. L. Bussard, "Trust Establishment Protocols for Communicating Devices," Ph.D. dissertation, Ecole Nationale Sup'erieure des T'el'ecommunications, France, 2004. [Online]. Available: http://www.geocities.com/laurentbussard/papers/phdThesisBussard04.pdf
10. M. Molle, K. Sohraby, and A. Venetsanopoulos, "Space-Time Models of Asynchronous CSMA Protocols for Local Area Networks," *IEEE J. Select. Areas Commun.*, vol. 5, no. 6, pp. 956 –968, July 1987.
11. B. R. Waters and E. W. Felten, "Secure, Private Proofs of Location," Department of Computer Science, Princeton University, Tech. Rep. TR-667-03, Jan. 2003. [Online]. Available: http://www.cs.princeton.edu/research/techreps/TR-667-03
12. R. J. Fontana, E. Richley, and J. Barney, "Commercialization of an Ultra Wideband Precision Asset Location System," in *IEEE Conference on Wideband Systems and Technologies*, Nov. 2003.
13. Y. Zhang, W. Liu, and D. Wu, "Secure Localization and Authentication in Ultra-Wideband Sensor Networks," *IEEE J. Select. Areas Commun.*, vol. 24, no. 4, pp. 829–835, Apr. 2006.
14. G. Welch and G. Bishop, "An Introduction to the Kalman Filter," University of North Carolina at Chapel Hill, Tech. Rep. TR 95-041, Apr. 2004.

CHAPTER 35

SAACCESS: SECURED AD HOC ACCESS FRAMEWORK

H. CHAOUCHI AND M. LAURENT-MAKNAVICIUS

CNRS Samovar UMR 5157, GET/INT/LOR, 9 rue Charles Fourier, 91011 Evry, France

Abstract: Ad hoc technology being developed for over a decade now has not yet succeeded to get into the telecommunication service value chain. This is due mainly to, the lack of network control, quality of service and security support. From a service provider's point of view, to use the ad hoc technology in the value chain, an efficient AAA framework is mandatory. This is not easy because of the self organising aspect of the ad hoc network. This paper presents a brief overview of security issues in ad hoc networks and introduces a new AAA approach in an ad hoc network in order to allow secure exchange of services, thus being chargeable. It is mainly based on decomposing the AAA service which is classically centralized, into three sub-services and distributing them securely in the ad hoc network. This will allow the ad hoc technology to securely extend the access network coverage and introduce new services exchange within the ad hoc network

1. INTRODUCTION

Ahoc network is a multi-hop network that is created by the mobile nodes when needed for their own communication purposes [1]. Typically this could mean that two hosts want to exchange some data. In an ad hoc network, mobile nodes come and go as they wish, so the topology of the network is changing quite rapidly. This creates new challenges for the protocols to be used in ad hoc networks. Most of the traditional protocols don't fit very well into ad hoc networks. An ad hoc network is quite a new concept, so there isn't any approved protocol yet, for example, for routing purposes.

An ad hoc network can be created in a number of ways. One solution is to run routing protocols in the mobile nodes. This approach requires careful attention, because the rate of change in an ad hoc network is quite rapid compared to the Internet, to which most of the current routing protocols are designed. Another approach would be to treat the ad hoc network as an incompletely connected physical medium [1].

H. Labiod and M. Badra (eds.), New Technologies, Mobility and Security, 425–436.
© 2007 *Springer*.

In the context of Always On era, ad hoc technologies integration with the infrastructure is without any doubt, the inevitable approach for extending at low cost the network access coverage. However a real and business oriented service deployment over ad hoc network requires firstly security of the communications and resource accounting. The lack of security and accounting mechanisms is the major issue that slows down the deployment of ubiquitous services. We believe that the integration of ad hoc and infrastructure-based technologies coupled with efficient security and accounting techniques is the answer for the urgent demand of network operators for appropriate architectures to host secure and large scale ubiquitous services.

There are several threats in ad hoc networks. First, those related to wireless data transmission such as eavesdropping, message replaying, message distortion and active impersonation. Second, those related to ad hoc construction of the network. This means that attacks can come also from inside the ad hoc network. Therefore we cannot trust one centralized node, because if this node would be compromised the whole network would be useless. Another problem is scalability. Ad hoc networks can have hundreds or even thousands of mobile nodes. This introduces important challenges to security mechanisms [2].

As most of the security issues in ad hoc networks are caused by trust less nodes, the authentication process is a strong solution to eliminate those misbehaving nodes. Nevertheless, ensuring authentication service in a self organized network is not easy to realize. We propose in this work to build a secured ad hoc infrastructure framework where the AAA service which is classically centralized in the infrastructure network is decomposed into three sub-services and partly executed by the infrastructure network. The Authentication service (Aaa), the Authorization and Accounting services (aAA). These services will be securely distributed in the servicing ad hoc nodes. For this purpose, a trust management framework is necessary.

One obvious and original consequence of the secured framework would be the integration of ad hoc technology in the service value chain by the introduction of a new service provider entrant (ad hoc network service provider), and a new network access provider (ad hoc network). The classical operator then will make profit by offering in addition to his classical services (access to Internet), new services for ad hoc nodes. For instance, it will act as a third party between the servicing ad hoc nodes, and the customers (local ad hoc nodes). This will be to guarantee the AAA service and a secured transaction for exchanged services (peer-to-peer, packet forwarding, resource consumption...).

2. SECURITY ISSUES AND CHALLENGES IN AD HOC NETWORKS

Ongoing research in ad hoc network security is mainly addressed in a pure ad hoc context and covers secure routing, key establishment, trust management, authentication and certification/revocation services. Assuring trustiness in ad hoc networks is a challenging issue due to the absence of communication infrastructures, and a fortiori third trust parties. Indeed, the infrastructure-less nature of ad hoc networks makes it difficult to adopt centralised or hierarchical trust models such as Public Key Infrastructures (PKI) [3]. Nevertheless, most of existing security paradigms for ad hoc networks assume the existence of public/private key pairs, and hence the existence of a key management infrastructure. For instance, secure routing protocols which are an essential security component for detecting and eliminating malicious nodes disrupting the network. That includes solutions such as ARAN [4], Ariadne [5], SAODV [6], and all of them are based on this very constrained prerequisite that nodes are configured with appropriate pre-shared key or public/private keys to support origin authentication.

As such, a number of works were conducted towards adapting certification and revocation services to ad hoc networks, and most of them identified the threshold cryptography [7,8] as a possible solution where k-out-of-n nodes collaboratively provide a certification service for other nodes in the network. The revocation service is operated when a minimum number of k nodes are accusing a node of misbehaving, and sometimes the action is balanced with the reputation of each accusing node.

Therefore, the threshold cryptography is adapted to ad hoc networks to establish keys that may be used next to secure the routing, authenticate nodes and exchange encrypted data. Another solution is the ID-based cryptography [9] where the node's identifier is part of its public key, so a public key is naturally bound to the node. Another solution [10] considers both cryptographies, the threshold one to initialize a private-key generation server for ID-based cryptography support.

3. AAA IN AD HOC NETWORKS

Typically Authentication, Authorization, and Accounting (AAA) are more or less dependent on each other. However, separate protocols are used to achieve the AAA functionality. IETF AAA working group is trying to design one AAA protocol that could be used in a variety of applications. AAAARCH is also trying to build a general architecture for AAA systems. Mobile Ad Hoc networking (MANET) brings new challenges to providing the AAA functionality. Ad Hoc networks are by their nature rapidly changing and dynamic. There isn't necessarily any network infrastructure present.

A number of research works were conducted on the classically centralized AAA functions [11,12], but few of them studied the possible interactions between AAA and ad hoc network. For instance, [13] focuses mainly on the authentication architecture for enabling distant users to access to services (like internet) through an ad hoc network [13] proposes to perform authentication based on EAP-TLS and PANA [14], but in a multi-hop network context. The EAP-TLS authentication phases end with the ad hoc node and the access network sharing a security association for next data exchanges to remain confidential.

3.1. Authentication

Some kind of authentication is needed in ad hoc networks. Because an ad hoc network is open in the way that mobile hosts can come and go, there is no way to know, which mobile hosts are present in the network. If some data for example are being transmitted, it is important to make sure that the communication is established with the right host.

One way to deal with low physical security and availability constraints is the distribution of trust [1]. Trust can be distributed to a collection of nodes. If all t + 1 nodes will be unlikely compromised, then a consensus of t + 1 nodes is trustworthy [2].

In [2] a distributed key management service is described. In this (n, t + 1) model there is not only one CA (*Certificate Authority*) but many. There are n special nodes which act as servers. As long as at most t servers are compromised the scheme works [1].

This key management approach is based on the *threshold cryptography*. A (n, t + 1) threshold cryptography scheme allows n parties to share the ability to perform a cryptographic operation, so that any t + 1 parties can perform this operation jointly whereas it is infeasible for t parties to do so, even by collusion [2].

In this key management service, n servers share the ability to sign certificates. A (n, t + 1) threshold cryptography scheme is used to make the service tolerate t compromised nodes. Service private key is divided to n shares, which are distributed to all servers. To sign a certificate each server signs the certificate with its share and transmits the certificate to a combiner. A combiner is able to sign the certificate if it gets t + 1 correct partial signatures. Compromised servers are not able to sign certificates, because there is at most t of them at any time [1].

After having built this kind of key management system a public key cryptography can be used to do the authentication [1].

3.2. Authorization

Authorization is also needed to avoid malicious host to be able to wreak havoc inside the network [1]. This can be prevented by keeping control of what hosts are allowed to do inside the ad hoc network. Authorization also needs some sort of distributed structure to avoid single point of failure. This is why the traditional way of using *access control lists* (ACL) in one central server isn't adequate in ad hoc networks [1].

3.3. Accounting

Accounting features are quite specialized in ad hoc networks. Because basically there is no network infrastructure that is providing the service, there isn't either the same kind of service provider concept as in traditional networks. In ad hoc networks individual mobile hosts are providing service to each others; hop by hop routing. There can be two kinds of situations in the charging point of view. The first one has no need to use charging. In this situation all the hosts have decided together that they want to form an ad hoc network for their own need to communicate with each other freely. This could mean they all belong to the same organization like in the case of military units or they are in the same place and want to communicate like in a meeting. So, this ad hoc network is most likely an infrastructureless intranet. In the second case, mobile nodes are just participating in the network to communicate with some of the other nodes only. In this situation, if some mobile node acts as a router in the network, providing connectivity between two nodes that are not within each others range, then it would be reasonable to charge some money for this routing service [1].

Accounting in ad hoc networks hasn't been studied very much. So there is no protocol to do the actual charging yet. This area is however quite interesting, because it is faced with questions like how mobile nodes can charge each others? We cannot assume connectivity to some central server that takes care of the charging in the ad hoc network, so there is a clear need for distributed charging protocols as well with the strong constraint that banks are accepting this new individual to individual charging [1].

3.4. AAA Systems

Ad hoc networks and general AAA systems can be seen as oxymoron. The biggest problem is related to the varying nature of the network. There are no home domains or foreign domains, because the networks are built in ad hoc way. Also the term service provider will have a different meaning than before. This does affect the AAA systems that the AAA working group is presenting, because some of the basic building blocks of their architecture are missing from the ad hoc networks [1].

The basic problem as we mentioned before is the model provided by the AAA working group which is a centralized trust model. This clearly doesn't fit well into ad hoc networks since it's decentralized. We need some other kinds of methods to achieve the AAA functionality.

One approach to provide authentication and authorization functionalities in ad hoc networks could be to use trust management based approaches like PolicyMaker or Keynote2 [1]. These are decentralized by nature and can provide the requested functionality in ad hoc networks quite easily. Also other protocols like SASL or ISAKMP/IKE could be used to provide the authentication functionality [1].

4. SECURED AD HOC ACCESS FRAMEWORK

Introduction of AAA into ad hoc environment is not an easy task due to the self organising aspect of the ad hoc network. The objective of this approach is to design a functional bridge (architecture) between the ad hoc network and the infrastructure network when it is available to support secured exchange of services between the ad hoc nodes. The designed architecture named AdIN (Ad hoc/Infrastructure) is represented in Fig. 1 below. It targets deploying several mechanisms such as

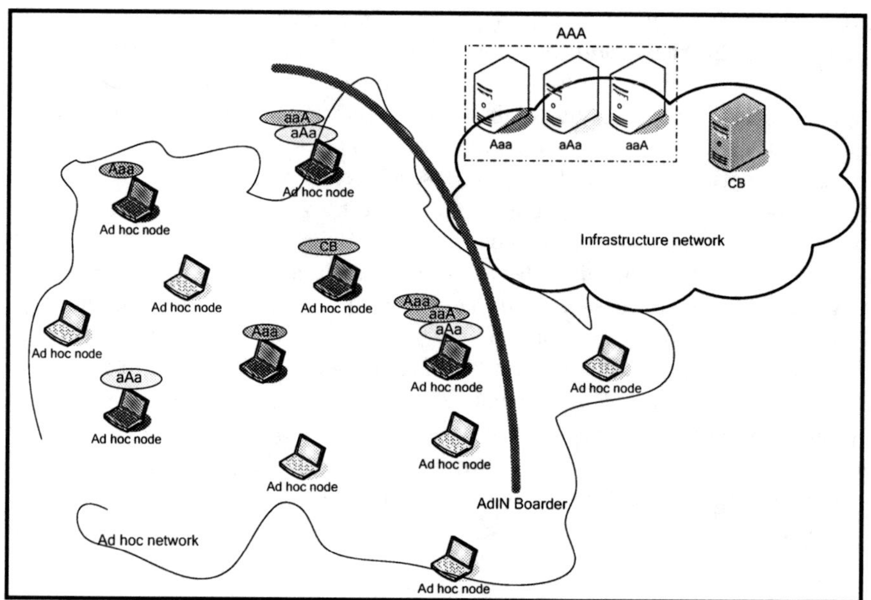

Figure 1. AdIN framework

authentication, authorization, accounting, key management. Neighbour and Service discovery mechanisms are also necessary to provide information for the ad hoc node in order to allow him get the appropriate service.

The first strong point of AdIN framework, is to decompose the AAA service in the infrastructure into Aaa, aAa, and aaA services in order to offer them fully or separately to the ad hoc network and this would be seen as a service rendered to the ad hoc node and hence chargeable. Another strong point of this architecture is the delegation of one or all of these services (Aaa, aAa, aaA) into certain nodes of the ad hoc network. These nodes are supposed to be secure and trusted by the infrastructure. For instance, there might be nodes that belong to the infrastructure network administration (i.e. airport buses equipped with ad hoc material). These nodes are specially set up by the operator willing to extend his infrastructure network to the ad hoc network these special nodes might be freely moving in the ad hoc network carrying with them the AAA services configured at the first place by the operator. The carried AAA services would be offered to the ad hoc network, when this one could not join the infrastructure network to benefit from the AAA service located in the infrastructure. This delegation of AAA services to the ad hoc network assumes a trusted relationship between those ad hoc nodes capable of offering the AAA services.

The aAA services could be implemented by the ad hoc nodes that are willing to provide services to the other nodes (content delivery or exchange, packets forwarding, Internet access...). So these services (aAa, aaA) might be easily distributed in the ad hoc network as long as an accounting and billing system is in place.

The Aaa service is more difficult to distribute totally since it authenticates users joining the ad hoc network. Those users are not known by the ad hoc network. That is why it is necessary to ensure an interdomain authentication between the ad hoc network and the infrastructure network. This interdomain signalling will be ensured by the AdIN boarder represented in the figure.

Finally the Charging and Billing (CB) service as represented in the figure could be offered by the infrastructure to the ad hoc nodes. It means that the infrastructure will be aware of the services exchanges between the ad hoc nodes and will charge the serviced ad hoc nodes for that. The servicing ad hoc node will get the payment for the service offered and will also pay the infrastructure network for supporting the CB on his behalf. As for the authentication (Aaa), authorisation (aAa) and accounting (aaA) services the infrastructure network will make profit on the usually not directly billed service which is here the CB.

4.1. Available Services Within Ad Hoc Nodes

There might be users in the ad hoc network offering services like in Peer to Peer (video, music records...). They might behave as very small operators offering at low cost contents Ad hoc level services such as peer-to-peer content exchange, packet forwarding, resource consumption...etc. In airport for instance, there might be also shops offering a number of services, free services like advertisement, charged services like download of pieces of music and so on. Furthermore, the classical AAA services could be offered by the infrastructure network to the ad hoc network and made profit from it.

In that sense, we differentiate in this architecture the network and user services. The network service is the service that the infrastructure network would offer to the ad hoc networks such as the authentication service needed by the ad hoc nodes. The user service is the service that an ad hoc node would offer to neighbour ad hoc nodes or to another ad hoc network such as ad hoc routing.

4.2. Neighbour and Service Discovery

Neighbour and Service discovery mechanisms are necessary to provide information for the ad hoc node regarding the service availability at the infrastructure boarder or inside the ad hoc network in order to allow him get the appropriate service.

The IETF has defined few protocols for service discovery. Service Discovery Protocol (SDP), and Service Location Protocol (SLP) [15] are designed for service discovery. SLP provides a scalable framework for the discovery and selection of network services. The protocol allows, with little or no static configuration, to discover network services for accessing network based application.

Another approach would be to use Web services (WS) based architecture. This permits to include a rich suite of specifications that provides complementary functions in the areas of security, reliability, and transaction-based messaging.

In the context of our AdIN architecture, the service discovery framework has to consider both user services (those offered by ad hoc nodes) and infrastructure services (those offered by the infrastructure network to the ad hoc network such as Aaa).

Special gateways could be used at the boarder of the ad hoc network to present the services that the infrastructure could offer to the ad hoc network and vice versa. The service discovery framework would be combined with the accounting and charging service which is tightly related to the AAA service.

4.3. User Identification and Anonymity

It is important to ensure user anonymity over the ad hoc network. So the servicing node and even spying nodes are not able to know who is connected and accessing some services. That is, the infrastructure provider must identify the user, but the ad hoc servicing nodes should not know the users identity. The only constraint for the servicing node is to be paid for the accessed services such as content delivery or ad hoc routing. As such, pseudonyms might be used within ad hoc networks while ordinary identification and authentication of the users to access the infrastructure is mandatory.

4.4. Authentication to the Infrastructure

The first 'A' in Aaa stands for authentication. It would be ideally distributed among ad hoc nodes. Due to the trust management problem, this would not be possible for now. However, the existing reliable and efficient Aaa system in the operator's network could be extended towards the ad hoc structure. Consequently, the extended Aaa service would be considered as a standalone service for ad hoc groups that would be billed for this authentication service.

The Aaa service will be offered by the infrastructure to the ad hoc network to ensure the deployment of services in the ad hoc network. The AAA service will operate in a self organized network (muti-hop) which is a challenging task. The PANA approach is a promising candidate for supporting Aaa service in a multi-hop context as it works over the IP level [13], and as such is independent of the underlying access technologies including ad hoc networks.

Moreover, an Inter-domain AAA will be necessary for the network operator to open its Aaa service to ad hoc subscribers from another operator. As such, any user might be able to join any ad hoc network and to ask for local available services.

4.5. Authorization and Accounting

aAA would be distributed within ad hoc nodes, and this is very challenging from a research point of view as there is a need to define how servicing ad hoc nodes are collecting accounting information of ad hoc customers. This accounting information would be used for supporting secure electronic transactions and charging within ad hoc nodes.

The Accounting service (aaA) will be adapted to services available in ad hoc network. The servicing ad hoc nodes will adopt a certain accounting policy per service (forwarded packets, service duration, and content value, bid...). This will solve the selfish behaviour of certain ad hoc nodes which is a very well known issue by encouraging them to participate in the ad hoc network communication.

4.6. AAA as a Basis for Securing Communications Between Ad Hoc Nodes

The idea is to benefit from the Aaa exchanges with the network operator to establish some security material (keys, algorithms...) within ad hoc networks and thus to ensure the protection of ad hoc communications. Some of the EAP authentication methods like EAP-TLS, EAP-TTLS... permit both authenticating parties to share a common key (called MSK for Master Session Key) at the end of EAP exchanges. In the context of ad hoc nodes, this EAP method might be very helpful for nodes to establish a common secret and use it to secure their connections. For instance, a group key might be pushed by the infrastructure to the ad hoc nodes. It is also possible for the infrastructure to generate and distribute attribute certificates to the ad hoc nodes.

4.7. Trust Management Within Ad Hoc Nodes

Trust management is necessary during the creation and the evolution of the ad hoc network. It is the fundamental premise to build a secured exchange of services where services such as authentication, authorisation, accounting, charging and billing are supported.

Trust management is mainly used to control that nodes are actively participating to the connectivity maintenance of the network, performing for instance packet forwarding. As secure routing protocols designed for ad hoc network are based on heavy cryptographic tools, deploying trust management might be advantageously used to ensure some control within ad hoc networks. Actually, trust management might be enough to ensure a certain security level without heavy crypto based protocols.

In an infrastructure network, network nodes (e.g. routers) are under the control of a certain administration (network operator, corporate network...), that is a strong trust basis. On the other hand, in an ad hoc network, there is no administration a priori to control the network. That is the reason why there is no intrinsic trust in an ad hoc network. It explains some security problems such as selfish behaviour of certain ad hoc nodes. When the ad hoc network can get access to the infrastructure network, it can benefit from a strong authentication offered by the infrastructure network and will then authenticate all the ad hoc nodes that will join the ad hoc network, thus ensuring the trust relation between all the ad hoc nodes.

The problem of trust creation and maintenance in ad hoc networks is worsen when this one cannot have access to a strong authentication like the one offered by the infrastructure network. There are few mechanisms that are dealing with this problem. For instance a reputation based mechanism which is very likely similar to trust building in a human community.

4.8. New Business Model

The resulted business model will be more complex with the interaction of four parties: operators, content and application/software providers, ad hoc nodes distributing contents to any nodes purchasing contents. The content provider's role will consist in providing the active ad hoc nodes with contents so a new market based on the peer-to-peer principle (with no deployment needs for the provider) is made possible. To bring new ad hoc node content distributors, remuneration of distributors may be envisioned.

Another point will be to introduce the application service provider role in the ad hoc network. However, the risk is high that the application providers are infected by a virus due to customers executing malicious programs. For such services to remain securely deployed, the ad hoc machine architecture should be carefully designed with disk partitioning for example.

Limited resource (UPC, bandwidth,...) networks like domestic networks might also benefit from the security AAA mechanisms defined in the AdIN approach. Domestic networks meet difficulties performing auto configuration, security initialization, and the AdIN approach may be of some help using the operator's access network as a third party. It might also be envisioned that videos digitally registered by individuals are billed to the neighbours including taxes for the authors' rights. However, there is a need to adapt AdIN mechanisms to such low resources devices.

5. FUTURE WORKS

As this paper introduces a new framework to allow the integration of ad hoc network with the infrastructure network, next step of our work is the implementation of the concepts introduced (AdIN boarder, decomposed AAA service,...) and the validation of the AdIN framework. In this architecture, the Aaa, and the aAA services will be distributed but will need somehow a certain control from the infrastructure network which is until now the sole trusted party that can guarantee the trueness of the information necessary for user authentication. The need for AAA distribution may happen when the ad hoc group is temporarily disconnected from the infrastructure and continuous service delivery is highly expected; especially in that case, the use of prepaid tools might be helpful for the serving service node to manage on its own the billing.

6. CONCLUSION

With the need for Always On, the coverage of current access networks infrastructures must be extended and the ad hoc network technology is an enabling technology for the always on technology at low cost, with auto configuration capabilities, network dynamicity management... In this new designed ad hoc/infrastructure architecture, new business oriented services will emerge within the ad hoc network with local ad hoc customers. However, until today, the ad hoc network technology is not finding its way to integrate the value chain of telecommunication. We propose in this paper to integrate the ad hoc technology in the service deployment value chain by decomposing the AAA service and executing it in the ad hoc network to ensure secured exchange of services. The infrastructure network would benefit from integrating the ad hoc technology in the access network since it will bring more users in the network. It will also benefit from converting the classically indirectly chargeable services such as authentication, authorisation, accounting, charging, billing,... etc directly chargeable to the ad hoc nodes using those services to build an ad hoc structure to exchange services.

REFERENCES

1. Levijoki S. (2000) Authentication, Authorization and Accounting in Ad Hoc networks, http://www.tml.tkk.fi/Opinnot/Tik110.551/2000/papers/authentication/aaa.htm#chap5
2. Zhou L., Haas Z. (1998), Securing Ad Hoc Networks, 1998. http://www.ee.cornell.edu/~haas/Publications/network99.ps
3. Housley R., Ford W., Polk W., Solo D.(1999), Internet X509 Public Key Infrastructure Certificate and CRL Profile, RFC 2459, January 1999.
4. Hu Y. C., Perrig A., Johnson D. B. (2002): Ariadne: A secure on-demand routing protocol for Ad Hoc networks, Proceedings of the 8th ACM International Conference on Mobile Computing and Networking, 2002.
5. Guerrero Zapatav M.(2002), Secure Ad hoc On-Demand Distance Vector Routing, ACM Mobile Computing and Communications Review (MC2R), vol. 6, no. 3, pp. 106–107, July 2002.
6. Luo H, Zerfos P., Kong J., Lu S., Zhang L. (2002) Self-securing Ad Hoc Wireless Networks, Seventh IEEE Symposium on Computers and Communications (ISCC'02), 2002.
7. Yi S., Kravets R. (2003), MOCA: Mobile Certificate Authority for Wireless Ad hoc Networks, in Proceedings of 2nd Annual PKI Research Workshop, NIST, Gaithersburg, MD, April 2003.
8. Boneh D., Franklin M. (2001) Identity-Based Encryption from the Weil Pairing. In J. Killian, editor, Advances in Cryptology, CRYPTO 2001, volume 2139 of Lecture Notes in Computer Science, pages 213-229. Springer Verlag, August 2001.
9. Khalili A., Katz J., Arbaugh W.A. (2003), Towards secure key distribution in truly ad hoc networks, IEEE Workshop on Security and Assurance in Ad hoc Networks, 2003.
10. IEEE Standard 802.1X-2004, Standard for Local and Metropolitan Area Networks: Port-Based Network Access Control, December 2004.
11. Lopez R. Marin, Bournelle J., Combes J.-M., Laurent-Maknavicius M., Gomez Skarmeta A. F. (2006), "Improved EAP keying framework for a secure mobility access service", International Wireless Communications and Mobile Computing Conference IWCMC 2006, Published in ACM Digital Library, Conference, Vancouver, Canada, July 2006.
12. Bournelle J., Laurent-Maknavicius M., Giaretta G., Guardini I., Demaria E., Marchetti L (2005)., Bootstrapping Mobile IPv6 using EAP, Joint IEEE Malaysia International Conference on Communications and IEEE International Conference on Networks, MICC-ICON 2005, Lumpur, Malaysia, November 2005.

13. Cheikhrouhou O., Laurent-Maknavicius M., Chaouchi H. (2006), Security architecture in a multi-hop mesh network, 5^{éme} conférence sur la Sécurité et Architectures Réseaux SAR 2006, Seignosse, Landes, France, juin 2006.
14. Parthasarathy M. (2005), Protocol for Carrying Authentication and Network Access (PANA) Threat Analysis and Security Requirements, RFC 4016, March 2005.
15. Service Location Protocol: http://www.openslp.org, IETF–Service Location Protocol: http://www.ietf.org/html.charters/OLD/svrloc-charter.html

CHAPTER 36

HIDING USER CREDENTIALS DURING THE TLS AUTHENTICATION PHASE

MOHAMAD BADRA[1], IBRAHIM HAJJEH[2], AND JACQUES DEMERJIAN[2]

[1] *LIMOS Laboratory, UMR 6158, CNRS France, badra@isima.fr*

[2] *ESRGroups, Security WG – France {Ibrahim.Hajjeh, Jacques.Demerjian}@esrgroups.org*

Abstract: TLS (Transport Layer Security) defines several ciphersuites providing authentication, data protection and session key exchange between two communicating entities. Some of these ciphersuites are used for completely anonymous key exchange, in which neither party is authenticated. However, they are vulnerable to man-in-the-middle attacks and are therefore deprecated. This article defines a set of ciphersuites to add client credential protection to the TLS protocol. This protection is essential in wireless infrastructures, in which it guaranties user's privacy and makes exchanges untraceable to eavesdroppers. We compare our proposition in terms of performance and cost to an ordinary TLS session

Keywords: Transport Layer Security (TLS), Public Key Infrastructures, and identity protection

1. INTRODUCTION

TLS [1] is widely deployed as a security protocol for securing exchanges. It provides end-to-end secure communications between two entities with authentication and data protection. TLS supports three authentication modes: authentication of both parties (if the client is able to present and to prove its ownership of a certificate), only server-side authentication, and anonymous key exchange. For each mode, TLS specifies a set of ciphersuites. However, anonymous ciphersuites are strongly discouraged because they cannot prevent man-in-the-middle attacks [1].

Client credential protection may be established by changing the order of the messages that the client sends after receiving ServerHelloDone [7]. This is done by sending the ChangeCipherSpec message before the Certificate and the CertificateVerify messages and after the ClientKeyExchange message. In other words, the client and the server start applying the negotiated ciphersuite, the session

H. Labiod and M. Badra (eds.), New Technologies, Mobility and Security, 437–445.
© 2007 *Springer.*

shared key and the encryption/decryption operations before sending date related to the client certificate. Unfortunately, this solution requires a major change to TLS machine state as long as a new TLS version.

Client credential protection may also be done through a DHE (Diffie-Hellman Ephemeral) exchange before establishing an ordinary handshake with credential information [5]. This would not however be secure enough against active attackers and would not be favorable for some environments (e.g. mobile), due to the additional cryptographic computations.

Client credential protection may be also possible, assuming that the client permits renegotiation after the first server authentication. However, this requires more cryptographic computations and increases significantly the number of rounds trips.

Also, client credential protection may be realized by exchanging a TLS extension that negotiates the symmetric encryption algorithm to be used for client certificate encrypting/decrypting [9]. This solution may suffer from interoperability issues related to TLS Extensions, TLS 1.0 and TLS 1.1 implementations, as described in [8].

This paper defines a set of ciphersuites to add client credential protection to TLS protocol. Client credential protection is provided by symmetrically encrypting the client certificate with a key derived from the:

- SecurityParameters.master_secret,
- SecurityParameters.server_random, and
- SecurityParameters.client_random.

The symmetric encryption algorithm is set to the cipher algorithm of the ServerHello.cipher_suite.

This paper is constructed as follows. Section 2 describes TLS and its integration to different contexts, such as wireless networks. Next in section 3, we discuss related works and in section 4 we present our solution to add credential protection to TLS. In section 5, we analyze our protocol and we discuss results obtained with an implementation of our proposition using OpenSSL and we compare its performance and cost to an ordinary TLS session. Finally, we give some concluding remarks.

2. THE TRANSPORT LAYER SECURITY PROTOCOL

TLS is a transaction security standard that provides connection security with the following basic proper-ties: mutual authentication, integrity-protected parameter negotiation and key exchange between two entities. This protocol consists of several sub-protocols, especially the Record and the Handshake protocol. The Record protocol provides basic connection security for various higher layer protocols through encapsulation. The Handshake protocol is used to allow peers to agree upon security parameters for the Record layer, authenticate themselves, instantiate negotiated security parameters, and report error conditions to each other. This protocol is used to negotiate the security attributes of a session. Once a transport connection is authenticated and a secret shared key is established with the TLS

Handshake protocol, data exchanged by application protocols can be protected with cryptographic methods by the Record layer using the keying material derived from the shared secret.

We illustrate TLS Handshake in three phases: the hello phase, the client authentication phase, and the finished and server authentication phase. During the hello phase, the client and the server negotiate the cryptographic option (asymmetric and symmetric encryption algorithm, hash function, key exchange method, etc.) and exchange two random values that will be used by the key computation process. The second phase consists of exchanging certificates and of proving the identity and the validity of these certificates. During this phase, the client generates a secret called pre_master_secret, which is sent encrypted using the server public key and applies its signature on handshake messages to prove its certificate ownership. During the third phase, the client and the server exchange the change cipher spec and the finished messages. The "change cipher spec" message is sent by both the client and the server to notify the receiving party that subsequent records will be protected under the newly negotiated cipher spec and keys. The finished message is immediately sent after the change cipher spec message to verify that the key exchange and authentication processes were successful. The finished message is the first message that is protected using the negotiated algorithms by the Record sub-protocol. When the client receives the server finished message, it will implicitly authenticate the server. In fact, the finished message is MACed [1] and encrypted using keys derived from the master secret, which is computed using the pre_master_secret. This latter is sent encrypted using the server public key, in which only the server is able to retrieve it in clear text.

2.1. TLS Deployment

TLS is being integrated in many infrastructures and environments due to its simplicity, extensibility and its security robustness. Originally, TLS has been designed to run over a reliable transport such as TCP. Due to the increasingly number of applications that are designed to operate over unreliable link such as UDP, Datagram TLS [10] has been developed and is actually under normalization by IETF. DTLS is designed to be as similar to TLS as possible; both to minimize new security invention and to maximize the amount of code and infrastructure reuse [10].

Actually, TLS is the most widely deployed protocol for authenticating WLAN entities. WLAN and 802.11 networks use EAP (Extensible Authentication Protocol) [6], which is a powerful umbrella that shelters multiple authentication methods and use PPP (Point to Point Protocol) [4] for transporting multi-protocol datagrams. However, WLAN and 802.11 do not indicate how EAP should be implemented within WLAN authentication frameworks. Consequently, many security protocols over EAP has been specified by IETF, in which majority of them are based on TLS; especially EAP-TLS [2], EAP-TTLS [11] and PEAP [13].

On the other hand, mobile communications are usually authenticated using mechanism based on the use of pre-shared keys (PSK). Nowadays, mobile and telecommunication communities are extremely interested in the use of shared secrets within TLS; particularly 3GPP2, which has agreed to adopt the TLS shared key concept as one of the authentication methods for 3GPP2-WLAN interworking. Consequently, we introduced the TLS-PSK standard protocol [3] and TLS Express [12] that support authentication based on PSK. This is because the use of PSK instead of certificate based-authentication could be an alternative when TLS is used in performance-constrained environments where limited CPU and bandwidth are the bottleneck.

However, the use of TLS in wireless LAN and 4G must not break the client's privacy, which is an elementary service for these architectures. Unfortunately, actual TLS specifications exchange certificate in clear text and then do not provide the ability for a client to negotiate credential protection without being re-negotiate the security parameters in an existing connection. Consequently, an eavesdropper is able to identify the communicating parties, since secret identifiers or certificates are sent in clear text. Thus, we propose to add credential protection to TLS, emphasizing interoperability, re-use and backward-compatibility with the existing TLS implementation.

3. RELATED WORKS

This section discusses other works related to credential protection integration to TLS. Current TLS specifications provide anonymous key exchange, allowing two entities to perform data encryption. In this case, the entities will not exchange their certificates and therefore, authentication or indication of the entity's credential will not be possible. Consequently, an intruder can easily perform a man-in-the-middle attack, intercepting the session exchanges and sends its public key instead of the server or the client ones. In fact, the attacker sniffs packets conveying entity public key, and then replaces it with its public key before sending the modified packets to the receiver. According to TLS, the client generates a pre_master_secret, which will be encrypted with the unauthenticated public key of the intruder. This latter will retrieve the pre_master_secret in clear text since it has the corresponding private key. Next, the intruder encrypts the pre_master_secret using the server public key.

Rescorla [5] proposed to resolve such attack by initiating a new TLS session once the anonymous TLS session has been established. The main idea of that approach is to re-handshake a TLS session with mutual authentication based on the use of certificates. All the messages of the second session, including certificates, are then encrypted using keys derived from the first session. Even though this solution provides anonymous exchanges, it increases latency and reduce throughput. In fact, asymmetric encryption/decryption operations are repeated during the re-handshake and that the re-handshake messages are symmetrically encrypted, which significantly increase the processing time on both client and server. Moreover, the number of re-handshake messages requires several round trips. Note that anonymous key exchange methods require the generation of ephemeral keys. Such a generation is too expensive to do; especially when entities repeat it for each session. Note also that this solution is similar to the case when the client and the server renegotiate their security parameters in an existing connection.

The third solution is proposed at the IETF-TLS Working Group meeting in Pittsburgh and discussed on the TLS mailing list [7]. It consists of changing the order of the TLS messages in a way the client sends the "change cipher spec" message before its "certificate" message. The "change cipher spec" is sent by the client to notify the receiving party that subsequent messages will be protected under the derived keys. That way the certificate is sent protected. However, TLS WG rejected this solution because it requires the definition of new version for TLS and it clutters up the TLS state-machine.

The client credential protection can be realized by exchanging a TLS extension that negotiates the symmetric encryption algorithm to be used for client certificate encrypting/decrypting [9]. As we introduced, this solution may suffer from interoperability issues related to TLS Extensions, TLS 1.0 and TLS 1.1 implementations, as described in [8]. Moreover, this solution helps in guessing the connected. In fact, the CertificateVerify conveys data related to the client public key signed by the client, including its public key. Since this message is send in clear text, an eavesdropper could easily deduce the client identity.

In [21], a review of existing anonymous authentication protocols and an alternative approach are presented. However, in this paper we only focus on the credential protection during their exchange over the network.

4. CREDENTIAL PROTECTION WITH TLS

We propose an enhanced way to avoid sending certificates or secret identifiers in clear text during the TLS handshake phase. Therefore, we specify a set of ciphersuites for TLS. These ciphersuites reuse existing key exchange algorithms that require based-certificates authentication, and reuse also existing MAC, stream and bloc ciphers algorithms from [1] and [16–19]. Their names include the text "CCP" to refer to the client credential protection. An example is shown below:

$$CipherSuite: TLS\_CCP\_RSA\_EXPORT\_WITH\_RC4\_40\_MD5$$

Key Exchange: RSA; Cipher: RC4_40; Hash: MD5

If the client has not a certificate with a type appropriate for one of the supported cipher key exchange algorithms or if the client will not be able to send such a certificate, it must not include any ciphersuite with client credential protection in the ClientHello.cipher_suites.

If the server selects a ciphersuite with client credential protection, it must request a certificate from the client. If the server selects one of the ciphersuites defined in this document, the client has to encrypt its TLS Certificate and CertificateVerify messages using the symmetric algorithm selected by the server from the list in ClientHello.cipher_suites and a key derived from the SecurityParameters.master_secret. For DHE key exchange method, the client always sends the ClientKeyExchange message containing its Yc (Diffie-Hellman public key value [2]).

In the case of DH_DSS and DH_RSA key exchange methods, the client certificate may already contain a suitable Diffie-Hellman key. Then the DHE public value (Yc) is implicit and does not need to be sent again. In this case, the client credential protection cannot be provided unless the client sent ClientKeyExchange carrying its Yc. On the other hand, the client certificate may not contain a suitable Diffie-Hellman key, and therefore the client has to send a ClientKeyExchange containing its Yc. In both implicit and explicit cases, the client has to then send a ClientKeyExchange message carrying its Yc but however, it is possible to correlate sessions by the same client. Consequently, the current document does not include the use of DH_DSS and DH_RSA key exchange methods.

Before sending its certificate, the client is able to compute the master secret and then the key_block. Thus, the client and the server derive from the key_block a key called credential_protection_key, with value set to the client_write_key (see [1] for key derivation and computation). This key is deployed by the client (respectively the server) to encrypt (respectively decrypt) the client's Certificate and the CertifateVerify messages.

Upon receiving the "client key exchange" message, the server decrypts it using its private key to retrieve the pre_master_secret. Next, the server computes the master_secret, and then it derives the encryption key in the same way as derived by the client. Afterwards, the server decrypts the Certificate and the CertificateVerify messages and verifies the validity of the client certificate. Next, the client and the server continue their session as shown in section 2.

5. DISCUSSION AND RESULTS

As we previously mentioned it, credential protection is a critical privacy issue; especially in wireless and mobile environment. Although it is widely use in today Wi-Fi networks, TLS does not support credential protection features, and therefore a malicious hacker can initiate an authentication session, and remotely collect users' identities.

The credential protection design is completely backwards compatible with the existing TLS implementation since it does not require any major changes to the TLS machine-state. In fact, TLS clients and servers can negotiate the credential protection service using the extra ciphersuites defined by this paper. Thus, TLS clients that support these extra ciphersuites can communicate with TLS servers that do not, and vice versa.

TLS defines a set of key exchange methods based on RSA and DH (Diffie-Hellman) algorithms [1]. When using a static DH based key exchange method (DH_DSS or DH_RSA), the ciphersuites introduced through this paper must not be used. In this case, the client certificate conveys the client DH parameters that will allow the client and the server to agree upon the same pre_master_secret. This latter will be used to compute the same symmetric encryption key. Consequently, if the client certificate is encrypted, the server will not be able to recover the client DH public key. It should be noted that this is not a major problem, since client certificates containing DH public keys are rarely used.

During the ordinary TLS authentication phase, the client executes the following operations:

- pre-master-secret encryption or computation, using the server public key,

On the server side, the following operations are executed

- pre-master-secret decryption or computation, using the server private key,

During the TLS authentication phase with the client credential protection, the client and the server executes the above operations and in addition, the client symmetrically encrypt its TLS Certificate and CertificateVerify messages, in which the server decrypts them using the same key.

Consider that the client and the server have negotiated the use of RSA-1024 as a key exchange method, DES as a cipher algorithm, and SHA-1 as a hash function. Consider also the client and the server execute TLS with mutual authentication based-certificate, which requires the mandatory submission of the client certificate over the wire

In order to evaluate the performance of our proposed protocol and to have a proof of concept, an implementation of our proposition was carried out using OpenSSL [15] tools. The implementation is tested and deployed in a Wi-Fi environment, whose security architecture is built according to the well-known IEEE 802.1X [14] standard. The TLS authentication within EAP/802.1X is quite straightforward. The TLS Handshake packets are encapsulated in an appropriate EAP form and transported between the station and the authentication server. Based on the successful TLS authentication dialog, the client will or will not get access to the network services.

In return to our benchmark, we performed it on a 1.2 GHz Intel processor PC. The throughput values used in estimating the execution time of EAP-TLS sessions (including EAP association [6]) with and without client credential protection are summarized in the table 1.

As an illustration, we obtain the following results for the client and the server (see Table 1). We repeated the test 50 times and we observed that the time for the TLS session execution without client credential protection is around 1.471 sec (we suppose that the network latency is the same for both TLS with and without

Table 1. TLS execution time with and without client credential protection

	Ordinary EAP-TLS	EAP-TLS with credential protection
Time	1.471 sec	1.701 sec

credentials protection and therefore it is ignored by our comparison). Credential protection adds an extra cost of about 0.23 sec, which is induced by two additional cryptographic operations: the calculation of a 128 bits encryption key using the PRF function, and the DES encryption/decryption of the client certificate. For more information on TLS performance evaluation, please refer to [20].

6. CONCLUSION

In this paper, we extended the famous TLS protocol with a new mechanism to protect client credentials and to make exchanges untraceable to eavesdroppers. Based on our experimental results, we demonstrated that the extra time required by our extension in comparison to a traditional TLS session does not exceed 0.25 second. Therefore, extra operations introduced by this paper have inconsiderable influence on the global performance of TLS, in comparison to other solutions based on tunnelling TLS or on TLS renegotiation that require double TLS session within encryption of the whole second one.

REFERENCES

1. T. Dierks, C. Allen, "The TLS Protocol Version 1.0", RFC 2246, January 1999.
2. B. Aboba, D. Simon, "PPP EAP TLS Authentication Protocol", RFC 2716, October 1999.
3. P. Eronen, Editor, et. al., "Pre-Shared Key Ciphersuites for Transport Layer Security (TLS)", RFC 4279, December 2005.
4. W. Simpson, Editor, "The Point-to-Point Protocol (PPP)", STD 51, RFC 1661, July 1994
5. E Rescorla, "SSL and TLS: Designing and Building Secure Systems", Addison-Wesley, March 2001.
6. B Aboba, et. al., "PPP Extensible Authentication Protocol (EAP)", RFC 3748, June 2004.
7. F. Corella, "adding client identity protection to TLS", message on ietf-tls@lists.certicom.com mailing list, http://www.imc.org/ietf-tls/mail-archive/msg02004.html, August 2000.
8. Y. Pettersen, "Clientside interoperability experiences for the SSL and TLS protocols", draft-ietf-tls-interoperability-00 (work in progress), October 2006.
9. P. Urien and M. Badra, "Identity Protection within EAP-TLS", draft-urien-badra-eap-tls-identity-protection-01.txt (work in progress), October 2006.
10. E. Rescorla and N. Modadugu, "Datagram Transport Layer Security", IETF Internet Draft, June 2004.
11. P. Funk, et. al., "EAP Tunneled TLS Authentication Protocol (EAP-TTLS)", IETF Internet draft (work in progress), August 2004.
12. M. Badra, et. al., "TLS Express", IETF Internet Draft, February 2005.
13. S. Josefsson, et. al., "Protected EAP Protocol (PEAP)", IETF Internet draft (work in progress), October 2005.
14. Institute of Electrical and Electronics Engineers, "Local and Metropolitan Area Networks: Port-Based Network Access Control", IEEE Standard 802.1X, September 2001.
15. OpenSSL WebSite, http://www.openssl.org.
16. RFC 4132, "Addition of Camellia Cipher Suites to Transport Layer Security (TLS)", July 2005.
17. RFC 3268, "Advanced Encryption Standard (AES) Ciphersuites for Transport Layer Security (TLS)", June 2002.
18. RFC 4492, "Elliptic Curve Cryptography (ECC) Cipher Suites for Transport Layer Security (TLS)", May 2006.
19. N. Modadugu, E. Rescorla, "AES Counter Mode Cipher Suites for TLS and DTLS", draft-ietf-tls-ctr-01.txt (work in progress), June 2006.

20. N. Sklavos, P. Kitsos, K. Papadopoulos, O. Koufopavlou, "Design, Architecture and Performance Evaluation of the Wireless Transport Layer Security (WTLS)", Journal of Supercomputing, Springer-Verlag, Vol. 36, No 1, 2006.
21. P. Persiano, I. Visconti, "A secure and private system for subscription-based remote services", ACM Transactions on Information and System Security (TISSEC), Volume 6 Issue 4, November 2003.

CHAPTER 37

ENHANCED SMART-CARD-BASED AUTHENTICATION SCHEME PROVIDING FORWARD-SECURE KEY AGREEMENT*

MAHDI ASADPOUR, BEHNAM SATTARZADEH, AND RASOOL JALILI

Department of Computer Engineering, Sharif University of Technology, Tehran, Iran
{asadpur@ce.,sattarzadeh@ce.,jalili@}sharif.edu

Abstract: Many smart-card-based remote authentication schemes have been proposed recently. In 2004, Yoon et al. presented an improved scheme which is the leading of a research track started from Sun, 2000. In this paper, we illustrate that Yoon et al.'s scheme is vulnerable to the parallel session attack and propose an enhancement of the scheme to resist that attack. In our scheme the parties further establish a forward-secure session key by employing only hash functions to protect the subsequent communications. We also demonstrate that our scheme has better security in comparison to other related works, while it does not incur much computational cost

Keywords: Authentication, Parallel session attack, Forward secrecy

1. INTRODUCTION

A password-based remote authentication scheme is the most widely used (and maybe the simplest) mechanism that authenticates a remote user to a server over an insecure communication. If adequate protection is not provided to the authentication protocol, an attacker, who has full control of all communication links, can launch several attacks such as forgery attack, server spoofing attack, and password guessing attack.

Since Lamport [10] presented his remote user authentication scheme based on one-way hash functions in 1981, several protocols have been proposed to improve

*This work is partially supported by Iran Telecommunication Research Center (ITRC), under grant No. 500/8478.

H. Labiod and M. Badra (eds.), New Technologies, Mobility and Security, 447–458.
© 2007 *Springer.*

its security, efficiency, and functionality. The server in Lamport's scheme needs to keep a password table for verifying the legitimacy of users. Hwang and Li [6], 2000, pointed out that this scheme suffers from the risk of modified password table and the cost of protecting and maintaining it. Moreover, they proposed an authentication scheme using smart cards based on public key cryptography which does not require the server to maintain a password table.

Besides the advantages of smart cards such as having some memory to store a long secret key and also a processor to perform cryptographic operations internally without exposing the secret information to outside, they have limited computational capabilities. Therefore, high computation cost operations such as public key cryptography of Hwang and Li's scheme are not suitable for this environment. In this respect, Sun [14] improved Hwang and Li's scheme, by using low computation cost (one-way hash) functions without employing the password table.

In 2002, Chien et al. [2] mentioned that the user in Sun's scheme does not freely choose his password and cannot authenticate the server. They further proposed a scheme to satisfy these capabilities. In 2004, Juang [8] integrated another feature into Chien et al.'s scheme. In his scheme, when mutual authentication is realized between the user and the server, they agree on a session key to be used for protecting their subsequent communications. Shieh and Wang [12], 2006, eliminated the cost of symmetric cryptography of Juang's scheme and also enhanced its security.

As Juang mentioned in his paper [8], one of the main criteria for session key agreement is provision of forward secrecy. The forward secrecy means that disclosure of long-term secret keys does not lead to exposure of session keys from earlier runs [5]. Neither Juang nor Shieh and Wang can satisfy this requirement, because each session key in their scheme only depends on the user's secret key and some other public parameters. So, an attacker knowing a compromised secret key can re-compute previous session keys. To the best of our knowledge, the smart-card-based remote authentication schemes with forward-secure key agreement (e.g. [11]), generally utilize the Diffie-Hellman key exchange algorithm [4] which requires costly exponential operations and contradicts with the low computational power of smart cards.

Ku and Chen [9], 2004, showed that Chien et al.'s scheme is vulnerable to the reflection attack. They presented an improved scheme that not only resists that weakness, but also is reparable once a user's permanent secret key is compromised. Recently, Yoon et al. [15] pointed out that the enhanced scheme is susceptible to the parallel session attack, and also its password changing phase is insecure. Accordingly, they proposed an improvement to resolve such weaknesses.

In this paper, we illustrate that Yoon et al.'s scheme still suffers from the parallel session attack. Then we modify their scheme in order to defend against that attack. We also make use of an efficient forward-secure key agreement scenario and apply this to the resulting scheme. We subsequently analyze the security and computation cost issues of our scheme compared to related works. These comparisons indicate the superiority of our scheme, as well.

The rest of this paper is organized as follows. In the following section, we review the Yoon et al.'s scheme. In section 3, we illustrate that their scheme is insecure against the parallel session attack. In section 4, an improved scheme is proposed to overcome this attack and provide session key agreement with forward secrecy, followed by its security and computation cost analyses. Finally, a concluding remark is given in section 6.

The following notations are used throughout this paper: S and U are the server and the user; ID and PW are identity and password of U; x is the permanent secret key of S; $h()$ is a one-way hash function; T_A is the timestamp of A; $\Rightarrow$ indicates a secure channel; $\rightarrow$ indicates a public channel; $\oplus$ and $\|$ are the exclusive-or, and the concatenation operators.

2. REVIEW OF YOON ET AL.'S SCHEME

In this section, we review the first three phases of Yoon et al.'s remote user authentication scheme. The fourth phase (*password change*) does not have any role in the proposed attack (next section).

2.1. Registration

Whenever U initially registers or re-registers to S, this phase is executed. In the following, the number of times when U re-registers to S is denoted by n.

1. U computes $h(b \oplus PW)$, where b is a random number.
2. $U \Rightarrow S: ID, h(b \oplus PW)$, while U uses a secure channel.
3. If it is the first time that U is registered by S, then S stores $n = 0$ in the account database of this user. Otherwise, S sets $n = n + 1$ in the existing entry for U and performs the following computations:

$$EID = (ID \parallel n)$$
$$V = h(EID \oplus x)$$
$$R = V \oplus h(b \oplus PW).$$

4. S issues a smart card containing V, R and $h()$ for U.
5. $S \Rightarrow U$: the smart card.
6. Upon receipt of the smart card, U enters the random number b into it.

2.2. Login

This phase is invoked whenever U wants to login S.

1. U inserts his smart card into a smart card reader and enters ID and PW.
2. The smart card computes c_1 and c_2, where T_U denotes U's current timestamp:

$$\begin{aligned} c_1 &= R \oplus h(b \oplus PW) \\ &= (V \oplus h(b \oplus PW)) \oplus h(b \oplus PW) \\ &= V = h(EID \oplus x) \\ c_2 &= h(c_1 \oplus T_U). \end{aligned}$$

3. $U \rightarrow S : ID, T_U, c_2$.

2.3. Verification

When the authentication request message $\{ID, T_U, c_2\}$ is received by S, the following verification steps are executed:

1. S checks ID and T_U to be valid and $T_U \neq T_S$, if not S rejects U's login request. The remote server S does $T_U \neq T_S$ checking to prevent the parallel session attack of Ku and Chen's scheme. Then S computes $c_1 = h(EID \oplus x)$ and $c_2 = h(c_1 \oplus T_U)$, and compares c_2 with the received one. If the comparison succeeds, S accepts U's login request and computes $c_3 = h(c_1 \oplus T_S)$, where T_S represents S's current timestamp.
2. $S \rightarrow U : T_S, c_3$.
3. In user side, U's smart card checks the validity of T_S and $T_S \neq T_U$. Next it computes $h(c_1 \oplus T_S)$ and compares the result with the received c_3. If equals, U authenticates S and the verification phase is successfully done.

3. PARALLEL SESSION ATTACK

In this section, we present a parallel session attack against Yoon et al.'s scheme. A parallel session attack occurs when two or more protocol runs are executed concurrently and messages from one run (the reference session) are used to form spoofed messages in another run (the attack session) [3]. So, we should prove that an intruder can login to the system by creating a valid login message from eavesdropped communications.

Suppose a valid user U sends the login request $\{ID, T_U, c_2\}$ to S, where T_U is the current timestamp of the user. If the request message is valid, U is authenticated and S responds $\{T_S, c_3\}$ to U, where T_S is its current timestamp.

Now, an attacker without knowing the user's password intercepts the above message. He sends $\{ID, T_S, c_3\}$ message back to S to start a new session, of course before the expiration of timestamp T_S. Login message $\{ID, T_S, c_3\}$ will pass the user authentication in S because:

1. ID and T_S are valid,
2. the current timestamp of the server T_S^* is different from its previous one sent to U $(T_S \neq T_S^*)$ and,
3. the computed result $h(c_1 \oplus T_S)$ equals the received c_3, which was originally generated by the server.

Generalizing the above attack, an adversary can pretend to be any legal user. Hence, Yoon et al.'s scheme cannot achieve the security requirement as they claim.

4. THE PROPOSED SCHEME

In this section, we explain our proposed scheme as an improvement on Yoon et al.'s scheme to protect against the mentioned parallel session attack. We then provide the session key agreement with forward secrecy in our scheme.

Like Yoon et al.'s scheme, the proposed scheme consists of four phases. But before their description starts, let highlight the main changes that our scheme makes.

4.1. Main Changes

In Yoon et al.'s scheme, due to the same structure of messages exchanged between the user and the server, the parallel session attack could succeed. Our scheme does change the structure and resist that attack.

Similar to the idea of Bellare and Yee [1] in utilizing hash functions to generate forward-secure random numbers, the forward secrecy in our scheme is achieved based on the one-way property of hash function $h()$. This property implies that given a message m, it is easy to compute its digest $d = h(m)$, but having the digest d in hand, it is hard to find m such that $m = h^{-1}(d)$ [13].

We take advantage of this one-way nature to make the past session keys irreversible. In this regard, S maintains a forward-secure parameter called FSP for each U in its database. The same parameter is also stored on the U's smart card. Every FSP is used in the generation of one session key, and after successful execution of the login and verification phases, it will be updated to a hash function of itself in both sides (S and U). The next forward-secure parameter takes part in the generation of the next session key and so on.

Now, if the FSP parameter of a user gets compromised, it may lead to disclosure of the current session key. But none of the previous session keys are revealed, because $h^{-1}(FSP)$ is unavailable and the previous FSP parameters cannot be re-computed. Therefore, the forward secrecy of session keys is maintained.

4.2. Registration

Registration phase depicted by Fig. 1 is executed whenever U registers or re-registers to S. The detail of this phase is as follows:

1. U selects his ID, PW, and a random number b. Then he generates $Mask = h(b)$ and computes $Mask \oplus PW$. $Mask$ is used to hide the user's password from the server.
2. $U \Rightarrow S: ID, Mask \oplus PW$.
3. S generates a random forward-secure number FSP_1 and sets $FSP_0 = 0$. If it is the first time of U's registration, S inserts the new record (ID, $n = 0$, FSP_0,

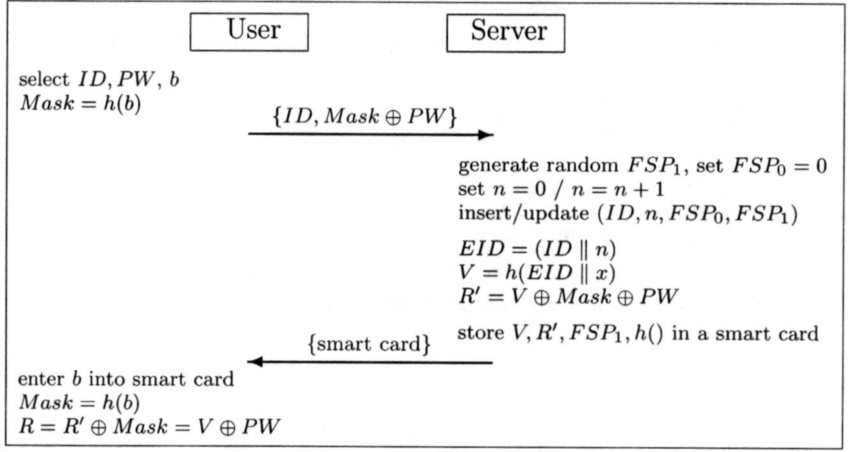

Figure 1. The proposed registration phase

FSP_1) in the account database, else it updates the corresponding record of U to (ID, $n = n + 1$, FSP_0, FSP_1). Here, n represents the number of times that U re-registers to S.

4. S performs the following computations:

$$EID = (ID \parallel n)$$
$$V = h(EID \parallel x)$$
$$R' = V \oplus Mask \oplus PW.$$

5. S issues a smart card containing V, R', FSP_1 and $h()$ for U.
6. $S \Rightarrow U$: the smart card.
7. U enters b into the received card. First the smart card computes:

$$Mask = h(b)$$
$$R = R' \oplus Mask$$
$$= (V \oplus Mask \oplus PW) \oplus Mask = V \oplus PW.$$

Then it replaces R' with R, so the smart card includes (V, R, FSP_1).

4.3. Login

This phase is invoked whenever U wants to login S (Fig. 2). Assume that initially the smart card stores (V, R, FSP_i), and the record of U in the account database is (ID, n, FSP_{i-1}, FSP_i).

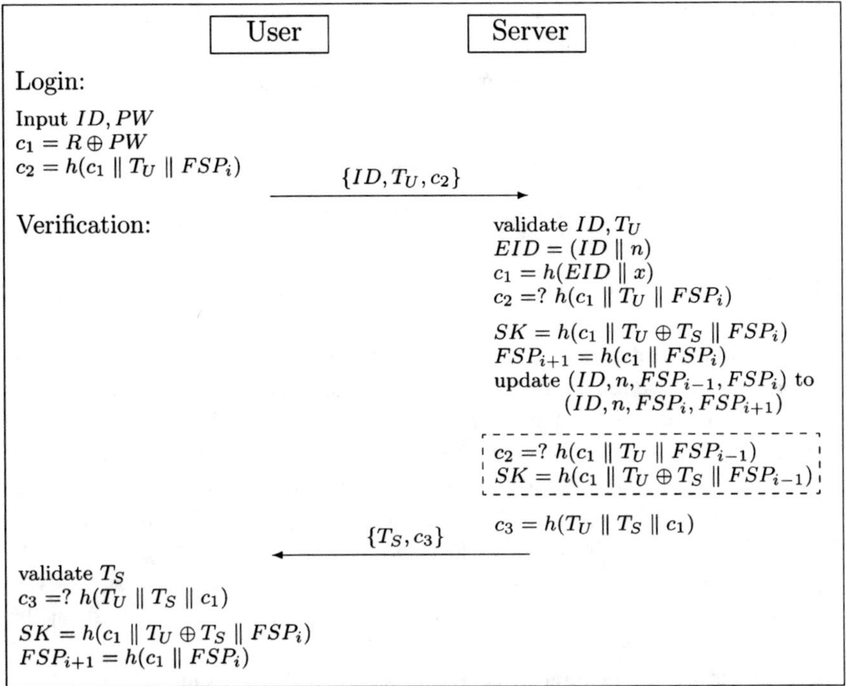

Figure 2. The proposed login and verification phases

1. U inserts his smart card into a smart card reader and enters his ID and PW. Using T_U, U's current timestamp, the smart card performs:

$$\begin{aligned} c_1 &= R \oplus PW \\ &= (V \oplus PW) \oplus PW \\ &= V = h(EID \parallel x) \\ c_2 &= h(c_1 \parallel T_U \parallel FSP_i). \end{aligned}$$

2. $U \rightarrow S: ID, T_U, c_2$.

4.4. Verification

After the authentication request message $\{ID, T_U, c_2\}$ is received, S and U execute the following operations:

1. S rejects U's login request if either ID or T_U is invalid, else it retrieves the corresponding record of U $(ID, n, FSP_{i-1}, FSP_i)$ and computes $EID = (ID \parallel n)$, $c_1 = h(EID \parallel x)$, and $c_2 = h(c_1 \parallel T_U \parallel FSP_i)$. Next S compares c_2 with the

received one. If equals, S authenticates U as the only party that knows c_1 (and FSP_i) thus can produce this term.
2. A session key and the next forward-secure parameter are calculated:

$$SK = h(c_1 \parallel T_U \oplus T_S \parallel FSP_i)$$
$$FSP_{i+1} = h(c_1 \parallel FSP_i)$$

where T_S denotes S's current timestamp. The record of U is also updated to (ID, n, FSP_i, FSP_{i+1})
There is one special case that can happen during the verification of c_2. If the previous execution of login and verification phases was not successfully finished, and U did not update the FSP_{i-1} to FSP_i while S did, then the forward-secure parameters would run out of synchronization. To remedy this problem and make U and S synchronized, S always stores the previous forward-secure parameter (FSP_{i-1}) along with the current one (FSP_i) is its database. Whenever the verification of c_2 by using FSP_i fails, it will be checked again by using FSP_{i-1} as shown in the dashed box of Fig. 2. In this case the session key is also computed through FSP_{i-1}, and no update operation is performed on the forward-secure parameter.
3. After successful verification of c_2 and generation of SK (using either FSP_i or FSP_{i-1}), S computes $c_3 = h(T_U \parallel T_S \parallel c_1)$ as its response to U.
4. $S \rightarrow U: T_S, c_3$.
5. If T_S is invalid, U terminates this session. U computes $h(T_U \parallel T_S \parallel c_1)$ and compares it with the received c_3. If the comparison succeeds, U authenticates S because c_1 is the shared secret key between them and no adversary without the knowledge of c_1 is able to generate a valid c_3.
6. The smart card computes $SK = h(c_1 \parallel N_U \oplus N_S \parallel FSP_i)$, then generates the next forward-secure parameter $FSP_{i+1} = h(c_1 \parallel FSP_i)$ and stores it in place of FSP_i which is not further needed.

Now, login and verification phases are successfully done, and U can securely communicate with S utilizing the established session key SK. Finally the smart card contains (V, R, FSP_{i+1}) and the account database maintains (ID, n, FSP_i, FSP_{i+1}).

4.5. Password Change

This phase is invoked when U wants to change his password and involves:
1. U enters ID and PW, and requests to change password.
2. The smart card computes $V^* = R \oplus PW$ and verifies whether it is equal to the stored V. If succeeds, U enters his new password PW_{new}.
3. The smart card generates R_{new} as $V^* \oplus PW_{new}$ and replaces R with R_{new}.

Table 1. Comparison of security aspects

Scheme\Feature	F_1	F_2	F_3	F_4	F_5	F_6	F_7	F_8
Sun [14]	No	No	No	No	No	No	No	Yes
Chien et al. [2]	No*	No	No	Yes	No	No	No	Yes
Juang [8]	Yes	Yes	No	Yes	No	No	No	Yes
Shieh and Wang [12]	Yes	Yes	No	Yes	No	No	No	Yes
Liao et al. [11]	Yes	Yes	Yes	Yes	No	No	Yes	Yes
Ku and Chen [9]	No*	No	No	Yes	No	Yes	Yes	Yes*
Yoon et al. [15]	No*	No	No	Yes	Yes	Yes	Yes	Yes*
Ours	Yes	Yes	Yes	Yes	Yes	Yes	Yes	Yes*

F_1: mutual authentication. F_5: securely change password.
F_2: key agreement. F_6: reparability.
F_3: forward secrecy. F_7: hide password from server.
F_4: freely chosen password. F_8: no password table.
No*: the scheme fails to provide, however it claims in the original.
Yes*: the scheme stores n for each user in the account database.

5. ANALYSIS

In this section, we examine our scheme in security and computation cost issues.

5.1. Security Analysis

Our improved scheme is similar to Yoon et al.'s scheme but it resists the parallel session attack and also supports the session key agreement with forward secrecy. Table 1 compares some selective features of our scheme with the related works, however it does not include the ones which all the schemes satisfy. In the following, we demonstrate that how the proposed scheme can fulfill security requirements and withstand different attacks:

- *Providing mutual authentication:*
  Since only the true user can generate a valid $c_2 = h(c_1 \parallel T_U \parallel FSP_i)$, by checking this term (step 1 of verification phase), the server can be sure that a user is really who he claims to be. Likewise, the user ensures that he is communicating with a legitimate server when checking $c_3 = h(T_U \parallel T_S \parallel c_1)$ (step 5 of verification phase). By the way, mutual authentication and hence impersonation prevention (no forgery and server spoofing attacks) are satisfied.
- *Parallel session attack protection:*
  An attacker cannot make a valid user's login message $\{ID, T_U, c_2\}$ by using the parts of previous server's verification message $\{T_S, c_3\}$. Because of difference between the format of $c_2 = h(c_1 \parallel T_U \parallel FSP_i)$ and $c_3 = h(T_U \parallel T_S \parallel c_1)$, no one can use c_3 instead of c_2 and vice versa. In other words and according to the mentioned attack, sending the message $\{ID, T_S, c_3\}$ won't pass the verification phase and of course won't lead to start a new session on the server. Totally, the *parallel session attack* is not feasible.

The reality of the matter is that in Yoon et al.'s scheme, the verification of $T_U \neq T_S$ in server's side was for preventing from the parallel session attack. But, in our scheme, with changing the structure of both c_2 and c_3 we omit this checking and do prevention from that attack.

- *Password guessing attack protection:*
  Since in the proposed scheme no password table is stored on the server and no parameter containing a password is transmitted over communication links during login and verification phases, so this attack cannot be applied to, at all.
  It is important to note that the password should never be revealed to the server, because practically it is probable that a user works with the same password to communicate with several servers. In this case, the administrator or any other insider of the server may try to impersonate the user to other servers. The $Mask = h(b)$ parameter can provide this requirement in our scheme due to the fact that it is only known to the user and when he submits the message including $Mask \oplus PW$ to the server in step 2 of registration phase, the server has no way to extract PW from $Mask \oplus PW$.
- *Reparability feature:*
  Using the same scenario mentioned in [9], if the $c_1 = h(EID \parallel x)$ parameter gets compromised e.g. when the smart card is stolen, the user can re-register to the server using the new parameters b_{new}, $Mask_{new}$ and PW_{new}. During the registration phase, the server updates $n_{new} = n + 1$ in the account database and issues a new smart card. Finally, the user owns that smart card containing: $EID_{new} = (ID \parallel n_{new})$, $V_{new} = h(EID_{new} \parallel x)$, and $R_{new} = V_{new} \oplus PW_{new}$. Hereafter, he can login to the server using his new smart card while the login request of the adversary who has obtained $h(EID \parallel x)$ will be rejected. So the reparability (see [7]) of the scheme is preserved.
- *Session Key Agreement with Forward Secrecy:*
  In order to prove the forward secrecy of our session keys, we shall prove that compromised long-term secret keys cannot expose session keys used before. The equations used in the computation of a session key were/are as follows:

$$SK = h(c_1 \parallel T_U \oplus T_S \parallel FSP_i)$$
$$c_1 = h(EID \parallel x) = V$$
$$FSP_{i+1} = h(c_1 \parallel FSP_i).$$

If either the secret data of U stored on his smart card (V, FSP_i), or the secret data maintained by S $(x, EID, FSP_{i-1}, FSP_i)$ get compromised, merely the current session key and the prior one are revealed. But as mentioned in subsection 4.1, the other values of forward-secure parameter that should be computed through the following equations:

$$c_1 \parallel FSP_{i-2} = h^{-1}(FSP_{i-1}),$$
$$c_1 \parallel FSP_{i-3} = h^{-1}(FSP_{i-2}),$$
$$c_1 \parallel FSP_{i-4} = \ldots$$

are still unavailable due to the one-way nature of hash function $h()$, and previous session keys cannot be re-computed. Hence the scheme provides the forward secrecy property for the session key agreement.

5.2. Computation Cost Analysis

We here evaluate the computation cost of the improved scheme and make comparison with the related schemes.

All phases of the proposed scheme only require limited number of hash computations, exclusive-or operations, and some other low-cost operations such as string concatenations. The hash operations can be performed efficiently and computation cost of the other operations is extremely low, so the efficiency of the user and the server are guaranteed in our scheme.

The computational costs of the phases of ours and the related schemes are compared in Table 2. It's worth noting that the login and verification phases play the main role in efficiency of authentication protocols. You can see that compared to Yoon et al. scheme [15] as our basis, the proposed scheme imposes $1T_{hash}$ to the user and $2T_{hash}$ to the server, and reduces $2T_{xor}$ from both of them in login and verification phases.

Needless to say that Table 1 and Table 2 should be considered both together so that it makes a sensible/reasonable comparison. For example, however Sun's scheme [14] has very low computational cost and is more efficient than ours, but can not support variety of features including mutual authentication and key agreement

Table 2. Comparison of computational costs

Scheme\Phase	Registration		Login and Verification	
	User	Server	User	Server
[14]	–	$1T_{hash}$	$1T_{hash} + 1T_{xor}$	$2T_{hash} + 1T_{xor}$
[2]	–	$1T_{hash} + 2T_{xor}$	$2T_{hash} + 3T_{xor}$	$3T_{hash} + 3T_{xor}$
[8]	–	$1T_{hash} + 1T_{xor}$	$3T_{sym} + 1T_{hash} + 1T_{xor}$	$3T_{sym} + 2T_{hash}$
[12]	–	$1T_{hash} + 2T_{xor}$	$3T_{hash} + 1T_{xor}$	$4T_{hash}+1T_{xor}$
[11]	$1T_{hash}$	$1T_{exp} + 1T_{hash}$	$4T_{exp} + 1T_{inv} +1T_{mul} + 3T_{hash}$	$3T_{exp} + 1T_{mul} + 3T_{hash}$
[9]	$1T_{hash} + 1T_{xor}$	$1T_{hash} + 2T_{xor}$	$3T_{hash} + 4T_{xor}$	$3T_{hash} + 3T_{xor}$
[15]	$1T_{hash} + 1T_{xor}$	$1T_{hash} + 2T_{xor}$	$3T_{hash} + 4T_{xor}$	$3T_{hash} + 3T_{xor}$
Ours	$2T_{hash} + 2T_{xor}$	$1T_{hash} + 1T_{xor}$	$4T_{hash} + 2T_{xor}$	$5T_{hash} + 1T_{xor}$

T_{exp}: time of modular exponentiation
T_{mul}: time of modular multiplication
T_{inv}: time of modular inversion
T_{sym}: time of symmetric encryption
T_{hash}: time of one-way hash function h()
T_{xor}: time of exclusive-or operation ⊕

(see Table 1). Another example is the Shieh and Wang's scheme [12], but it can not specially satisfy the forward secrecy feature while our scheme can.

6. CONCLUSION

In this paper, we illustrated that Yoon et al.'s smart-card-based authentication scheme is vulnerable to the parallel session attack. Then we proposed a scheme to solve that attack, and be involved with the forward-secure key agreement feature. All of our improvements are made by employing only (low cost) one-way hash functions.

Subsequently, we examined our scheme in the security issue where we evaluated its different features in comparison to the related works, and in computational cost issue where we compared its cost versus the others, specially the base one.

As the result, we concluded that the proposed scheme not only achieves all the advantages of Yoon et al.'s scheme, but also withstands the parallel session attack and supports key agreement with forward secrecy. As it came out, providing these facilities does not impose much more computational cost on our scheme.

REFERENCES

1. Bellare M, Yee B (2003) Forward-security in private-key cryptography. Topics in Cryptology–CT-RSA, Lecture Notes in Computer Science 2612:1–18
2. Chien HY, Jan JK, Tseng YM (2002) An efficient and practical solution to remote authentication: smart card. Computers & Security 21(4):372–375
3. Clark J, Jacob J (1996) Attacking authentication protocols. High Integrity Systems 1(5):465–474
4. Diffie W, Hellman M (1976) New directions in cryptography. IEEE Transactions on Information Theory 22(6):644–654
5. Diffie W, Oorschot PCV, Wiener MJ (1992) Authentication and authenticated key exchanges. Designs, Codes and Cryptography 2(2):107–125
6. Hwang MS, Li LH (2000) A new remote user authentication scheme using smart cards. IEEE Transactions on Consumer Electronics 46(1):28–30
7. Hwang T, Ku WC (1995) Reparable key distribution protocols for Internet environments. IEEE Transactions on Communications 43(5):1947–1949
8. Juang WS (2004) Efficient password authenticated key agreement using smart cards. Computers & Security 23(2):167–173
9. Ku WC, Chen SM (2004) Weaknesses and improvements of an efficient password based remote user authentication scheme using smart cards. IEEE Transactions on Consumer Electronics 50(1):204–207
10. Lamport L (1981) Password authentication with insecure communication. Communications of the ACM 24(11):770–772
11. Liao IE, Lee CC, Hwang MS (2006) A password authentication scheme over insecure networks. Journal of Computer and System Sciences 72(4):727–740
12. Shieh WG, Wang JM (2006) Efficient remote mutual authentication and key agreement. Computers & Security 25(1):72–77
13. Stallings W (2005) Cryptography and network security. 4th ed. New Jersey, Prentice Hall
14. Sun HM (2000) An efficient remote use authentication scheme using smart cards. IEEE Transactions on Consumer Electronics 46(4):958–961
15. Yoon EJ, Ryu EK, Yoo KY (2004) Further improvement of an efficient password based remote user authentication scheme using smart cards. IEEE Transactions on Consumer Electronics 50(2):612–614

CHAPTER 38

A MATHEMATICAL FRAMEWORK FOR RISK ASSESSMENT

MARCO BENINI AND SABRINA SICARI

Dipartimento di Informatica e Comunicazione, Universit degli Studi dell'Insubria, via Mazzini 5, IT-21100, Varese, Italy {marco.benini, sabrina.sicari}@uninsubria.it

Abstract: Risk assessment is an important step in the development of a secure system: its goal is to identify the possible threats to a system, their impact and, henceforth, to evaluate the connected risks. Although several systematic approaches have been developed to perform a risk assessment task, the current methodologies rely on the quantitative evaluations of experts in a substantial way. This paper addresses the problem of detaching the methodology results from the subjective judgements of experts, by formalising a risk assessment methodology in an appropriate mathematical framework that reduces the subjective aspects in experts' evaluations

1. INTRODUCTION

Despite the fact that risk assessment is a well-established engineering practice to evaluate the security status of a complex system, the significance of the obtained results is often debated [1] since it depends on the quantitative judgements of one or more human experts and, thus, the results are said to be influenced by the subjective views of the experts.

The present work wants to cope with this problem by developing a mathematically formalised risk assessment procedure whose dependence on the judgements of human experts is greatly reduced and controlled.

The core of our proposal lies in considering the quantitative measures given by the experts as *relative*: the measures are considered to be part of a *metric*, i.e., an organised system of possible values, and the meaning of a single value lies in its relationship with the *structure* of the whole metric.

In this way, a sensible risk assessment methodology, like the one we are going to discuss, relies only on the structure of the metrics. The immediate effect is that the

H. Labiod and M. Badra (eds.), New Technologies, Mobility and Security, 459–469.
© 2007 *Springer.*

evaluations of different experts using each one a different metric, can be compared and integrated as far as the metrics are compatible.

Henceforth, the need for a mathematical formalisation becomes clear: a mathematical framework allows to define the notion of metric, its structure and the idea of compatible metrics, and, furthermore, it allows to prove that a risk assessment does not depend on the choice of a metric, but, instead, essentially the same result is obtained by using any compatible metric.

Therefore, in Section 2, we introduce the risk assessment methodology and, in Section 3, we describe the mathematical framework allowing to prove its correctness, i.e., its ability to obtain a result. In Section 4, we show an example that clarifies our claim that the risk assessment procedure is independent from the values in the metric and that compatible metrics will produce equivalent results. Finally, in Section 5, a comparison with the existing approaches is shown, allowing the reader to position our result in the existing research streams.

2. THE RISK ASSESSMENT METHODOLOGY

In general, the goal of risk assessment is to determine the likelihood that the identifiable threats of a system will harm, weighting their occurrence with the damage they possibly cause. Therefore, a risk assessment methodology is a procedure whose outcome is an estimation of the risk connected with the occurrence of one or more threats.

In this work, we analyse the risk assessment methodology introduced in [2]; specifically, in this Section, we illustrate the methodology, briefly discussing its foundations, while in Section 3, we will introduce its main properties and we will use them to derive some interesting facts about the quality of the risk evaluations the methodology produces.

In general, the risk is measured by a function r of two variables: the damage potential of the hazard and its level of exploitability. The damage potential is often defined as the average loss of money an attack may cause, although, in our approach, any other kind of sensible measure can be used. Moreover, the level of exploitability is a measure of the difficulty to make an attack including both the easiness and the reproducibility of an attack, as defined, e.g., in the STRIDE/DREAD theory [3].

The methodology evaluates the total risk of a threat by means of a sequence of steps described as follows:

1. The threat to the system under examination is modelled by using an attack tree [4,5]: the attack goal is the root node and its children nodes represent possible ways of achieving it. Recursively, the children can be alternative subgoal, each one satisfying the father goal (`or` subtrees) or partial subgoals, whose composition satisfies the father goal (`and` subtrees). The leaves of the tree are the *potential* vulnerabilities that should be matched with the *actual* vulnerabilities of the system. To each vulnerability v is associated an index $E_0(v)$, called its *exploitability*, which measures the difficulty to exploit v in order to perform a successful attack.

2. The dependencies among the identified vulnerabilities are introduced: a vulnerability A depends on a vulnerability B if, when B is already exploited, then A becomes easier to exploit. Moreover, each dependency is weighted by an exploitability value, using the same metric as E_0.
3. The final exploitability of each single vulnerability is calculated taking into account its initial value, which does not consider dependencies, and its dependencies, according to the algorithm described in Section 2.1.
4. The risk associated to the threat under examination is finally computed by recursively aggregating exploitabilities along the attack tree. The exploitability of an `or` subtree is the easiest exploitability of its children, and the exploitability of an `and` subtree is the most difficult exploitability of its children. The aggregated exploitability measures the level of feasibility of the attack and is combined with the damage potential to finally assess the risk of the threat.

The *metric* employed in the evaluation of exploitabilities and their dependencies is the set of possible values for $E_0(v)$. We require this set to be a partial order: this choice reflects the difficulty to compare an arbitrary pair of vulnerabilities in order to decide their relative difficulty; usually, similar vulnerabilities are easily compared, while different vulnerabilities may be compared only to some extent, e.g., saying that both are easier or more difficult to exploit than a third one.

Evidently, it is safe to suppose that the metric contains a finite number of elements, since the vulnerabilities of the system are always finite, and, similarly, it is safe to assume that the partial order contains a maximum, denoted as 1, and a minimum, denoted as 0. This is justified since every actual vulnerability is easier to exploit than to violate the ideal perfectly secure component, while each vulnerability is harder to exploit than the ideal perfectly insecure component.

In practice, a meaningful assessment of $E_0(v)$ is a matter of both experience and ingenuity, but, as explained in Section 3, just the *relative* exploitability has to be estimated.

2.1. Exploitability of Dependent Vulnerabilities

The system is formalised as a graph $\mathcal{A} = \langle C, L \rangle$ where C is the set of *components* and L is the set of *links* between components. The components and the links are exposed to the set of vulnerabilities V_C and V_L, respectively, where an element $(u, v) \in V_C$ means that the component u is susceptible to be subverted thanks to the flaw v. Therefore, the set of the system's vulnerabilities is $V = V_C \cup V_L$.

Initially, during the step 1 in our methodology, an expert assesses how easy and repeatable is to exploit every single vulnerability to gain the control of a component or a link in the given architecture. We call this the initial exploitability $E_0(v)$ of the vulnerability v in the system $\mathcal{A}$.

In general, the functions $E_i \colon V \mapsto \mathcal{O}$ map vulnerabilities to $\mathcal{O}$, a partial ordered set of degrees of exploitability. The functions are indexed by a step number (details later), thus the E_0 function generates the exploitability values $E_0(v)$. The $\mathcal{O}$ set, modelling the expert's metric, is a finite, partially ordered set containing two distinct elements, 0, its minimum, and 1, its maximum, as already explained in the preceding Section.

However, the architecture of the system imposes dependencies among vulnerabilities. For example, we need to understand if it is easier to exploit a vulnerability of a component given that an input link attached to it has already been compromised or a component attached to any of its input links has already been violated. We denote with $E(v|w)$ the exploitability of v given that the vulnerability w has already been exploited.

The dependencies among vulnerabilities are represented in the *dependency graph* $\mathcal{D} = \langle V, D \rangle$, whose nodes are the vulnerabilities and the edge (w, v) is in D iff $E(v|w) \geq E_0(v)$, i.e., an edge (w, v) means that it is easier to compromise an element suffering the v vulnerability when one has already compromised an element affected by the w vulnerability.

The number of exploitability evaluations is bounded since the number of edges in the $\mathcal{D}$ graph is, at most, $|V|(|V|-1)$. However, in practice, most of the vulnerabilities are usually independent, and the evaluations the expert has to *guess* is typically closer to $|V|$ than to $|V|^2$.

Initially, the value associated to each node v in the dependency graph $\mathcal{D}$ is $E_0(v)$, that is, the initial measure of how difficult is to exploit the vulnerability. The conditional exploitabilities are used to label the edges they belong to. The conditional values are derived in step 2 of our methodology.

The initial assessment depicted in the graph $\mathcal{D}$ does not take into account that each vulnerability could be exploited thanks to the previous exploitation of one of the vulnerabilities on which it depends. Therefore, the labels of the nodes are iteratively updated by considering the easiest way, i.e., the maximum value, to exploit an incoming vulnerability in the dependency graph. In turn, each incoming vulnerability could be exploited by controlling the affected element or leveraging on the dependency itself: the most difficult, i.e., the minimum, constraints the value. Therefore, the update rule for labels is defined by the following formula:

$$(1) \qquad E_{i+1}(v) = \sup(\{E_i(v)\} \cup \{\inf\{E(v|w), E_i(w)\} : (w, v) \in D\}).$$

Thus, the third step of the methodology consists in iteratively applying (1) for each vulnerability, until the system converges to an equilibrium after a suitable number n of steps. Then, the values of $E_n(v)$ represent the final exploitability of each vulnerability v considering also its dependencies.

3. THE MATHEMATICAL FRAMEWORK

As the reader may expect, the choice of the metric $\mathcal{O}$ influences the results obtained in the application of the methodology. On a more subtle level, the mathematical characters of the methodology depend on the *structure* of the ordering $\mathcal{O}$, as we are going to clarify in this Section.

The first property of the method is its *convergence*; we want to prove that the methodology guarantees to reach an equilibrium in the computation of the exploitability values. A side effect of the proof is that the equilibrium is a reached in a number of steps bounded by $|\mathcal{O}||V|$, being V the set of vulnerabilities.

Theorem 1: Given a dependency graph $\mathcal{D} = \langle V, D\rangle$, there is a number k such that, for every $v \in V$, $E_{k+1}(v) = E_k(v)$.

Proof. We notice that, for any number i and for any $v \in V$, $E_{i+1}(v) \geq E_i(v)$ because of the definition of exploitability. Moreover, we know that the exploitability values form a finite partial order $\mathcal{O}$: let n be the number of elements in $\mathcal{O}$.
Therefore, before reaching the equilibrium, for every step i, there is a vulnerability v such that $E_{i+1}(v) > E_i(v)$. This situation can occur only n times per vulnerability, since, after at most n updates, the value $E_i(v)$ reaches the maximum of the metric, thus it cannot be updated anymore. Hence, after at most $n|V|$ steps, every vulnerability must be stable. □

Corollary 1: Given a dependency graph $\mathcal{D} = \langle V, D\rangle$, there is a number k such that, for every $v \in V$ and $i, j \geq k$, $E_i(v) = E_j(v)$.

It is important to remark that Theorem 1 provides an upper bound to the number of iterations: as obvious by the use of the pigeon hole principle, this bound is unnecessarily large, and in practice, convergence is usually obtained in a few steps.

In our observation, it resulted that the methodology enjoys another interesting property, that reveals how the iterative calculation depends only on the structure of the ordering of metrics. Specifically,

Theorem 2: Given a dependency graph $\mathcal{D} = \langle V, D\rangle$ and two metrics $\mathcal{O}_a = \langle O_a, \leq_a, 0_a, 1_a\rangle$ and $\mathcal{O}_b = \langle O_b, \leq_b, 0_b, 1_b\rangle$, if $g\colon \mathcal{O}_a \to \mathcal{O}_b$ is a morphism[1] from $\mathcal{O}_a$ to $\mathcal{O}_b$ such that $g(E_0^a(v)) = E_0^b(v)$ for every $v \in V$ and $g(E^a(v|w)) = E^b(v|w)$ for every $(w, v) \in D$, then, for any $v \in V$ and for any i, $g(E_i^a(v)) = E_i^b(v)$, where E^a and E^b are the exploitability functions using, respectively, $\mathcal{O}_a$ and $\mathcal{O}_b$ as metrics.

Proof. By induction on i, the number of steps; the base step is obvious by hypothesis, while the induction step is as follows:

$$\begin{aligned} g(E_{i+1}^a(v)) &= g(\sup_a(\{E_i^a(v)\} \\ &\quad \cup\{\inf_a\{E^a(v|w), E_i^a(w)\colon (w, v) \in D\})) \\ &= \sup_b(\{g(E_i^a(v))\} \cup \{\inf_b\{g(E^a(v|w)), g(E_i^a(w))\colon (w, v) \\ &\quad \in D\}) \\ &= \sup_b(\{E_i^b(v)\} \cup \{\inf_b(E^b(v|w), E_i^b(w)\colon (w, v) \in D\})) \\ &= E_{i+1}^b(v) \end{aligned}$$ □

[1] A morphism is a structure-preserving function. In particular, a function f between $\mathcal{O}_a$ and $\mathcal{O}_b$ is a morphism iff for every $x, y \in O_a$ such that $x \leq_a y$, it holds that $f(x) \leq_b f(y)$.

Some comments are due:

- The hypothesis "$g(E_0^a(v)) = E_0^b(v)$ for every $v \in V$ and $g(E^a(v|w)) = E^b(v|w)$ for every $(w, v) \in D$" encodes the fact that the initial situation the method is applied to, is the same, modulo the g morphism.
- The g is a morphism, i.e., a function respecting the relation and the constants of the order. The meaning of this requirement is that the metrics are compatible, that is, any comparable pair of values in the first metric is associated with a pair of values in the second metric that gets compared in the same way.

3.1. An Illustrating Example

This section tries to clarify our findings by means of an abstract example that evidences the core of our results without the complexities of a real case study.

The scenario is as follows: we have two security experts, Alice and Bob, working together to evaluate the risk of a network attack to a complex system. They developed a suitable attack tree (not shown) and they agree both on the set of vulnerabilities affecting the system, and on the way they depend one on each other. Hence, our experts produce the system dependency graph, whose nodes are the identified vulnerabilities and whose arcs are the dependencies.

In practice, the depicted scenario is common: the possible ways to conduct an attack, the identification of the vulnerabilities and, finally, the dependencies among the identified vulnerabilities are subjects on which experts can easily integrate their knowledges, thus producing a common, agreed picture of the security status of a system.

Differently, when the experts are asked to quantify the risks connected to the identified vulnerabilities, their evaluations may diverge because of the application of different metrics. In our example, Alice adopts the metric a while Bob uses the metric b; both of them are represented in Fig. 1. The drawing shows the minima (0_a and 0_b) at the bottom, the maxima (10_a and 10_b) at the top, and a value x is less than y if x is below y and connected to. The supremum of two elements x and y is the minimal point above x and y, connected to both of them, and, dually, the infimum of x and y is the closest connected point below them.

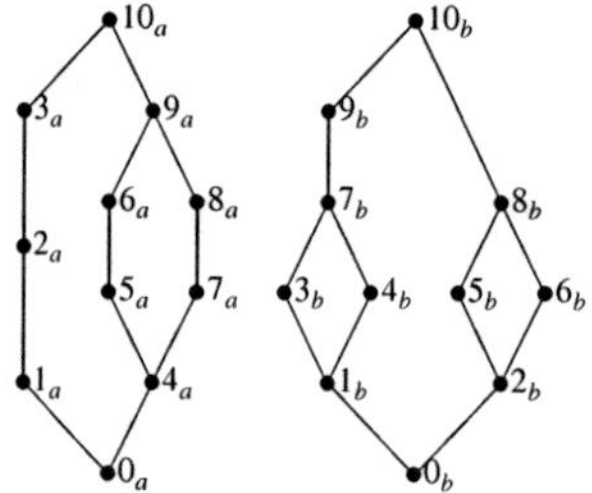

Figure 1. The metrics a and b

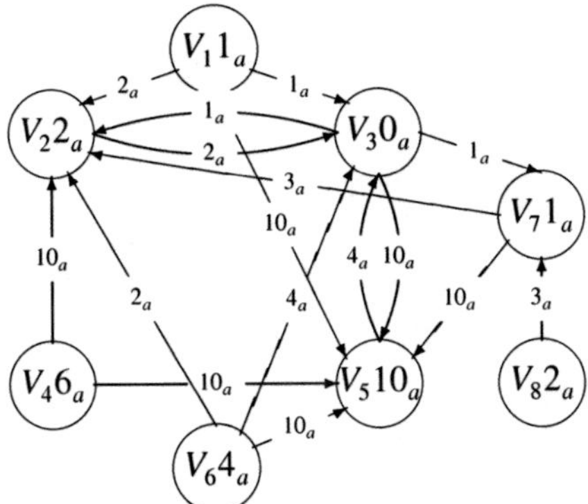

Figure 2. The initial evaluation of Alice

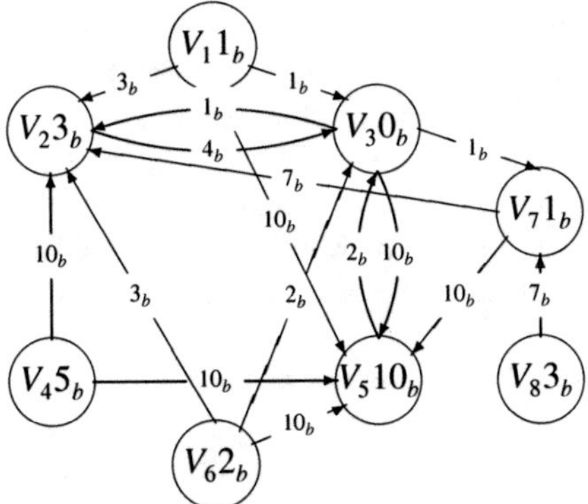

Figure 3. The initial evaluation of Bob

In the scenario, Alice develops an initial evaluation of the exploitability values, synthesised in Fig. 2; Bob does the same, as illustrated in Fig. 3. These evaluations are the result of the application of the experts' experience and judgement, thus, at least to some extent, the values are subjective.

Applying our methodology, Alice and Bob can calculate the final risk assessment, considering also the role of dependencies: after a few iterations of the application of (1), Alice derives the following risk vector

$E^a(V_1)$	$E^a(V_2)$	$E^a(V_3)$	$E^a(V_4)$	$E^a(V_5)$	$E^a(V_6)$	$E^a(V_7)$	$E^a(V_8)$
1_a	10_a	10_a	6_a	10_a	4_a	2_a	2_a

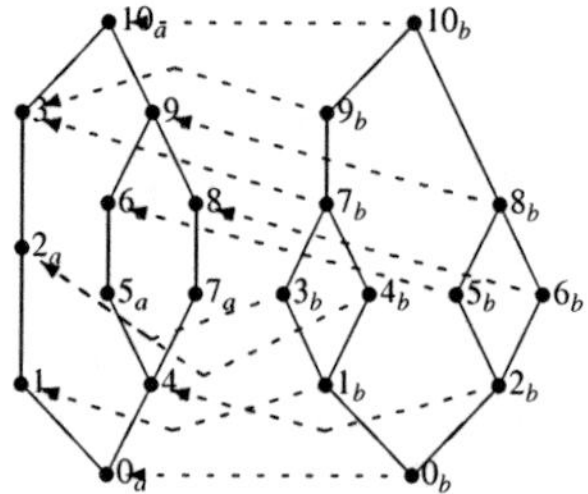

Figure 4. The morphism from the metric b to a

while Bob obtains as his final result

$E^b(V_1)$	$E^b(V_2)$	$E^b(V_3)$	$E^b(V_4)$	$E^b(V_5)$	$E^b(V_6)$	$E^b(V_7)$	$E^b(V_8)$
1_b	10_b	10_b	5_b	10_b	2_b	3_b	3_b

It is evident that the derived evaluations are different, so we expect the measured risk to differ. Nevertheless, following the Theorem 2, if the metrics are compatible, then the results coincide, modulo a *renaming* of the values in the metrics.

The *renaming* function is the morphism relating the metrics of our experts, and its existence is the criterion to say that the metrics are compatible. For example, the metric b used by Bob can be mapped in the metric a of Alice via the morphism g shown in Fig. 4: it is immediate to check that, if $x < y$ in the metric b, then $g(x) < g(y)$ in the metric a.

If one transforms the resulting risk vector of Bob by means of the g morphism, the result is

$g(E^b(V_1))$	$g(E^b(V_2))$	$g(E^b(V_3))$	$g(E^b(V_4))$	$g(E^b(V_5))$	$g(E^b(V_6))$	$g(E^b(V_7))$	$g(E^b(V_8))$
1_a	10_a	10_a	6_a	10_a	4_a	2_a	2_a

that is exactly the result of Alice.

Henceforth, Theorem 2 says that, the values in a metric have a conventional meaning which is determined up to morphisms between compatible metrics.

4. RELATED WORKS

Even though the application of risk management methodologies has been widely discussed and analysed, see, e.g., [6–9] among information security experts there appears to be no agreement on the best or the most appropriate method to assess the probability of computer incidents [10].

In literature there are many attempts to face the risk assessment problem; some of them define systematic approaches while others provide more ad-hoc methods to evaluate the likelihood of (a class of) violations.

In particular, we have found of interest Baskerville's description [11] of the evolution of various ad-hoc methods to measure risk that sometimes could be combined to improve the accuracy of the security evaluation.

On the side of systematic approaches, O. Sami Saydjari et al. [12] present a system security engineering methodology to discover system vulnerabilities, and to determine what countermeasures are best suited to deal with them: the paradigm that synthesises this work is *analysing information systems through an adversary's eyes*.

With respect to the previous works, our approach, starting from its initial definition in [2], has been based on the structured evaluation of single vulnerabilities along with their mutual dependencies. In this respect, the results in [12,13] are similar to ours, although they do not propose a formal methodology based on mathematical arguments. In fact, the distinctive aspect of our work with respect to the discussed ones is the mathematical formalisation of the risk assessment method in order to derive its characterising properties.

In this respect, there are more formalised approaches, employing a graph-based representation of systems and their vulnerabilities, that provide methodologies whose properties are, at least partially, mathematically analysed. Among those approaches, of prominent interest are those based on attack graphs [14,15], where state-transition diagrams are used to model complex attack patterns. The extreme consequence of this family of approaches is to use model-checking techniques to simulate attacks, like in [15].

In comparison, our approach is simpler both in the methodology and in its formalisation. Despite its simplicity, our results are stronger on the mathematical side and some experimentation [16] make evident the practical value of the method in real-world situations.

In fact, we use the attack tree model [4,5] to evaluate the security threats combining them with the dependency graph, a formalisation of an essential piece of knowledge of experts. This combination is the subject of our mathematical analysis, and dealing with a richer structure than the simple attack trees, we are able to derive stronger properties.

As a matter of fact, independently from their application areas, the risk assessment methodologies have a core weakness: the use of subjective metrics. In fact, in the scientific community the main criticism to these methodologies is about the fact that values assigned on the basis of a personal knowledge and experience are regarded as *guessed* values, making the total risk evaluation process unreliable.

It is a fact that the evaluation metric behind exploitability deeply influences the risk evaluation. But, at least in our treatment, what matters is the *structure* of the metric rather than its absolute values, thus limiting the previous criticism.

Generalising, in many field of ICT there is the need to define an objective metric. In the abstract, a metric is defined [17] as the instrument to compare and to measure a quantity or a quality of an observable. Our treatment of metrics follows the work of N. Fenton, in particular [18].

In agreement with him, we consider measurement as the process by which numbers or symbols are assigned to attributes of entities, in our case to the exploitability of a vulnerability. Therefore, even though there is no widely recognised way to assess risks and to evaluate the induced damages, there are various approaches that provide methodologies by which the risk evaluation becomes more systematic.

Looking towards risk assessment as a decision support tool, Fenton [19] proposed the use of Bayesian networks. Differently, our approach towards objective risk assessment is based on the abstraction over values, thus what matters is the *structure* of the metrics. Hence, objectivity is gained by considering values in the metric not as *absolute measures of risk*, but, instead, as *relative evaluations of risks*. Therefore, in agreement with [20,12,19,21], the information computed by our model can be used as a decision support.

5. CONCLUSIONS

This work addressed the problem of formalising a risk assessment procedure. The problem behind the formalisation is that a risk assessment depends on the quantitative evaluation of exploitabilities, a task performed by human experts, and, thus, subjective.

The idea developed here to cope with this problem, is to abstract over the metrics, i.e., the values used to quantify the threats, in order to force the risk assessment methodology to depend on the structure of a metric, instead of depending on its absolute values. In this way, two experts adopting different but compatible metrics will produce similar evaluations, that is, evaluations that are different in their form, but equivalent in their meaning.

In this respect, we formalised in an appropriate mathematical framework a simple, yet effective risk assessment methodology, and we have proven inside the framework its correctness, i.e., its ability to produce the expected evaluations, in Theorem 1. Moreover, we claimed that the methodology depends only on the structure of the employed metrics, and we gave a precise mathematical formalisation of this statement. As a side effect, we have shown a formal characterisation of the notion of compatible metrics in terms morphisms between partial orders.

REFERENCES

1. Redmill, F.: Risk analysis: A subjective process. Engineering Management Journal **12**(2) (April 2002) 91–96
2. Sicari, S., Balzarotti, D., Monga, M.: Assessing the risk of using vulnerable components. In Gollmann, D., Massacci, F., Yautsiukhin, A., eds.: Quality of Protection. Security Measurements and Metrics, New York, NY, USA, Springer-Verlag (June 2006) 65–78
3. Howard, M., Leblanc, D.: Writing Secure Code. Microsoft Press (2003)
4. Moore, A., Ellison, R.: Survivability through intrusion-aware design. Technical Report 2001-TN-001, CERT Coordination Center (2001)
5. Schneier, B.: Modelling security threats. Dr. Dobb's Journal (December 1999)
6. Alberts, C., Dorofee, A., Stevens, J., Woody, C.: Introduction to the Octave approach (October 2003)

7. den Braber, F., Dimitrakos, T., Gran, B., Lund, M., Stølen, K., Aagedal, J.: The CORAS methodology: Model-based risk management using UML and UP. In Favre, L., ed.: UML and the Unified Process. IRM Press (2003) 332–357
8. Jenkins, B.: Risk analysis helps establish a good security posture; risk management keeps it that way (1998) White paper.
9. Siu, T.: Risk-eye for the IT security guy (February 2004)
10. Sharp, G., Enslow, P., Navathe, S., Farahmand, F.: Managing vulnerabilities of information system to security incidents. In: ICEC '03: Proceedings of the 5th International Conference on Electronic Commerce, New York, NY, USA, ACM Press (2003) 348–354
11. Baskerville, R.: Information system security design methods: Implications for information systems development. ACM Computing Survey **25**(4) (1993) 375–412
12. Evans, S., Heinbuch, D., E.Kyle, Piorkowski, J., J.Wallener: Risk-based system security engineering: Stopping attacks with intention. IEEE Security & Privacy Magazine **2**(6) (2004) 59–62
13. Moskowitz, I., Kang, M.: An insecurity flow model. In: NSPW '97: Proceedings of the 1997 Workshop on New Security Paradigms, New York, NY, USA, ACM Press (1997) 61–74
14. Noel, S., Jajoidia, S., O'Berry, B., Jacobs, M.: Efficient minimum-cost network hardening via exploit dependency graphs. In: ACSAC '03: Proceedings of 19th Annual Computer Security Applications Conference, IEEE Computer Society (2003) 86–95
15. Sheyner, O., Haines, J., Jha, S., Lippmann, R., Wing, J.: Automated generation and analysis of attack graphs. In: SP'02: Proceedings of the 2002 IEEE Symposium on Security and Privacy, Washington, DC, USA, IEEE Computer Society (2002) 273–284
16. Benini, M., Sicari, S.: Risk assessment: Intercepting VoIP calls. In: Proceedings of the VIPSI 2007 Venice Conference. (March 2007) To appear.
17. Arshad, S., Shoaib, M., Shah, A.: Web metrics: The way of improvement of quality of non web-based systems. In Arabnia, H.R., Reza, H., eds.: SERP'06: Proceedings of the International Conference on Software Engineering Research and Practice. Volume 2., CSREA Press (2006) 489–495
18. Fenton, N.: Software measurement: A necessary scientific basis. IEEE Transactions on Software Engineering **20**(3) (1994) 199–206
19. Fenton, N., Neil, M.: Making decisions: Bayesian nets and mcda. Knowledge-Based Systems **14**(7) (November 2001) 307–325
20. Biswas, G., Debelak, K., Kawamura, K.: Application of qualitative modelling to knowledge-based risk assessment studies. In Ali, M., ed.: IEA/AIE'89: Proceedings of the Second International Conference on Industrial and Engineering Applications of Artificial Intelligence and Expert Systems. Volume 1., New York, NY, USA, ACM Press (1989) 92–101
21. Sahinoglu, M.: Security meter: A practical decision-tree model to quantify risk. IEEE Security & Privacy **3**(3) (May/June 2005) 18–24

CHAPTER 39

AN EFFICIENT HYBRID CHAOTIC IMAGE ENCRYPTION SCHEME

ASIM, M. AND JEOTI, V.

Department of Electrical & Electronic Engineering, Universiti Teknologi PETRONAS, Malaysia

Abstract: In recent years chaotic cryptography has attracted significant attraction for multimedia security. However, most of these chaotic ciphers still suffer from excessive computation for a given level of security [1–3]. In view of these, a novel scheme is proposed in this paper to encrypt digital images that has far smaller computational load. The proposed scheme is based on multiple-piecewise linear chaotic maps (m-PLCMs) and S-box, where the R, G and B channels of digital images are first segmented into low correlated frames of the appropriate length. Identical chaotic key stream is used for the various frames of the R, G and B channels of the segmented image. The reuse of the chaotic stream reduces the computational complexity by a higher factor and the combination of S-box makes it infeasible to extract the chaotic stream due to its low differential and linear probabilities. The analysis of the proposed scheme proves that it efficiently fulfills the tradeoff between security and encryption speed

1. INTRODUCTION

In today's digital world, secure transmission of images and videos have become an important issue for the applications such as Pay TV, video conferencing and medical imaging systems etc. Many cryptographic algorithms have been proposed and are widely used today such as DES, IDEA, Triple-DES, AES, and RSA etc.

Properties of the chaotic systems such as sensitivity to initial conditions and control parameters, ergodicity, mixing and exactness can be usefully used for in cryptography and also for generating pseudo-random codes. Due to these attention grabbing properties of the chaotic systems, a number of cryptographic algorithms have been proposed recently for the secure transmission of digital images and videos [1–3] [7], but many of them have been reported to be insecure against the known-plaintext and chosen plaintext attacks [8].

Along with the existing security problems of the chaotic ciphers, their speed has also come under great scrutiny due to their low throughput compared to that of AES,

H. Labiod and M. Badra (eds.), New Technologies, Mobility and Security, 471–480.
© 2007 *Springer.*

DES etc. However like other researchers [12], we believe that both conventional and chaotic cryptography can benefit from each other.

In this paper, a Hybrid Chaotic Image Encryption Scheme (HyChIES) is proposed. HyChIES is based on m-PLCMs and S-box. HyChIES achieves a good performance both from speed and security analysis point of view.

This article is organized as follows. In section 2, detailed description of HyChIES is presented. Section 3 presents detailed security analysis of HyChIES. In section 4, we talk about the efficiency of HyChIES and compare its computational load with the recently proposed chaotic ciphers for digital images. This article is closed with conclusion.

2. HYBRID CHAOTIC IMAGE ENCRYPTION SCHEME—HYCHIES

Hybrid Chaotic Image Encryption Scheme (HyChIES) is a complete image encryption scheme which operates on pixel values of the image. In HyChIES, digital image is first divided into less correlated frames of the appropriate length. After dividing image into less correlated frames, same chaotic stream is used for masking the corresponding pixel values in all frames. In order to make chaotic key stream extraction infeasible, an S-box is introduced. Input and output of S-box are masked with the output of chaotic stream. In what follows, we describe each and every step of the proposed HyChIES in some detail.

2.1. Scrambling

The ideal way to scramble the 2-D R, B and G channels of an image is to transform each pixel value to a different position in an unpredictable dynamic fashion by a 2-D or 3-D chaotic cat map like [7]. However, it entails high computational load as calculation of new coordinate values require large number of multiplications and additions. Hence to reduce the computational load, R, G and B channels are divided into multiple non-overlapping blocks of the size 4×4, 8×8 or 16×16 and scrambled.

In HyChIES, R, G and B channels are divided into multiple non-overlapping pixel blocks of dimension 16×16. If the size of the R, G and B channels is not a factor of 16, it is padded with extra zeros. When the image is decrypted, these pad pixels are chopped off.

After dividing into 16×16 non-overlapping pixel blocks, these blocks are permuted or scrambled by an efficient random pattern. The pattern for transforming the 16×16 pixel blocks to the new position is generated from generalized 2-D cat map. The pattern should be invertible in order to reproduce the original map correctly. The classical and generalized cat maps are given by (1).

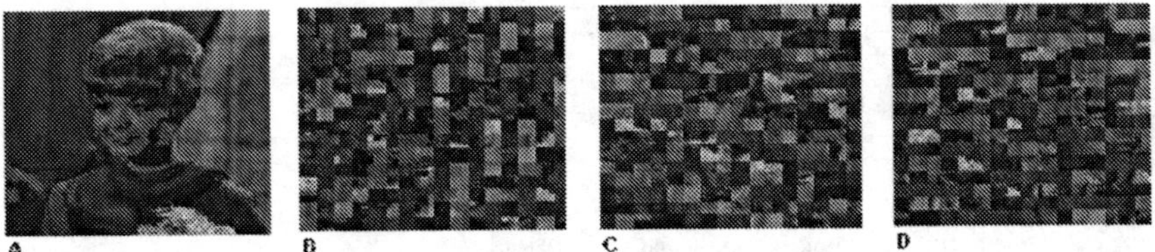

Figure 1. Scrambling 16 × 16 pixel blocks of R, G and B channels: (A) Original pain image (B) Output of 1st (C) Output, of the 2nd, (D) Output of 3rd—round

$$\begin{bmatrix} x_{n+1} \\ y_{n+1} \end{bmatrix} = \begin{bmatrix} 1 & 1 \\ 1 & 2 \end{bmatrix} \begin{bmatrix} x_n \\ y_n \end{bmatrix} \bmod 1$$

$$\begin{bmatrix} x_{n+1} \\ y_{n+1} \end{bmatrix} = \begin{bmatrix} 1 & a \\ b & ab+1 \end{bmatrix} \begin{bmatrix} x_n \\ y_n \end{bmatrix} \bmod N \qquad (1)$$

The scrambling process is shown in Fig. 1 for an image taken from [10]. Scrambling three times is sufficient enough to achieve the appropriate level of randomness.

2.2. Division into 1-D Frames

In this phase, 2-D R, G and B channels are mapped into 1-D matrices R^{1D}, G^{1D} and B^{1D} by an invertible scanned pattern. The scanned pattern is reading values column wise as shown in Fig. 2. After arranging into 1-D matrices, R^{1D}, G^{1D} and B^{1D} are divided into frames of the size "N", such that frames are as uncorrelated as possible.

In our experimental analysis, R^{1D}, G^{1D} and B^{1D} matrices of an image database obtained from [10] are divided into frames of the length 2500, which is quite reasonable value for the frames length. After the frames creation process, number of different pixels (NDP) at the corresponding positions in the first frame and the rest of the frames of R^{1D}, G^{1D} and B^{1D} matrices are calculated by using (2) and (3).

$$\begin{array}{lll} if\ f_1^R(i) \neq f_j^R(i) \text{ then} & if\ f_1^B(i) \neq f_j^B(i) \text{ then} & if\ f_1^G(i) \neq f_j^G(\text{i}) \text{ then} \\ D^R(i) = 1 \text{ else} & D^B(i) = 1 \text{ else} & D^G(i) = 1 \text{ else} \\ D^R(i) = 0 & D^B(i) = 0 & D^G(i) = 0 \\ where\ i = 1, \ldots..2500 & where\ i = 1, \ldots.2500 & where\ i = 1, \ldots..2500 \\ j = 1, \ldots., L & j = 1, \ldots., L & j = 1, \ldots., L \end{array} \qquad (2)$$

here f_j^R, f_j^R *and* f_j^R represents the jth frame of R^{1D}, G^{1D} and B^{1D} matrices and L represents the total number of frames. After computing $D^R(i)$, $D^B(i)$ and $D^G(i)$, the NDP ratios for the frames of R^{1D}, G^{1D} and B^{1D} matrices are calculated by (3).

Figure 2. Scan pattern for reading R^{1D}, G^{1D} and B^{1D} matrices

(3) $$NDP^R = \frac{\sum_{i=1}^{2500} D^R(i)}{2500} \times 100\%, \quad NDP^B = \frac{\sum_{i=1}^{2500} D^B(i)}{2500} \times 100\%, \quad NDP^G = \frac{\sum_{i=1}^{2500} D^G(i)}{2500} \times 100\%$$

Through analysis, it is founded that the average value for NDP^R, NDP^B and NDP^G is between 90–96% for images obtained from [10]. In other words, 90–96% pixels of frame 1 are different from all other subsequent frames and the average correlation coefficient between the first frame and the rest of the frames is negligible. Hence the use of the same chaotic stream for the multiple frames will not degrade the security level.

2.3. Initial Condition Extraction from External Secret Key

In this step the initial condition IC for the m-PLCMs are extracted from a secret key of 16 characters. The secret key in ASCII form is denoted by (4).

(4) $$K = K_1\ K_2\ K_3 \cdots K_{16}$$

here K_i denotes the 8-bit block of the secret key. The initial condition is derived in the following fashion. Initial condition for the m-PLCMs used in the extraction of the chaotic key stream is derived by employing 16-PLCMs with different control parameters. The PLCM is defined by (5).

(5) $$CM(x,p) = \left\{ \begin{array}{cc} x/p & 0 \le x \le p \\ (x-p)/(1/2-p) & p \le x \le 1/2 \\ CM(1-x,p) & 1/2 < x \le 1 \end{array} \right\}, \ \textit{where}\ 0 < p < 1/2$$

The initial condition for the ith PLCM is extracted from the ith 8-bit character (ASCII value) of the secret key by (6).

(6) $$IC_i = K_i/256$$

here IC_i serves as the initial condition for the ith PLCM in the following way:

(7) $$R = \sum_{i=1}^{16} CM_i^{K_i}(p_i, IC_i)$$

$$IC = R \bmod 1$$

here R refers to the intermediate variable. K_i refers to the number of iteration of the ith chaotic system with corresponding initial condition IC_i and control parameter p_i. IC refers to the initial condition value for m-PLCMs that are used in the extraction of chaotic key stream in HyChIES.

2.4. Extraction of Chaotic Key Stream

HyChIES employs three PLCMs for the encryption of the frames of R^{1D}, G^{1D} and B^{1D} matrices. The selected map is one of the simplest PLCM and posses the following properties: 1) It is ergodic, mixing; 2) It has uniform density function $f(x) = 1/(\beta - \alpha)$; 3) The autocorrelation function of a particular chaotic orbit $\tau(n) = \delta(n)$. The three PLCMs are denoted as follows:

$$CM_1 = CM_1(IC, p_1), CM_2 = CM_2(IC, p_2), CM_3 = CM_3(IC, p_3)$$

Here $p_1 = 0.15$, $p_2 = 0.27$ and $p_3 = 0.43$ are three different control parameters and PLCM is highly sensitive for considered parameters [11] [13]. The chaotic key stream from m-PLCMs is extracted by (8).

$$\phi_k = \lfloor CM_k * (10^7) \rfloor \,(\text{mod}\, 256) \;\; where\ k = 1, 2, 3 \tag{8}$$

ϕ_k represents the extracted 8-bit value from k^{th} PLCM

2.5. Substitution by S-box

Substitution by S-box is a nonlinear transformation step that plays significant role in fighting against the differential and linear cryptanalysis. HyChIES uses S-box of AES which has maximum differential and linear probabilities of 2^{-6} and 2^{-3} respectively [4]. The combination of S-box and chaotic key stream makes the overall system secure and makes the extraction of the chaotic key stream infeasible. Detail regarding the functionality and its design philosophy of S-box can be found in [4]. In HyChIES, input and output of S-box are masked with the chaotic key stream, so that the straight forward use of the inverse S-box can not recover the input data. HyChIES utilizes the same chaotic key stream in two stages in an efficient way such that each byte is masked with the output of the two different maps in one iteration. Section 2.6 describes this step in more detail.

2.6. Add Chaotic Key Stream

The extracted chaotic key stream ϕ_k is XORed with the corresponding pixel value in each frame of R^{1D}, G^{1D} and B^{1D} matrices by using (9).

$$\begin{array}{llll} f_j^R(i) \oplus \phi_k, & \text{where} \;\; k = 1, 2\ or\ 3, & j = 1, 2\ldots, L \text{ and } i = 1, 2\ldots, 2500 \\ f_j^G(i) \oplus \phi_k, & \text{where} \;\; k = 1, 2\ or\ 3, & j = 1, 2\ldots, L \text{ and } i = 1, 2\ldots, 2500 \\ f_j^B(i) \oplus \phi_k, & \text{where} \;\; k = 1, 2\ or\ 3, & j = 1, 2\ldots, L \text{ and } i = 1, 2\ldots, 2500 \end{array} \tag{9}$$

The value of ϕ_k is selected dynamically by using another PLCM in the following fashion:

$$\begin{aligned} CM_c &= CM(IC, p_c) \\ I &= \left(\lfloor CM_c{}^{*}10^7 \rfloor \text{ mod } 16\right)+1 \\ F &= K(I) \text{ mod } 3 \end{aligned} \tag{10}$$

Here CM_c represents the output of the PLCM with initial condition IC and control parameter p_c that controls the three chaotic maps used in the extracting chaotic key stream. I represents the index of a particular character in the secret key K_i that determines the value of the control variable F. The control variable helps in the selection of ϕ_k. In HyChIES, the key addition step is performed twice – once before and once after the substitution by S-box.

3. SECURITY ANALYSIS

A good cipher should be robust against all kinds of cryptanalytic, statistical and brute-force attacks. In this section, the security analysis of the proposed image encryption scheme is discussed such as statistical analysis, sensitivity analysis with respect to the key etc. to prove that the proposed cryptosystem is secure against the most common attacks.

3.1. Histogram Analysis

An image-histogram illustrates how pixels in an image are distributed by graphing the number of pixels at each color intensity level. In Fig.3, the histograms of plain and ciphered images are shown. It is clear that histograms of the encrypted image are fairly uniform and significantly different from the respective histogram of the original image and hence do not provide any clue to employ any statistical attack on the proposed image encryption procedure.

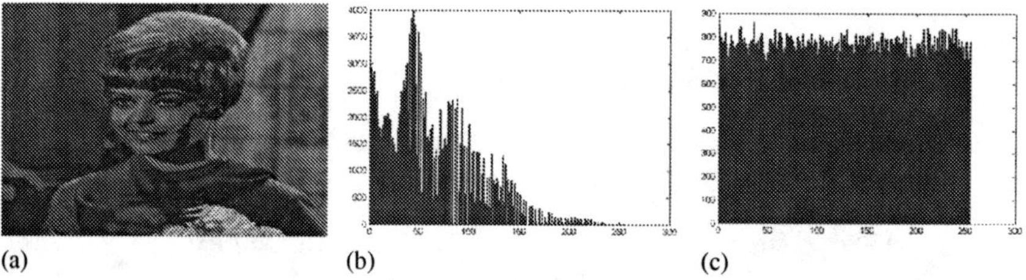

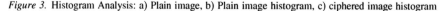

Figure 3. Histogram Analysis: a) Plain image, b) Plain image histogram, c) ciphered image histogram

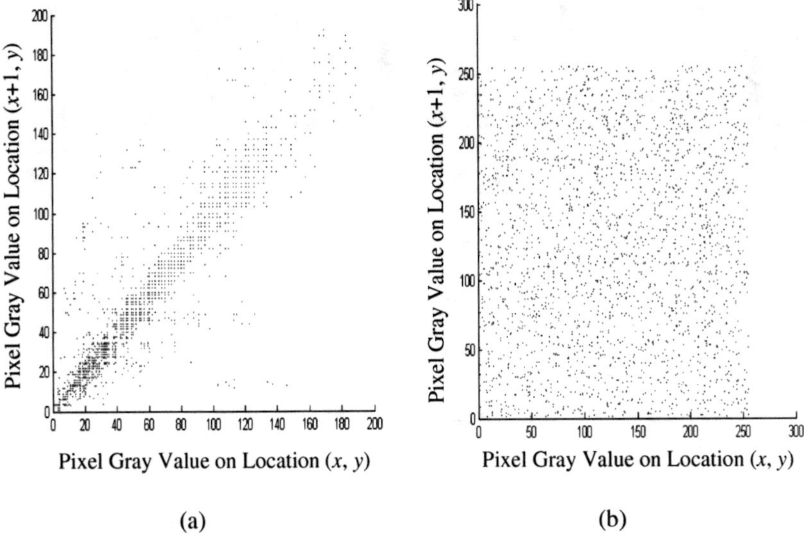

Figure 4. Correlation of two adjacent pixels: a) distribution of horizontally adjacent pixels in an original frame, b) distribution of horizontally adjacent pixels in an encrypted frame

3.2. Correlation Coefficient Analysis

In this section, the correlation between the various plain images obtained from [10] and their corresponding cipher images are extensively studied using HyChIES, where all of images were encrypted using the secret key '*abcdefghijklmnop*'. The correlation coefficients for some images calculated by (11).

From the Table 1 and Fig. 4, it is evident that correlation coefficients are very small which implies that no correlation exists between original and its corresponding cipher image.

Table 1. Correlation coefficient between the image and corresponding cipher image for a number of images obtained from [10]. The encryption has been done using the secret key '*abcdefghijklmnop*'

File Name	File Description	Type	Size	Correlation coefficient
4.1.01	Girl	Color	256×256	0.0025
4.1.05	Tree	Color	256×256	−0.00099853
4.1.06	Jelly beans	Color	256×256	−0.00070419
4.2.01	Splash	Color	512×512	0.00023768
4.2.03	Baboon	Color	512×512	−0.00082352
4.2.04	Girl(Lena)	Color	512×512	0.000071325
4.2.05	Airplane(F-16)	Color	512×512	0.00094581
4.2.06	Sailboat on Lake	Color	512×512	−0.0003127

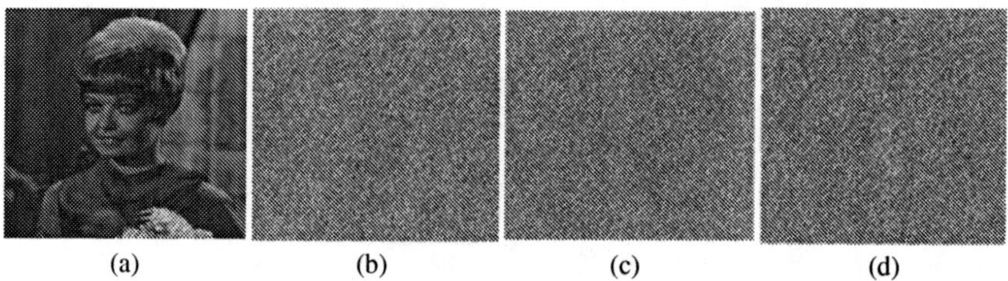

Figure 5. Key sensitivity test: a) Plain image, b) Encrypted Image with '*abcdefghijklmnop*', c) Decrypted image with wrong key '*abcdefghijklmnoo*', d) decrypted image with wrong key '*abcdefghijklmmop*'

$$r = \frac{\sum_{m}\sum_{n}(A_{mn} - \overline{A})(B_{mn} - \overline{B})}{\sqrt{\left(\sum_{m}\sum_{n}(A_{mn} - \overline{A})^2\right)\left(\sum_{m}\sum_{n}(B_{mn} - \overline{B})^2\right)}} \tag{11}$$

where $\overline{A}$ and $\overline{B}$ represents the corresponding mean values.

3.3. Key Sensitivity Analysis

The cipher image obtained by HyChIES is extremely sensitive to the change in secret key i.e. a single bit of difference b/w encrypting and decrypting key should make it unable to decrypt the ciphered image. Alternatively, it can be said that a single bit of change in an encrypting key should result in totally different encrypted image. For testing the key sensitivity of the proposed HyChIES, we have performed the following steps:

- An Original image in Fig. 5(a) is encrypted using the secret key '*abcdefghijklmnop*', that is shown in Fig. 5(b).
- Tried to decrypt the encrypted image Fig. 5(b) with slightly wrong key *abcdefghijklmno**o***', that resulted in image Fig. 5(c).
- Ttried to decrypt the encrypted image Fig. 5(b) with another slightly wrong key '*abcdefghijklm**m**op*', that resulted in image Fig. 5(d).

It is evident that it is unable to decrypt an image with a slightly different key and HyChIES is highly key sensitive.

4. EFFICIENCY OF HYCHIES

The proposed HyChIES is based on m-PLCMs as chaotic cryptosystems based on single chaotic systems can be easily broken by estimating the parameters of chaotic system [5]. PLCM is selected due to its perfect cryptographic properties and its statistical properties are investigated in [11] [13].

HyChIES is a hybrid scheme, which posses the properties of chaos and S-box. The inclusion of S-box makes it infeasible to extract the chaotic key stream [8]. The

Table 2. Computations for encryption of equivalent one byte required by considered ciphers

Operations	[1]	[2]	[7]	HyChIES
XORing	—	90	2	2
Multiplication	3.33	60	12	0.1017
Shifting	—	—	—	1
Substitution	—	30	—	1
Addition	3	—	7	0.0127
Subtraction	1	—	1	[a]
Modulo	3	[b]	4	0.00636

[a] May or may not be performed due to the unpredictable behavior of m-PLCM

[b] May not be performed and calculations are for the simple case of the proposed cipher

particular S-box used herein is borrowed from AES which has very low differential and linear probabilities of 2^{-6} and 2^{-3} respectively [4]. S-box coupled with the chaotic stream cipher will provide new directions towards exploring better ways of providing security.

HyChIES employs an external secret key, which is mapped to an *IC* by a sub-algorithm. The sub-algorithm retains the positional significance of each character in the external secret key unlike [1] [3] [8,9]. The algorithm has a large key space and passes the sensitivity test with respect to the secret key.

In HyChIES, R^{1D}, G^{1D} and B^{1D} matrices are first divided into low correlated frames of the appropriate length. The reuse of the same chaotic key stream for multiple frames reduces the computational load of HyChIES significantly and achieves the same performance parameters which are shown in section 3. In Table 2, the computational load of the recently proposed chaotic ciphers for the encryption of equivalent 8-bit is shown. In case of [2], we assumed the number of iterations is equal to 90, while computation of [1] does not include the multiplications and additions, resulting due to iterations of chaotic systems by Runge-Kutta method. In HyChIES, calculations are for an image of the size 256*256*3.

5. CONCLUSIONS

In this article a novel hybrid chaotic image encryptions scheme–HyChIES is proposed for real time digital image encryption. HyChIES fulfills the trade off between a given security level and encryption speed more efficiently than the recently proposed chaotic ciphers. HyChIES passes all the statistical, sensitivity tests and has a large key space. HyChIES has a very less computational load and is very suitable for the real-time encryption of digital images. AES–S-box is introduced along with chaotic scheme that makes it infeasible to extract the chaotic key stream. 2–4 rounds of HyChIES will increase the security level with a very little cost of computational load.

6. ACKNOWLEDGEMENTS

The authors would like to gratefully acknowledge the financial support from Universiti Teknologi PETRONAS for presenting the paper to NTMS 2007 and anonymous reviewers for their valuable comments.

REFERENCES

1. Fei et al. (2005) An image encryption algorithm based on mixed chaotic dynamic systems and external key, In Proc. Of IEEE International Conference on Communications, Circuits and Systems, Vol 2, pp 1135–1139.
2. Pareek et al. (2006) Image Encryption using Chaotic Logistic Map, Image and Vision computing.
3. Pareek et al. (2005) Cryptography using Multiple One Dimensional Chaotic Maps, Communications in Nonlinear Science and Numerical Simulation, Vol. 10, Issue 7, pp 715–723.
4. J. Daemen and V. Rijmen (2002) The design of Rijndael–AES–The Advanced Encryption Standard, Springer-Verlag.
5. Peitgen et al. (2002) Chaos and Fractals-New Frontiers of Science, Springer-Verlag.
6. M. I. Sobhy and A.-E.R. Shehata (2001) Methods of Attacking Chaotic Encryption and Countermeasures, IEEE International Conference on Acoustics, Speech, and Signal Processing, Vol 2, pp 1001–1004.
7. Chen et al. (2004) A Symmetric Image Encryption Scheme Based on 3d Chaotic Cat Maps Chaos, solutions, and fractals 21(2004), pp 749–761.
8. M. Asim and V. Jeoti (2007) On the Security of a Recent Chaotic Cipher, To appear in the 3rd International. Colloquium on Signal Processing and its Applications.
9. M. Asim and V. Jeoti (2007) M. Asim and V. Jeoti, Comments on Fei-Qui-Min Chaotic Cipher", To appear in the 3rd International. Colloquium on Signal Processing and its Applications.
10. http://sipi.usc.edu/database/
11. Li et al. (2001) Statistical Properties of Digital PLCMs and their Roles in Cryptography and Pseudo Random Coding, In Cryptography and Coding: 8th IMA International Conference pp 205–221.
12. Kocarev et al. (1998) From Chaotic Maps to Encryption Schemes, IEEE International Symposium on Circuits and Systems, Vol 4, 31 pp 514–517.
13. Li et al. (2001) Pseudo-Random bit Generator based on Couple Chaotic Systems and its Applications in Stream Cipher Cryptography, 2nd International Conference on Cryptology in India, Chennai, pp 316–329.
14. Tao et al. (1998) Perturbance-based Algorithm to Expand Cycle Length of Chaotic Key Stream. Electronics Letters, pp 873–874.

CHAPTER 40

A MODEL BASED ON PARALLEL INTRUSION DETECTION SYSTEMS FOR HIGH SPEED NETWORKING SECURITY

SOUROUR MEHAROUECH, ADEL BOUHOULA, AND TAREK ABBES

Department of Computer Technology and Networks, Higher School of Telecommunications SupCom, 2083 Cit El Ghazala, Tunisia

Abstract: During this time when the Internet provides essential communication between an infinite numbers of people and is being increasingly used as a tool for commerce, security becomes a tremendously important issue to deal with. It is also important to note that, recently, the intrusion detection systems (IDS) have been unable to provide an effective security mechanism for defending high speed networks. Existing networks intrusion detection systems (NIDS) can barely keep up with bandwidths of some hundred Mbps whereas, nowadays, the network speed presses forward 10 Gbps. So in order to protect high speed networks, we propose a new approach aiming at accelerating the intrusion detection operation. The approach is based on three main steps: traffic classification, load balancing and a high availability mechanism. This paper describes the above mentioned approaches and presents an experimental evaluation of their effectiveness

1. INTRODUCTION

One essential aspect for network security is the network intrusion detection system; their use is so widespread. The 2001 CSI/FBI Computer Crime and Security Survey [8] shows that of 61 % respondents use intrusion detection system to protect their networks.

Actually, there are two main intrusion detection approaches: the behaviour approach, based on statistical description of the normal behavior of users or applications. And the misuse approaches, based on collecting attack signatures in order to store them in an attacks base.

Commonly, the performance of a network intrusion detection system is characterized by the probability that an attack is detected in combination with the number of false alerts. However, equally important is the system's ability to process traffic at the maximum rate offered by the network with a minimal packet loss. Significant

H. Labiod and M. Badra (eds.), New Technologies, Mobility and Security, 481–491.
© 2007 *Springer*.

packet loss can leave a number of attacks undetected and degrades the overall effectiveness of the system [11].

In order to solve this challenging problem, we propose to divide the high-speed network traffic into several classes. Each class is fed into a balancer, which carries out load balancing between a numbers of analysers. We propose also the application of a high availability mechanism because NIDSs behave as fail open devices. That's why skilled hackers try first to disable NIDSs before attacking other hosts in the network. The high availability mechanism controls the NIDS's activities in order to quickly detect any decrease in the performance. Moreover, it discharges an overloaded NIDS, and sends its traffic to another NIDS.

The remainder of this paper is organized as follows. In the next section, we discuss related work. Section 3 describes the architecture of our proposed system and the details of each component. Then, we describe in section 4, 5, 6 the application of the traffic classification, the load balancing and the high availability in the intrusion detection. We present in section 7 the results of the evaluation of the solution. Finally, in the last section we summarize the paper and outline future work.

2. RELATED WORK

Currently there are mainly two approaches to increase the performance on NIDSs. On the one hand, some researchers attack the problem by increasing the efficiency of detection algorithms, e.g. AC-BM algorithm [12], SSPP-BM algorithm [10] and sekar et. al approach [3]. Research work in [3] presents a fast pattern matching algorithm whose runtime is insensitive to the number of attack signatures. This solution enables the IDS to support real-time performance at up to 500 Mbps. Research works in [12,10,4] are mainly concentrates on designing efficient parallel multi-pattern algorithms for the NIDS. On the other hand, some researchers introduce distributed framework in data collection by dividing the high-speed network traffic load between multiple intrusion detection engines [1,2]. The work of Kruegel et al. [1] is pioneering in the domain, as it is the first one to subdivide the network traffic. The method allows a further stateful analysis. However, this work did not take into account the enterprise security policies the IDS characteristics, and it uses a very simple algorithm (Round Rabin) for balancing the traffic. The evaluation of this work shows that the system is able of real-time detection at 170 Mbps. Research works in [2] use also multiple sensors operating in parallel, the evaluation of this work shows an increase 20 % in sensors performance.

The commercial world attempted also to respond to this problem and a number of vendors [5–7] now claim to have intrusion detection system that can operate on high-speed ATM or Gigabit Ethernet links. For example, ISS [5] offers Net-ICE Gigabit Sentry, a system that is designed to monitor traffic on high-speed links. The tool claims to be "the industry's first gigabit intrusion detection system that can accurately and effectively monitor full gigabit speed links in real-time. No other intrusion detection system in the world can match Sentry Gigabit for robustness, reliability, and accuracy within the gigabit environment". However, the evaluation of the tool states, "GigaSentry handles a full Gigabit in lab conditions, but real-world performance will likely be less... Customers should expect at least 300 Mbps real-world performance [1].

3. HIGH SPEED INTRUSION DETECTION ARCHITECTURE

The overall goal of this work is to perform a stateful real time intrusion detection system for high-speed networks. This solution is characterized by the following requirements:

- The system implements a misuse detection approach where signatures representing attack scenarios are matched against a stream of network events.
- Intrusion detection is performed by a set of sensors; each sensor is autonomous and does not interact with other sensors.
- Traffic classification for intrusion detection must be done with regard to the traffic types and properties.
- Load balancing for intrusion detection has to be performed in a stateful way that guarantees the detection of all the threat scenarios.
- The architecture should result in a scalable design where we can add sensor as needed to match increased network throughput with the least reconfiguration of the system.

As we see in Fig. 1, the system consists of several components. Each one is responsible of doing a very determined task.

- Classifier: it captures packets circulating in the network and divides the overall network load into several classes. For example we can put the HTTP traffic in a class, FTP traffic in another class, and the other types of traffic can be putted in one single class. Packets that belong to the some class are sent to the same balancer.
- Back-up classifier: the presence of only one classifier might result in a single point of failure. For example, if the classier is breakdown or if it crash, the disturbance on the network security operation would be considerable. To overcome this problem, we add in our architecture a back-up classifier that has the some functionality that the classifier.
- Balancer: every balancer collects the class of traffic destined to it by the classifier and carries out load balancing between a number of analysers .
- Analysers (NIDSs): they are responsible for determining if an intrusion has occurred. The output of this component is an alert forwarded to the manager. In each set of analysers, responsible for analysing the same class of traffic, we put only a subset of all detectable intrusion scenarios with respect to the class that they analyse.
- Manager: it enables a user to view the output of the analysers and controls the behaviour of the system.
- Database: it stores alert records, session logs and system parameters.

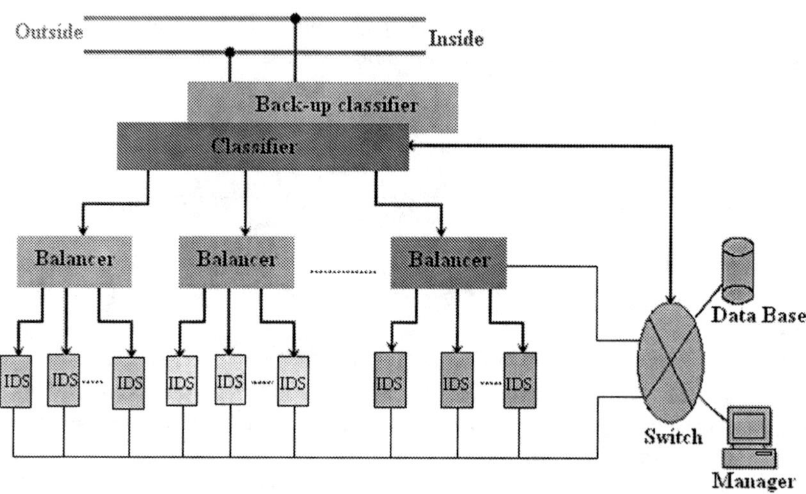

Figure 1. Parallel intrusion detection systems

The crucial problem of ever increasing high speed traffic can be tackled by disturbing the analysis among several NIDS. In addition, our solution offers other benefits:

- Reducing false negative rate: Our solution distributes the analysis among several NIDSs; this operation reduces considerably the number of dropped packets (not analyzed in time) by following the false negative rate.
- Intrusion prevention: when classifying packets, a splitter engine can stop suspected flows. Besides, it can inspect the load towards some services. In case of an excess in use, the splitter hushes up busy addresses. The classifier can also sends suspect traffic to honeypots.
- Faster analysis of alerts: Our solution separates the analysis of each traffic class; this separation allows a fast analysis of alerts since each group of alerts corresponding to a type of traffic can be analyzed separately. Our solution permits also to obtain shorter and independent log files.

4. TRAFFIC CLASSIFICATION FOR INTRUSION DETECTION

Classification algorithms are well investigated in many network applications. In the same way, intrusion detection systems can benefit from this technique in order to solve the challenging problem of supporting the high-speed traffic. We present in this section forms of our classification rules, after we explain our approach to classify the network traffic.

- Classification rules

  We express the classification operations by means of rules. The first rule to be satisfied will forward the traffic to the corresponding balancer. A rule can be divided into 2 parts. The left hand side called header contains a set of parameters to be matched by the packet and a direction parameter to indicate the source of the traffic. The right hand side of the rule specifies the action to be taken and a priority to resolve the case where many rules are satisfied at the same time by one packet. Below is the schema of a traffic classification rule.

(1) R: (field1, … , field n) $_{\text{Direction}}$ → <action, priority>
Where direction {in, out, bi}

We give below the different types of our traffic classification rule:
Type 1: (adresses, port) $_{\text{Direction}}$ →<action, priority>
We frequently employ this model to supervise the services provided by a server.
Type 2: (*, port) $_{\text{Direction}}$ →<action, priority>
It defines rule used to supervise particular services. Generally these rules are indexed by a direction in order to restrict the supervision to only incoming flow.
Type 3: (address, *)$_{\text{Direction}}$ →<action, priority>
This category of rules is generally used to tune the level of verification and protocol analysis exploration depending on the remote networks.

We explain now our approach to classify the network traffic. We consider that the classification rules have 4 dimensions: source and destination addresses and source and destination ports. The number of these parameters can be extended without any change in the method. The classification proceeds in 2 steps. We first perform a geometric search in a 2 dimensional space in order to cluster rules by the port criteria. We construct thus several sets of candidate rules that fall in the same ranges of source and destination ports. The second step processes each group of rules in order to represent them in the form of graph. At runtime, classifying the traffic is an easy task since it suffices to look for the suitable directed graph to traverse.

- Classification with the Port Criteria
  The first step is to construct the sets of candidate rules with the same ranges of source and destination ports. The port definition in a classification rule can have simple integer values of 16 bits (from 0 to 216-1), or a more complex format given by sets and ranges. This leads to the overlap problem between classification rules. To solve it, we divide the possible values and ranges in elementary ranges. Note that a range can be represented by its 2 endpoints. Hence to find a range, it suffices to consider one of the 2 boundaries. For instance if we choose the lower endpoint, we can look for the greatest endpoint smaller than the port value. We accomplish this search in sublinear time. After finding the elementary intervals of the source and the destination ports, we point to the graph used to classify the packet by address criteria.
- Classification with the Address Criteria
  At the end of the first phase, we obtain several groups of rules. The second phase builds a directed graph for each group. Then traversal of the graph defines the suitable balancer (or action) to forward the traffic. The IP addresses used in the classification rules present some peculiarities that make the search operation a complex task. In fact, they can be defined on a number of bits lower than 32 bits.

In this case, they represent a range of addresses which leads to rules overlap. The algorithm that we propose postpones the processing of these overlaps. In fact, during the graph construction, we divide every dimension to a set of bytes. The constructed graph looks like a multi-bit graph with a stride size equal to 8. An IPv4 address contains 4 bytes so that a host is represented by 4 complete bytes while a network is assigned less complete bytes and only one partial byte. The partial byte can be viewed as a couple of integer, the first one being the byte value and the second one is the number of masked bits. We concatenate all the dimensions and we modify the bytes order. We first put together all complete bytes and we complete them with the partial bytes. The number of rules that match the same complete bytes sequence will be low and therefore we minimize the number of overlapping partial bytes.

The graph construction consists in transforming complete bytes into labeled links between nodes. After, we label the final nodes by the classification rules having the highest priority. This priority follows the apparition order in the rule list. Once the winning classification rule is obtained for a packet, we forward it to the suitable balancer.

- Fragmentation Problem

  The classifier must ensure the flow integrity especially in the case of fragmented packets. In fact, the IP fragment does not usually include the TCP header but only the first one contains this information. Therefore the traffic splitter cannot apply the classification rules that use the ports parameters. To solve this problem, we keep in the splitter the IP packet identifier of the first fragment as well as the winning rule to classify this packet. All fragments can now be routed to the some balancer.

5. LOAD BALANCING FOR INTRUSION DETECTION

By classifying the network traffic, we divide the load into several classes. But it is still the case that the load of one class is important to be analysed by one single NIDS. That is why adding a load balancing is important.

We design a new load-balancing algorithm, which divides the data of each class based on the current value of each analyser's Load function. The incoming data packets are forwarded to the analyser that has least load. The analysers' Load function models are described as follows:

Suppose the number of balancers is M and the number of analysers by balancer is N_M. To define the Load function, we use the following functions:

Mi,j (t) : defined as analysers' memory function. The value of this function stands for the percent utilization of an analyser's memory at time t.

Ui,j(t) : defined as analysers' CPU function. The value of this function stands for the percent utilization of an analyser's CPU at time t.

Ci,j(t): defined as connection function. The value of this function stands for the relative number of connections being processed by an analyser at time t and not achieved.

Pi,j(t): defined as packet function. The value of this function stands for the relative number of packets not yet analysed by the IDS responsible of them at time t. Hence, the Load function models for each analyser can be expressed as follows:

$$Li,j\,(t) = ai\ Mi,j\,(t) + bi\ Ui,j(t) + ci\ Ci,j(t) + di\ Pi,j(t) \quad i = 1,2,\ldots, M \text{ and } j = 1, 2, \ldots, N_M \tag{2}$$

Where ai, bi, ci and di are weight coefficients, which represent the relative impact of different parameters on the Load function value. The sum of all weight coefficients should be equal to 1. However, the same parameter may have different impact on Load function value in different network traffic environments. Hence, we can adjust the weight coefficients to optimise the algorithm based on specific network traffic class. Therefore each balancer has its own coefficients with respect to the class of traffic that it balances.

To collect the Mi,j and Ui,j parameters, the manager sends requests to each NIDSs. In fact each analyser knows these parameters, so it sends back a response with the necessary information. To collect Ci,j and Pi,j parameters, we consider two counters for every NIDS: one for Ci,j and one for Pi,j. Following the arrival of a new connection or a new packet we increase the values of the corresponding counters of the NIDS responsible for their analysis. Once a packet or a connection is analysed we decrease the values of the counters.

The question, now, is when the load balancing needs to be carried out? If the balancers carry out load every time they receive a new packet or a new connection, they become overloaded. So to solve this problem, we program the balancers to carry out load balancing periodically. The manager consults the analysers to have information on their parameters only at the end of the $P_{\text{load balancing}}$ period. These parameters will be useful to take the adequate decisions of balancing of the next period.

We are interested now in ensuring a stateful analysis. For this, we propose to record the result of balancing, i.e. the NIDS that have been elected to analyse the connection with the identifying of the connection, in an arborescent structure. For each new connection we create a new node. When we receive a packet putting an end to a connection we destroy its node.

6. HIGH AVAILABILITY FOR INTRUSION DETECTION

Nowadays, guaranteeing the high availability is only limited to services like File server, Domain Name service, Web mail. Our idea is to apply this technique to intrusion detection systems. Our solution is based on two steps: first, the manager exchanges with the NIDSs heartbeats messages in a definable interval ($P_{\text{Heartbeat}}$). If during a certain time the manager doesn't receive a signal from a NIDS, it declares that the NIDS in question is in dysfunction and informs the balancer responsible for this element. In the second step, the balancer stops the traffic toward this analyser, separates it from the group, and redirects the load toward other active entities. To detect failed analysers, our algorithm is based on the utilities PING: echo request (echo (8)), echo response (echo (0)) and it uses the following parameters:

- Heartbeat period: the number of seconds that must elapse before the next ping echo request is sent.
- Ping fail period: the manager marks the NIDS as not running if a NIDS stops responding for a period longer than the Ping fail period.
- Ping success period: the manager marks the NIDS as active if one single echo response was received.

The parameters below show the most important results:

- Status: active or in dysfunction.
- Pings sent: the total number of pings sent, during the current session.
- Pings received: the total number of pings received, during the current session.
- Pings failed: the total number of pings that were sent but no response was received during current session.
- Ping success rate: the ping success rate, calculated as the fraction of received pings of the number of pings sent. This parameter is very important because it shows if there is any decrease in the performance of NIDSs.

To redirect the traffic, we implement in our architecture a fail-over mechanism that stops the traffic toward failed analysers, separates them from the group and redirects the load toward other active ones. Our redirection of traffic is based on a warm stand-by fail-over. This technique offers a faster redirection without the intervention of humans.

We are interested now in the architecture of the redirection, so we propose an "active-active" configuration where each analyser can be the back-up NIDS of a broken down one. That means that each analyser does its task and if we have a problem, each one is able to receive the load of the other. The advantage of this architecture is that we take benefit from all the analysers, in contradiction with the case where we have back-up NIDSs used only in the event of incidents.

7. PERFORMANCE EVALUATION

The main goal of the experiments described in this section is to get a preliminary evaluation of the practicality and effectiveness of our approach. The first experiment aims at measuring the performance of the classifier against different traffic speed. We define a performance index for the classifier as follows:

(3) $\text{Performance} = \text{Rateclassifier}/\text{Ratenetwork}$

Ratenetwork is defined as the network traffic speed. Similarly, Rateclassifier is defined as the amount of traffic handled by the classifier at a given network traffic speed. It is calculated by measuring the total system time taken to classify a fixed amount of network traffic generated at Ratenetwork by tcpreplay. Beside the classifier, the Libpacp/Libnet solution is kept to capture and redirect the traffic. We implement the classifier on a standart laptop Pentium 4 (1 GB of RAM), 1 Gbps

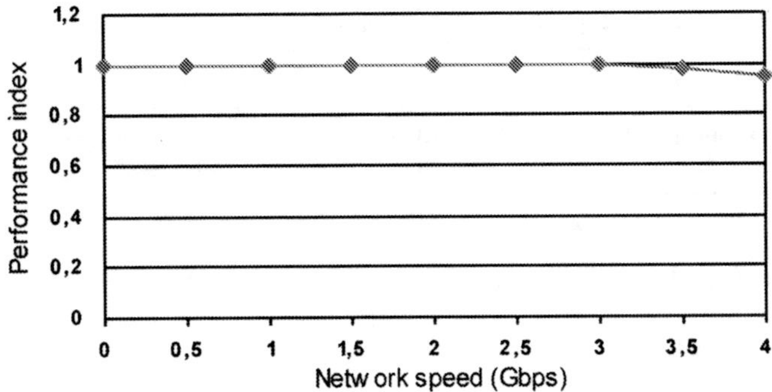

Figure 2. Performance of the classifier

Ethernet card running Linux Redhat 7.2. We observe in Fig. 2 that performance index starts falling when network traffic exceeds 3,5 Gbps. In fact the degrading performance of the classifier can be attributed to packet loss happening because of decrease in performance of Libpcap library.

Our second experiment aims at measuring the performance of our architecture to detect attacks. To run our experiments we used traffic produced by MIT Lincoln Labs. The traffic log was injected using tcpreplay. To achieve high-speed traffic we had to "speed up" the traffic. We chose Snort 2.6 as our "reference" NIDS. Each host was equipped with Intel Pentium 4 (1 GB of RAM), 1 Gbps Ethernet card running Linux Redhat 7.2. We use a number of tools, as ARP0c and APsend, permitting the simulation of attacks. We must verify that the signatures of the generated attacks are present in the database of the analysers, so if they don't detect the attacks, we will be sure that this is caused by the incapacity to keep up with network speed. The test results in Table 1 show that the 100% attacks detection is about 10 Mbps for 1 analyser, 300 Mbps for 2 analysers and 700 Mbps for 4 analysers. So the percentage of attack detection increases with the number of analysers, therefore even if network speed becomes faster, we can add more analysers to keep up with the speed. This result shows the perspectives of our approach and its capacities in the future.

Our third experiment measures the received improvement with the implementation of the high availability mechanism. The test results in Table 2 show that the percentage of attacks detection increases to 77 % for 1 Gbps. This improvement is

Table 1. Probability of attacks detection for increasing network speed

Network speed (Mbps)	100	300	500	700	800	900	1000
% of attacks detected (1 analyser)	82	28	13	7	5	2	1
% of attacks detected (2 analyser)	100	100	79	65	62	53	44
% of attacks detected (4 analyser)	100	100	100	100	90	85	76

Table 2. Improvement of the high availability (HD) mechanism

Network speed (Mbps)	100	300	500	700	800	900	1000
% of attacks detected without HD	100	100	100	100	90	85	76
% of attacks detected with HD	100	100	100	100	94	89	77

due to automatic detection of a collapsed analyser and the faster redirection of its traffic to an active one.

8. CONCLUSION AND FUTURE WORK

Intrusion Detection Systems (IDS) are an essential tool for monitoring safety in computer systems and networks. But the constant increase in network speed and throughput pose new challenges to these systems. Current NIDSs are designed to 100/500 Mbps, nevertheless large network installations are Gigabit Ethernet, so the task of detection becomes increasingly difficult with only one NIDS. In order to address the problem of intrusion detection analysis in high speed networks, the paper presented the design, implementation, and experimental evaluation of a scalable solution that combines a traffic classification technique, a load balancing device and a high availability mechanism. The results obtained allow us to conclude that our solution is effective for improving the performance of NIDS to keep up with the increased network throughput.

Future work will include a mechanism of alerts correlation permitting to reduce the number of false alerts and to detect attacks when they are performed across multiple sessions or against multiple hosts.

REFERENCES

1. Kruegel Christopher, Valeur Fredrik, et al. Stateful intrusion detection for high-speed networks. Santa Barbara: Reliable Software Group, University of California. In Proceedings of the IEEE Symposium on Research on Security and Privacy, Oakland, CA, May 2002. IEEE Press.
2. I. Charitakis, K. Anagnostakis, and E. Markatos. An active traffic splitter architecture for intrusion detection. In Proceedings of 11th IEEE/ACM International Symposium on Modeling, Analysis and Simulation of Computer and Telecommunication Systems (MASCOTS 2003), Orlando, pages 238_241, October 2003.
3. R. Sekar, V. Guang, S. Verma, and T. Shanbhag. A High-performance Network Intrusion Detection System. In Proceedings of the 6th ACM Conference on Computer and Communications Security, November 1999.
4. M. Fisk, G. Varghese, Fast content-based packet handling for intrusion detection, Technical Report CS2001-0670, Department of Computer Science and Engineering, University of California, San Diego, June 2001.
5. ISS. BlackICE Sentry Gigabit. http://www.networkice.com/products/sentry gigabit, November 2001.
6. CISCO. CISCO Intrusion Detection System. Technical Information, Nov 2001.
7. Inc. Top Layer Networks. IDS balancer with etrust intrusion detection. http://www.toplayer.com/pdf/IDSBdsCA_pdf.pdf. Jan 2004.

8. R. Power, 2001 CSI/FBI Computer Crime and Security Survey, Computer Security Institute, Computer Security Issues and Trends VII (1) (2001) http://www.goci.com/prelea_000321.htm
9. M. Roesch. Snort – Lightweight Intrusion Detection for Networks. In Proceedings of the USENIX LISA '99 Conference, November 1999.
10. Wu Yanga, Bin-Xing Fanga, Bo Liub, Hong-Li Zhanga. Intrusion detection system for high-speed network. Computer Network and Information Security Technique Research Center, Harbin Institute of Technology, Harbin 150001, China (March 2004).
11. Lambert Schaelicke, Thomas Slabach, Branden Moore and Curt Freeland. Characterizing the Performance of Network Intrusion Detection Sensors. Department of Computer Science and Engineering University of Notre Dame.
12. J. McAlerney, C. Coit, S. Staniford, Towards faster string matching for intrusion detection or exceeding the speed of snort, DARPA Information Survivability Conference and Exposition, Anaheim, California (June 2001) 367–373.

CHAPTER 41

A DYNAMIC POLICY BASED SECURITY ARCHITECTURE FOR MOBILE AGENTS

MISBAH MUBARAK[1], ZARRAR KHAN[1], SARA SULTANA[1], HAJRA BATOOL ASGHAR[1], H. FAROOQ AHMAD[2], HIROKI SUGURI[2], AND FAKHRA JABEEN[1]

[1]*National University of Sciences and Technology (NUST)*

[2]*Communication Technologies, 2-15-28 Omachi Aoba-ku, Sendai, Japan, hajra-mcs@nust.edu.pk, fakhra@niit.edu.pk*

Abstract: Mobile agents are undoubtedly the upcoming trend in the agent community. As their use is increasing, security problems are coming forth. A mobile agent must address several security issues. An agent may be malicious, the platform on which it executes may carry on sinister activities on it or the agent may be harmed by another malicious agent. Most of the security approaches do not provide any flexibility or dynamic decision making. Therefore, security threats increase further. In our paper, we suggest an approach that uses dynamic ontology based policies to enforce security in platforms as well as in mobile agents. The approach is a hybrid one which utilizes other security measures as well. We have also evaluated the approach on software quality parameters other than security like portability, flexibility and efficiency. The results have been satisfactory

Keywords: Mobile agents, encryption, digital signatures, ontologies, reconfigurable policies

1. INTRODUCTION

Distributed computing has undergone several fundamental changes in the past few years. Initially, the resources of a system were bound to that particular system only. Mobile agents [1] have changed the trend. They offer a lot of flexibility to the distributed computing applications. In mobile agent system, the agent moves from one host to another in order to utilize the services and resources that are available locally at the systems. An agent is an autonomous active entity that acts on behalf of its owner in order to accomplish a certain goal. It performs tasks that are favorable for the users but on the other hand it may also perform malicious tasks depending on its goals.

H. Labiod and M. Badra (eds.), New Technologies, Mobility and Security, 493–505.
© 2007 *Springer.*

The advent of mobile agents was hailed as a solution to the ever increasing demand for greater bandwidth requirement. However, technology of mobile agents did not take off as it was foreseen in the beginning solely due to security issues. Although mobile agents greatly increase the efficiency in a distributed system but the risk of security threats becomes far too great. The mobile agent may over utilize system resources, steal important information and use the system as a point of attack to other systems. Security threats due to mobile agents are classified in three basic types

- Confidential information retrieval
- Denial of service attacks
- Corruption of information

The problem specially becomes serious once mobile agents are used for mission critical and real world applications [1]. Another issue that may possibly occur is that the systems the mobile agent accesses may be malicious. It may be possible that the system tries to take important information from the mobile agent or modify the agent by adding or removing its code [2]. Therefore in general four categories of threats for the mobile agents are possible

- Threats from a malicious agent that may harm the system
- A malicious system may attack the agent
- A malicious agent may attack a useful agent
- Any other entity may attack the system

There may be several solutions to the security issues in mobile agents.

- Cryptographic authentication of the owner on behalf of which the agent is acting.
- Security in the execution environment of the languages.
- Policy decisions that are based on the owner's identity by the software that runs the agents.

A policy is a course of action, guiding principle, or procedure considered expedient, prudent, or advantageous. In terms of MultiAgent systems they are declarative rules governing choices in a system's behavior. They are able to restrict system's behavior. They can be used for flexibility, adaptability and security of a system. Policies may be expressed using formal policy languages, rule based policy notations and attribute table representations [3]. They loosely couple the code so that the runtime behavior of applications can be easily adapted according to the requirements. A misconception about policies is that they impose certain constraints on the execution of an application. Policies actually are rules that aid an agent in achieving its social goals resolving security issues. Policies require strong support of middleware. One such middleware is Java Virtual Machine (JVM) in which the high level language policies are compiled into byte codes. In the paper we use Java as the language to express policies and the underlying MultiAgent system.

An advanced variation of policies is dynamically reconfigurable policies. By the use of such policies, an application can adapt according to the requirements at runtime. However, in order to achieve this, the policies need to be semantically rich which is only possible with the help of ontologies. These policies are able to perform the operation without changing the source code of the application thus decreasing security issues.

Mobility can be classified into two basic types i.e. weak mobility and strong mobility. Weak mobility is the movement of code and data from one computational unit to another. Strong mobility is the movement of code, data and execution state from one point to another. Java due to its strong network programming capabilities, offers strong support for achieving mobility. It provides the features of dynamic class loading, serialization and machine independence which are very helpful in achieving mobility [4]. However, due to security reasons, Java does not allow any access to its run time execution state. The reason for this is that a malicious platform may harm the agent severely by modifying the execution state. Therefore, security becomes a big issue once strong mobility is implemented in the case of mobile agents.

With the help of mobility, dynamic customization and configuration of internet applications is also possible. Mobile agents also have the ability to adapt functionality according to user needs with the help of dynamically reconfigurable policies. Therefore, it is vital to deal with the security aspect of mobile agents.

The paper first discusses different possible approaches for implementing security. It then analyzes the draw backs of these approaches and argues how our proposed approach is better than those discussed in Section 2. After that, it gives the basic idea and concept of the proposed approach with the help of detailed modeling in Section 3. It then proposes the implementation of the approach in a strongly mobile FIPA compliant MultiAgent system. Conclusion and future work is given in Section 5.

2. RELATED WORK

In this section we discuss different approaches for security and prove that these approaches, when used along with policies provide a better and highly flexible solution to the problem.

We first discuss some approaches that provide security to the platforms against malicious mobile agents.

2.1. Sand Boxing

This approach is used for the protection of the platform from malicious mobile agents. In this case, local code has got access to the critical system resources. However, the remote code or the mobile agent has got restricted access rights. The remote code executes inside a restricted area called 'Sand Box'. It may affect operations like

- Interacting with local file systems
- Creating a network connection
- Accessing system properties on a local system
- Invoking programs in a local system

The approach is commonly implemented in Java since it has got a class loader, verifier and security manager that enforce security. The problem is that if any of the components fails, it leads to security violation. For instance a remote class may be wrongly classified as a local class. Our system removes this deficiency since it has got a dual authentication mechanism which leads to two pass verification procedure [10].

2.2. Code Signing

This is another approach that is used to protect platforms against malicious mobile agents. It uses digital signature which verifies that the code of the mobile agent has not been modified since it has been signed by its creator. This method verifies the identity of the owner by checking it against a list that is maintained on the system. However, it only verifies the owner and does not guarantee the safety of the code. Once the mobile agent is verified, it is given full privileges to system resources. The drawbacks of the approach are

- The approach assumes that all the entities mentioned in the list are clean
- The mobile agent is granted complete access to the resources. It may then change the trusted list and invite other malicious agents [11,12].

Our architecture solves the problem by dynamically deciding about the nature of the mobile agent based on certain parameters. After a close inspection, the architecture allows a certain percentage of access to the system resources depending upon the authenticity. Complete access to system resources is granted in highly exceptional cases.

2.3. State Appraisal Function

The state appraisal function is another approach that protects the platform. It keeps a check on the execution state of the mobile agent. In this case, the author who creates the mobile agent writes a state appraisal function in it. The function is composed of the access rights that the agent has on the platform, depending upon the current state of the agent. Therefore, the sender, instead of the platform, decides the set of permissions that are to be requested. The approach solves the problem in case the platform is malicious. However, if the mobile agent itself is malicious, it may create serious security threats. Moreover, it's very difficult to write an appraisal function every time one sends the mobile agent. Our architecture solves the problem as the dynamic policies are already encoded which may be used by the agent at runtime. Moreover, our architecture provides protection against malicious agents as well [13].

We now discuss approaches that provide security to the mobile agent against malicious platforms.

2.4. Execution Tracing

This is an approach that provides protection to the mobile agent. In this case, the mobile agent keeps a log of the action that it performs at each platform. The log consists of an identifier that identifies all the statements that the agent has executed on any platform. In case, any information is required from an external execution environment, the platform's digital signatures are mandatory. The messages that are attached to the mobile agents consist of unique identifier of the message, identity of the sender platform, time stamp, a finger print and the final state of the mobile agent. If the owner of the agent processes the execution trace of the mobile agent and suspects that a certain platform has been malicious with the mobile agent, he may ask the platform to reproduce the trace. A comparison is made between the finger prints of previous and reproduced trace. This helps in identifying any sinister activity. Disadvantage of the approach is the heavy logs that the agent has to maintain while traveling. Our proposed architecture solves the problem as the mobile agent carries only policies along with it [14].

Since the use of mobile agents poses security threats as mentioned in Section 1, **encryption of mobile agents** is an easy way to get protection against these threats. The mobile agent along with its messages is transformed into cipher text by a function that is parameterized using a public key. The cipher text is then transmitted over the network via Java RMI in the form

$C = E_K(M)$

C refers to the cipher text of the mobile agent

K represents the public key that parameterizes the function

M is the mobile agent to be encrypted

E refers to encryption of the mobile agent

The enemy may copy the entire cipher text. He is, however, unaware of the key. In some of the cases, the intruder may inject fake messages and later send them back or they may break the ciphers by cryptanalysis [8]. Therefore, making an intelligent choice for the key is mandatory. Nevertheless, if the encrypted key is found by the enemy, he may inject malicious code in the mobile agent which can then cause a serious damage to the platform. Therefore, a strong security mechanism is not provided in this case. Our dual layer security architecture solves this problem by enforcing policies on the use of system resources as mentioned in Section 3. Moreover without introducing policies, the entire agent has to be encrypted which requires a lot of computation thus lacking efficiency. Our approach solves the problem by encrypting only the confidential parts of the agent [8].

2.5. Cooperating Agents

It is an approach in which a critical task is carried out by two mobile agents. These mobile agents carry out their operations in entirely separate platforms. There is a secret and confidential communication channel between the two agents. With the help of these channels, the agents transfer their confidential information to each other like

- Agent itinerary
- List of messages sent,
- List of all the platforms that have been recently visited current one and the ones that are still pending to be visited.

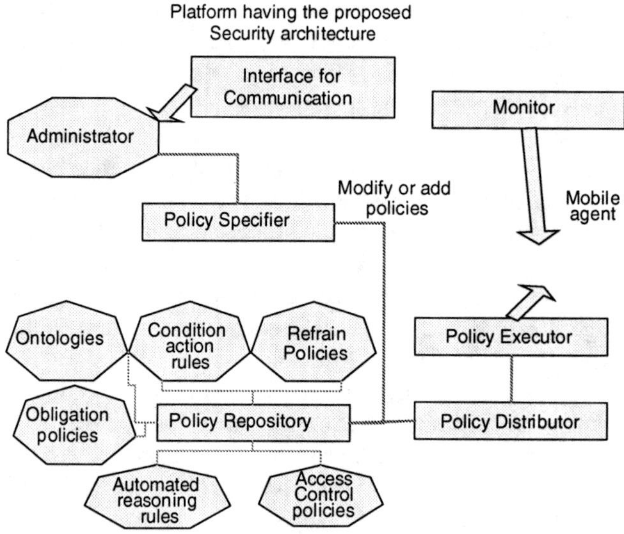

Figure 1. Proposed Security architecture

In this way if malicious platform changes the route of the agent, the cooperating agent will immediately get to know. It will then notify its sender.

The approach assumes that there are very few platforms that are actually malicious. Moreover, setting up a reliable communication channel between agents is not cost effective. Our proposed architecture poses no such problem as it does not require a separate communication channel. The security architecture software, once installed, protects the platform and all the outgoing mobile agents from the platform.

3. PROPOSED ARCHITECTURE

We propose dual layer security architecture in the use of mobile agents. Policies provide a most promising approach to control access to system resources. However, one may question what guarantee do we have that the policies are correct? Simple policies are enforced automatically without any reasoning. Therefore, if any incorrect policy is enforced, it may lead to security violation. That is the reason due to which we have used 'Ontology based semantically rich policies'. Ontologies [15,17], formally specify a concept and the inter relationship between those concepts. They can provide automated reasoning using certain inference procedures. This helps to enforce a meaningful runtime policy after automated reasoning with the help of ontologies. Ontologies also aid in resolving the heterogeneity of platforms. The condition however is, that there should be a single shared ontological representation. For an explanation see [15].

3.1. Policy Enforcement Architecture

The architecture for policy enforcement consists of the following main components
Policy Storage area: This is the storage area where all the policies and ontologies are stored.
Policy Distributor: This is the module that distributes all the policies to the desired agents.
Policy Implementer: This module maps the policies to its correct action. It uses automated reasoning and inference procedure in order to size up the situation.
Monitor: It monitors the agent execution state and its environment.
Policy specifier: It helps to edit policies and ontologies. The module can only be used by a system administrator.
Figure 1 demonstrates the lay out of security architecture.

3.2. Policies Enforced in the Architecture

The categories of policies that have been dealt in our architecture are

- Authorization policies
- Obligation policies
- Refrain policies

In the end we discuss how all these policies give security to strongly mobile agents as shown in Figure 2.

3.3. Authorization Policies in the Architecture

These are access control policies that control the access of the mobile agents to the platform resources. Policies are added in the form of event condition rules in this case. The policies of this category are further divided into

- Encryption policies
- Percentage of access to system resources

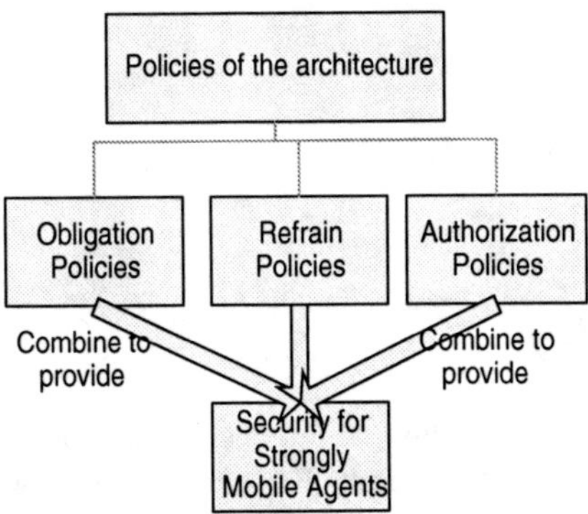

Figure 2. Policies of the architecture

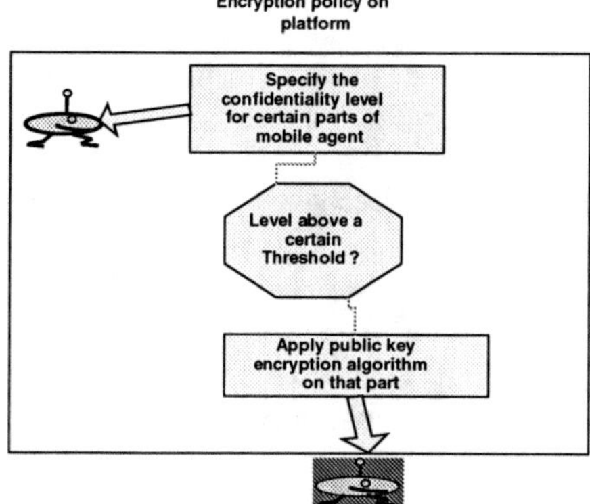

Figure 3. Encryption policy on specific part of the mobile agent

- Monitoring the agent in case of complete access
- Monitoring the behavior of the platform on which the agent is executing.

We discuss all the enforcement of the above mentioned policies in detail.

Encryption policies: In order to enforce encryption policies, the creator of the agent identifies the part that he/she feels is sensitive. That sensitive part is encrypted using a public key encryption algorithm. This saves a lot of computation effort that is wasted in case the entire agent is encrypted. The creator specifies a certain confidentiality level in integer format which if above a certain threshold (the threshold is also in integer format) will encrypt the agent automatically as shown in Figure 3.

An example of encryption policy for a mobile agent is

```
domain a=/OrganisationID/
inst authriz policyName (
on (confid_level. 90)
subject s=E/siteA/compC
target t=a/siteB/compC
do
t.go((deviceName.getSite()).toStrin
g(). "run()");
when(Creator.checkStatus(MAgent(
)==true)
)
```

Example 1: Example of an encryption authorization policy

The above example represents an authorization policy which checks the confidentiality level assigned by the creator of the mobile agent. If it is above 90 for example, the subject that is the mobile agent is encrypted. After encryption, it is sent to the destination by the *'t.go()'* function.

Policies granting access to system resources: There are other policies in the systems that check the characteristics of incoming mobile agents by using ontology based automated reasoning techniques. In other words, the credibility of the mobile agent is deduced based on certain factors like

- Sender identity
- Operation to be performed by the agent
- Check against the list of harmful threats in the code

After making all these checks using automated reasoning and inference procedures, the system decides which percentage of access is to be granted to the mobile agent.

```
domain a=/OrganizationID/
inst msrResource Percentage {
on applyInference(input);
subject s=a/siteA/check_Percentage();
target t = a/siteA/grant_Percentage();
do t.grant(grant_Percentage());
when Monitor.check(grantAccess==true);
}
```

Example 2: Policy specifying percentage of resource usage for the mobile agent

The above policy is an example of one that grants access to system resources based on the dynamic decision about the mobile agent. applyInference(input) is the name of the function that applies the inference procedure using ontologies and automated reasoning on the mobile agent. Check_Percentage() is a function that is based on the result of applyInference(input) function. It checks the level of resources that should be granted to the mobile agent. GrantAccess (Boolean flag) is the function that grants access to the system resources.

Policies monitoring agent execution in a platform: Monitoring policies are implemented by the 'Monitor' module in the architecture. The purpose of these policies is to dynamically monitor the actions of the agent. It keeps a track of whatever the agent executes. If the agent performs any malicious activity on the resources, it has the right to block the agent and suspend its execution at that very instant.

```
domain a=/OrganizationID/
inst monitor M_Execution {
on traceExecution(m_agent);
subject s=a/siteA/check_Execution();
target t = a/siteA/grant_suspendExecution();
do t.grant(check_Platform());
when Monitor.check(executionMal==true);
}
```

Example 3: monitor policy

This policy keeps a track on the execution of the agent by a function traceExecution(m_agent). The argument m_agent specifies the mobile agent. The function check_Execution() checks for malicious activity by comparing the agent activities against a list of malicious activities. Whenever the condition is found to be true, the execution of the agent is suspended using grant_suspendExecution() function.

Policies protecting the agent from malicious activities in the platform: There is another policy that monitors the platform in case if there is some virus program hidden in the system which may harm the execution of the mobile agent. Whenever there is some malicious activity going on the platform, the execution of all the programs is suspended until the malicious program is repaired or quarantined.

```
domain a=/OrganizationID/
inst monitor P_Execution {
on traceExecution(platform);
subject s=a/siteA/check_Execution();
target t = a/siteA/grant_suspendExecution();
do t.grant(Stop_Functions(argl.arg2...argn));
when Monitor.check(executionMal==true);
}
```

Example 4: Policy catering with harmful platform

The function 'traceExecution (platform)' keeps a check on the executions that are performed on it. If Monitor.check(Boolean flag) returns true which means there is malicious activity going on the platform, the execution of all programs is suspended.

3.4. Obligation Policies in the Architecture

Obligation policies are event condition rules. They specify what actions are to be performed as a result of an event. In the proposed architecture, these policies are stored in the policy repository and whenever a condition meets, the policy is enforced by the policy executor. For example, in case when the creator of the mobile agent tries to access the agent results and execution state in case of strong mobility by decrypting it, if he enters the wrong key, results are blocked.

```
inst oblig loginFailure {
on loginfail(key);
subject s = /Block;
target t = {userid};
do t.disable((s.log(userid));
}
```

Example 5: Login policy

The obligation policy 'login failure' blocks access to the mobile agent in case if wrong decryption key is entered. Loginfail(key) is the function which checks if there has been a login failure condition. The user id is blocked and disabled using the functions t.disable((s.log(userid)).

3.5. Refrain Policies in the Architecture

Refrain policies allow specifying actions that the user must not perform under certain conditions. These policies control the subject (a mobile agent in our case) to act even if it has got full access to the target [16]. For example, a normal user of a platform may not access the sensitive code, data and execution state information from the mobile agents when refrain policies are enforced. However, the creator of the mobile agent may be able to access it. The verification may be made using digital signatures.

```
inst refrain NonDisclosure {
subject s = /normalUsers;
action HideProjectInformation();
target t = /people;
when t.signedNDA = false
}
```

Example 6: Disclosure of information policy

With the help of this policy, the information is not disclosed to local users as they cannot verify the digital signature process. The function HideProjectInformation() hides the project or the application information whenever the user does not confirm the digital signatures. Test for digital signature verification is made by t.signedNDA function [3].

3.6. Policies for Strongly Mobile Agents

In case of strong mobility, the agents carry their execution state along with them as mentioned in Section 1. Though carrying the execution state of an agent increases the efficiency as the overhead of restarting the mobile agent at the destination is eliminated. However, it poses severe security threats since if the execution state of the agent is modified maliciously, it may lead to severe damage to all the platforms that the agent visits. Therefore, the Monitor component of the security architecture monitors the agent execution all the time. Whenever the agent's execution state is tried to be modified by some external source, this module performs a check policy and enforces blockage of that external source. Policies in Example 5 and 6 are instances of strong mobility's security policies. Therefore, the architecture solves the security problems that are posed by strong mobility. See Figure 4. It shows whenever a malicious program tries to access the execution state of an agent, it gets terminated.

3.7. Evaluation of the Architecture Against Quality Parameters

We evaluate the software against the following quality parameters

Reliability: The security architecture enforced is hybrid architecture. It employs a manifold technique of security including policies, encryption and digital signature.

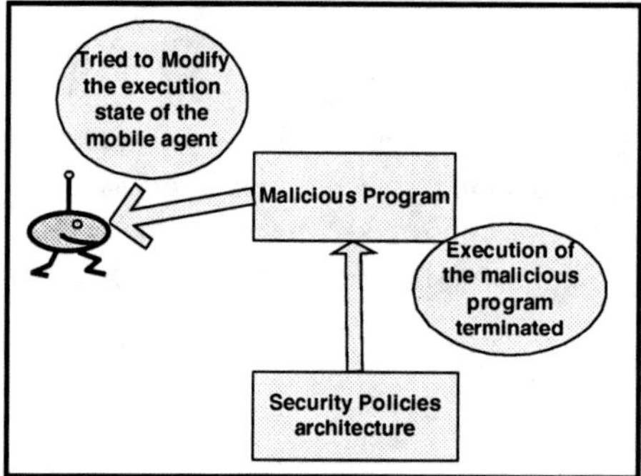

Figure 4. Malicious program trying to execute in security architecture

It deals with all issues which may harm the platform or the mobile agent. At present, there is no security loop hole. However, we are working further on the architecture for the identification of any other security problems.

Efficiency: The architecture imposes such policies which encrypt only the sensitive portion of the agent. This requires less computation and makes the encryption process efficient. The mobile agent does not carry logs along with it which require more bandwidth and flood the network. The architecture is therefore efficient.

Portability: the policy enforcer is a pluggable component which may get plugged into any system independent of the underlying architecture or MultiAgent system.

Security: As proved in the previous section, the architecture implements manifold security mechanism.

4. CONCLUSION AND FUTURE WORK

In the paper we have proposed a security framework for strongly mobile agents. The security plug in once installed on any platform has the ability to protect all the incoming and outgoing mobile agents. It also makes sure that no malicious activity is carried out on the agent platform. Moreover, the architecture is a hybrid approach which implements policies on the top of other security mechanisms like encryption, digital signature etc. This adds flexibility in the system as the security mechanisms can be utilized using dynamic and reconfigurable system policies. Without these policies, the security mechanisms included a single pass security check. Moreover, the system lacked flexibility as the mechanisms were hard coded. The dynamic policies introduce the element of intelligence and runtime decision making in the system by inference procedures. Currently, we are working on Strong mobility in any FIPA compliant MultiAgent system. In the future we intend to plug in this security architecture in the Strong mobility system.

REFERENCES

1. "Mobile Agents White Paper", General Magic, 1998. <URL:http://www.genmagic.com/technology/techwhitepaper.html>
2. Robert S. Gray and David Kotz and George Cybenko and Daniela Rus: "D'Agents: Security in a multiple-language,mobile-agent system"
3. Rebecca Montanari, Emil Lupu, Cesare Stefanelli: "Policy based dynamic reconfiguration of mobile code"
4. Misbah Mubarak, Sara Sultana, Zarrar Khan, Hajra Batool Asghar, H Farooq Ahmad, Fakhra Jabeen: "A review of mobility techniques". *In proceedings of 19th Assurance Systems Symposium, Tokyo Institute of Technology, Japan*
5. Wolfgang Nejdl, Daniel Olmedilla, Marianne Winslett, and Charles C. Zhang: "Ontology-Based Policy Specification and Management", Research Center and University of Hannover, Germany
6. Todd Papaioannou : "On the structuring of distributed systems, the argument for mobility", doctoral thesis
7. Wayne Jansen, Tom Karygiannis National Institute of Standards and Technology: NIST Special Publication 800-19 – "Mobile Agent Security"
8. Andrew S Tenanbaum: "Computer networks", 4th edition, vrije universiteit Amsterdam, Netherlands
9. "Fast software encryption": 4th International workshop, FSE, Haifa, Israel
10. R. Wahbe, S. Lucco, T. E. Anderson, and S. L. Graham, "Efficient software-based fault isolation," In Proceedings of the 14th ACM Symposium on Operating Systems Principles, pages 203–216, Dec. 1993.
11. "Signed Code," (n.d.). Retrieved December 15, 2003, from James Madison University, IT Technical Services Web site: http://www.jmu.edu/computing/infosecurity/engineering/issues/signedcode.shtml
12. "Introduction to Code Signing," (n.d.). Retrieved December 15, 2003, from Microsoft Corporation, Microsoft Developer Network (MSDN) Web site: http://msdn.microsoft.com/library/default.asp?url=/workshop/security/authcode/intro_authenticode.asp
13. W. M. Farmer, J. D. Guttman, and V. Swarup, "Security for mobile agents:Authentication and state appraisal," In Proceedings of the European Symposium on Research in Computer Security (ESORICS), pages 118–130, Sep. 1996.
14. H. K. Tan and L. Moreau, "Extending Execution Tracing for Mobile Code Security," In K. Fischer and D. Hutter (Eds.), Proceedings of Second International Workshop on Security of Mobile MultiAgent Systems (SEMAS'2002), pages 51-59, Bologna, Italy.2002.
15. Misbah Mubarak, Sara Sultana, Zarrar Khan, Hajra Batool Asghar, H Farooq Ahmad, Fakhra Jabeen: "An approach to ontological interoperability", In proceedings of 2nd IEEE international conference on emerging technologies. (ICET 2006)
16. John A. Knottenbelt: A report on Policies for Agent Systems, Imperial College of Science, Technology and Medicine
17. M Andrea Rodriguez : "Similarity based ontology integration". In Proceedings of the 1st International Conference on Geographic Information Science, October 2000

CHAPTER 42

A SCHEME FOR INTRUSION DETECTION AND RESPONSE IN AD HOC NETWORKS

MARIANNE A. AZER[1], SHERIF M. EL-KASSAS[2], AND MAGDY S. EL-SOUDANI[3]

[1]*National Telecommunication Institute, Computer Department, Cairo, Egypt*
Email: marazer@nti.sci.eg,
[2]*American University in Cairo, Computer Science Department, Cairo Egypt*
Email: sherif@aucegypt.edu,
[3]*Cairo University, Faculty of Engineering, Electronics and Communications Department, Cairo, Egypt*
Email: melsoudani@menanet.net

Abstract: The dynamic and cooperative nature of ad hoc networks present substantial challenges in securing and detecting attacks in these networks. In this paper, we propose three schemes for intrusion detection in ad hoc networks and demonstrate their effectiveness by applying them to the wormhole attack. The first scheme is based on attack graphs, the second is based on the theory of the diffusion of innovations and the third is based on the aggregation lists of events. The advantages and disadvantages of each scheme are identified and a combined model for intrusion detection is presented. Furthermore, a response module is proposed to augment the intrusion detection functions

Keywords: Ad hoc networks, intrusion detection, security, wormhole attack

1. INTRODUCTION

The dynamic and cooperative nature of ad hoc networks present substantial challenges in securing these networks. There are recent research efforts in providing various attack prevention schemes, e.g. authentication and encryption schemes, to secure the ad hoc routing protocols. Such security measures are not sufficient all the time, and intrusion detection remains an important security goal (to complement the various existing detection mechanisms). Intrusion detection techniques can be mapped into three categories [1]: signature-based detection, anomaly detection, and specification-based detection. However, the special characteristics of ad hoc

H. Labiod and M. Badra (eds.), New Technologies, Mobility and Security, 507–516.
© 2007 *Springer.*

network, like the lack of support infrastructure, dynamic network topology, distributed operation, bandwidth constraints, variable capacity links, use of low power devices, limited CPU and memory, limited physical security, and complexity of design of network protocols impose lots of limitations to the intrusion detection system needed for those networks.

In this paper we propose three intrusion detection schemes which may be classified as anomaly detection schemes and demonstrate their application to the detection of the wormhole attack. The first is a centralized scheme based on attack graphs, the second and the third scheme are decentralized and are based on the theory of the diffusion of innovations and the aggregation of lists of events, respectively. The advantages and disadvantages of each scheme are identified and a combined model for intrusion detection is presented. Furthermore, a response module is proposed to augment the intrusion detection functions.

The remainder of this paper is organized as follows. Section 2 contains the basic assumptions and a brief introduction to the routing protocol considered throughout this paper and the wormhole attack. In section 3, a novel centralized intrusion detection scheme is proposed, whereas in section 4 two decentralized intrusion detection schemes are presented. In section 5 the advantages and disadvantages of the proposed intrusion detection schemes are identified and a combined model with an added response module is proposed. Finally in section 6, conclusions and future works are given.

2. BACKGROUND

In this section we present our basic assumptions and give a brief overview on the Ad Hoc on Demand Distance Vector (AODV) routing protocol and the wormhole attack.

2.1. Assumptions

We assume that there are Network Monitors NMs that monitor all the suspected parameters for the wormhole attack such as the route request (RREQ) and route reply (RREP) packets, the back off, the transmission power the number of packets in and out from a node. When a transmission occurs from a source to a destination, those parameters are used to make a central decision about the network's behavior according to the chosen decision scheme. Our intrusion detection architecture allows cooperative NMs to promiscuously monitoring all suspected parameters, and exchange their local data if necessary through secure channels and will not be compromised [2]. Our intrusion detection model is subject to the following assumptions:

1. The MAC and IP addresses of all mobile nodes are registered in NMs and remain unchanged.
2. MAC addresses cannot be forged.
3. Every NM and its messages are secure and authenticated.

4. Every node must forward or respond to the messages according to the protocol within some finite period of time.
5. NMs are well selected to be able to cover all nodes and perform all required functionality.
6. If a node is out of range of a NM, it must be in the range of neighboring monitors.
7. If some nodes do not respond to broadcast messages, this will not cause serious problems.
8. The (AODV) routing protocol is the only network protocol in the network and each node has only one network interface.

2.2. The Ad Hoc on Demand Distance Vector Routing Protocol

The AODV [3] builds and maintains routes between nodes only as needed by source nodes. When a source node desires a route to a destination for which it does not already have a route it broadcasts RREQ packet across the network. Nodes receiving this packet update their information for the source node and set up backwards pointers to the source node in the route tables. In addition to the source node's IP address, current sequence number, and broadcast ID, the RREQ also contains the most recent sequence number for the destination of which the source node is aware. A node receiving the RREQ may send a RREP if it is either the destination or if it has a route to the destination with corresponding sequence number greater than or equal to that contained in the RREQ. If this is the case, it unicasts a RREP back to the source. Otherwise, it rebroadcasts the RREQ. Nodes keep track of the RREQ's source IP address and broadcast ID. If they receive a RREQ, which they have already processed, they discard the RREQ and do not forward it. As the RREP propagates back to the source, nodes set up forward pointers to the destination. Once the source node receives the RREP, it may begin to forward data packets to the destination. If the source later receives a RREP containing a greater sequence number or contains the same sequence number with a smaller hop count, it may update its routing information for that destination and begin using the better route. As long as the route remains active, it will continue to be maintained. A route is considered active as long as there are data packets periodically traveling from the source to the destination along that path. Once the source stops sending data packets, the links will time out and eventually be deleted from the intermediate node routing tables. If a link break occurs while the route is active, the node upstream of the break propagates a route error (RERR) message to the source node to inform it of the now unreachable destination(s). After receiving the RERR, if the source node still desires the route, it can reinitiate route discovery.

2.3. The Wormhole Attack

A particularly severe security attack, called the wormhole attack, has recently been introduced in the context of ad hoc networks. During the attack [4], a malicious node captures packets from one location in the network, and "tunnels" them to another malicious node at a distant point, which replays them locally. The tunnel can be established in many different ways, such as through an out-of-band hidden channel (e.g. a wired link), packet encapsulation, or high powered transmission. This tunnel makes the tunneled packet arrive either sooner or with less number of hops compared to the packets transmitted over normal multihop routes. This creates the illusion that the two end points of the tunnel are very close to each other. The wormhole attack can affect network routing, data aggregation and clustering protocols, and location-based wireless security systems. Several solutions have been proposed in the literature for the wormhole attack, the solutions can be categorized into location and time based approaches such as in [5], [6], [7],[8], [9] and [10], key based approaches such as in [11], statistical based approaches such as in [12], and [13] and graph based solutions as in [14], and [15].

3. CENTRALIZED INTRUSION DETECTION SCHEME

In this section we propose a novel scheme for intrusion detection using attack graphs. Network attack graphs represent a collection of possible penetration scenarios in a computer network. The graph can focus on the extent to which an adversary can penetrate a network to achieve a particular goal, given an initial set of capabilities. They represent not only specific attacks but categories of attacks. They can detect previously unseen attacks which have common features with attacks in graphs. Based on the wormhole attack description in section 2, we have developed an attack graph for this type of attacks; it is depicted in Fig. 1. Separating attack nodes based on functionality enhances the expressiveness of attack trees for automated vulnerability & threat analysis. We shall adopt the stratified node topology suggested in [16]. It has three primary layers: event level nodes, state level nodes, and top level nodes. Event level nodes correspond to events resulting from a hacker's attempt to exploit system vulnerabilities & are used in determining the ultimate goal of an attacker; state level nodes define conceptual steps an attacker takes to reach a goal, whereas top level nodes represent the ultimate intentions of an attacker. In our wormhole attack graph, event nodes, state nodes and top level nodes are represented by the symbols T_1, T_2, and T_3 respectively. For example, in the graph shown below, if node A receives the RREQx forwarded by the source node, this state is not suspicious and is not marked as a triggering event. However, after having received the RREQx, node A did neither send a RREP to the source node to declare having a route to the destination nor forward the RREQx to its neighbors; but rather it sent a normal packet. Hence this event is suspicious and should be considered as an event level node and marked as the first triggering event T_1. Similarly, if a node B receives the RREQx, apparently there is no problem as the RREQx was already sent by a source node. However this forwarded RREQx followed a normal packet reaching node B that should have been forwarded as a normal packet to node B's neighbors or followed by a RREQ different than RREQx. Hence this state should be considered as a state level node and marked as T_2. Finally a RREP which contains

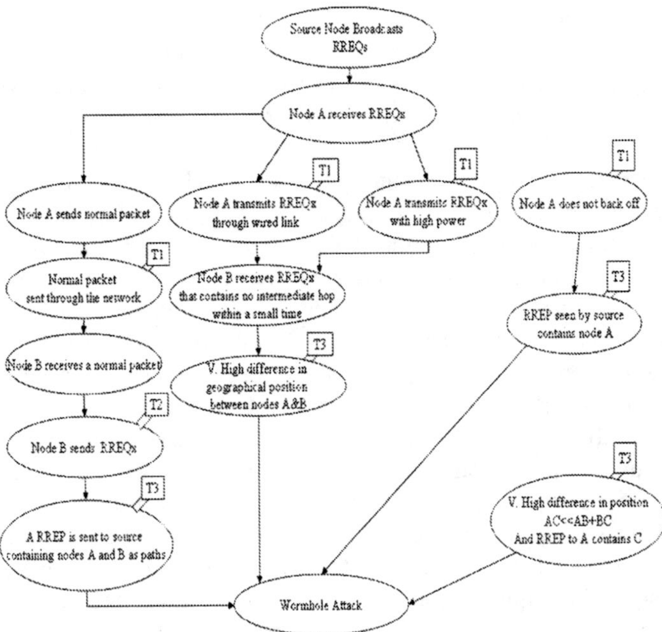

Figure 1. The wormhole attack graph with stratified nodes

as an optimum route two nodes who had suspicious behavior should be considered as a top level node and marked as T_3. Different alarms are sent to the nodes by the central authority according to the triggering event's level. Information regarding the packets entering and exiting from a node are sent to the central authority from the distributed network monitors. The central authority marks the suspected events as triggering events in the described stratified way, allocates the suspected nodes in the network and chooses a proper intrusion response.

4. DECENTRALIZED INTRUSION DETECTION SCHEMES

In this section we propose two decentralized intrusion detection schemes. The first is based on the theory of diffusion of innovations and will be presented in section 4.1 and the second is based on the aggregation of lists of events and will be presented in section 4.2.

4.1. The Diffusion of Innovations

In [17], a decision making scheme was suggested for scenarios in which ad hoc nodes must reconfigure when that network-wide consensus is needed before any change can take place. This scheme is based on the social science theory of the diffusion of innovations. We propose applying this scheme to our intrusion detection problem with a customization to make it suitable for our detection model as follows. We call the intrusion detection in this scheme the persuasion. The persuasion has two main phases, namely the NM's observations and the NM's judgment as it is shown in Fig. 2. First, a NM attempts to form its own observations based on the collected parameters, mentioned in section 2, by comparing them to a threshold. For example if the number of RREPs containing the same nodes in the selected optimum path and the power that was used to transmit their corresponding RREQs within this repeated path exceed a certain threshold, a NM marks those nodes as suspected ones in its set of observations. Strong evidence can be defined as high correlation between a NM's observations and its neighbors' observations. Four actor types are selected to form a decision model to be used in the second phase of the persuasion, when NMs attempt to make a judgment. If the NM succeeds to make a decision, it is labeled an early adopter (EA). If the NM cannot make evidence based decision because there is low correlation between its observations and its neighbors' observations, a NM attempts to make a weaker judgment based on what its peers think. The first type of weaker judgment is labeled as early majority (EM) if a NM has more than a certain threshold of neighbors holding the same particular judgment who are themselves EAs. Failing this, the node attempts to make a late majority (LM) decision. The criterion for this is to have a certain threshold of neighbors holding a common judgment who are themselves either EAs or EMs. If no judgment can be made, a node is labeled a laggard (LD). This judgment-making process continues in a parallel fashion throughout the network. All of the judgments that are made are soft state decisions. Back to our example, during the observation phase each node had marked some nodes as suspicious. In the judgment phase each monitor checks if suspicious nodes in his list are suspicious in other NM's lists. By trying to make a judgment and adopt one of the actor roles described above, to become EA, EM, LM, or LD in deciding that a node is malicious. For the network

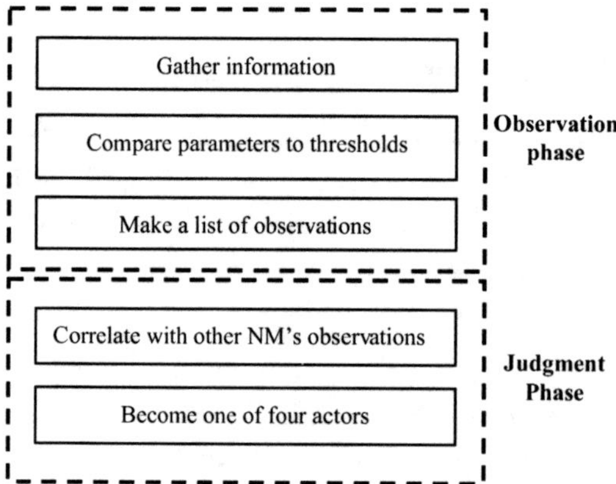

Figure 2. The two persuasion phases in the proposed intrusion detection scheme

to reach global consensus, a critical mass of EAs must emerge pointing out that certain nodes are malicious and need to be isolated.

4.2. Aggregated Lists of Events

The aggregated lists of events scheme was proposed in [18]. A similar scheme can be used for our intrusion detection problem as follows. The basic idea is to set up a monitor at each node in the network to produce evidences, and then share them among all nodes. Evidence is a set of relevant information about the network state. For each captured packet, the monitor displays a complete view of the packet and payload, and adds some general statistics as a timestamp, frame number and length in bytes. Together with the captured packets, relevant statistics like counters for transmission, signal strength, carrier sensing time, and packet transmission time are added. In this way a list of events at each NM is built, and an algorithm is used to match two lists of events. It starts from the first list and for every event it tries to find a matching event on the second list for the same packet. The final output is a single list of events which combines lists. A few packets disappearing in the two lists are index of channel failures, while a sequence of disappearing packets is considered as an attack. The idea is to merge one list at a time with the result of the previous merge. In other words, $list_1$ and $list_2$ are merged, and the result is matched with $list_3$, until every list is processed. In this way an aggregated list of all events which happened in the network in a given time frame is obtained. As an example, one of the events could be a RREQx that was sent from a source node for route discovery. While matching with the NM of the node to which the RREQ was sent, neither a corresponding RREPx nor a forwarded RREQx were found in the output packets of the node in question. Hence there is a mismatch that occurred while aggregating the two lists. The lists of the whole NMs are treated similarly to get an aggregated list of all events which happened in the network in a given time frame. Each node can raise a vote about what happened in the network, majority rule is used and a clash between the honest and cheaters reveals the situation.

5. COMBINED INTRUSION DETECTION AND RESPONSE MODEL

In the previous sections, we have proposed three intrusion detection schemes. Each of the proposed schemes has its advantages and disadvantages. The attack graph is applicable to other types of attacks, which are not only routing attacks but also attacks at the application level. It also helps in identifying the malicious nodes, and expecting the incoming actions in some cases. Nevertheless, there is no clear technique for its automated generation, and it needs a central authority which is, although dynamically chosen, prone to attacks and not practical for ad hoc networks. The decentralized diffusion of innovations benefits from being completely distributed and not needing lots of resources, however it needs strong evidence for convergence and cannot reach a stable solution easily with the dynamic network topology change. The aggregated lists of events scheme has the advantage of allowing each node doing the merge to make its own decision and construct its opinions with its own measures without needing a global trust policy for the network. Nevertheless, it suffers from the difficulty of matching the lists and the storage requirements for the list. The three schemes also need a post detection action to be taken after an intrusion is taken. In [19], a testbed for post intrusion detection actions called ADEPTS was presented. ADEPTS focuses on containment, i.e. restricting the effect of the intrusion to a subset of the entire set of services which may allow users access to limited functionality of the system. Furthermore, it estimates the likely path of spread of the intrusion from the alarms & the structure of the attack graph and then determines the appropriate responses to be taken. Another approach was proposed in [10] where a TRACE packet is used to identify the location of hop count change by a malicious node and isolate it. Still after isolating malicious nodes, the network needs to be reconfigured properly. In [20], a cut set algorithm builds a map of available nodes within the network and finally Iterations are done through the resultant cut set to reestablish a path and guarantee network normal operation.

To benefit form each of the proposed intrusion detection schemes and have a complete model that possesses intrusion detection and response capabilities we propose the combined scheme depicted in Fig. 3. In the combined scheme, the NMs collect some parameters and compare them with thresholds. If any of the parameters exceeds its threshold, a first solution uses the wormhole attack graph to allocate the attack's region and then ADEPTS estimates the likely path of spread of the intrusion from the alarms & the structure of the attack graph and then determines the appropriate responses to be taken. A second solution is the use the diffusion of innovations in which a NM adopts one of four actor types, EA, EM, LM, or LD and the network reaches global consensus when a critical mass of EAs emerges and then a decision that a node is malicious is taken. The third scheme can also be used to form an aggregated list of events to figure out the suspicious events and a voting scheme is used to decide that a node is malicious. The two decentralized schemes

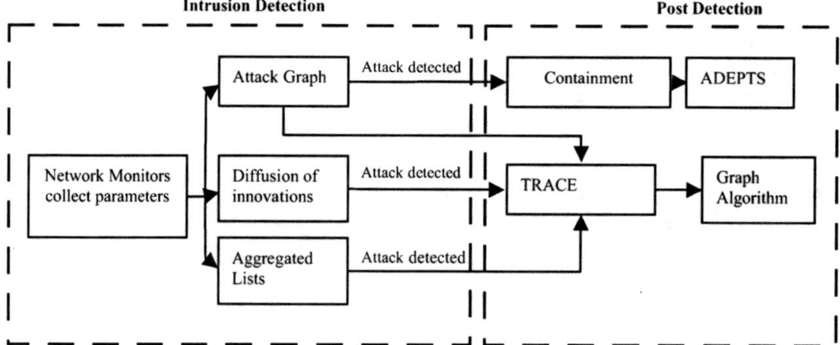

Figure 3. Proposed combined model for intrusion detection and response

can also be merged together. By aggregating lists using the aggregated list of events method and then using the diffusion of innovations, the NMs can correlate between aggregated lists to become one of the four actors. Afterwards, the attack graph can be used to locate the attack's region. Once an attack is declared a TRACE packet is used to detect the malicious nodes, the graph algorithm for recovery and failover in ad hoc networks can then be used to determine the minimum cut set and reestablish connections between network nodes. It should be mentioned that the TRACE and graph algorithm based techniques can also be used in conjunction with the attack graph detection technique.

6. CONCLUSIONS AND FUTURE WORK

In this paper, we focused on the intrusion detection in ad hoc networks. NMs were used for monitoring certain routing parameters and comparing them with thresholds. Three different schemes for decision making were proposed, the first was a novel centralized scheme based on attack graphs. An attack graph for the wormhole attack was developed and the scheme's effectiveness was demonstrated based on this attack graph. Two distributed schemes were then proposed, one scheme based on the diffusion of innovations and the other on list aggregation. The advantages and disadvantages of each scheme were identified and a combined scheme for intrusion detection and response was proposed. In the combined scheme, three post intrusion detection actions were presented, the first can be used in conjunction with the attack graph detection technique and the second and third can be used to detect, isolate and reestablish network connections with the three detection schemes. For the future we plan to evaluate build an elaborate simulation environment to experiment with and evaluate our proposed schemes' performance through simulations. We also plan to design a complete combined scheme from the three detection and response techniques. The threshold values for the NMs' parameters should be investigated through simulations as well.

REFERENCES

1. Anantvalee T, Wu J (2006) *A Survey on Intrusion Detection in Mobile Ad Hoc Networks.* Wireless/Mobile Network Security: Springer 2006, Ch7, pp. 170–196.
2. Tseng C, Song T, Balasubramanyam P, Ko C, Levitt K (2005) "A Specification-Based Intrusion Detection Model for OLSR". RAID 2005, pp. 330–350.
3. Perkins CE, Royer EM (2000) *The Ad hoc On-Demand Distance Vector Protocol.* In C. E. Perkins editor, Ad hoc Networking, Addison-Wesley, 2000, pp. 173–219.
4. Khalil I, Bagchi S, Shroff NB (2005) "LITEWORP: A Lightweight Countermeasure for the Wormhole Attack in Multihop Wireless Networks." *dsn,* 2005 International Conference on Dependable Systems and Networks (DSN'05), pp. 612–621.
5. Hu YC, Perrig, A, Johnson DB (2003) "Packet leashes: A defense against wormhole attacks in wireless network,." *INFOCOM.* (2003).
6. Capkun S, Buttyan L, Hubaux J (2003) "SECTOR: Secure Tracking of Node Encounters in Multi-hop Wireless Networks. "*ACM Workshop on Security of Ad Hoc and Sensor Networks (SASN),* pp. 1–12 Washington, USA, Oct 2003.

7. Wang, W, Bhargava, B. Lu, Y. and Wu, X. "Defending against Wormhole Attacks in Mobile Ad Hoc Networks," *preprint,* http:/www.cs.purdue.edu/homes/wangwc/papers/MC2R sample101104.pdf
8. Hu L, Evans D (2004) "Using Directional Antennas to Prevent Wormhole Attacks." *Network and Distributed System Security Symposium (NDSS),* San Diego, Feb 2004.
9. Vassilaras S, Vogiatzis D, Yovanof G (2005) "Misbehavior Detection in Clustered Ad-hoc Networks with Central Control," *itcc,* pp. 687–692, *International Conference on Information Technology: Coding and Computing (ITCC'05)* - Volume II, 2005.
10. Wang X (2006) "Intrusion Detection Techniques in Wireless Ad Hoc Networks." compsac, oo. 347–349, *30th Annual International Computer Software and Applications Conference (COMPSAC'06),* 2006.
11. Zhang Y, Liu W, Lou W, Fang Y (2005) Securing sensor networks with location-based keys, *WCNC 2005 - IEEE Wireless Communications and Networking Conference,* no. 1, March 2005, pp. 1909–1914.
12. Song N, Qian L, Li X (2005) "Wormhole Attacks Detection in Wireless Ad Hoc Networks: A Statistical Analysis Approach," *ipdps,* p.289a, *19th IEEE International Parallel and Distributed Processing Symposium (IPDPS'05)* - Workshop 17, 2005.
13. Buttyan L, Dora L, Vajda I (2005) "Statistical wormhole detection in sensor networks," Hungary, July 2005.
14. Wang W, Bhargava B (2004) "Visulization of wormholes in sensor networks," *Proceeding of the ACMWorkshop onWireless Security (WiSe),* pages pp. 51–60, 2004.
15. Azer M, El-Kassas S, El-Soudani M (2006) "Using Attack Graphs in Ad Hoc Networks - For Intrusion Prediction Correlation and Detection.," *SECRYPT 2006,* pp. 63–68.
16. Daley K, Larson R, Dawkins J (2002) "A Structural Framework for Modeling Multi-Stage Network Attacks," icppw, p. 5, *2002 International Conference on Parallel Processing Workshops (ICPPW'02),* 2002.
17. Forde T, Doyle L, O'Mahony D (2006) Ad Hoc Innovation: Distributed Decision Making in Ad Hoc Networks. IEEE *Communications Magazine,* vol. 44, no. 4, April 2006.
18. Aime M, Calandriello G, Lioy A (2006) "A Wireless Distributed Intrusion Detection System and a New Attack Model," iscc, pp. 35–40, *11th IEEE Symposium on Computers and Communications (ISCC'06),* 2006.
19. Foo B, Wu Y, Mao Y, Bagchi S, Spafford E (2005) "ADEPTS: Adaptive Intrusion Response Using Attack Graphs in an E-Commerce Environment," dsn, pp. 508–517, *2005 International Conference on Dependable Systems and Networks (DSN'05),* 2005.
20. Watkins D, Scott C, Randle D (2006) "A Graph Alorithm Based Approach to Recovery and Failover in Tactical Manets," pp. 253–260, *Seventh ACIS International Conference on Software Engineering, Artificial Intelligence, Networking, and Parallel/Distributed Computing (SNPD'06),* 2006.

CHAPTER 43

QUERY ANSWERING IN DISTRIBUTED DESCRIPTION LOGICS

FAISAL ALKHATEEB AND ANTOINE ZIMMERMANN
{faisal.alkhateeb,antoine.zimmermann}@inrialpes.fr
INRIA Rhône-Alpes and LIG, France

Abstract: This paper describes the notion of query answering in a distributed knowledge based system, and gives methods for computing these answers in certain cases. More precisely, given a distributed system (DS) of ontologies and ontology mappings (or bridge rules) written in Distributed Description Logics (DDL), distributed answers are defined for queries written in terms of one particular ontology. These answers may contain individuals from different ABoxes. To compute these answers, the paper provides an algorithm that reduce the problem of distributed query answering to local query answering. This algorithm is proved correct but not complete in the general case

1. INTRODUCTION

The emergence of the Semantic Web has focused the attention on developing systems in which knowledge can be shared in a distributed environment. Besides, query answering in expressive knowledge base is a difficult task. Therefore, when knowledge description is separated into different knowledge bases, query answering becomes an even more tedious problem. The present paper investigates a new approach to the problem, where local knowledge bases (ontologies) may represent heterogeneous domains, but are related with directional mappings that express how one can interpret foreign knowledge from a given peer's point of view. Queries are posed in terms of one local ontology (the target), and answers are given in the context of the target ontology, while taking advantage of the overall distributed knowledge. For such an approach, Distributed Description Logics (DDL [1]) is an appropriate knowledge representation language, but it has currently no supports for queries with variables. So, this paper defines the notion of distributed answers to a query. Then, to evaluate queries, we propose to reduce the problem of distributed

H. Labiod and M. Badra (eds.), New Technologies, Mobility and Security, 517–528.
© 2007 *Springer.*

query answering to a local query answering problem, assuming that there already exist local query evaluation algorithms (*e.g.*, [2,3]).

Several peer-to-peer (P2P) data management systems have been proposed recently, which are divided into two main categories: centralized systems (*e.g.*, [4]) and decentralized systems (*e.g.*, [5,6,7]). [5] presents a relational P2P data management system, and the mappings between relational peer schemas are inclusion and equivalence statements of conjunctive queries. The Piazza system [8] is a P2P data management system that relies on a tree based data model: data in XML and XQuery-based mapping language for mediating between peers. PEPSINT [4] supports interoperation of both XML and RDF data sources, using a hybrid architecture with a super peer containing the global ontology. EDUTELLA [6] provides an RDF-based metadata infrastructure for P2P networks. [9,7] describe the SomeWhere semantic P2P data management system that promotes a small vision of the Semantic Web based on simple ontologies distributed at a large scale, and logical mappings between ontologies make possible the creation of a web of people.

Most of the existing systems are assumed to work with rather homogeneous data, and they might prove useless if different ontologies are developed for a different context of application. This is the particularity of our work: DDL can handle distributed knowledge where each ontology may provide different context, and mappings between domains are handled through the use of the so-called bridge-rules. Another peculiarity of our approach is that, even though the query is posed in terms of one particular local ontology, a single answer may contain individuals from several knowledge bases.

The paper is organized as follows: we present the syntax and the semantics of DDL in Section 2. Section 3 presents the syntax and the semantics of the query language that we consider. In Section 4, we prove a theorem that provides a guideline for a query evaluation procedure over DDL. The concluding remarks and further work are presented in Section 5.

2. DISTRIBUTED DESCRIPTION LOGICS

Our work is based on DDL [1]. DDL serves to describe a distributed knowledge base (DKB) composed of several local KBs (written in standard DL) and of "bridge rules" that serve to connect terms from different local KBs.

2.1. Ontology Language: DL

Syntax. Our local KBs, that we will refer to as *ontologies*, are written in Description Logics (DL). The basic elements in DL are concepts, roles and individuals. Concepts (class of individuals) and roles (relations between individuals) are either primitive (named concepts or roles) or complex (recursively defined with constructors and other concepts or roles). Individuals can only be described by a name. Constructors are given in Table 1.

Table 1. Syntax and semantics of DL constructors

Construct name	Syntax	Semantics
individual	a	$a^I \in \Delta^I$
atomic concept	A	$A^I \subseteq \Delta^I$
universal concept	$\top$	$\top^I = \Delta^I$
empty concept	$\bot$	$\bot^I = \emptyset$
conjunction	$C \sqcap D$	$(C \sqcap D)^I = C^I \cap D^I$
disjunction	$C \sqcup D$	$(C \sqcup D)^I = C^I \cup D^I$
negation	$\neg C$	$\Delta^I \setminus C^I$
exists restriction	$\exists R.C$	$\{x \mid \exists y.\langle x, y\rangle \in R^I \wedge y \in C^I\}$
value restriction	$\forall R.C$	$\{x \mid \forall y.\langle x, y\rangle \in R^I \Rightarrow y \in C^I\}$
number	$\leq nR$	$\{x \mid \sharp\{y.\langle x, y\rangle \in R^I\} \leq n\}$
restrictions	$\geq nR$	$\{x \mid \sharp\{y.\langle x, y\rangle \in R^I\} \geq n\}$
nominals	$\{a_1, \ldots, a_n\}$	$\{a_1^I, \ldots, a_n^I\}$
atomic role	R	$R^I \subseteq \Delta^I \times \Delta^I$
role conjunction	$R \sqcap S$	$(R \sqcap S)^I = R^I \cap S^I$
role disjunction	$R \sqcup S$	$(R \sqcup S)^I = R^I \cup S^I$
role complement	$\neg R$	$(\Delta^I \times \Delta^I) \setminus R^I$
transitive closure	R^+	transitive closure of R^I
inverse role	R^-	$\{\langle y, x\rangle \mid \langle x, y\rangle \in R^I\}$
role composition	$R \circ S$	$\{\langle x, z\rangle \mid \exists y.\langle x, y\rangle \in R^I \wedge \langle y, z\rangle \in S^I\}$
TBox axioms	**Syntax**	**Interpretation constraints**
subsumption	$C \sqsubseteq D$	$C^I \subseteq D^I$
role inclusion	$R \sqsubseteq S$	$R^I \subseteq S^I$
role transitivity	Trans(R)	$R^I = (R^+)^I$
ABox axioms	**Syntax**	**Interpretation constraints**
class membership	$C(a)$	$a^I \in C^I$
role membership	$R(a_1, a_2)$	$\langle a_1^I, a_2^I\rangle \in R^I$
idendity	$a_1 = a_2$	$a_1^I = a_2^I$

Ontologies are composed of axioms asserting truth about a knowledge domain. We distinguish terminological axioms, *i.e.*, statements about concepts and roles, and assertional axioms, *i.e.*, facts about individuals. All possible axioms are summarized in Table 1.

Definition 1 (Ontology) *An ontology (or local knowledge base) O is a pair* $\langle T, A\rangle$ *where T is a set of terminological axioms called TBox, and A is a set of assertional axioms called ABox.*

Given an ontology O, we denote by Sig(O) the set of all terms (primitive concept/role and individual names) appearing in the axioms of O and we call it the *signature* of O. Moreover, given a set of terms Σ, we denote by $\mathcal{A}(\Sigma)$ the set of all possible axioms inductively built out of terms of Σ and constructors from Table 1.

Semantics. The semantics of DL define the notion of interpretation, and specify when an interpretation satisfies an axiom or an ontology.

Definition 2 (Interpretation of a set of terms) *Given a set of terms Σ, an interpretation I of Σ is a pair $\langle \Delta^I, \cdot^I \rangle$ where Δ^I is a non-empty set called the domain of I, and $\cdot^I$ is an interpretation function that maps individuals of Σ to elements of Δ^I, concepts of Σ to subsets of Δ^I and roles of Σ to subsets of $\Delta^I \times \Delta^I$. The interpretation function is extended to complex concepts or roles by applying interpretation rules of Table 1 recursively.*

Satisfiability allows one to identify interpretations that are consistent with the statements expressed in axioms.

Definition 3 (Axiom satisfiability) *Let Σ be a set of terms, and $\phi \in \mathcal{A}(\Sigma)$ an axiom in terms of Σ. An interpretation I of Σ satisfies ϕ iff it fulfills the constraint associated with ϕ in Table 1. In this case, we denote it $I \models \phi$.*

Given an ontology O, an interpretation I of $\text{Sig}(O)$ satisfies O iff it satisfies all the axioms of O. In this case, we write $I \models O$ and call I a *model* of O. Moreover, we denote by $\text{Mod}(O)$ the set of all models of O. Additionally, an ontology is *satisfiable* if it has at least one model.

The next section explains how knowledge from different ontologies can be related, and how distributed systems are interpreted and satisfied.

2.2. Distributed Systems

Basically, a distributed (knowledge-based) system (DS for short) is a structure composed of ontologies and ontology mappings interconnecting them. Since there are now several ontologies, it is convenient to identify the provenance of terms. To do so, we add a prefix to terms or axioms denoting the ontology whence they come. Additionally, we will consistently use K to denote a set of indexes.

Syntax. In DDL, ontology mappings are expressed by *bridge rules*.

Definition 4 (Bridge rule) *Let O_i and O_j be two ontologies. A bridge rule from O_i to O_j ($i \neq j$), is an expression of one of the following two forms:*
– $i{:}C \xrightarrow{\sqsubseteq} j{:}D$ an into-bridge rule;
– $i{:}C \xrightarrow{\sqsupseteq} j{:}D$ an onto-bridge rule;
– $i{:}x \mapsto j{:}y$ a (partial) individual correspondence;
– $i{:}x \overset{=}{\mapsto} j{:}\{y_1, \ldots, y_n\}$ a complete individual correspondence;
where $i{:}C$ and $j{:}D$ are either two concepts or two roles of O_i and O_j respectively, $i{:}x$ is an individual of O_i and $j{:}y, j{:}y_1, \ldots, j{:}y_n$ are individuals of O_j.

Informally, a distributed system is a set of ontologies interconnected with ontology mappings.

Definition 5 (Distributed System) *A Distributed System (DS) is a pair $S = \langle (O_i)_{i\in K}, (B_{ij})_{i\neq j}\rangle$ where for all $i, j \in K$, O_i is a DL ontology as defined in the previous section, and B_{ij} is a (possibly empty) set of bridge rules between O_i and O_j.*[1]

The notion of distributed system can capture P2P knowledge-based systems, the Semantic Web, or multi-agent systems.

Semantics. In a distributed system in DDL, each ontology is interpreted according to the local DL semantics. Since interpretation domains of different ontologies can be heterogeneous, a distributed interpretation also describe how two different domains are interrelated. This done via domain relation, as shown in the following definition.

Definition 6 (Distributed interpretation) *Let $S = \langle \mathbf{O}, \mathbf{B}\rangle$ be a DS. A distributed interpretation. $\mathcal{I} = \langle \mathbf{I}, \mathbf{r}\rangle$ of S assigns to each ontology O_i an interpretation $I_i = \langle \Delta^{I_i}, \cdot^{I_i}\rangle$ of $\mathrm{Sig}(O_i)$ and for all pairs $i \neq j$, a domain relation $r_{ij} \subseteq \Delta^{I_i} \times \Delta^{I_j}$.*[2]

Informally, a domain relation r_{ij} serves to express what objects from Δ^{I_i} represent from the O_j's point of view. A distributed interpretation satisfies a DS when (1) its local interpretations locally satisfy their respective ontology, and (2) satisfies constraints imposed by bridge rules as defined in the following definition.

Definition 7 (Distributed satisfiability) *Let $\mathcal{I} = \langle \mathbf{I}, \mathbf{r}\rangle$ be a distributed interpretation. $\mathcal{I}$ satisfies a bridge rule b_{ij} (written $\mathcal{I} \models_d b_{ij}$) when the following constraints hold:*

- *if b_{ij} is $i{:}C \xrightarrow{\sqsubseteq} j{:}D$ then $\mathcal{I} \models_d b_{ij} \Leftrightarrow r_{ij}(C^{I_i}) \subseteq D^{I_j}$;*
- *if b_{ij} is $i{:}C \xrightarrow{\sqsupseteq} j{:}D$ then $\mathcal{I} \models_d b_{ij} \Leftrightarrow r_{ij}(C^{I_i}) \supseteq D^{I_j}$;*
- *if b_{ij} is $i{:}x \mapsto j{:}y$ then $\mathcal{I} \models_d b_{ij} \Leftrightarrow y^{I_j} \in r_{ij}(x^{I_i})$;*
- *if b_{ij} is $i{:}x \overset{=}{\mapsto} j{:}\{y_1, \ldots, y_n\}$ then $\mathcal{I} \models_d b_{ij} \Leftrightarrow r_{ij}(x^{I_i}) = \{y_1^{I_j}, \ldots, y_n^{I_j}\}$.*

A distributed interpretation satisfying a DS is called a model of the DS.

Definition 8 (Model of a DS) *A model of a DS $\langle \mathbf{O}, \mathbf{B}\rangle$ is a distributed interpretation $\mathcal{I} = \langle \mathbf{I}, \mathbf{r}\rangle$ such that for all $i \in K$, $I_i \models O_i$ and for all $i, j \in K, b_{ij} \in B_{ij}$, $\mathcal{I} \models_d b_{ij}$. This is denoted by $\mathcal{I} \models_d S$.*

An axiom (resp. a bridge rule) α (resp. β) is a semantic consequence of S if all models of S satisfy α (resp. β).

Next section defines the notion of targeted query over a DS, and specify the problem we address in this paper: answering queries targeted on a specific ontology within a DS.

---

[1] When there is no ambiguity, we will write $\langle \mathbf{O}, \mathbf{B}\rangle$ to denote $\langle (O_i)_{i\in K}, (B_{ij})_{i\neq j}\rangle$.

[2] For all $d \in \Delta^{I_i}$, we use $r_{ij}(d)$ to denote $\{d' \in \Delta^{I_j} | \langle d, d'\rangle \in r_{ij}\}$, for any $D \subseteq \Delta^{I_i}$, we use $r_{ij}(D)$ to denote $\bigcup_{d\in D} r_{ij}(d)$, and for any $R \subseteq \Delta^{I_i} \times \Delta^{I_j}$ to denote $\bigcup_{\langle d,e\rangle\in R} r_{ij}(d) \times r_{ij}(e)$.

3. QUERY ANSWERING IN DDL

We are interested, in this section, in defining answers to queries posed in terms of one ontology in DDL. So the query is specified in one specific context, and answers are given *w.r.t.* this context, even though the overall DS is used to determine these answers. We first define the syntax of such queries, and then define the set of possible answers to a given query over a DDL knowledge base.

3.1. Syntax

Informally, a query as we consider here is simply a set of axioms that can be constructed from the terms of one specific local ontology as well as variables. So, we will consider the following formal definition of a query.

Definition 9 (Query) *A query Q is a tuple of the form $\langle X, Y, \Sigma, F\rangle$, where X and Y are sets of variables, Σ is a set of terms, X, Y and Σ are pairwise disjoint and $F \subseteq \mathcal{A}(\Sigma \cup X \cup Y)$ is a set of DL axioms where variables from X and Y are used as terms.*

In the above definition, variables of X are called *distinguished variables* and variables of Y are existentially quantified and called *non-distinguished variables*. A very common notation for queries is $q(\overline{x}) \leftarrow body(\overline{x}, \overline{y})$, where $\overline{x}$ represents the distinguished variables, $\overline{y}$ the non-distinguished ones, and *body* represents the axioms. The set of terms upon which the axioms are defined is made implicit. We will use this more convenient notation in our practical examples.

Example 1: The query $q(?x) \leftarrow \texttt{Student}(?y) \wedge \texttt{hasStudent}(?x, ?y)$ corresponds to the tuple $\langle X, Y, \Sigma, F\rangle$, s.t. $X = \{?x\}$, $Y = \{?y\}$, $\Sigma = \{\texttt{Students}, \texttt{hasStudent}\}$, $F = \{\texttt{Student}(?y), \texttt{hasStudent}(?x, ?y)\}$.

Definition 10 (Targeted query on an ontology) *Given an ontology O, a targeted query Q on O is a tuple $\langle X, Y, \mathrm{Sig}(O), F\rangle$.*

In this paper, we are only interested in answering targeted queries, which means we only want to get answers that are related to the knowledge domain of the target ontology, yet in accordance to all the knowledge described in every ontologies and bridge rules.

3.2. Semantics

As in distributed system semantics, there are two levels of query satisfiability: local and global. Local satisfiability corresponds to the usual query satisfiability over an ontology language. Global satisfiability necessitates a more elaborate definition since it has to deal with the presence of bridge-rules.

Definition 11 (Assignment) *An assignment is a mapping $\alpha : X \rightarrow \Sigma$ from a set of variables X to a set of terms Σ.*

Given a query Q and an assignment α, $\alpha(Q)$ denotes a the axioms of Q in which all variables x are replaced by $\alpha(x)$. Consequently, $\alpha(Q)$ is a set of axioms with no distinguished variables.

Definition 12 (Interpretation extended to variables) *Let Σ be a set of terms, Y be a set of variables disjoint from Σ and I be an interpretation of Σ. Then, an extension I' of I to a set of variables X is an interpretation of $\Sigma \cup X$ such that $\forall x \in \Sigma\ x^{I'} = x^{I}$.*

This extension of interpretations serves to define satisfaction of axioms with variables, as shown below.

Definition 13 (Satisfied query) *Let $Q = \langle X, Y, \Sigma, F\rangle$ be a query, $\alpha : X \rightarrow \Sigma$ an assignment and I be an interpretation of Σ. Then, I satisfies $\alpha(Q)$ if there exist an extension I' of I to Y such that $I' \models \alpha(Q)$. In this case, we simply write $I \models \alpha(Q)$.*

This definition makes sense since $\alpha(Q)$ is a set of axioms built out of $\Sigma \cup Y$, and I' interprets it.

Definition 14 (Local answer to a query) *Let $Q = \langle X, Y, \mathrm{Sig}(O), F\rangle$ be a query targeted on ontology O. An answer to Q over O is an assignment $\alpha : X \rightarrow \mathrm{Sig}(O)$ such that $\forall I \in \mathrm{Mod}(O)$, $I \models \alpha(Q)$. We denote by $\mathrm{Ans}(Q, O)$ the set of all local answers to a query Q targeted on O.*

Now we must define a distributed answer to a targeted query. The difficulty lies in the fact that there might be answers with terms from different ontologies. In this case, the resulting assignment α may be such that $\alpha(Q)$ has terms from different ontologies. Yet, none of our currently defined interpretations can interpret axioms where terms from different ontologies appear. So, we will provide a new definition for that matter.

The idea of the following definition is to extend a local interpretation to an interpretation of all the terms that appear in a DS. However, since we only want to answer the query in the target ontology's context, we want this extended interpretation to stay within the same domain of interpretation, hence, to be local. To this extent, we first define the extension of an ontology signature *w.r.t.* a DS.

Definition 15 (Extended signature) *Let $S = \langle \mathbf{O}, \mathbf{B}\rangle$ be a DS, with O_t a particular ontology of S. The extended signature $\mathrm{eSig}_S(O_t)$ of O_t w.r.t. S is the smallest set such that:*
- $\mathrm{Sig}(O_t) \subseteq \mathrm{eSig}_S(O_t)$;
- *for all $e \in \mathrm{Sig}(O_i)$, there exists a distinct term $e^{i \rightarrow t} \in \mathrm{eSig}_S(O_t)$.*

Now our goal is to be able to interpret the extended signature in the local domain of the target ontology. Such an interpretation is defined by taking advantage of a given distributed interpretation.

Definition 16 (Combined interpretation) *Let $\mathfrak{I} = \langle \mathbf{I}, \mathbf{r} \rangle$ be an interpretation of a DS, O_t be an ontology of the DS. The combined interpretation of I targeted on O_t is an interpretation $\mathfrak{I}_t^c$ of the extended vocabulary* eSig$_S(O_t)$ *defined by:*
- $\mathfrak{I}_t^c = \langle \Delta_t, \cdot^{\mathfrak{I}_t^c} \rangle$;
- $\forall e \in \mathrm{Sig}(O_t)$, $e^{\mathfrak{I}_t^c} = e^{I_t}$; *and*
- $\forall i \neq t, \forall e \in \mathrm{Sig}(O_i)$, $(e^{i \to t})^{\mathfrak{I}_t^c} = r_{it}(e^{I_i})$.

It must be remarked that when e is an individual of ontology $O_i \neq O_t$, then $r_{it}(e^{I_i})$ is a set of elements of the domain of interpretation. So, foreign individuals are interpreted as subset of Δ_t, not as elements. However, our goal is to obtain a standard DL interpretation of foreign terms. We envisage two possibilities to overcome this problem: either we restrict the combined interpretation to foreign concepts and roles and ignore foreign individuals; or we add an explicit constraint on individual terms in order to make their interpretation a singleton. More precisely, this second choice imposes that a bridge rule $i{:}x \overset{=}{\mapsto} t{:}\{x^{i \to t}\}$ is added to the DS for each foreign individual term $x \in \mathrm{Sig}(O_i)$. While this does not fully solve the problem, it at least ensures that if ; is a model of the DS, then $\mathfrak{I}_t^c$ is a well defined DL interpretation. This is actually all we need to have a valid definition of query answer.

Our approach works for these two options, and we will present the differences that are implied by the two approaches.

Definition 17 (Distributed answer to a query) *Let $S = \langle \mathbf{O}, \mathbf{B} \rangle$ be a distributed system. Let $Q_t = \langle X, Y, \mathrm{Sig}(O_t), F \rangle$ be a query targeted on O_t in S. A distributed answer to Q_t is an assignment $\alpha : X \to \mathrm{eSig}_S(O_t)$ such that for all models $\mathfrak{I}$ of S, $\mathfrak{I}_t^c \models \alpha(Q_t)$. We denote by* dAns$(Q_t, O_t, S)$ *the set of all distributed answers to a query Q_t targeted on O_t in S.*

***Example** 1: To illustrate our approach, we consider the following DDL system that contains only two ontologies $\mathbf{O}_1$ and $\mathbf{O}_2$, and the bridge-rules $\mathbf{B}_{21}$. In this system* `Teacher` *and* `Student` *are concepts, and the roles are* `teachesTo` *and* `hasStudent`*:*

$\mathbf{O}_1$	1:∃`teachesTo` ⊑ `Teacher` 1:`funct(teachesTo)` 1:`teachesTo(John, Bob)`
$\mathbf{B}_{21}$	2:`hasStudent` $\overset{\sqsubseteq}{\to}$ 1:`teachesTo`$^-$ 2:`John-Doe` $\overset{=}{\mapsto}$ 1:`John`
$\mathbf{O}_2$	2:`Student` ⊑ ∃`hasStudent`$^-$ 2:`hasStudent(John-Doe, Robert)` 2:`Student(Larry)`

Consider now the following query targeted on $\mathbf{O}_1$:

$$Q_1(?x, ?y) \leftarrow \texttt{Teacher}(?x), \texttt{teachesTo}(?x, ?y)$$

the answer to the above query is:

$$\langle ?x, ?y\rangle \in \{\langle\texttt{John}, \texttt{Bob}\rangle, \langle\texttt{John-Doe}, \texttt{Robert}\rangle, \langle\texttt{John-Doe}, \texttt{Bob}\rangle, \langle\texttt{John}, \texttt{Robert}\rangle\}$$

Note that our approach is capable of assigning individuals in the same answer from different domains (*e.g.*, $\langle\texttt{John}, \texttt{Robert}\rangle$), which is not provided by previous approaches.

Consider now the following query with the same body as the previous one:

$$Q_2(?y) \leftarrow \texttt{Teacher}(?x), \texttt{teachesTo}(?x, ?y)$$

the answer to the above query is:

$$?y \in \{\texttt{Bob}, \texttt{Robert}, \texttt{Larry}\}$$

Although both of the above two queries have the same body, `Larry` appears in the answer of the second query while it does not appear in the answer of the first one. This is because we know that there is someone who teaches to `Larry` but we do not know whom, so there is no assignment to the variable $?x$ in the first query.

4. EVALUATING A TARGETED QUERY OVER DDL

Though our definitions for local and distributed answers are given for general queries, we are interested in answering conjunctive queries over DDL knowledge base. Conjunctive queries only allow variables in replacement of individuals, and only have axioms of the form $C(\tau)$ and $R(\tau_1, \tau_2)$ where C is a concept, R is a role and τ, τ_1, τ_2 are either individual terms or variables.

In order to present our method for computing answers, we first show an important theorem that includes properties satisfied by the combined interpretation.

***Theorem** 1: Let ; $= \langle \mathbf{I}, \mathbf{r}\rangle$ be a distributed interpretation of a DS, and O_t a target ontology. The following properties holds:*

- if $\mathcal{J} \models_d i{:}B \overset{\sqsubseteq}{\rightarrow} t{:}C$ then $\mathcal{J}_t^c \models t{:}B^{i\rightarrow t} \sqsubseteq C$;
- if $\mathcal{J} \models_d i{:}S \overset{\sqsubseteq}{\rightarrow} t{:}R$ then $\mathcal{J}_t^c \models t{:}S^{i\rightarrow t} \sqsubseteq R$;
- if $\mathcal{J} \models_d i{:}D \overset{\sqsupseteq}{\rightarrow} t{:}C$ then $\mathcal{J}_t^c \models t{:}C \sqsubseteq D^{i\rightarrow t}$;
- if $\mathcal{J} \models_d i{:}T \overset{\sqsupseteq}{\rightarrow} t{:}R$ then $\mathcal{J}_t^c \models t{:}R \sqsubseteq T^{i\rightarrow t}$;
- if $I_i \models i{:}B \sqsubseteq D$ then $;_t^c \models t{:}B^{i\rightarrow t} \sqsubseteq D^{i\rightarrow t}$.

Moreover, if we consider the second choice mentioned after Definition 16, the following holds too:

- if $I_i \models i{:}B(a)$ then $;_t^c \models t{:}B^{i\rightarrow t}(a^{i\rightarrow t})$;
- if $I_i \models i{:}S(a, b)$ then $;_t^c \models t{:}S^{i\rightarrow t}(a^{i\rightarrow t}, b^{i\rightarrow t})$;

where C is any concept, R is any role and B, D, S, T are defined according to the following grammar:

$$B ::= A \mid \exists S.\top \mid B \sqcup B \mid [\{a_1, \ldots, a_n\}]$$
$$D ::= A \mid D \sqcap D \mid D \sqcup D \mid [\{a_1, \ldots, a_n\}]$$
$$S ::= P \mid S^- \mid S \sqcup S$$
$$T ::= P \mid T^- \mid T \sqcup T \mid T \sqcap T$$

and A is a primitive concept, P a primitive role and $a_1, \ldots, a_n$ are individuals.[3]

The proof for this theorem is not very difficult but long and may be hard to follow because of the many variables and notations involved.[4]

When using DL constructs in axioms or bridge rules other than the ones used in Theorem 1, there is no guarantee that a corresponding axiom is satisfied by the combined interpretation $\mathcal{I}_t^c$. Counter examples for all constructs not in the list above are found in the online appendix.

4.1. Query Evaluation

In order to compute the distributed answers to a query targeted on an ontology, we build a new ontology which extends the targeted one with terms from foreign ontologies and axioms deduced from foreign axioms and bridge rules, according to Theorem 1. This ontology corresponds to a kind of deductive closure of the target ontology with respect to a DS. This new ontology is thus called the "targeted distributed closure". More precisely, the target ontology signature is extended to $e\text{Sig}_S(O_t)$ and all axioms satisfied by all combined interpretations are added to it. Building the closure ontology O_t' serves to transform the distributed query answering problem into a local query answering problem. Indeed, the following property holds:

$$\text{Ans}(Q_t, O_t) \subseteq \text{Ans}(Q_t, \text{Closure}(O_t, S)) \subseteq \text{dAns}(Q_t, O_t, S)$$

So this gives a correct algorithm for computing distributed answers, on the condition that we have (1) a correct and complete deduction procedure for DDL, computing all semantic consequences of a DS in finite time; (2) local query answering facilities. The first point is of course very optimistic, but it can give a criteria on the type of DL and/or ontologies that permits computable answers. Moreover, it is not needed to have the complete closure in order to have the above

[3] Individuals are allowed only if the second choice mentioned after Definition 16 is taken.

[4] The reader can refer to the online appendix at the following url: `http:// www.inrialpes.fr/ exmo/people/zimmer/NTMS2007proof.pdf`.

Algorithm 1: Closure(O_t, S)

Data: A distributed system $S = \langle \mathbf{O}, \mathbf{B} \rangle$, and an ontology $O_t \in S$.
Result: a new ontology O_t'.
begin
  $O_t' ::= O_t$;
  for each bridge rule b_{ti} such that $S \models_d b_{ti}$ **do**
    if b_{ti} is of one of the forms given in Theorem 1 **then**
      add the corresponding axiom to O_t'
  for each axiom ϕ_i in O_i **do**
    if ϕ_i is of one of the forms given in Theorem 1 **then**
      add the corresponding axiom to O_t'
  return O_t'
end

property. Any ontology extending O_t with axioms obtained by applying the rules in Theorem 1 satisfies this property. So, it is possible to extend the ontology progressively, while computing new answers step by step. This method, which corresponds to a lazy evaluation, permits to give answers at anytime during the process. Finally, in the favorable case when the closure is computable and the query has only distinguished variables, the set of local answers to the query over the closure is most likely equal to the set of distributed answers over the DS.

***Example** 2:* According to Algorithm 1, the closure of ontology $\mathbf{O}_1$, which includes all possible deductions that can be made by the bridge-rules and foreign axioms, is:

$1{:}\exists\mathtt{teachesTo} \sqsubseteq \mathtt{Teacher}$
$1{:}\mathtt{funct(teachesTo)}$
$1{:}\mathtt{Student}^{2\to 1} \sqsubseteq \exists(\mathtt{hasStudent}^{2\to 1})^-$
$1{:}\mathtt{hasStudent}^{2\to 1} \sqsubseteq \mathtt{teachesTo}$

$1{:}\mathtt{teachesTo(John, Bob)}$
$1{:}\mathtt{Student}^{2\to 1}(\mathtt{Larry}^{2\to 1})$
$1{:}\mathtt{hasStudent}^{2\to 1}(\mathtt{John\text{-}Doe}^{2\to 1}, \mathtt{Robert}^{2\to 1})$
$1{:}\mathtt{John\text{-}Doe}^{2\to 1} = \mathtt{John}$

5. CONCLUSION AND FUTURE WORK

We have investigated the query answering problem over DDL, and showed that, in some specific cases, the problem can be reduced to answering query over local DL ontology by constructing the closure ontology. The closure ontology extends a local ontology with all possible axioms that can be deduced from other ontologies in the system using bridge-rules between different domains, which allow to deduce additional information from other ontologies, and thus find answers that were not given by the sole local knowledge base.

Our approach guarantees correctness as soon as there exists a sound algorithm for querying DL ontologies. However, it does not guarantee completeness in the general case. Nonetheless, we conjecture and strongly believe that with ontologies in DL-Lite and minimal constraints on bridge rules, the algorithm we propose would prove to be complete, and it is part of our future investigations to do so.

Future work will also regard four main directions: optimization, implementation of this approach, test of the performance of our approach in terms of scalability (*i.e.*, the number of peers in the system) and study of the algorithmic complexity. Comparison with existing approach in terms of expressivity and completeness of results is also envisaged. Especially, a possible optimization would consist in taking into account the query in the construction of the closure, such that it would not be needed to deduce all possible axioms, and in particular, reduce the set of needed axioms to a finite one.

REFERENCES

1. Borgida, A., Serafini, L.: Distributed Description Logics: Assimilating information from peer sources. Journal of Data Semantics **1** (2003) 153–184
2. Calvanese, D., De Giacomo, G., Lembo, D., Lenzerini, M., Rosati, R.: DL-Lite: Tractable description logics for ontologies. In: Proceedings of the 20th National Conference on Artificial Intelligence (AAAI 2005). (2005) 602–607
3. Calvanese, D., De Giacomo, G., Lembo, D., Lenzerini, M., Rosati, R.: Data complexity of query answering in description logics. In: Proceedings of the Tenth International Conference on Principles of Knowledge Representation and Reasoning (KR 2006). (2006) 260–270
4. Cruz, I.F., Xiao, H., Hsu, F.: Peer-to-Peer Semantic Integration of XML and RDF Data Sources. In: Third International Workshop on Agents and Peer-to-Peer Computing (AP2PC 2004). (2004)
5. Halevy, A., Ives, Z., Suciu, D., Tatarinov, I.: Schema mediation in peer data management systems. In: Proceedings of ICDE. (2003)
6. Nejdl, W., Wolf, B., Qu, C., Decker, S., Sintek, M., Naeve, A., Nilsson, M., Palmer, M., Risch, T.: EDUTELLA: A P2P Networking Infrastructure Based on RDF. In: Proceedings of the 11th international conference on World Wide Web (WWW2002). (2002)
7. Goasdoue, F., Rousset, M.C.: Querying Distributed Data through Distributed Ontologies: A Simple but Scalable Approach. IEEE Intelligent Systems **18** (2003) 60–65
8. Halevy, A.Y., Ives, Z.G., Mork, P., Tatarinov, I.: Piazza: data management infrastructure for semantic web applications. In: Proceedings of the 12th international conference on World Wide Web (WWW2003), New York, NY, USA, ACM Press (2003) 556–567
9. Adjiman, P., Chatalic, P., Goasdoué, F., Rousset, M.C., Simon, L.: SomeWhere in the Semantic Web . In: International Workshop on Principles and Practice of Semantic Web Reasoning. (2005)

CHAPTER 44

WSRANK: A NEW ALGORITHM FOR RANKING WEB SERVICES

XIAODI HUANG

School of Mathematics, Statistics and Computer Science, The University of New England, Armidale, NSW, Australia, huang@turing.une.edu.au

Abstract: Web services are one of important components in the service-oriented computing paradigm. As the increasing number of Web services is available for service requests, it is demand to develop an algorithm for ranking them with respect to some criteria. In addition, the algorithm should be capable of taking into account the fact that the user requirements for a service may be diverse. This paper presents approaches for ranking Web services in different scenarios. Two algorithms are developed for this purpose

1. INTRODUCTION

The Web is evolving from a collection of Web pages to a collection of services. In the last few years, Web service technologies have received much attention as they aim to provide a flexible mechanism for interaction over the Web between one machine and another. Web services, accessible via standardized protocols, are viewed as a fundamental component of the service-oriented computing paradigm. With the increasing number of Web services available, there is a growing need for efficient ways of ranking services. It often happens that there are many available services matching a service request. A ranking algorithm is therefore employed to position the range of candidate Web services. It produces a list of matched services in decreasing order of suitability to particular requests against a number of criteria related to the attributes of services. Effective mechanisms for ranking Web service play a critical role for organizations to engage in business collaborations and service compositions, as well as to identify potential service partners. There is no much literature on ranking WSs. The notion of ranking, however, is applied to Web search applications, as well as to query engines of relational databases. Many ranking algorithms for Web pages have been developed. The most popular among them are the HITS algorithm proposed by Kleinberg [1] and the PageRank algorithm by

H. Labiod and M. Badra (eds.), New Technologies, Mobility and Security, 529–539.
© 2007 *Springer.*

Brin et.al [2,3]. The idea of PageRank has been applied to a graph structure derived from the contents of tuples of a database [4–7]. XRANK [4] ranks XML elements using the link structure of a database. Employing the similar notion, ObjectRank [6] transfers edge bounds in order to distinguish between containment and IDREF edges. Huang et al. [8] propose a way of ranking the tuples of a relational database using PageRank, where their connections are determined by the query workload dynamically, rather than by the schema statically. Rank aggregation is reported in the literature [8–10]. Recently, [11,12] propose context and constraint ranking.

Our focus in this paper is on ranking Web services. All above-mentioned approaches are based on the link structure of objects being ranked; that is, the structural or reference relationships among objects. In contrast, our approach relies on the inherent characteristics of Web services, which are characterized by a feature vector.

The contributions of this paper are that we formalize the problem of ranking Web services, providing two algorithms for ranking them in four cases.

The remainder of this paper is organized as follows. The approaches for ranking Web services in several contexts are presented in the following section. Section 3 describes the algorithms, followed by describing preliminary experiments. Section 5 concludes this paper and points out our future work.

2. RANKING WEB SERVICES

In the section, we present the approaches that rank Web services (WSs) with respect to different criteria.

The WS ranking face new challenges. First, each WS is characterized by many attributes so that ranking should be based on each attribute or a combination of the attributes. The result of ranking is probably derived from one attribute or a subset of the attributes. Rather, it is always from the aggregation of all attribute scores. One, for example, may rank all candidate WSs that are able to provide a particular service over their attribute of *security*. The services that have the higher *security* level are selected, instead of those that have the higher quality of services (QoSs). In general, ranking WSs should also cater for the diversities of user requirements.

Second, a number of WSs are usually related to a workflow in the business world. A workflow consists of a set of subsequent subtasks. Some subtasks rely on the successful completion of previous subtasks. While ranking WSs, we should give more weights to the reliabilities of these prerequisite subtasks. Different subtasks are of varied concerns on different aspect of WSs. One subtask, for example, may be concerned more about the responding time while others pay more attention to the security. A WS ranking differs from others in that how many attributes the ranking is based on, and that how many objects are compared for each ranking list. The search engines such as Google usually rank documents in terms of one attribute, i.e., their hyperlinked structure [1–3], while the WS ranking considers the multiple attributes of a WS. Google compares one document against another in terms of their importance in a collection. In contrast, WS ranking in the context of workflow compares a group of related WSs against another group. Each group includes a number of WSs that are candidates for providing the services specified in a workflow. The purpose of ranking is to order the optimal combination of a number of WSs as opposed to only one WS. A WS ranking for a workflow selects an optimal sequence of WSs from a number of candidate WSs that fulfill different subtasks of the workflow. The difficulty lies in the fact that a number of WSs associated with a workflow may be ranked differently with respect to diverse attributes. Furthermore, computing rankings of a sequence is complicated by the fact that the local optimization of WSs is possibly different from a global one of a combination of sequential WSs. As a consequence, techniques for computing rankings founded solely on hyperlinks [1–3] are not directly applicable for ranking WSs, particularly not for those in the context of a workflow.

We characterize a WS as an attribute vector derived from an attribute set. This set consists of some common attributes such as *trustiness, security*, and *response time*. In this paper, we are not concerned about how to obtain the attribute vector of a WS. Instead, we suppose that the scores of all given attributes take on values in [0, 1], and that the higher the value, the better. For example, WS_1 with the *response time* value of 0.9 is better than WS_2 with that value of 0.3. In the following, we will discuss how to rank a number of candidate WSs with respect to their attributes. Formally, suppose each candidate WS in a set $C = \{W(1), W(2), \cdots, W(m)\}$ is characterized by an attribute vector $W(k) = (w_1(k), w_2(k), \cdots, w_d(k))$, in which each element denotes the score of an attribute in an attribute set A, and $d = |A|$. We suppose $A = \{security,\ trustiness,\ responsetime,\ price\}$. Each WS is able to accomplish at least one of tasks in a task set T associated with a workflow. The notation of $Rank(C,\ A,\ T)$ returns an order set of WSs in C with respect to the attribute(s) in A for the task(s) in T.

2.1. Ranking Web Services for a Task

We rank WSs for a particular task in relation to one specific or multiple attributes. The attribute-specific WSRank attempts to find the most suitable candidate WS for a particular task t in T with respect to the score of attribute a_j in A. That is, we have $Rank(C, a_j, t) = \tau_j$ where τ_j is an ordered list, $a_j \in A (j = 1, \ldots, d)$, and $t \in T$. Denoting the position of service $W(k)$ in order τ_j as $\tau_j(W(k))$, we have $w_j(k_1) \geq w_j(k_2)$ if $\tau_j(W(k_1)) < \tau_j(W(k_2))$ where $w_j(k)$ is the score of the j-th component of the attribute vector $W(k)$ in C. In other words, the ranking returns a WS that has the highest score of attribute j among all the candidates regardless of other attributes. According to their *trustiness* scores, for example, we rank all available WSs that are able to carry out an activity in a process, and then select the service with the highest score.

How to rank WSs with respect to more than one attribute? One WS may be ranked in varied positions in accordance with different attributes. WS_2, for example, is ranked as the first place with respect to the *security* level, and the fourth to the *response time*. What is the combinational ranking for WS_2 with consideration of both attributes of the *security* level and *response time*? In principle, this question is a kind of the rank aggregation problem.

Definition 1 *(Rank aggregation): a global ranking from multiple input ranking lists. Given a d-dimensional attribute vector* $\mathbf{W}$*, m candidate WSs, and d ranking lists* $\tau_1, \tau_2, \cdots, \tau_d$ *of these candidates corresponding to each attribute in A, rank aggregation finds an optimal list of these d ranking lists.*

The above definition does not answer the question of how to define the optimal criterion. Several distance functions for the criterion have been proposed in the literature. We focus on two well-known ones: the *Kendall tau* and *Spearman footrule*.

The *kendall tau* distance counts the number of pairwise disagreements between any two of d lists:

$$s_{KT}(\tau_i, \tau_j) = \frac{2}{m(m-1)} \sum_{k=1;k<l}^{m} |\{(\mathbf{W}(k), \mathbf{W}(l)) | \tau_i(\mathbf{W}(k)) < \tau_i(\mathbf{W}(l)) \text{ and } \tau_j(\mathbf{W}(k)) > \tau_j(\mathbf{W}(l))\}|$$

where $\tau_i(\mathbf{W}(k))$ is the position number of the k-th WS in order list τ_i.

The *Spearman footrule* distance is defined as the sum over all m candidates of the differences between their positions in two lists.

$$s_{SF}(\tau_i, \tau_j) = \frac{2}{m(m-1)} \sum_{k=1}^{m} (\tau_i(\mathbf{W}(k)) - \tau_j(\mathbf{W}(k)))$$

Note that the two above distances have been normalized so that the distances take on a value in [0, 1].

Rank aggregation produces an optimal ranking list from d respective lists of m candidate WSs for task t.

$$\begin{aligned} Rank(C, A, t) &= \tau_A^* \\ &= \arg\min_{\tau'_A} \sum_{k=1}^{d} s(\tau'_A, \tau_k) \end{aligned}$$

where s can be either s_{KT} or s_{SF}.

In fact, the attribute-specific WSRank is regarded as a special case of rank aggregation.

Example 1: given two WSs being characterized by $\mathbf{W}(1) = (0.6, 0.8, 0.2, 0.1)$, and $\mathbf{W}(2) = (0.5, 0.9, 0.3, 0.76)$, we obtain

The *security* ranking list is $\tau_1 = [\mathbf{W}(1) > \mathbf{W}(2)]$ (because of $0.6 > 0.5$);
The *trustiness* ranking list is $\tau_2 = [\mathbf{W}(2) > \mathbf{W}(1)]$
The *response time* ranking list is $\tau_3 = [\mathbf{W}(2) > \mathbf{W}(1)]$
The *price* ranking list is $\tau_4 = [\mathbf{W}(2) > \mathbf{W}(1)]$
The aggregation ranking is $\tau_A^* = [\mathbf{W}(2) > \mathbf{W}(1)]$

Note that the score of the *price* attribute of a WS has been converted so that the bigger it is, the better.

2.2. Ranking Web Services in the Context of Workflows

A workflow is an abstraction of a business process. It comprises a number of logic steps (known as tasks or activities), dependencies among tasks, routing rules, and participants. Unlike ranking by only one task, ranking WSs for a workflow involves in positioning the WSs for a number of tasks in terms of different criteria.

In a workflow, a task represents either a human activity or a software system. The emergent need of workflows to model e-service applications makes it essential that workflow tasks be associated with Web services. Assume a composite service S that finishes the tasks required by a workflow includes T component services. Each component service will be selected from a number of candidate WSs. The selected WSs form an execution plan vector, the component of which implements one sub-task, respectively. We attempt to find an optimal execution plan vector that maximizes the score.

We formalize the problem as follows.

Definition 2 *(Ranking WSs for a workflow): a workflow is represented as the task vector* $\boldsymbol{T} = (t_1, t_2, \cdots, t_T)$, *with each sub-task* t_i *associated with a set of candidate WSs* $W^{t_i} = \{W^{t_i}(1), W^{t_i}(2), \cdots, W^{t_i}(k_{t_i})\}$ *where* $W^{t_i}(i) = (w_1^{t_i}(i), w_2^{t_i}(i), \cdots, w_d^{t_i}(i))$ *denotes the i-th WS that is able to finish the sub-task t. A set of all candidate WSs is denoted as* $C = \bigcup_{i=1}^{T} W^{t_i}$ *for the workflow.*

Note that $W(i)$ may be as the same as $W(j)(i \neq j)$. This means that one candidate WS possibly implement two or more sub-tasks of the given workflow. We attempt to find an optimal permutation of the number of T WSs in C such that the score is maximized. $Rank(C, A, T) = \underset{W^{t_1}, W^{t_2}, \ldots, W^{t_T}}{\arg\max}\ score\,[W^{t_1}, W^{t_2}, \cdots, W^{t_T}]$ where $score[W^{t_1}, W^{t_2}, \cdots, W^{t_T}] = \sum_{j=1}^{d} [w_j^{t_1}(k) w_j^{t_2}(l) \cdots w_j^{t_T}(o)]$. Also, in the equation $i, k, \cdots, o \in \{1, 2, \cdots, n_{t_1}, \cdots, n_{t_T}\}$ denote the index of the selected candidate WSs, and $n_{t_1}, \cdots, n_{t_T}$ are the number of available candidate WSs for sub-tasks $t_1, \ldots, t_r$, respectively.

The multiplication of attribute scores may result in a very small number. For analytical purposes, it is usually easier to work with the logarithm of the attribute score than with the attribute score itself. Because the logarithm is monotonically increasing, we are able to re-write the above equation as:

$$
\begin{aligned}
Rank(C, A, T) &= \underset{W^{t_1}, W^{t_2}, \ldots, W^{t_T}}{\arg\max} \sum_{j=1}^{d} [w_j^{t_1}(k) w_j^{t_2}(l) \cdots w_j^{t_T}(o)] \\
&= \underset{W^{t_1}, W^{t_2}, \ldots, W^{t_T}}{\arg\max} \sum_{j=1}^{d} [\ln w_j^{t_1}(k) + \ln w_j^{t_2}(l) + \cdots + \ln w_j^{t_T}(o)]
\end{aligned}
$$

In other words, we first select one WS from C for sub-task t_i, and then find the combination of the number of T WSs such that their score is maximized.

In summary, the sub-problems of WS ranking for a workflow are of two types:

- An optimal WS for each task with respect to an attribute
- An optimal sequence of WSs for a workflow with respect to the various attribute for different tasks.

Problem 1 is equivalent to the problems presented in Section 2.1.

Example 2: given a workflow with two subtask $T=\{t_1, t_2\}$, there are one candidate WS for $t_1 (n_{t_1} = 1)$ represented as $W^{t_1}(1) = (0.2, 0.25, 0.65, 0.53)$, and three candidate WSs for t_2 $(n_{t_2} = 3)$ as $W^{t_2}(1) = (0.23, 0.45, 0.37, 0.5)$, $W^{t_2}(2) = (0.7, 0.3, 0.6, 0.24)$, and $W^{t_2}(3) = (0.6, 0.2, 0.46, 0.14)$.

The scores of their permutations are given by:

$$\begin{aligned} score[W^{t_1}(1), W^{t_2}(1)] &= (0.2 \times 0.23 + 0.25 \times 0.45 + 0.65 \\ &\quad \times 0.37 + 0.53 \times 0.5) \\ &= 0.664 \end{aligned}$$

$score[W^{t_1}(1), W^{t_2}(2)] = 0.7322$, and $score[W^{t_1}(1), W^{t_2}(3)] = 0.5432$

Therefore we have $Rank(C, A, T) = [W^{t_1}(1) > W^{t_2}(2)]$.

2.3. Conditional Ranking Web Services

In some cases, we rank only candidate WSs that satisfy some conditions. For illustration, the clients may want to find a travel WS that has the highest *trust* reputation, as well as whose *price* does not exceed a threshold. Selecting candidate WSs with the highest scores of the *trust* attributes, we rank only those whose *price* is below a given threshold.

We now formalize this problem. As presented previously, each candidate WS in a set C is associated with an attribute set A. The set $A$is further divided into two subsets; namely the set of R consists of specified attributes for conditional selection, and the set of A-R contains unspecified attributes for ranking. The attributes for specifying conditions are denoted as $R = \{r_1, r_2, \cdots, r_s\} \subset A$. A condition usually consists of relational algebra operators, the attributes from R, as well as values in their domain. The select operation, for example, examines the attribute values in R of each candidate WS for task t_k, and generates a subset of WSs, containing only those WSs that satisfy the specified predicate. The produced subset will be ranked. The predicate may be simple, involving the comparison of an attribute of WS with either a constant value or another attribute value. The predicate may also be composite, involving more than one condition, with conditions combined using the logical connectives such as "and", and "or".

All those WSs that satisfy the constraints are kept for ranking while those whose scores of specified-attributes are not within the predicates will be filtered. Using the approach presently before, we are able to rank all remaining WSs in accordance with their un-specified attributes scores. We call this model as "Filter+Ranking".

In practice, the conditions cannot completely be satisfied. In such a case, we need to consider the degree of satisfaction while ranking WSs. The partial satisfaction is related to attributes and conditions. The question is how to quantify the approximation rate of a given WS to the conditions. Suppose the filtering condition involves in attribute r, and its threshold score p. We take into account three cases of quantifying the satisfaction rate for a particular WS.

Case 1: Given a predicate $P(r) = p$, the satisfaction degree is given by $M(r, W(i)) = 1 - |w_r(i) - p| / \max_j\{|w_r(j) - p|\}$. That is, we normalize the difference between the attribute r score of the i-th WS and the threshold. The smaller this value is, the better the approximation.

Case 2: Given a predicate $P(r) > p$, the satisfaction degree is given by:

$$M(r, W(i)) = \begin{cases} (w_r(i) - p) / (\max_j\{w_r(j)\} - p) & if \max_j\{w_r(j)\} > p \\ 0 & otherwise \end{cases}$$

Case 3: Given a predicate $P(r) < p$, the satisfaction degree is given by:

$$M(r, \mathbf{W}(i)) = \begin{cases} (w_r(i) - p) / (\min_j\{w_r(j)\} - p) & if \max_j\{w_r(j)\} > p \\ 0 & otherwise \end{cases}$$

Note that the value of p is within [0, 1] by means of normalization.

As for a condition using the composite predicate, we choose the *min* function for "and", and *max* for "or".

The score of a WS for conditional ranking contains two parts, namely, the score for unspecified attributes, and the score for matching constraints. In other words, those WSs whose unspecified attributes have higher scores, and that satisfy the specified constraints well will be ranked in higher positions in the list.

In Sections 2.1 and 2.2, we suppose that the ranking candidate WSs completely satisfy the conditions. As such, the position of a particular WS in a list is solely determined by its score of the attribute corresponding to the list. The conditional ranking WSs, however, differ from this in twofold. First, only unspecified attributes in a set *A-R* are used for ranking. Second, the scores of these attributes of a WS are weighted by the degrees of conditional satisfaction before they are ranked. The conditional ranking of WSs is now ready to be formalized as:

$$Rank(C, A, t|R) = \tau^*_{A-R}$$

where the order list τ^*_{A-R} is of the form:

The equation $w_i(k_1)M(j, W(k_1)) \geq w_i(k_2)M(j, W(k_2)$ holds, if $\tau^*_{A-R}(W(k_1)) < \tau^*_{A-R}(W(k_2))$ where $i \in A - R$ and $j \in R$.

Example 3: Suppose we have R={*price*}, the selection condition given by $0.5 < price < 0.8$, and three candidate WSs denoted as $\boldsymbol{W}(1)$=(0.5, 0.54, 0.5, 0.6), $\boldsymbol{W}(2)$=(0.2, 0.34, 0.7, 0.75), and $\boldsymbol{W}(3)$=(0.25, 0.64, 0.38, 0.55).

For the condition $P(price) > 0.5$, the degrees of satisfaction are calculated as

$$M(4, \boldsymbol{W}(1)) = (0.6 - 0.5)/(0.75 - 0.5) = 0.4$$

$$M(4, \boldsymbol{W}(2)) = (0.75 - 0.5)/(0.75 - 0.5) = 1$$

$$M(4, \boldsymbol{W}(3)) = (0.55 - 0.5)/(0.75 - 0.5) = 0.2$$

Similarly, for the condition $P(price) < 0.8$, we have

$$M(4, \boldsymbol{W}(1)) = (0.8 - 0.6)/(0.8 - 0.55) = 0.08$$

$$M(4, \boldsymbol{W}(2)) = (0.8 - 0.75)/(0.8 - 0.55) = 0.2$$

$$M(4, \boldsymbol{W}(3)) = (0.8 - 0.55)/(0.8 - 0.55) = 1$$

For the "and" condition $0.5 < P(price) < 0.8$, we therefore have

$$M(4, \boldsymbol{W}(1)) = 0.08,$$

$$M(4, \boldsymbol{W}(2)) = 0.2, \text{ and } M(4, \boldsymbol{W}(3)) = 0.2.$$

For the purpose of rank aggregation, the above WSs are first being converted into:

$$\boldsymbol{W}(1) = (0.5{*}0.08, 0.54{*}0.08, 0.5{*}0.08)$$

$$\boldsymbol{W}(2) = (0.2{*}0.2, 0.34{*}0.2, 0.7{*}0.2)$$

$$\boldsymbol{W}(3) = (0.25{*}0.2, 0.64{*}0.2, 0.38{*}0.2)$$

Then we apply the previous approach to ranking them.

2.4. Ranking Web Services with Considering User Preference

In some cases, we should take into account the user's constraints and preference for WS ranking. Each client has preferences that bias him towards certain attributes, thereby weighing them higher scores than others for a particular task.

As an illustration, we take a travel agent WS as an example. Web Services need to be accessed one after another, e.g., booking air travel, booking hotel room and so forth. Different paths for WSs sequential consumption will exist as companies will compete by offering WSs with similar functionalities, e.g., different bus companies offering the same connection between two cities. A user is concerned with different attribute scores of these WSs. The user could pay more attention to the *security* level of the "booking air travel", than to the *price* of the "hotel room". As such, the ranking algorithm should be capable of ranking the WSs according to the user-specified preference over different attributes of each candidate WS for a workflow.

We provide the definition for the user preference vector.

Definition 3 *(user Preference Vector): a d-dimensional vector denoted by* $\boldsymbol{P} = (p_1, p_2, \ldots, p_d)$*, where d is the total number of attributes of a WS and* p_i *represents the user's degree of interest in the i-th attribute. The vector* $\boldsymbol{P}$ *is normalized such that* $\sum_{i=1}^{d} p_d = 1$.

The user preference–specific WSRank is described as
$Rank(C, A, t|\boldsymbol{P}) = \tau_A^*$
where the order list τ_A^* *is of the form: The equation* $w_i(k_1)p_i \geq w_i(k_2)p_i$ *holds, if we have* $\tau_A^*(W(k_1)) < \tau_A^*(W(k_2))$ *where* $i \in A$.

Example 4: Given two WSs of $\boldsymbol{W}(1) = (0.6, 0.8, 0.2, 0.8)$, and $\boldsymbol{W}(2) = (0.5, 0.9, 0.3, 0.6)$ as in the previous example, a client is concerned particularly with the *security* level of the first WS by specifying the preference vector $\boldsymbol{P}_1 = (0.8, 0.4, 0.1, 0.6)$, as well as with the *price* of the second WS by $\boldsymbol{P}_2 = (0.5, 0.4, 0.1, 0.7)$. We obtain

The *security* ranking list is $\tau_1 = [\boldsymbol{W}(1) > \boldsymbol{W}(2)]$ (because of 0.6*0.8 > 0.5*0.5);
The *price* ranking list is $\tau_4 = [\boldsymbol{W}(1) > \boldsymbol{W}(2)]$ (because of 0.8*0.6 < 0.6*0.7).
The rests remain the same as those of *Example 1*.

3. ALGORITHM FOR RANKING WEB SERVICES

It is straightforward to give the algorithm for ranking WSs for one task. We rank all WSs in decreasing order of their corresponding attribute values. In this section, we present the algorithms for rank aggregation, as well as for ranking WSs in the context of a workflow.

Rank aggregation: Rank aggregation is an efficient way of producing a global rank from multiple input ranking lists. It can be achieved through various techniques. The simplest technique is positional ranking or Borda's method [8], since it is easy to compute in linear time. We employ a modified version of Borda's method [8] for rank aggregation. Each WS is assigned a score corresponding to the position in a ranked list of preference over an attribute of the WS. One WS is therefore associated with the number of $|A|$ ($|A - R|$ for conditional ranking) scores for different attribute lists. All candidate WSs are sorted by their total scores summing over the attribute lists.

The procedure is as follows. Given the attribute lists $\tau_1, \tau_2, \cdots, \tau_d$, we assign a score for each attribute of a candidate WS. The total score of the j-th WS is calculated. All the WSs are then sorted in decreasing order of the total score.

Algorithm for rank aggregation
Input: a rank list $\tau_1, \tau_2, \cdots, \tau_d$ of the number of m candidates
Output: an optimal list WS()

Begin initialize

1: Sort respective lists $\tau_1, \tau_2, \cdots, \tau_d$
2: Assign a position score for every WS in each of d lists

For $j = 1$ to m
For $i = 1$ to d
$B_i(j)$ = the number of candidates whose score of attribute i is ranked below j in τ_i
end for
$WS(j) = \sum_{i=1}^{d} B_i(j)$ // total score for the j-th WS
end for
sort WS(j) in decreasing order where j=1, …, m

Borda's method can be implemented in linear time. Recently, more sophisticated methods have been proposed. The reader is referred to [8–10] for more details on the various proposed algorithms.

Rank in the context of a workflow: As presented before, we consider enumerating every possible path and calculating the score of all candidate WSs for each subtask in a workflow. The possible combination of a d-dimensional WS vector for the length T of sub-tasks is as high as $\prod_{i=1}^{T} n_{t_i}$ where n_{t_i} is the number of candidate WSs available for subtask t_i. This calculation is expensive and prohibitive in practice.

A computationally simpler algorithm for ranking WSs is as follows. We can calculate the score recursively. By doing so, we define

$$\alpha_i^t = \begin{cases} 1 & if \quad t = 0 \\ \sum_{i=1}^{d} \alpha_i^{t-1} w_i^t(j) & otherwise \end{cases}$$

Algorithm for ranking Web services for a workflow

Input: a set of candidate Web services W^{t_k} for each sub-task t_i in a task vector T=(t_1, t_2,…, t_T)

Output: a set of optimal candidate O={}

Begin initialize

O={}
$\alpha^0 = (1, 1, \cdots, 1)_d$
for t = 0 to T
for j=1 to n_t / /all W_{t_j} for sub-task t
$j^* = \arg\max_{W_{t_j}} \sum_{i=1}^{d} \alpha_i^{t-1} w_i^t(j)$
append W^{j*} to O
for i=1 to d
$\alpha_i^t = \alpha_i^{t-1} w_i^t(j^*)$
end for
end for
end for
return O

In the presence of the user-preference vector, the values in Lines 5 and 8 are weighted by corresponding weights, and then ranked.

4. EXPERIMENTS

Our preliminary experiment is encouraging. We empirically evaluate the algorithms for the ranking problem. As a dynamic algorithm, our ranking algorithm is tested on synthetic and real datasets. We will plot the performance of our algorithms with respect to different number of attributes, to various number of candidates, to different number of specified attributes for conditional ranking, and so on. We will examine the different ranking results with and without the user preference vector.

5. CONCLUSION

Web services are becoming increasingly important part of the WWW. How to select most suitable Web services for a service request is a worthwhile research topic. This paper have presented the formalization of the problem of ranking Web services in four scenarios, proposing approximation optimal algorithms for ranking a number of candidate Web services. Our future work includes more extensive experiments on the proposed approaches.

REFERENCES

1. Kleinberg J (1999) Authoritative sources in a hyperlinked environment. Journal of ACM, 46: 604–632
2. Page L, Brin S, Motwani R, Winograd T (1999) The pagerank citation ranking: Bringing order to the web, SIDL-WP-1999-0120, Stanford University
3. Page L, Brin S (1998) The anatomy of a large-scale hypertextual web search engine. Computer Networks 30:1–7
4. Guo L, Shao F, Botev C, Shanmugasundaram J (2003) XRANK: Ranked Keyword Search over XML Documents. In: ACM SIGMOD, San Diego, pp 16–27
5. Bhalotia G, Hulgeri A, Nakhe C, Chakrabarti S, Sudarshan S (2002) Keyword searching and browsing in databases using BANKS. In: 18th International Conference on Data Engineering (ICDE), pp 431–440
6. Balmin A, Hristidis V, Papakonstantinou Y (2004) Objectrank: Authority-based keyword search in databases. In: Proceedings of the 30th VLDB Conference, Toronto, pp 564–575
7. Huang X, Xue Q, Yang J (2003) Tuplerank and implicit relationship discovery in relational databases. Advances in Web-Age Information Management LNCS, 2762/2003, pp 445–457
8. Dwork, Kumar R, Naor M, Sivakumar D (2001) Rank aggregation methods for the web. In: Proceedings of the 10th International World Wide Web Conference, pp 613–622
9. Fagin R, Lotem A, Naor M (2003) Optimal aggregation algorithms for middleware. Journal of Computer and System Sciences 66 : 614–656
10. Fagin R, Kumar R, Mahdian M, Sivakumar D, and Vee E (2006) Comparing partial rankings, SIAM J. Discrete Mathematics 20:628–648
11. Agrawal R, Rantzau R, Terzi E (2006) Context-sensitive ranking. In: ACM SIGMOD, Chicago, IL, USA, pp 383–394
12. Zhang Z, Hwang S, Chang KCC (2006) Boolean+Ranking: querying a database by K-constrained optimization. In: ACM SIGMOD, Chicago, IL, USA pp 359–370

CHAPTER 45

AUTHENTICATED WEB SERVICES: A WS-SECURITY BASED IMPLEMENTATION*

VINCENZO AULETTA[1], CARLO BLUNDO[1], STELVIO CIMATO[2], EMILIANO DE CRISTOFARO[1], AND GUERRIERO RAIMATO[1]

[1]*Dipartimento di Informatica e Applicazioni, Università degli Studi di Salerno, Via Ponte Don Melillo - I-84084 Fisciano (SA), Italy., {auletta, carblu, emidec, raimato}@dia.unisa.it*

[2]*Dipartimento di Tecnologia dell'Informazione, Università di Milano cimato@dti.unimi.it*

Abstract: Web Services technology provides software developers with a wide range of tools and models to produce innovative distributed applications. After the initial diffusion of the standard technology the attention of the developers has focused on the ways to secure the information flows between clients and service providers. For this purpose several standards have been proposed and adopted. Another important issue is how to count the number of accesses to a given service in order to develop standard business models, in which the providers get paid for the offered resources. In this paper we propose an implementation, based on WS-Security, of an existing framework for authenticated Web metering, and compare it with an *ad-hoc* implementation. Our analysis shows that WS-Security is mature enough to provide a flexible and dynamic layer to underlie complex and interactive applications which require security management, without the need of developing ad-hoc solutions for each provided feature

1. INTRODUCTION

Web Services technology is often presented as an efficient solution to the increasing request for flexible connection and dynamic cooperation of distributed systems over the Internet. More in general, such technology offers a new set of standards and techniques which has having a very significant impact on IT organizations and is collecting a large amount of interest concretized in common research projects and money investments.

---

*The work of the authors has been supported in part by the European Commission through the IST program under Contract IST-2002-507932 ECRYPT and under contract FP6-1596 AEOLUS

H. Labiod and M. Badra (eds.), New Technologies, Mobility and Security, 541–553.
© 2007 *Springer*.

Indeed, interoperability, cross platform communication, and language independence are only a part of the appealing characteristics of Web Services. The standardization and the flexibility introduced by Web services in the development of new applications translate into increased productivity and gained efficiency. Development costs are also reduced since the use of standard interfaces, enabling the integration among applications and the reuse of existing solutions, may shorten the total development cycle.

Actually there are many examples in which Web Service technology has been employed, typically in enterprise application integration, development of B2B integration initiatives and Service Oriented Architecture projects. As soon as companies provide business services and release new products and tools Web services enabled, a problem which in our opinion will become more and more challenging, is the need of methodologies to accurately measure the number of accesses to a given service. Metering techniques have been developed in the context of measuring accesses to Web pages, where it is important to have a statistics on the exposure of advertising [13,10]. Recently, several directions for designing efficient and secure metering schemes have been proposed. Many proposals are based on various cryptographic techniques, as secure function evaluations, threshold cryptography, and secret sharing. For instance, see [14,7,11].

In the context of Web services usage, such metering techniques can be helpful in the development of new business models, where an *audit agency* is in charge of registering *clients* accessing the services, and of paying the *servers* claiming for payment, after that a valid *proof* of the number of serviced requests is presented. In metering schemes, a *proof* is a value that the server can compute only if a fixed number of clients have visited it or a client has visited it a certain number of times. Such a value is sent to the audit agency at fixed time intervals.

The widespread use of Web Service for providing business level solutions put a great emphasis on security issues of the developed applications. The goal is to provide software developers with standard security techniques, enabling the creation of business models in which service providers get paid for the offered resources. In this sense, it is important to combine authentication techniques with mechanisms which allow to count the number of accesses for a given service in a secure way. Up to a few years ago, the most diffused approach for handling authentication credentials was to use custom SOAP headers to transmit user credentials. In last years, many supplemental standards for guaranteeing security features in Web Services have been proposed and collected under the WS-Security specification [12]. This specification (whose last version has been released at the beginning of the 2006) is the result of the work by the OASIS Technical Committee [4] and proposes a standard set of SOAP extensions that can be used when building secure Web services to provide *confidentiality* (messages should be read only by sender and receiver), *integrity and authentication* (the receiver should be guaranteed that the message is the one sent by the sender and has not been altered), *non repudiation* (the sender should not be able to deny the sending of the message, and *compatibility* (messages should be processed according to the same roles by any node on the path message). Such mechanisms provide a building block that can be used in conjunction with other Web service extensions and higher-level applications to accommodate a wide variety of security models and security technologies. Some other related standard proposals are also available, such as SAML [5], XKMS [6], and so on. Each proposal refers to a particular aspect of securing the information flow and protecting the data exchange between the client and the service provider. Although WS-Security (as well as the other secure related specification) is relatively new, only recently, it has been widely adopted and (almost) completely supported by multiple toolkits.

In this work, we consider the problem of implementing a secure, authenticated, and efficient *metering system*. In [8] a Web Services-based framework is described that implements the metering scheme presented in [9] for Authenticated Web Services. This implementation is characterized by static invocations of the Web Services using custom SOAP headers to provide authentication. In this paper, we describe a novel implementation of the framework given in [8], based on WS-Security. The SOAP header contains the authentication token which can be used when accessing a restricted web service, where the authentication token is embedded in the message using the mechanisms provided by WS-Security. We compare our implementation with the one presented in [8] and analyze its performance. This analysis shows that WS-Security is mature enough to provide a flexible and dynamic layer to underlie complex and interactive applications which require security management, without the need of developing ad-hoc solutions for each provided feature. Moreover, the available implementations of the standard allow to solve standardization and compatibility issues, achieving a completely standard and dynamic solution. In particular, we show how WS-Security can be used to overcome the addition of customized information inside the SOAP header to provide authentication in metering schemes.

Furthermore, we present a thorough performance analysis that display that the implementation based on WSS performs better than the one presented in [8], despite the latter provides a dedicated solution.

The rest of the paper is organized as follows. In Section 2, we present the scheme for Authenticated Metering. In Section 3, we recall the framework and its implementation presented in [8]. In Section 4, we propose our implementation based on WS-Security. Finally, in Section 5, we give a thorough performance analysis to compare the two implementations.

2. AUTHENTICATED METERING SCHEME

Metering systems monitor services accesses either for statistical or billing goals. An *audit agency* is in charge of counting the number of clients accessing the services, so that service providers can be payed according to the number of visits received. Because it measures the interactions between many servers (service providers or Web site hosts) and clients (Web surfers), the audit agency needs a mechanism that ensures the validity and accuracy of usage measurements by preventing fraudulent actions by servers or clients.

In this work we refer to the software infrastructure presented in [9] and the framework given in [8] that implements this infrastructure. The metering system consists of n clients, say $C_1, ..., C_n$ interacting with an audit agency A and a server S. The audit agency is in charge of keeping trace of the number of times the clients accesses a service provided by the server S. The scheme is tightly based on registration and authentication of clients and it relies on three phases:

Initialization The initialization phase starts when a client $\mathcal{C}$ contacts the audit agency $\mathcal{A}$ and requests access to the provided service. A number of subsequent operations are started by each of the players:

- The audit agency $\mathcal{A}$ calculates a random seed, say w_0, and computes the value of the k-th application of a cryptographic hash function $\mathcal{H}$, that is $w_k = \mathcal{H}^k(w_0) = \mathcal{H}(\mathcal{H}^{k-1}(w_0))$, where $\mathcal{H}^2(w_0) = \mathcal{H}(\mathcal{H}(w_0))$. Then, $\mathcal{A}$ stores the tuple $[id_C, k, w_0]$ holding the client identifier $id_{\mathcal{C}}$, the number of guaranteed accesses k, and the seed w_0. Finally, it sends the two tuples $[id_{\mathcal{C}}, k, w_0]$ and $[id_{\mathcal{C}}, w_k]$ to the client $\mathcal{C}$ and the server $\mathcal{S}$, respectively.
- The client $\mathcal{C}$ stores the tuple sent by $\mathcal{A}$ and retrieves the initial seed w_0 and the number k of accesses it registered for. Then, $\mathcal{C}$ generates and stores the k values[1] $w_1 = \mathcal{H}(w_0), w_2 = \mathcal{H}(w_1), \ldots, w_k = \mathcal{H}(w_{k-1})$. The client will use the token w_{k-j} to access an authenticated service the j-th time. The client keeps trace of the last access and maintains an access counter.
- The server $\mathcal{S}$ stores the tuple received from $\mathcal{A}$ in its database of registered clients, associating it with a counter $L_{\mathcal{C}}$ initially set to 0.

Interaction Interaction happens when the client $\mathcal{C}$ visits the server site $\mathcal{S}$ and wants to access the restricted services offered by $\mathcal{S}$. Suppose that $\mathcal{C}$ is accessing the service for the j-th time.

- The client $\mathcal{C}$, within a service request, sends the token $w_{k-j} = \mathcal{H}^{k-j}(w_0)$ to $\mathcal{S}$, and it updates the access counter for the authenticated service decrementing its value.
- The server $\mathcal{S}$, on the reception of the token, performs an access control, verifying that the value resulting from the application of the hash function $\mathcal{H}$ to the received token matches the last stored value for the client $\mathcal{C}$, that is $\mathcal{H}(w_{k-j}) = w_{k-j+1}$. If the two values match, then he updates its stored values with the new received token w_{k-j} and increments the client counter $L_{\mathcal{C}}$.

Verification The verification phase begins when the server $\mathcal{S}$ claims the payment after a certain number of visits received by a client $\mathcal{C}$ in a certain period of time previously agreed with the agency $\mathcal{A}$.

- The server $\mathcal{S}$ sends to the audit agency $\mathcal{A}$ the tuple $(id_C, W, L_{\mathcal{C}})$, where $id_{\mathcal{C}}$ is the client's identifier, W is last stored authentication token for $\mathcal{C}$, and $L_{\mathcal{C}}$ is the client's counter.
- The audit agency $\mathcal{A}$ verifies that the value sent by $\mathcal{S}$ for $\mathcal{C}$ equals to $\mathcal{H}^{k-L_C}(w_0)$. In this case, $\mathcal{S}$ may claim a payment from $\mathcal{A}$ for the services granted to $\mathcal{C}$.

[1]The values $w_0, w_1, \ldots, w_n$ are referred to as *hash chain* with seed w_0.

3. THE FRAMEWORK

In [8], a framework has been designed to carry out the operations discussed in Section 2. In this framework the three entities, client, server, and audit agency, expose their services as Web Services. The time diagram of the interactions is presented in Figure 1. As depicted, the four interactions are carried out in the following way:

- *Client/Audit Agency.* Client registers himself with the audit agency in order to get a valid authentication token. To this aim, the audit agency exposes the `returnClientToken` service. The SOAP request contains the number of requested accesses; while, the SOAP response is composed by a univocal client identifier $id_{\mathcal{C}}$, the number of obtained accesses k, and the initial seed w_0 used to calculate the authentication tokens (i.e., the values in the hash chain).
- *Audit Agency/Server.* The server exposes the `clientRegister` service in order to receive from the audit agency the data associated with the newly registered client. The SOAP request invoked by the audit agency contains the client identification $id_{\mathcal{C}}$ and the last value of the hash chain $\mathcal{H}^k(w_0)$.
- *Client/Server.* This stage handles the client access to the restricted service. Therefore, a dedicate handler must come into play in order to provide authentication. The server uses the authentication information ($id_{\mathcal{C}}$, $hashValue$) sent by the client $\mathcal{C}$ and its stored information associated to $\mathcal{C}$ to verify whether the client $\mathcal{C}$ has the right to access the requested service. If so, it grants access, otherwise it raises a SOAP fault. A `Synchronization` service is also provided in case the client sends a wrong hash value through having the right one. The server, as challenge for the client $\mathcal{C}$, provides the hash value currently associated to $\mathcal{C}$, say w_j, as well as the counter $L_{\mathcal{C}}$. The client replies with the pre-image of the hash value, namely w_{j-1}. Due to the one-way property of the hash function, this is possible only if the client knows the hash chain. Indeed, from $L_{\mathcal{C}}$ and w_0 the client can compute w_{j-1} as $\mathcal{H}^{k-L_{\mathcal{C}}-1}(w_0)$.
- *Server/Audit Agency.* The last interaction is performed when the Server has to require to the Audit agency the payment for the $\mathcal{C}$'s accesses. In this case, it has

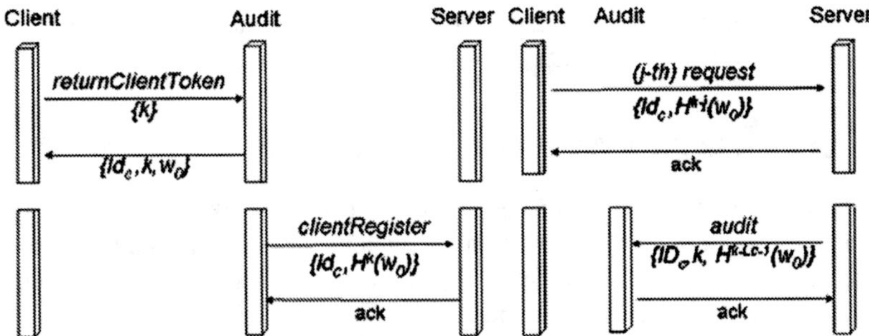

Figure 1. Architecture of the framework in [8]

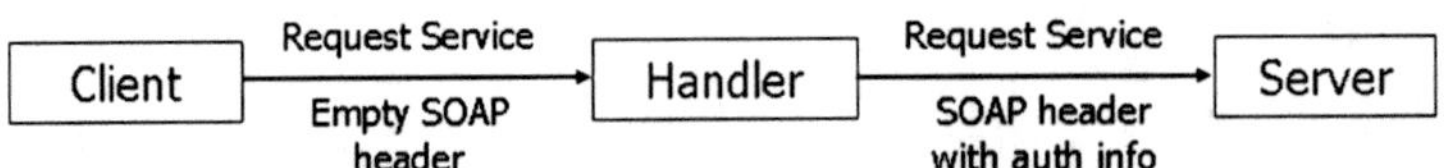

Figure 2. Intervention of the handler to add authentication information

to provide to the audit agency a *proof* of the number of served requests. The audit agency exposes the `Audit` service. Server's SOAP requests contain the client's identifier, the last stored authentication token for that client, and the number of registered accesses. The audit verifies that the values sent by $\mathcal{S}$ match the stored values, and acts consequently.

3.1. Implementation Based on Ad-hoc SOAP Headers

The implementation of the framework presented in [8] has been developed within the Java platform, using Tomcat as the Servlet Container, and Axis [1] as the SOAP engine.

One of the crucial points of the framework is the submission of the authentication credentials of the Client to the Server. The Client sends an authentication token (composed of $id_{\mathcal{C}}$, *hashValue*) in a SOAP Header. This header is created by a SOAP Handler, which is in charge of intercepting SOAP requests, adding client's authentication information to the header and attach it to the SOAP message before sending it to the Web Service, as showed in Figure 2. The Java class has to implement the *java.xml.rpc.handler.Handler* interface and, in particular, the *handleRequest* method.

The core of this method can be summarized as follows. Recover message envelope. Create a SOAP header named *authHeader*. Attach the *idClient* and the *hashValue*.

On the server-side, there is a dedicated handler (*AuthenticateHandler*) that is in charge of intercepting the Authentication Header and verifying the correctness of the authentication token before the service is invoked.

This implementation was based on *Java2WSDL* and *WSDL2Java* tools. Starting from the Java classes of the services, they obtain the correspondent WSDL files using the *Java2WSDL* tool. After that, the tool *WSDL2Java* creates from the WSDL files all the stubs, skeletons, and data types needed to the framework.

4. IMPLEMENTATION BASED ON WS-SECURITY

In this section, we give a new implementation of the framework proposed in [8]. The main differences of this implementation are dynamic invocations of the services and the use of WS-Security to execute the authentication operations.

In static invocation, the clients generate proxy stubs at development time, using the WSDL of the Web Service. The clients invoke the methods in the proxy stub, which in turn invokes the methods in the Web Service. Therefore, if the WSDL changes, these proxy stubs have to be re-generated. Then, our implementation instead

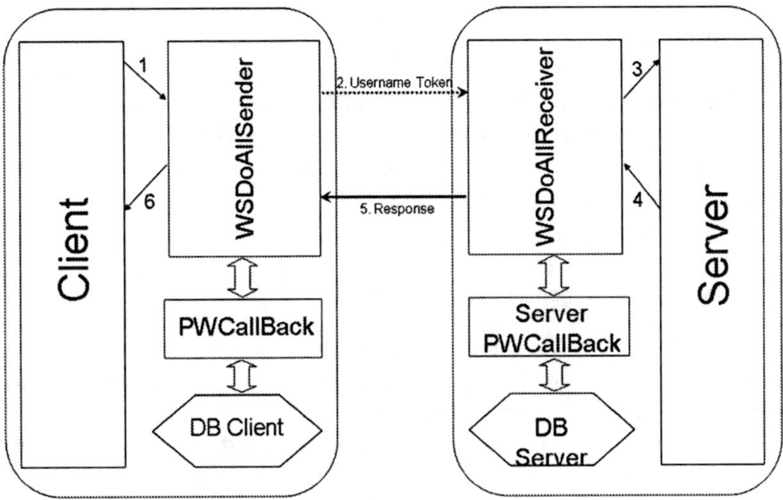

Figure 3. The interaction scheme of the UsernameToken profile as in WSS4J

is based on dynamic invocations. Proxy stubs are generated at run-time, using the dynamic invocation APIs. The advantage is that any change to the WSDL file will not affect the Web Service invocation. To provide client authentication, we use *WSS4J* [2], the standard implementation of WS-Security provided by the Apache Software Foundation. This package implements WS-Security mechanisms and also achieves interoperability between JAX-RPC [3] based and .NET based client/servers. WS-Security has been designed as a flexible set of mechanisms that can be used to build several security protocols. WS-Security gives an appropriate profile to provide authentication: the *User Name Token* allows users to authenticate themselves by sending as credentials a *Username* and eventually a *Password.* Together with the password, some extra information can be set, such as the type of the password (*PasswordText* or *PasswordDigest*), a *Nonce* to avoid replay attacks or a *TimeStamp.* In this way, the management of authentication information is no longer carried out by ad-hoc and customized handlers, but through *WSS4J built-in* handlers, named *WSDoAllSender* and *WsDoAllReceiver.* These handlers are transparent to user applications and operate according to standard specifications. Figure 3 shows the interaction scheme of the *UsernameToken* profile as implemented by WSS4J.

1. The SOAP message created by the client application is intercepted by *WSDoAllSender* and processed before it is sent to the server. Programmers do not have to modify this handler, since it gets the information needed to build the *UsernameToken* header by the security header in the SOAP message through a standard class, named *PWCallBack.* Application developers take care of this class, which must implement the *javax.security.auth.CallBackHandler* interface, according to the application needs.

2. The SOAP message is sent to the service. The *WSDoAllReceiver* intercepts the message and processes the authentication information stored in the message. Once again, the handler makes use of a callback class, named *ServerPWCallback*. This must implement the *javax.security.auth.callback.CallbackHandler* interface to verify whether the client has the right to access the service.
3. If the verification is passed, the handler returns the control to the application otherwise a *WSSecurityException* is thrown.

4.1. Framework Implementation with WSS4J

As seen in Section 2, authentication tokens are used in the *interaction* phase, when clients have to send credentials to the server to access the service. We have seen in the previous section that this can be achieved by using the WSS4J *UsernameToken*, where the *idClient* is the username and the *hashValue* is the password.

On the server-side, the *UsernameToken* profile is processed by the *WSDoAllReceiver* handler, which has to be included in the Axis service request flow chain. This is specified in the *Web-Service Deployment Descriptor* file and done using the Axis services deployment tool. On the client-side, instead, the *UsernameToken* profile is built by the *WSDoAllSender* handler, which has to be included in the Axis service request flow chain. To do this, application developers could perform a static deploy of the service on the client-side. However, we discard this solution for two reasons. First, the *username* field is assigned dynamically by the audit agency to any newly registered client and changes if the logged user changes. Second, the handler would be globally inserted in the client's requestFlow and be called for *any* invoked service.

We decided to dynamically provide to WSS4J the username of the *UsernameToken* profile, using the dynamic configuration of the Axis Engine. The code for the *org.apache.axis.configuration.SimpleProvider* class was written to implement the *org.apache.axis.EngineConfiguration* interface.

This solution is graphically presented in Figure 4 and has been implemented in a dedicated infrastructure that hides all the details related to the handler and to the dynamic configuration. To this aim, a Java package, **CHM** (Client Handler Manager) has been released.

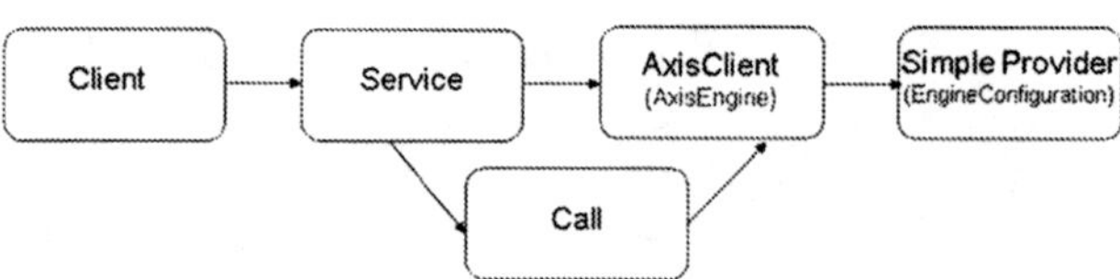

Figure 4. The dynamic configuration of Axis Engine

Users just have to dynamically provide username and password, by querying them from a database. We have implemented repositories as MySQL databases, accessed through the JDBC API. The data held by the Server, the Audit, and the Client have been stored in separate databases holding one table for each subject. MySQL has been chosen, being open-source and cross-platform.

Classes contained in the CHM package extend the Axis Client API, thus achieving transparency and compatibility. The core class of the infrastructure is the *Service* class. It implements the dynamic configuration which allows to use the *Username-Token* profile by providing dedicated methods to set username and password (in our case the client's identifier and the hash value, respectively) to bind to the configuration. It is within this classthat the dynamic configuration is carried out through the use of the *EngineConfiguration* interface. Other two classes have been extended: the *Call* and the *PWCallback* classes. The former essentially sets the password property with a dynamic input value and sets the callback class *PWCallBack* that will be invoked by *WSDoAllSender*. This is in charge of providing to *WSDoAllSender* the password associated to username given to WSS4J.

5. PERFORMANCE ANALYSIS

In this section we compare performances of our dynamic WS-security based implementation against the static ad-hoc solution given in [8].
The experiments were held on the following test bed:

- PC IBM ThinkCentre 50 - Pentium 4 2,6 GHz with 760 MB RAM.
- Linux Suse 10 - Kernel 2.6.16.
- Sun Java VM version 1.4.2_13 SDK.
- Apache Tomcat 5.5.28 as Servlet Container.
- Axis 1.3 as the SOAP Engine.

Our first experiment compared times taken by the client to build the header with the security information. Tests were repeated 100 times with 100 different accesses. Figure 7 shows times for the 100 requests in both the implementations. With the ad-hoc SOAP headers based on WSDL4Java, we obtained an average construction time of 93,69 ms. Whereas, with the WSS4J-based implementation, the average time to build the header for the SOAP request was 51,65 ms. In both tests we witness almost uniform times, with some periodical peaks basically due to internal operations of the Java virtual machine, such as garbage collection. The difference between the two implementations is remarkable. The WSS4J-based implementation performs better thanks to the smaller number of involved objects in order to build the header. With WSDL2Java a set of stubs are requested in order to insert the handler in the chain and to serialize information.

In the second test, we measured times taken by the server to process the header with the security information. Figure 6 shows processing times for each request with both the implementations. Also in this case the WSDL4Java-based implementation has worse performances. In fact, the implementation described in [8] reports an

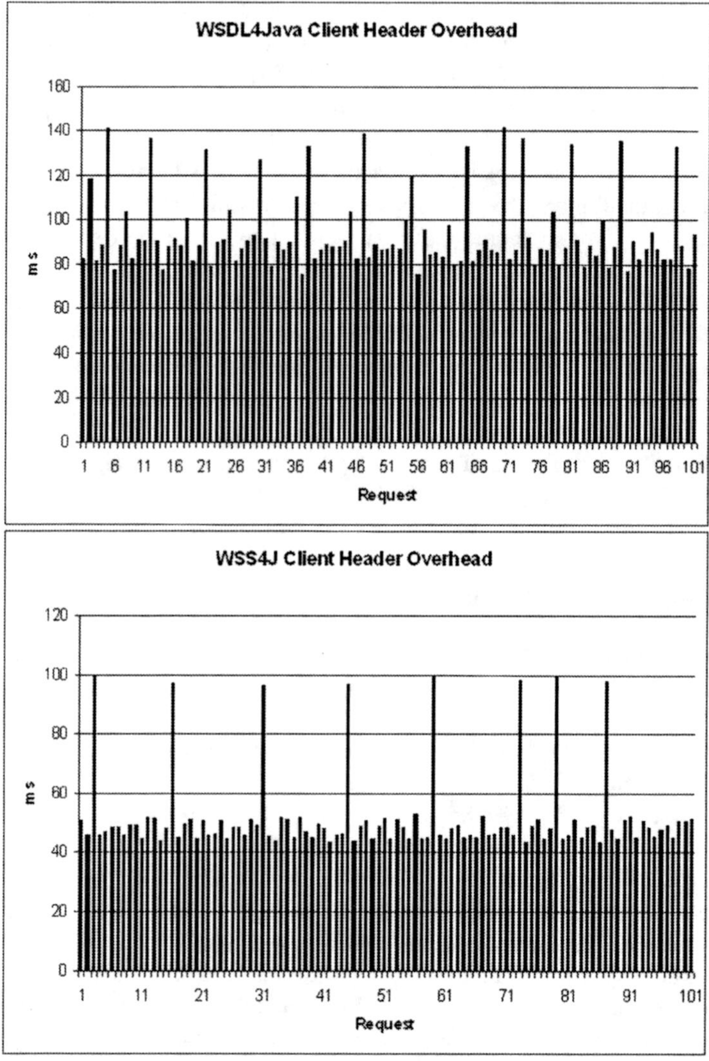

Figure 5. Time to build client header with the WSDL4Java and WSS4J implementations

average response time of 81 ms for 100 accesses. Our implementation instead obtains an average response time of 61 ms.

In the third test, we have compared the entire round trip times, i.e. the total times required for an invocation of a service. To this aim, we used a very simple application (`Echo`), so that performance analysis is not affected by computation or overhead bounded to the service. The comparison between the two implementations should only be related to the managing of authentication information. Results of this test are summarized in Figure 7. Also in this case, the WSS4J-based implementation has better performances.

As discussed in section 4, besides relying on the WS-Security standard, our implementation makes the programmer task easier by the dynamic invocations of deployed Web Services for all the provided features. Despite this greater work assigned to the SOAP engine, performance experiments showed that no increase occurs in running times of any service. The client, in the Initialization

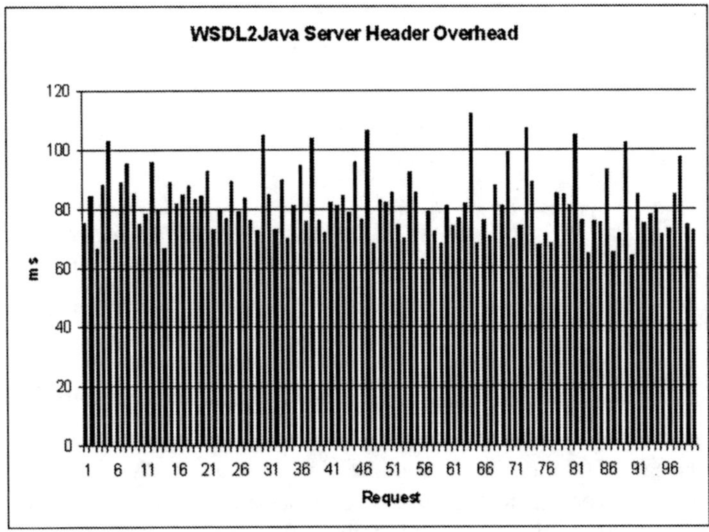

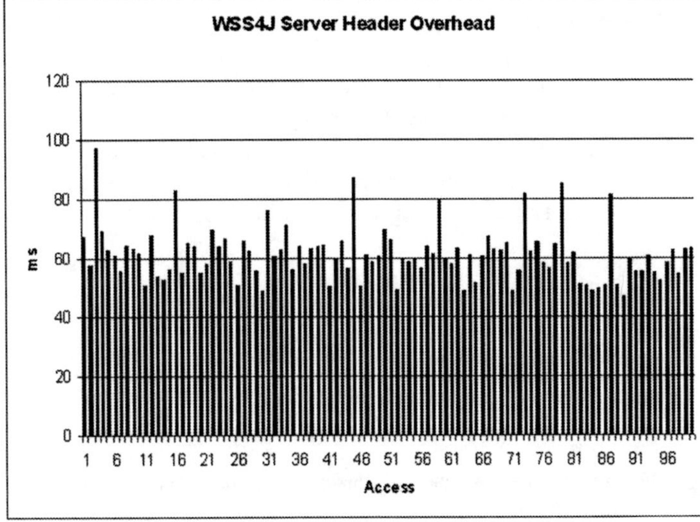

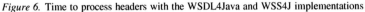

Figure 6. Time to process headers with the WSDL4Java and WSS4J implementations

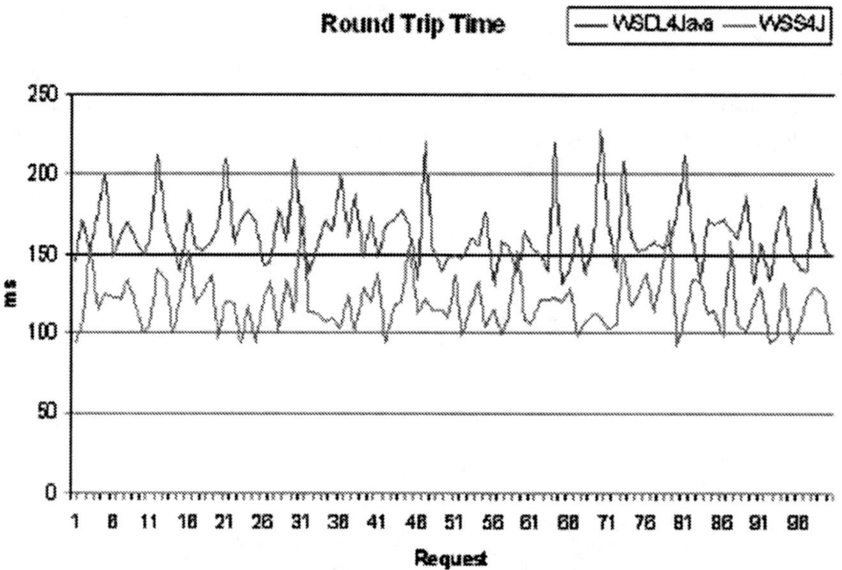

Figure 7. Round trip time for the two implementations

phase, contacts the audit agency to request a certain number of accesses. Tests were ran with a growing number of requested accesses, in order to evaluate different performances with more and more hashing operations. We saw that times grow gracefully according to the number of requested accesses. The time required for the client's registration of the two implementations is essentially equivalent. Therefore, achieving dynamic invocations does not lead to a performance worsening.

In conclusion, performance evaluations showed that WS-Security can be used to implement the scheme of Authenticated Web-Services. Therefore, we can conclude that WS-Security is mature enough to provide a flexible and dynamic layer to underlie complex and interactive applications which require security management, without the need of developing ad-hoc solutions for each provided feature.

REFERENCES

1. Apache Axis. *http://ws.apache.org/axis/.*
2. Apache Web Services Security For Java. *http://ws.apache.org/wss4j/.*
3. Java API for XML-based Remote Procedure Call (JAX-RPC). *http://java.sun.com/webservices/jaxrpc/.*
4. Organization for the Advancement of the Structured Information Standards (OASIS). *http://www.oasis-open.org/home/index.php.*
5. SAML. *http://www.oasis-open.org/committees/security/.*
6. XML Key Management Specification (XKMS). *http://www.w3.org/TR/xkms/.*

7. S. G. Barwick, W. Jackson, and K. Martin. A general approach to robust web metering. *Designs, Codes, and Cryptography*, 36(1):5–27, 2005.
8. C. Blundo and S. Cimato. A framework for authenticated web services. In *Proceedings of Europen Conference on Web Services (ECOWS 04), Lecture Notes in Computer Science.*
9. C. Blundo and S. Cimato. A software infrastructure for authenticated web metering. *IEEE Computer*, 37(4):28–33, 2004.
10. M. K. Franklin and D. Malkhi. Auditable metering with lightweight security. *Journal of Computer Security*, 6(4):237–256, 1998.
11. S. S. Kim, S. K. Kim, and H.-J. Park. New approach for secure and efficient metering in the web advertising. In *Proceedings of International Conference on Computational Science and Its Applications (ICCSA 2004), Lecture Notes of Computer Science*, volume 3043, pages 215–221. Springer-Verlag, Berlin, 2004.
12. A. Nadalin, C. Kaler, P. Hallam-Baker, and R. Monzillo. Web Services Security: SOAP Message Security 1.1. *OASIS. http://www.oasis-open.org/committees/download.php/16790/wss-v1.1-spec-os-SOAPMessageSecurity.pdf*, 2006.
13. M. Naor and B. Pinkas. Secure and efficient metering. In *Proceedings of Advances in Cryptology – Eurocrypt '98, Lecture Notes in Computer Science*, volume 1403, pages 576–590, 1998.
14. W. Ogata and K. Kurosawa. Provably secure metering scheme. In *Proceedings of ASIACRYPT 00, Lecture Notes in Computer Science*, volume 1976, pages 388–398. Springer-Verlag, Berlin, 2000.

CHAPTER 46

AN ARCHITECTURAL MODEL FOR A MOBILE AGENTS SYSTEM INTEROPERABILITY

ZEGHACHE LINDA, BADACHE NADJIB,
AND ELMAOUHAB AOUAOUCHE

Research Centre in Scientific and Technological Information (CERIST), Algiers, Algeria
l.zeghache@dtri.cerist.dz, badache@mail.cerist.dz, elmaouhab@wissal.dz http://www.cerist.dz

Abstract: An important goal in mobile agent technology is interoperability between various agent systems. A way of achieving this goal would be to envisage a standard to be imposed on these various "agents systems" in order to allow the inter-working of various architectures of mobile agents. During the past years, different scientific communities proposed to different standardization actions, such as the Foundation for Physical Intelligent Agents (FIPA) and the Object Management Group's MASIF (Mobile Agent System Interoperability Facilities). Although, they finally share some major targets, the OMG and FIPA current results show their distinct origins, particularly for interoperability between or within distributed systems. In this paper, we first analyze the similarities and differences, advantages and disadvantages of the Object Management Group (OMG) mobile agent and the Foundations for Intelligent Physical Agents (FIPA) intelligent agent approaches. Based on this analysis, we try to integrate these two standards to propose an architectural model for mobile agents system interoperability

Keywords: Mobile agent, Mobile agent system, Interoperability, Standard, MASIF, FIPA

1. INTRODUCTION

Mobile agents provide many benefits for the development of new generation Internet systems, such as great capabilities for distributed Internet system programming, in which there is the need for different kinds of integrated information. Also the mobiles agent paradigm ensures satisfactory performance for distributed access to Internet databases, for distributed retrieving and filtering of information and for minimizing network workload. Finally, mobile agents have been proved very effective in supporting the asynchronous execution, weak connectivity and disconnected operations[15].

H. Labiod and M. Badra (eds.), New Technologies, Mobility and Security, 555–566.
© 2007 *Springer.*

Mobile agents reside in a highly heterogeneous environment; this heterogeneity appears in many dimensions. Mobile agents migrate to a host where an execution environment is set up for them; upon arriving there, they might execute code, make remote procedure calls (RPCs) in order to access the resources of the host, collect data and eventually might initiate another process of migration to another host [17]. A problem arises when mobiles agent platforms are different.

In this paper, we focus on mobile agent interoperability. We investigate the existing standards, examine their different origins and explore their possible integration in order to bring an architectural model for a mobile agents system interoperability.

2. AGENT SYSTEM INTEROPERABILITY

Mobile agent is a relatively new paradigm, but there is already a number of mobile agent system implementations, such as AgentTcl, Aglets, and Odyssey. Unfortunately, these systems are usually incompatible with each other. Thus, agents built for one platform cannot be used in another, if that does not pose problems in a closed universe, it is differently in the case of Internet. [11]

During the past years, different scientific communities proposed different standardization actions :

- **M**obile **A**gent **S**ystem **I**nteroperability **F**acility (**MASIF**) defined by the Object Management Group (OMG), based on CORBA and is intended for mobile agents systems.
- **F**oundation of **I**ntelligent **P**hysical **A**gents (**FIPA**). It standardizes an agent platform for inter-working agents systems.

Although, they finally share some major targets, the MASIF and FIPA current results show their distinct origins.

2.1. MASIF Specification

The Object Management Group's first effort in the agent field resulted in the MASIF specifications adopted in 1998, [11]. It focused on defining a conceptual framework, services and interfaces for CORBA-based interoperability between heterogeneous mobile agent platforms. The re-use of existing CORBA services [4], such as naming service, life cycle, security and externalization templates, is also analysed by MASIF [3].

MASIF's framework defines the concepts of agent (either mobile or stationary), hosted by places, run by agent systems, belonging to regions. MASIF specifies two interfaces :

- The MAFFinder interface defines operations for place, agent and agent systems lookup;
- The MAFAgentSystem interface deals with the management of an agent system, places and agents (creation, destruction, agent migration, etc.).

There should be at least one server implementing the MAFFinder interface per region, and each agent system should implement the MAFAgentSystem interface. [11]

MASIF suffers from many weaknesses: How can regions be interconnected? How can an agent system receive an incoming agent of a different agent system type? How may heterogeneous agents communicate? Moreover, it appears that mobility makes the reuse of today's CORBA services neither simple (e.g. naming), nor sufficient (e.g. security).

Finally, the fact that only one implementation is available today (Grasshopper) makes it impossible to assess the actual support for interoperability. Nevertheless, MASIF has to be regarded as a basic infrastructure and conceptual framework for further specification and revision.

2.2. FIPA Specification

The Foundation for Intelligent Physical Agents was officially created and registered as a non-profit association in 1996. FIPA aims at producing "specifications that maximise interoperability across agent-based applications"[6].

FIPA conceptual model defines the following concepts [7]:

- Agent Platform (AP): is a physical infrastructure in which agents are deployed.
- Agent: is the fundamental actor on an AP.
- The Agent Management System (AMS): is a "white pages" server. It manages the creation, deletion and general status of the agents.
- Directory Facilitator (DF): is a "yellow pages" server. Agents may register with the DF or query the DF to find out the services offered by other agents.
- A Message Transport Service (MTS): is the default communication method between agents.

Moreover, Agent communication is one of the core components of the FIPA's conceptual model for agent systems. Agents can pass semantically meaningful messages to one another in order to accomplish the tasks required by the application. These messages are expressed in the FIPA-ACL language. This later is a speech-act based Agent Communication Language developed by FIPA to allow communication between heterogeneous agents. It defines Protocols for complex message interaction and semantic languages for complex content expression [5].

However, in FIPA standards there is no specification about agent mobility.

2.3. MASIF and FIPA Comparison

Several similarities and differences between FIPA and MASIF specifications can be drawn (Table 1).

MASIF and FIPA standardisation efforts are not completely different; they both focus on interoperability between agent systems. However, each one has its own religion to achieve interoperability; MASIF uses agent mobility while FIPA concentrates on agent communication.

Table 1. MASIF and FIPA Comparison

Criterion	MASIF	FIPA
Agent	Mobile	Stationary
Agent Management	Supported by the MAF Agent System interface	Assured by Agent Management System
Agent Registration	Supported by the MAF-Finder interface	Managed by Agent Management System
Service Registration	No specification	Supported by the Directory Facilitator
Agent Location	Sequential search algorithm	Federation of DFs to discover services offered by agents
Mobility	Enable agent migration between Agent Systems of the same profile (language,authentication and serialization methods) via standardized CORBA IDL interfaces	No specification
Communication	No specification	Based on a high-level speech act communication language and a predicate logic based content language
Interoperability	Based on agent migration and require homogeneous platforms	Based on agents communication in heterogeneous environments
Error recovery	No specification	No specification

Regarding these characteristics of FIPA and MASIF, we can notice that the two standardisations are complementary.

3. RELATED WORK

Mobile Agent Interoperability has been an active field of research. Various solutions have been proposed.

One approach to mobile agent interoperability has been suggested by [13]. The basic idea consists of dividing agents into platform specific and platform independent parts into the body and the head. The body handles the agent-programming interface of each agent platform and a head can be placed on top of it. The agent-head can migrate via a Gateway.

The solution described by [12] is based on generative mobility. In generative mobility, a blueprint of an agent's functionality is transported, together with information on the agent's state. At its destination, an agent factory regenerates the executable code of the agent on the basis of its blueprint. An agent may then restore its state and resume execution.

Several approaches relying on MASIF and FIPA specifications have been considered.

The developers of Grasshopper [2]provides a FIPA add on to Grasshopper platform which comply with MASIF standard. The main components of a FIPA compliant platform are provided in form of stationary Grasshopper agents.

Another approach described by [1]propose agent migration using FIPA-ACL message, which means that transmission of mobile agents between two agents systems will be carried out using the message system between agents.

4. OUR PROPOSAL

The aim of this paper is to propose an architectural model in order to promote interoperability between various mobile agents systems and take into account the advantages of MASIF and FIPA specifications. The architecture we propose is based on MASIF conceptual model to ensure agent management and mobility. In other side, it takes from FIPA its agent's communication specifications.

4.1. Reference Model

The reference model of our architecture is composed of a set of conceptual elements managed by common operations defined within interfaces (Fig.1).

Agent : is a computer program that acts autonomously on behalf of a person or organization and it can be mobile or stationary. It is characterised by a state, authority, name and location.

Place : is a context within an agent system in which an agent can execute. It must contain three components: language interpreter, state module and security module.

Space Information : is an abstraction of communication mean that allows agent communication in independent way from the languages. It provides different style of interaction.

Agent System : is a platform that can create, interpret, execute, transfer, and terminate agents. It is associated with an authority that identifies the person or organization for whom the agent system acts.

Region : is a set of agent systems that have the same authority, but are not necessarily of the same type.

4.2. Agent Management

In order to ensure agent management (create, suspend, resume, terminate) and agent transfer (migration) in a standard way, we use MAFAgentSystem Interface proposed by MASIF.

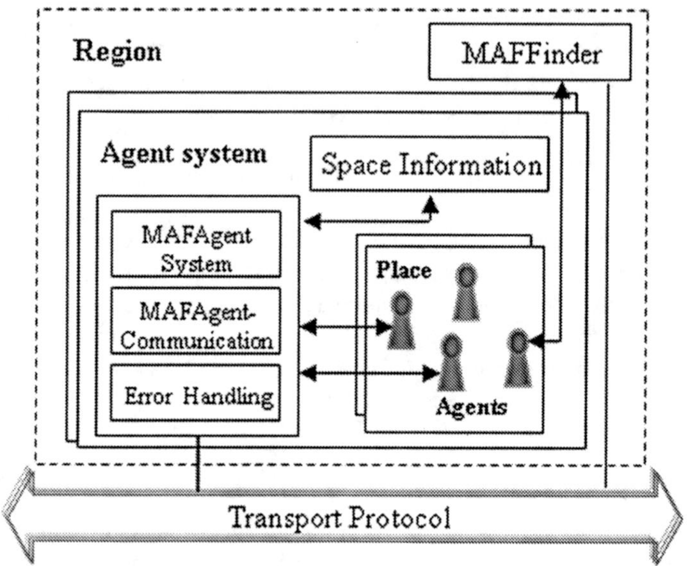

Figure 1. Proposed Architecture

We have extended this interface with new methods to allow the control of an agent directory which maintains an index of all the agents that are currently resident on an Agent System. It offers white pages services to other agents. Each agent must register with the agent directory after creation to have a global unique identifier.

4.3. Agent Localization

To allow agent localization in an efficient way, we propose to use the service advertisement. We suggest creating a directory service on each region called MAFFinder. This MAFFinder provides yellow pages services to other agents. Agents may register their services with the MAFFinder or query the MAFFinder to find out the services offered by other agents.

The MAFFinder has the following logical structure (Fig.2):

- To each region we associate a MAFFinder called Regional MAFFinder.
- To each type of service we associate a MAFFinder called Service-MAFFinder. Every agent that wishes to publish its service to other agents, should find an appropriate Service-MAFFinder and request the registration of its agent description.
- We define System-MAFFinder to maintain descriptions of all agent systems residents in the region.

Service-MAFFinders and System-MAFFinders are registered within the Regional MAFFinder. In order to allow agents to access the MAFFinder, we redefine the MAFFinder interface of MASIF.

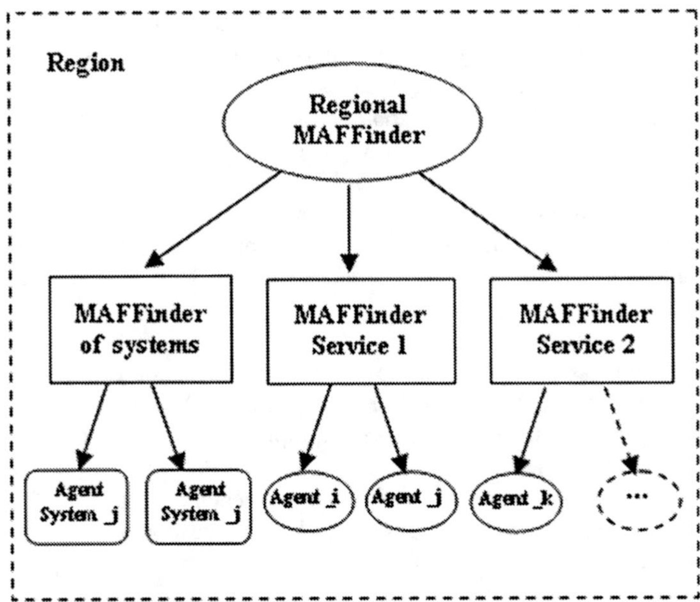

Figure 2. MAFFinder logical structure

It contains a set of methods permitting the registration of agent system, advertisement of agent services and localisation of services offered by other agents. In this proposition, we use a mechanism that searches first locally and then extends the search to other MAFFinders, if allowed. For this purpose, we envisage to federate MAFFinders.

The federation of MAFFinders can be achieved by registering Regional MAFFinders with each other with federate method defined in MAFFinder interface.

4.4. Agent Communication

Mobile agents are expected to be autonomous and interact with their environment and other agents in order to solve complex distributed problems. They need a common knowledge core, so that negotiation and cooperation can be managed.

Three basic problems need to be addressed for agents to effectively share knowledge : First, how can we translate from one knowledge representation language to another? Second, how can we guarantee that meaning of concepts, objects and relationships is the same across different agents? And third, how is this potentially sharable knowledge going to be shared and communicated between agents?

The tool that follows this layered abstraction is Agent Communication Language (ACL). [16]

Agent Communication Language : In this proposition, we have used FIPA agent communication language (FIPA-ACL) to package messages in standard way.

FIPA-ACL is based on the speech act theory: messages are actions, or communicative acts, as they are intended to perform some action by virtue of being sent[8]. The specification consists of a set of message types and the description of their pragmatics. The specification provides also a set of high level interaction protocols.

FIPA-ACL is based on synchronous exchange of messages. However, mobile agents are not permanently reachable because of their mobility. It is desirable to build a communication system above an asynchronous support. Thus we see great interest of using space information approach.

Space Information Approach : According to [9], space information consists of set of tuples in the form (K, A, V), where K is a key (identity), A is an access control list that specify all agents allowed access the tuple; and V consists of the information itself (FIPA-ACL message), (Fig. 3).

Agent communication is done indirectly through elementary operations for writing WRITE (K, A, V), reading READ (K, A, V) and destructive reading DREAD (K, A, V) in public information space.

It is not tied to a particular programming language and can be used for asynchronous "one-to-many" and "many-to-many" style communication.

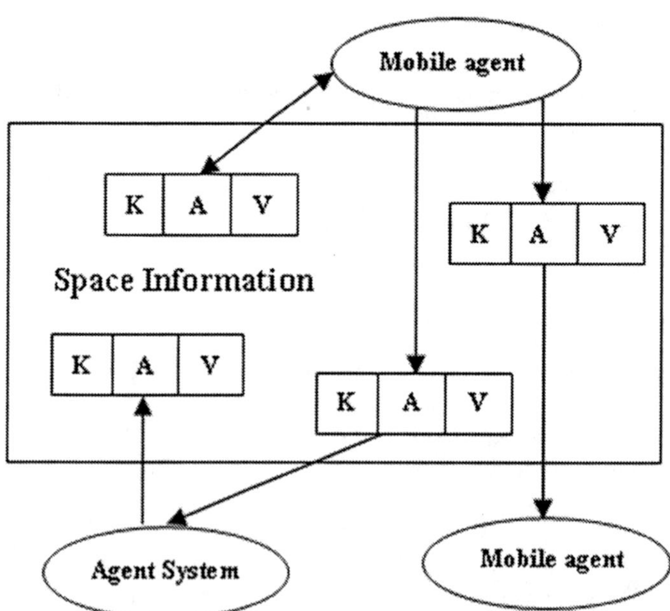

Figure 3. Space information approach

Agent Communication Interface : In order to manage FIPA-ACL messages exchange within information space in standard way, we define agent communication interface called **MAFAgentCommunication**. It contains data definitions and methods necessary to create, send, receive, notify and destroy FIPA-ACL messages.

4.5. Error Detection and Recovery

Mobile agent execution can fail due to a number of reasons. So the error Handling is important not only because mobile agents can carry sensitive data, but also in order to have information the interpretation of which is vital for determining the cause of the failure [14].

We define, in our architecture, ErrorHandling interface the role of which consists in generating reasonable error information and, if possible, to retrieve the agent state.

In case of a failure on mobile agent level, the procedure depicted in Fig. 4 is initiated. It is described in the following:

1. If a failure on agent has occurred, an SOS message is passed to MAFAgentSystem.
2. MAFAgentSystem stores the SOS message, makes agent state at failed and send an error message to the agent owner.
3. The agent system will go ahead collecting the remainders of the mobile agent and send it to EerrorHandling service.
4. Upon reception of the mobile agent, the ErrorHandling Service persistently writes the remainder of the agent to a local database.
5. The owner of a mobile agent can invoke the MAFAgentSystem in order to recover this agent.
6. The MAFAgentSystem returns the error message to the agent owner.
7. The owner of a mobile agent invokes the ErrorHandling in order to retrieve the remainder of this agent.

4.6. Agent Security

In this architecture, we consider the following risks of security relating to the agents mobility and the messages exchange.

1. Messages security :There are two basic potential security risks when sending a message from one agent to another:
   - The primary risk is that a message is intercepted, and modified in some way.
   - The secondary risk is that the message is read by another entity. We propose to use digital signing and encryption to bypass these risks respectively.
2. Mobile agent security : Because a mobile agent is a computer program that can travel among agent systems, it is often compared to a virus. So, it is imperative for agent systems to identify and screen incoming agents.

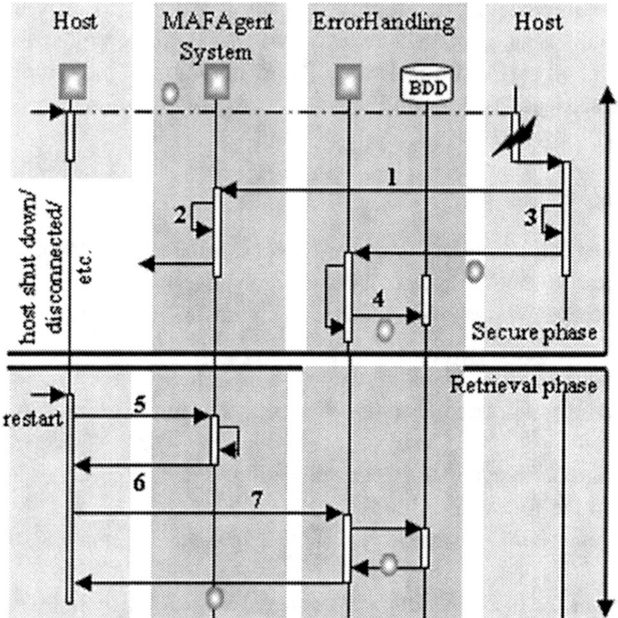

Figure 4. Error recovery

To ensure the safety of agents and agents systems in the proposed architecture, we use current CORBA security specifications [11]:

- Agent identity: an agent system can identify and verify the authority that sent the agent. An agent system can provide information within a credential object about an agent that it is hosting :
  - The agent's name and principal authentication,
  - The authenticator (algorithm) used to evaluate the agent's authenticity.
- Client Authentication for Remote Agent Creation: CORBA security specification offer client authentication services via the PrincipalAuthenticator interface.
- Mutual Authentication of Agent Systems: CORBA security allows the mutual authentication of agent systems by setting the following association options: EstablishTrustInClient and EstablishTrustInTarget.
- Agent Authentication and Delegation: When possible, it is desirable to propagate the agent's credentials along with the agent as it moves between agent systems. This may only be possible using composite delegation.

5. CONCLUSION

This paper focused on the heterogeneous agent interoperability issue. During the past years, different scientific communities proposed different standardization actions, such as the Foundation for Physical Intelligent Agents (FIPA) and the Object Management Group's MASIF.

In this paper, we have first analyzed the similarities and differences, advantages and disadvantages of MASIF and FIPA approaches :

- MASIF specification does not address agent communication;
- FIPA specification does not address agent mobility;

Based on this analysis, we have tried to integrate these two standards to propose an architectural model to enable mobile agent system interoperability. We have used MASIF specification to standardize agent management, agent transfer and localisation. We have introduced mobile agents communication via FIPA-ACL specification. Doing so, make a new range of interoperability options available to mobile agents.We have also proposed a simple solution to error detection and agent recovery.

Work is continuing to identify infrastructure and design elements that facilitate interoperability.

REFERENCES

1. Ametller J., Robles S., Borrell J.: Agent migration over FIPA ACL messages. In: Proceedings of Fifth International Workshop on Mobile Agents for Telecommunication Applications, Springer, Berlin, Germany (2003), pp. 210–219.
2. Baumer C., Breugst M., Choy S., Magedanz T.: Grasshopper A universal agent platform based on OMG MASIF and FIPA standards. In the First International Workshop on Mobile Agents for Telecommunication Applications (MATA'99), Ottawa, Canada (1999), pp. 1–18.
3. Bruno Dillenseger, Huan Tran Viet: Towards full agent interoperability. In 2nd International ACTS Workshop on Advanced Services in Fixed and Mobile Telecommunications Networks, Centre for Wireless Communications, Singapour(1999)
4. David Acremann, Gilles Moujeard, Laurent Rousset:Développer avec CORBA en Java et C++. Edition Compus Press,(1999)
5. Dogac Asuman, Cingil Ibrahim : Book of the course: B2B e-Commerce Technology: Frameworks, Standards and Emerging Issues. Middle East Technical University Ankara, Turkey(2003)
6. Foundation for Intelligent Physical Agents (FIPA): FIPA Abstract Architecture Specification.http://www.fipa.org, (2000)
7. Foundation for Intelligent Physical Agents (FIPA): Agent Management Specification. http://www.fipa.org, (2000)
8. Foundation for Intelligent Physical Agents (FIPA): FIPA ACL Message Structure Specification. http://www.fipa.org, (2000)
9. Lingnau A., Drobnik O.:Making mobile agents communicate: A flexible approach. In the 1st annual Conference on Emerging Technologies and Applications in Communication (etaCOM'96), Portland, Oregon(1996)
10. Laurent Magnin: Internet, Complex environment for situated agents. Informatique Research Centre of Montreal, Canada(1999)
11. Object Management Group: Mobile Agent System Interoperability Facilities specification. Object Management Group TC Document orbos/00-01-02, (2000)
12. Overeinder B.J., De Groot D.R.A., Wijngaards N.J.E., Brazier F.M.T.: Generative Mobile Agent Migration in Heterogeneous Environments. Proceedings of the ACM symposium on Applied computing, P. 101–106, Madrid, Spain (2002)
13. Pauli Misikangas, Kimmo Raatikainen: Agent Migration Between Incompatible Agent Platforms. 20th IEEE International Conference on Distributed Computing Systems (ICDCS'00,), P. 4, (2000)

14. Steffen Richard: Mobile Agent Support Services. PhD Thesis, Aachen (2002)
15. Stéphane Perret: Mobile Agent for nomadic information access in large scale networks. PhD Thesis, Joseph Fourier university, France (1997)
16. Yannis Labrou, Tim Finin, Yun Peng:The Interoperability Problem: Bringing together Mobile Agents and Agent Communication Languages. Proceedings of the Thirty-second Annual Hawaii International Conference on System Sciences, Vol. 8, P.8063, (1999)
17. Yannis Labrou, Tim Finin, Yun Peng:Mobile agents can benefit from standards efforts on inter-agent communication. IEEE Communications Magazine, Vol. 36, No. 7, pp. 50–56, July (1998)(special issue on Mobile Software Agents for Telecommunications).

CHAPTER 47

P2PNET: A SIMULATION ARCHITECTURE FOR LARGE-SCALE P2P SYSTEMS

LECHANG CHENG[1], NORM HUTCHINSON[2], AND MABO R. ITO[3]

[1]*Department of Electrical and Computer Engineering, University of British Columbia*
lechangc@ece.ubc.ca
[2]*Department of Computer Science, University of British Columbia*
norm@cs.ubc.ca
[3]*Department of Electrical and Computer Engineering, University of British Columbia*
mito@ece.ubc.ca

Abstract: Simulation of P2P systems at the scale of millions of nodes is important because some problems with the protocols or their implementations might not appear at smaller scales. In this work, we propose a parallel message-level simulator, P2PNet, which can simulate P2P systems with up to millions of nodes. P2PNet applies the technique of time expansion and uses real time to synchronize the processing of events among the participating processors. Simulation results show that P2PNet has small overhead compared with a single-processor event-driven simulator, a large speedup when multiple computers are used and no late events

Keywords: Event-driven, peer-to-peer parallel, simulation

1. INTRODUCTION

Today a lot of effort is being put into developing large-scale Internet applications based on P2P technology [1,2,3,4]. As P2P systems utilize the resources of end user systems, they usually involve thousands or even millions of nodes. For systems on such a scale, it is impossible to test the designed protocols with real, large-scale implementations in the Internet. Therefore, simulation is critical to the building and understanding of these systems. Furthermore, as errors in the protocols or their implementations might only appear at scales of millions of participating nodes, simulation at that scale is important for the thorough evaluation of P2P systems before their deployment [5].

H. Labiod and M. Badra (eds.), New Technologies, Mobility and Security, 567–581.
© 2007 *Springer.*

There are two challenges to simulating new Internet protocols at large scales: the size and complexity of the topology of the Internet, and the complexity of the layered Internet protocols (TCP, UDP, IP etc.) that form the underpinning of the new application protocol. It is extremely difficult to evaluate any Internet protocol at both the full complexity of the Internet topology and the complexity of the underlying Internet protocols. Simulations are normally forced to trade off accuracy for complexity by modeling either the Internet topology or the underlying Internet protocols in a very simple way. Packet-level simulation can evaluate the performance of the application protocol with detailed models of the underlying Internet protocols but with small scale models of the Internet's topology. Message-level simulations contain more precise and large-scale models of the Internet's topology but are able to do so only by using a very simple application model of message delivery.

Most of the current simulation tools utilize a centralized event-driven scheduler which is not scalable in simulating large scale networks. In order to facilitate simulations for large-scale networks, parallel distributed event-driven simulations (PDES) have been developed. The central problem for PDES is to synchronize the event schedulers and thus guarantee that events are processed in chronological order. Existing PDES systems can be divided into two types: conservative, where simulation time moves in lock-step on all the processors, or optimistic, where each processor advances simulation time in isolation but must retain sufficient state to roll-back should it get too far ahead of its peers. Both methods experience high overheads: the conservative method exhibits high communication costs for the synchronization of simulation time and the optimistic method has high state saving and rollback costs.

In this paper we present the design of a message-level simulator which can simulate P2P systems at a scale of a million nodes. We take advantage of a few characteristics of message-level simulations of P2P systems to improve the performance of the simulation. 1) In message-level simulations, the message delay (the delay between the logical time at which the message is generated and the logical time at which it should be processed) is primarily caused by latency in the Internet, which is on the order of 10s or 100s of milliseconds. Therefore, the precision of the synchronization of logical time can be less exact. 2) P2P systems usually are designed to accommodate fluctuations in Internet latency and packet loss. Therefore, messages that occasionally arrive late will not significantly affect the simulation results.

In this work, we propose a new architecture (P2PNet) for simulating P2P applications. The main idea of P2Pnet is to use real time to synchronize all the participating processors so as to reduce the synchronization overhead. P2PNet executes on a cluster of computers, each of which runs an event-driven scheduler. Each event scheduler uses real time to calculate an estimate of the global minimum simulation time in the system which we call the lower bound time-stamp (LBTS). The scheduler only processes events that have timestamps less than LBTS. LBTS has a piece-wise linear relationship with real time and the ratio between real time and the rate at which LBTS changes is dynamically adjusted so as to eliminate or minimize the occurrence of late messages while minimizing the total simulation execution time. In this paper, our novel contributions are as follows:

- We develop a new mechanism for parallel distributed event-driven simulation which uses real time to synchronize the event processing of the participating processors.
- We propose and implement an algorithm to adjust the rate at which logical time advances relative to real time which is adaptive to the varying workload experienced by the processors.

The structure of this paper is as follows. Section 2 discusses some related work. Section 3 presents the details of P2PNet. Section 4 presents some preliminary simulation results. Section 5 concludes and discusses future work.

2. RELATED WORK

As P2P systems usually are intended to involve millions of nodes, it is important that the simulations used to verify their correctness scale to millions of nodes. Parallel distributed event-driven simulations (PDES) [6,7,8,9] have been developed for many years and some PDES tools have been developed for simulation of large-scale Internet applications. In PDES, instead of having a single event scheduler, there are multiple event schedulers which interact with each other. The core engine of a PDES can be classified as using either a conservative or an optimistic approach. In the conservative approach, simulation is carried round by round. In each round, processors synchronize with each other to determine a look-ahead window and process only the events in that window. In the conservative approach, the chronological order of the processing of events is guaranteed. The main problem of the conservative method is the high communication overhead involved in synchronization. In the optimistic approach, each processor processes its events and advances its logical time greedily. When a late event arrives (an event with a timestamp less than the current logical time), the processor rolls back to the event's time by restoring previously check-pointed state. The main problem of the optimistic approach is the high cost of saving the state necessary for recovery and restoring this saved state when events arrive late. In addition, most of the current PDES simulation packages including pdns [6], SSF [7], and GloMoSim [9] are designed for packet level simulation and are not able to scale to millions of nodes.

Lin. et al. [5,10] proposed a simulation architecture for P2P systems which was the first to take advantage of the features of P2P systems. It is based on the conservative method and introduces a slow-message relaxation optimization which trades simulation accuracy for speed.

Emulation [11] is another mechanism to study large-scale systems. Our work can be regarded as a combination of emulation and PDES. It differs from previous PDES systems in that it relies on real time to synchronize the event processing.

3. P2PNET ARCHITECTURE

The P2PNet simulator is designed to simulate P2P protocols on network topologies with millions of nodes. As one single computer lacks the physical memory and processing power to simulate protocols in such large-scale networks, multiple computers must be used. It can be either a cluster of computers or a distributed memory multiprocessor connected through a high throughput network. Figure 1 illustrates the architecture of P2PNet. The simulator applies a two-tier parallel distributed event-driven simulation (PDES) architecture. One computer is the coordinator of the simulation and all the remaining computers are workers.

Figure 2 shows the structure of a P2PNet worker. In P2PNet, each worker hosts a logical process (LP) and a number of simulated protocol instances. An LP consists of three components: 1) an event scheduler, 2) a simulation network component, and 3) a message delivery component. Each of the protocol instances is denoted a virtual node (VN).

The event scheduler is in control of the execution and the execution of the virtual nodes is driven by events. The event scheduler contains an event queue. Each event is assigned a time stamp which indicates the logical time at which the event should be processed. Events are stored in the event queue in increasing time stamp order. In the simulation, the event scheduler iteratively fetches events from the event queue and calls the event handler to handle each event. Event processing can trigger other events. Periodically the event scheduler will send and receive synchronization messages to the coordinator to synchronize the event processing process.

The simulation network component delivers events between virtual nodes. It abstracts the underlying Internet protocols up to the transport layer. It contains the topology of the simulated network. Based on the latency between virtual nodes in the underlying network topology, it builds a routing table. It also provides the mapping between virtual nodes and LPs. To deliver an event, the simulation network calculates the time stamp of the event and find out the LP which hosts the destination VN. If the destination LP is the same as the current LP, it will schedule the event into the event scheduler. If the destination LP is in another machine, it will call the message delivery

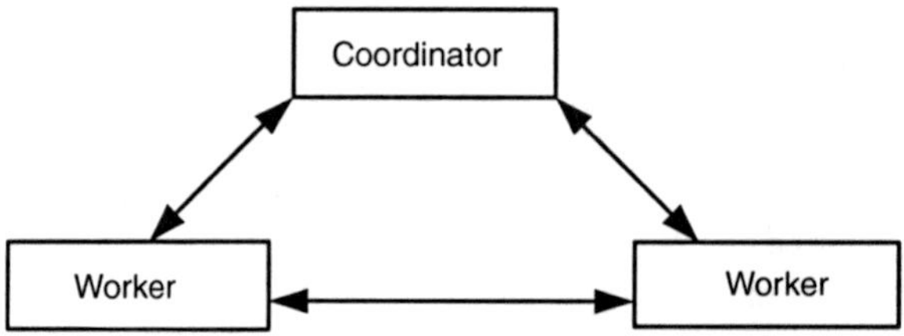

Figure 1. Two-tier structure of P2PNet

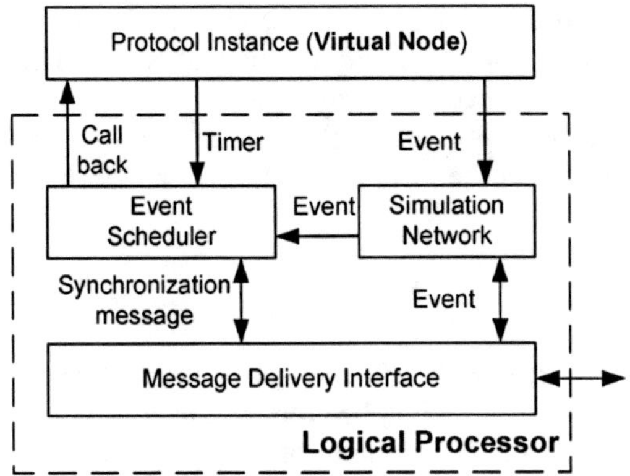

Figure 2. Structure of P2PNet

component to send the event out. When the destination LP receives the event, the simulation network component will insert it into the event queue of the event scheduler.

The message delivery component deals with message transmission between LPs. It can utilize socket, RMI, message passing interface (MPI) or any other message passing service. It delivers both the events from the simulation network component and the synchronization messages from the event scheduler. It can also provide message buffering and message retransmission should reliable message delivery be needed.

3.1. Time Systems

In PDES, each LP processes events and adjusts its local logical time independently. Each LP sets its local logical time to the time stamp of the current event it is processing. In order to synchronize the event processing of LPs, a global lower bound time stamp (*LBTS*) is calculated. At any time, LPs only process events that have time stamps which are less than *LBTS*. Therefore, for any LP, the local logical time is always less than *LBTS*.

In this work, we use real time to synchronize the event processing of the LPs. There are three time systems in P2PNet, local logical time $\hat{t}$, real time t, and LBTS $\bar{t}$. Real time is the real execution time of the simulator since the simulation starts. We assume that LPs have their physical clocks synchronized, and so the real time is the same for all LPs (within the precision of the clock synchronization algorithm).

In P2PNet, the LBTS has a (piece-wise) linear relationship with real time.

$$\bar{t} = (t - t_0)/K + \bar{t}_0 \tag{1}$$

When real time advances Δt, LBTS advances Δ/K. K is called the time expansion factor (TEF). As the application behavior changes during the simulation, the TEF is adjusted dynamically. The pair $(t_0, \bar{t}_0)$ is called a time reference point; each time the TEF is adjusted, a new time reference point is established. As long as the time reference point and TEF are the same for all the LPs, the LBTS will be the same as well.

3.2. Time Management Algorithm

The primary task of a PDES is to guarantee the chronological order of event processing since events which happen in one LP might be processed in another LP. An event E is processed in *chronological order* if its time stamp is larger than the local logical time of the LP when it is scheduled for processing. Otherwise, it is called a late event.

Supposes that LP1 starts to process event D when the real time is t_1, local logical time is $\hat{t}_1$ and the LBTS is $\bar{t}_1$. After some processing time Δt_p (in real time), it triggers event E and the time stamp of event E is $\hat{t}_1 + \Delta\hat{t}_l$ ($\Delta\hat{t}_l$ is the latency between the source and destination of event E in the simulated network topology). Suppose that event E has to be processed by LP2 and LP1 sends event E to LP2 right after it is generated. Suppose that LP2 will receive event E after a communication delay Δt_c. When LP2 receives the event, the *LBTS* is $\bar{t} = \bar{t}_1 + (\Delta t_p + \Delta t_c)/K$.

If order to guarantee that event E is processed in chronological order, the local logical time $\hat{t}_2$ of LP2 when it receives event E must be less than the time stamp of event E ($\hat{t}_2 \leqslant \hat{t}_1 + \Delta\hat{t}_l$). Recall that $\hat{t}_2 \leqslant \bar{t}_2$. Therefore, if

$$(2) \qquad \bar{t}_2 = \bar{t}_1 + \frac{\Delta t_c + \Delta t_p}{K} \leq \hat{t}_1 + \Delta\hat{t}_l$$

we will have $\hat{t}_2 \leqslant \hat{t}_1 + \Delta\hat{t}_l$ and event E will be processed in chronological order. Inequality (2) is equivalent to

$$(3) \qquad \bar{t}_1 - \hat{t}_1 \leq \Delta\hat{t}_l - \frac{\Delta t_c + \Delta t_p}{K}$$

Taking the minimum of the right side of inequality (3) over all events generated by LP1, we obtain:

$$(4) \qquad (\bar{t}_1 - \hat{t}_1) \leqslant \min(\Delta\hat{t}_l) - \frac{\max(\Delta t_c) + \max(\Delta t_p)}{K}$$

where $\min(\Delta\hat{t}_l)$ is the minimum network latency between any virtual node in LP1 and any virtual node not in LP1. $\max(\Delta t_c)$ is the maximum transmission delay between LP1 and other LPs. $\max(\Delta t_p)$ is the maximum event processing time in LP1. In the following, $\bar{t}_1 - \hat{t}_1$ is denoted the local logical time delay. $T_D = \min(\Delta \hat{t}_l) - \frac{\max(\Delta t_c)+\max(\Delta t_p)}{K}$ is denoted the logical time delay threshold of LP1. For LP1, if inequality (4) is true, then all events triggered by it will be processed in chronological order.

The local logical time delay $\bar{t}_1 - \hat{t}_1$ is the delay between the *LBTS* and the local logical time of LP1. It becomes positive when the time expansion factor is too small and consequently the LP is unable to process all the events it should within a logical time interval. It is zero if the time expansion factor is large enough. Therefore, if the *LBTS* is the same for all LPs, and for each LP $(\bar{t}_1 - \hat{t}_1)$ is controlled within the threshold defined in the right side of inequality (4), all messages will be processed in chronological order. The time synchronization problem becomes:

1. To adjust the time expansion factor K of each of the LPs so that $\bar{t}_1 - \hat{t}_1$ can be within the threshold defined in the right side of inequality (4).
2. To synchronize the time expansion factor and time reference point periodically so that the *LBTS* is the same for all LPs.

The event processing time and message transmission time are on the order of 100s of microsecond. With careful partition of the simulated network topology, $min(\Delta \hat{t}_l)$ can be on the order of 10s or 100s of millisecond. As the time expansion factor is always large than 1, the logical time delay threshold T_D is mainly determined by $min(\Delta \hat{t}_l)$ and can be on the order of 10s or 100s of millisecond. The threshold T_D provides a lot of flexibility in adjusting TEF. Insufficient TEF for certain amount of time will not cause any late events.

3.3. Adaptation of TEF for a Single LP

The primary goal of synchronization is to find an appropriate TEF so that $(\bar{t}_1 - \hat{t}_2)$ will be within the threshold. In fact, the time expansion factor represents the ratio of the execution time taken by the simulation versus the simulation logical time.

In the simulation, the event arrival rate and the event processing rate may be different for different LPs, and may change during the simulation. Therefore, the TEF calculation algorithm needs to be adaptive. In this section, we describe our time expansion factor adaptation algorithm. Periodically, the event scheduler measures the following three factors:

- **Logical time adjustment ratio**: This is calculated as the logical time adjustment versus the *LTBS* adjustment in the monitoring period. If this is smaller than 1, then it means that the time expansion factor is not large enough. If this is equal to 1, this means that the time expansion factor is sufficiently large.
- **Logical time delay**: It is the difference between LBTS and logical time. This delay will be caused by an insufficient time expansion factor in this or a previous monitoring period.
- **Empty loop rate**: During the simulation, the scheduler iteratively fetches a suitable event from the event queue and processes it. There are loops in which no suitable events are found, which are called empty loops. The empty loop rate

is the ratio of observed empty loops to the maximum number of empty loops a scheduler can execute in the monitoring period. This indicates the percentage of time that is wasted in a monitoring period because the scheduler found no suitable event to process. With this percentage we will be able to estimate the lowest time expansion factor that is necessary for the scheduler to process the events in the monitoring period.

The algorithm to adjust the time expansion factor is as follows. In any monitoring period, if the time adjustment ratio is less than some threshold α_l, or the logical time delay is larger than some threshold β_h, the time expansion factor will be multiplied by 2. However if the time adjust ratio is larger than α_h, and the logical time delay is less than threshold value β_l, we calculate the new TEF as follows:

$$TEF_{new} = TEF * E_L * (1 + \theta)$$

where E_L is the empty loop rate. We use a TEF that is $(1 + \theta)$ times of the minimum TEF to accommodate the fluctuation of the simulation conditions. In other cases, the TEF remains unchanged. The thresholds α_h, α_l, β_h, β_l are used to accommodate the fluctuation and calculation errors of the logical time adjustment ratio and logical time delay. θ is used to force the setting of the TEF to be slightly conservative.

3.4. Synchronization of TEF of LPs

The algorithm for synchronizing TEF is performed periodically. At the end of each monitoring period, the coordinate sends a TEF inquiry message to all the workers. Upon receiving the inquiry message, each LP will calculate its new minimum TEF with the algorithm mentioned in Section 3.4. The workers will send an inquiry reply message to coordinator. After collecting all the reply messages from the worker, the coordinator LP will find the maximum of TEF of all LPs and broadcast a factor adjustment message which contains the global minimum TEF. The worker will adjust the TEF according to the factor adjustment message.

3.5. Discussion

In this part, we will discuss some issues related to the applicability of P2PNet: adaptability, performance and the scalability.

3.5.1. Adaptability

P2PNet is designed to be a general purpose message-level simulator for P2P systems. In the real situations, different P2P applications have the different event arrival rate and the event processing rate. The event arrival and processing rate may also change during the simulation. We argue that with the TEF adjustment algorithm, P2PNet is able to adapt to the various applications with different event arrival rates and event processing rates.

The simulation of P2PNet can be divided into two states: start up state and adaptive state. In the start up state, P2PNet tries to find a TEF that is larger than necessary so as to reduce late event. Currently, P2PNet starts with a default TEF. The TEF can be set as a very large number (such as 1000 or even 10000) so that it will be large enough for most simulation situations. Furthermore, the algorithm can exponentially increase TEF if the logical time falls behind the LBTS (increase by 2 for every monitoring period). Therefore, P2PNet can adapt to the applications with a broad range of simulation workload. As the TEF will be adjusted down when a larger than enough TEF is found, this large initial value of TEF only have a small effect on the overall simulation execution time if the simulation duration (logical time) is large enough. Further work might be to investigate adjustment algorithm which can adjust TEF more rapidly. We can either choose a higher increasing factor (large than 2) or adjust the TEF according to the logical time adjustment ratio.

In the adaptive state, the adjustment algorithm tries to adjust the TEF according to simulation situations. P2PNet can exponentially increase the TEF when the logical time falls behind LBTS. By default, P2PNet can change the TEF by 2 for every monitoring period (20ms). As discussed in Section 3.2, the logical time delay threshold T_D can be on the order of 100s of millisecond. The P2PNet can run with insufficient TEF for 5 monitoring periods without causing any late event. Therefore, the adjustment algorithm can gradually increase TET by 32 times in 100ms without causing any late event. It would be unusual for the average event arrival rate and event processing rate of large-scale simulations to change rapidly in the time scale of tens of milliseconds. Therefore, the adjustment algorithm can work for reasonable simulation workload. In the future work, we will also investigate proactive adjustment algorithm which predict the future event arrival rate to deal with abrupt change in the event arrival rate.

3.5.2. *Performance Analysis*

For parallel distributed event simulations, communication cost involved in exchanging events among LPs is the main overhead compared with sequential event-driven simulation. In a sequential event–driven simulation, one can send an event from one virtual node to another by inserting it back to the event queue. In PDES, if the destination virtual node of an event is in a different LP from the source virtual node, the event has to be physically sent from one computer to another. Suppose there are N total virtual nodes in the simulation and each virtual node generates R events every millisecond. Let a be the average processing time for an event and b be the communication cost of an event (including the cost of sending and receiving an event), the simulation duration be P. The total execution time with sequential event driven simulation is $NRPa$. With W workers, the total simulation execution time is

$$NRP(a+\frac{W-1}{W}b)/W \approx NRP(a+b)/W \tag{5}$$

The minimum time expansion factor is

$$(6) \qquad NR(a+\frac{W-1}{W}b)/W \approx NR(a+b)/W$$

The speed up is

$$(7) \qquad Speedup = \frac{W}{1+(W-1)/W*(b/a)}$$

Where W is large, the speedup rate is

$$(8) \qquad Speedup \approx \frac{W}{1+(b/a)}$$

The overhead rate is about b/a. One can see from equation (8) that when the number speedup linearly related to the number of workers, restrained by the relative communication cost. Given the number of workers and the relative communication cost b/a, equation (8) gives the maximum speed up of any parallel simulators.

3.5.3. Scalability

P2PNet is designed to be scalable to simulate P2P applications on network topologies with millions of nodes. A regular PC can usually host thousands and tens of thousands of application instances. Thus, P2PNet should also scale to hundreds of and a few thousands of processors.

For each of the worker, it processes events and sends and receives events from other workers. The workload of a worker depends on the average event arrival rate R and the number of the virtual nodes hosted by the worker. Therefore, we can simulate larger networks by simply adding more workers. Furthermore, each worker and sends and receives 2 synchronization messages for every monitoring period. The overhead of synchronization is both constant and negligible.

The possible bottleneck of P2PNet architecture is the coordinator as it communicates with all workers for synchronization. The coordinator processes $3W$ synchronization messages for every monitoring period T. Recall when W is large, the time expansion factor is approximately $NR(a+b)/W$. Therefore, the message processing rate for the coordinator in real time is

$$\frac{3W}{(N/W)R(a+b)T}$$

When the number of average virtual nodes per worker is fixed, the work load of the coordinator is proportional to the number of workers.

When $N = 10^6$, $W = 1000$, $R = 0.1ms$, $a+b = 0.2ms$, the event processing rate is 7.5/ms. This is an affordable work-load by regular PCs. We can further reduce the work load of each coordinator by introducing a two-tire coordinator architecture and adding more coordinators.

The internal network bandwidth is another issue related to the scalability P2PNet. The total number of messages generated during the simulation is NRP. Suppose that the average message length is L bytes. Since the minimum simulation execution time is $NRP(a+b)/W$. Therefore, the average bandwidth needed for P2PNet is

$$\frac{NRPL}{NRP(a+b)/W} = \frac{WL}{a+b}$$

The required internal bandwidth does not depend on the size of the simulated network. It increases linearly with the number of workers. When $W = 1000$, L is 40 bytes, $a+b$ = 0.2ms. The bandwidth required is 200 MB/s. Therefore, P2PNet can scale to thousands of worker with moderate requirement on the internal bandwidth.

4. EXPERIMENTS

In this section, some preliminary simulation results are presented. The above algorithm has been implemented in Java and evaluated with a simple overlay application in which each node has a ping client. In the simulation, the ping client periodically picks a random node and sends a ping message to it. In receiving a ping message, the destination client performs some calculation (to simulate the event processing time) and replies to the ping message. The delay between two consecutive ping messages sent by a ping client follows a uniform distribution. The processing time of the event also follows a uniform distribution. As the topology has no effect of the simulation, we use a simple star topology and network size varies from 100 to 1000000.

4.1. P2PNet Versus Regular Event Scheduler

First, we compare the performance of P2PNet with one worker versus a sequential event-driven simulator (SEDS). As the sequential event-driven simulator will experience no communication overhead and thus takes the minimum time to perform a simulation, it is used as the benchmark. Compared with sequential event-driven simulator, the overhead caused by using real time to guide the event processing includes:

- Empty loops due to larger than necessary TEF.
- Monitoring of the execution of the event scheduler, especially the system calls to get the system time.

In the experiment, the simulation is run for 1000ms (logical time). In order to eliminate the effect of the initial value of TEF toward the simulation execution time, the initial value of the TEF is set as 1. The other parameters are set as: α_l = 0.9, α_h = 0.95, β_l = 1.0, β_h= 1.0, θ=0.2. The monitoring period is set as 20 ms.

Table 1, 2, 3 shows the total simulation run time (real time) for both P2PNet and a sequential event-driven simulator (SEDS) in millisecond. It shows that the ratio between the execution time of P2PNet and that of a sequential event-driven simulator decreases as the simulation load increases, approaching 1.35 in high work

Table 1. Network Size 1000, average between-message delay 50 ms

Average message processing time	Simulation run time (SEDS)	Simulation run time (P2PNet)	Ratio
0.01	848	1432	1.689
0.02	1214	2035	1.677
0.05	2481	3549	1.430
0.1	4331	6606	1.525
0.2	8317	10877	1.308
0.5	18766	26310	1.402
1	36107	50031	1.386
2	71822	93551	1.303

Table 2. Network Size 1000, average event processing time 0.1 ms

Average between message delay	Simulation run time (SEDS)	Simulation run time (P2PNet)	Ratio
200	950	1872	1.971
100	2360	3237	1.371
50	4523	6540	1.446
20	10762	13877	1.289
10	20930	30368	1.451
5	43083	57911	1.344
2	105600	131352	1.244

Table 3. Average between-message delay 50 ms, average message processing time 0.1ms

Network size	Simulation run time (SEDS)	Simulation run time (P2PNet)	Ratio
100	527	1537	2.917
200	808	1874	2.319
500	2138	2984	1.396
1000	4457	6540	1.467
2000	8546	12008	1.405
5000	20981	30400	1.449
10000	42672	57591	1.350

load situations. As P2PNet is aimed at high work load situations, an overhead of 35% is acceptable. Future work will be investigating techniques to reduce the overhead.

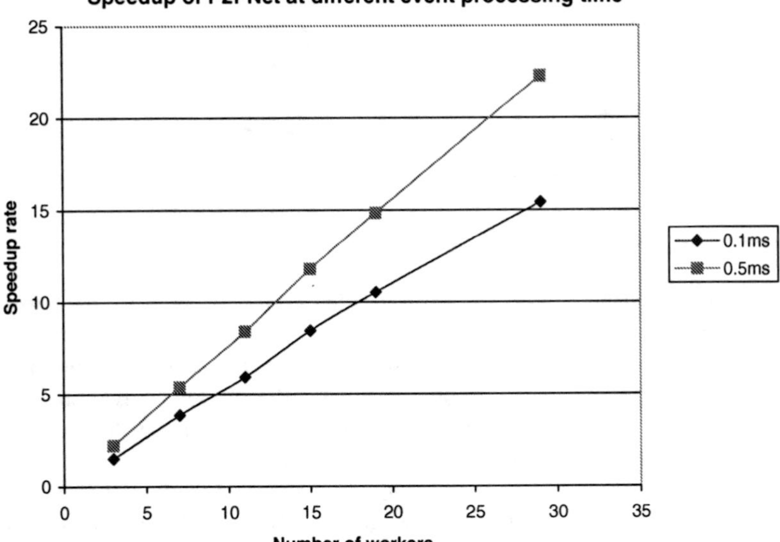

Figure 3. Speedup of the P2PNet versus regular event scheduler (network size 1000000, average event arrival rate 1/100ms)

4.2. Speedup of P2PNet

Second, we compare the performance of P2PNet with multiple workers to that of a sequential event-driven simulator. The simulation was conducted on a cluster of distributed-memory multiprocessors. In this simulation, the network size is 1000000 and the average event arrival rate is 1/100ms. As discussed in Section 3.5, the speedup rate is mainly determined by the number of workers and the relative communication cost (the cost of sending and receiving an event divided by the event processing time). In this experiment, the average event processing time is 0.1ms and 0.5ms. The cost of sending and receiving a message is around 0.35ms. We use a buffer size of 5. Figure 3 shows the speedup of P2PNet versus the number of workers with different average event processing time. At both situations, the speedup increases linearly with the number of workers. In all the simulations, P2PNet has no late events.

The slopes of the speed up are different. Table 4 shows the estimated speed up slope versus the measured slope. The estimated slope is calculated as $1/(a+b/a)$. Compared with the estimated speed up slope, P2PNet has a low overhead of around 10% in either case.

Table 4. Estimated speed up slope versus measured speed up slope

Average event processing time	0.1 ms	0.5 ms
Speed up slope	0.537	0.774
Estimated speed up slope	0.588	0.877
Overhead	9.54%	13.3%

5. CONCLUSIONS

In this paper, we describe a message-level simulator for P2P systems that uses real time to synchronize the event processing of the processors. The technique dynamically adapts to different simulation situations. Simulation results show that the simulator does not have large overhead compared with a sequential event-driven simulator. When used in a cluster of computers, it provides high speedup in high work load situations without introducing any late events.

There is much more work to be done on the foundation that we have laid. First, we observe some oscillation in the value of the TEF during simulation, which leads to performance degradation. We would like to understand the cause of this oscillation and find a better algorithm for adjusting the TEF. Second, more extensive evaluation needs to be done. So far, we have only tested the simulator with a simple application. We will be testing the simulation on various overlay applications such as Pastry and Chord, and using it to evaluate some optimization to their protocols. Third, because of lack of computer facilities, we have to date only performed simulations on a small cluster of 30 processors. We plan to test the simulator in a larger cluster to evaluate its extensibility to hundreds of processors.

REFERENCES

1. I. Stoica, R. Morris, D. Karger, F. Kaashoek, and HariBalakrishnan, "Chord: A scalable Peer-To-Peer lookup service for internet applications," in *Proceedings of the 2001 ACM SIGCOMM Conference*, pp. 149–160, 2001.
2. A. Rowstron and P. Druschel, "Pastry: Scalable, decentralized object location, and routing for large-scalepeer-to-peer systems," in *IFIP/ACM International Conference on Distributed Systems Platforms (Middleware)*, pp. 329–350, 2001.
3. B. Y. Zhao, L. Huang, S. C. Rhea, J. Stribling, A. Joseph, and J. D. Kubiatowicz, "Tapestry: A global-scale overlay for rapid service deployment," *IEEE J-SAC*, vol. 22, pp. 41–53, January 2004.
4. S. Ratnasamy, P. Francis, M. Handley, and R. K. andScott Schenker, "A scalable content-addressable network," in *Proceedings of the 2001 conference on applications, technologies, architectures,and protocols for computer communications*, pp. 161–172, ACM Press, 2001.
5. S. Lin, A. Pan, R. Guo, and Z. Zhang, "Simulating large-scale p2p systems with the wids toolkit," in *Proceedings of the 13th IEEE International Symposium on Modeling, Analysis, and Simulation of Computer and Telecommunication Systems*, 2005.
6. G. F. Riley, R. Fujimoto, and M. H. Ammar, "A generic framework for parallelization of network simulations," in *MASCOTS*, 1999.
7. J. Cowie and H. Liu, "Towards realistic million-node internet simulations," in *Proceedings of the 1999 International Conference on Parallel and Distributed Processing Techniques and Applications*, 1999.

8. D. M. Rao and P. A. Wilsey, "Simulation of ultra-large communication networks," in *MASCOTS*, 1999.
9. X. Zeng, R. Bagrodia, and M. Gerla, "Glomosim: A library for parallel simulation of large-scale wireless networks," in *Workshop on Parallel and Distributed Simulation*, 1998.
10. S. Lin, A. Pan, Z. Zhang, R. Guo, and Z. Guo, "Wids: An integrated toolkit for distributed system developmen," in *Proceedings of the 10th USENIX Workshop on Hot Topics in Operation System*, June 2005.
11. A. Vahdat, K. Yocum, K. Walsh, P. Mahadevan, D. Kostic, J. Chase, and D. Becker, "Scalability and accuracy in a largescale network emulator," in *Proceedings of 5th OSDI*, 2002.

CHAPTER 48

A MIDDLEWARE FOR MANAGING SENSORY INFORMATION IN PERVASIVE ENVIRONMENTS

COSTAS PONTIKAKOS AND THEODORE TSILIGIRIDIS

Division of Informatics Mathematics and Statistics, Agricultural University of Athens, 75 Iera Odos, 11855, Athens, Greece, tsili@aua.gr

Abstract: Context and context-aware computing have attracted remarkable attention in recent years. Research into wireless sensor networks is rapidly moving from simulations to realistic testbeds. The work presented here investigates the architecture and design of an agent-based sensor network for monitoring and spraying control of olive fly in an ubiquitous precision farming environment. It contributes by providing a generic, flexible and extensible model for handling heterogeneous sensor data, which can be managed using a simple user interface. The innovative aspect stands for the development of a multi-agent system and focuses on the communication aspects of the proposed middleware architecture. The modeling allows the encapsulation of different modules as agents and integrates them, by utilizing a standardized XML schema in order to hide complexity to the users. The proposed design adopts a layered architecture, which is based on some software agents that solve different tasks and communicate among themselves their results and requests

1. INTRODUCTION

Ambient Intelligence (AmI) relies on the areas of ubiquitous computing, ubiquitous communication and intelligent user interfaces aiming at the seamless delivery of services and applications. The vision suggests an environment of a potentially large number of embedded mobile devices, or software components, interacting to support user goals and activities. It also suggests a component-oriented view, in which the components are independent and distributed. This environment is characterized by the autonomy, reactivity, distribution, collaboration and adaptation of its artifacts, which, in this sense, share the same characteristics as agents [8]. AmI requires these agents to be able to interact with numerous other agents in the environment around them so as to achieve their goals.

H. Labiod and M. Badra (eds.), New Technologies, Mobility and Security, 583–595.
© 2007 *Springer.*

Agent technology has been considered as an alternative for enhancing pervasive computing environments [2]. The environment provides the infrastructure that enables AmI scenarios to be realized. In relation to pervasiveness, it is important to note that scalability, by means of ensuring that large numbers of agents and services are accommodated, as well as heterogeneity of agents and services must be facilitated by the provision of appropriate ontologies [3]. In most of the cases, relevant context is assumed as known in advance and agents just learn how to adapt their decision to it. However, agent-based architectures that make an explicit separation between a selection of relevant context from the decision making process of the agent have also been proposed [1]. Addressing all of these aspects will require efforts to provide solutions to issues of context architecture, operation, integration and visualization of distributed sensors, ad-hoc services, and network infrastructure.

Agricultural Information Systems (AIS) that provide access to electronic agro-environmental (e.g biological, climatic, meteorological, etc.) records are a step towards the direction of providing accurate and timely information to farmers in support of decision-making. This has motivated the introduction of ubiquitous computing technology in precision farming, based on designs which respond to particular conditions and demands. It is therefore important to realize how autonomous agents can cope with the complexities associated with implementing AmI environments for precision farming settings. Note that farms are distinguished from the distributed nature of the information, the intensive collaboration and mobility of their personnel, as well as their need to access agro-environmental information occasionally. We suggest an iterative user-centered development method to understand the farm activities and to envision farm's AmI scenarios, in which the main systems components have been identified as agents that respond autonomously in accordance with the context surrounding the activities performed at the farm. Specifically, based on the envisioned scenarios and the easy integration of components represented by autonomous agents, we have conceived a flexible middleware, called *Dacus_oleae*, for spray control of olive trees. In this work, we focused on the communication aspects of the middleware architecture, allowing different context-aware applications to query, retrieve and use sensor data in a way that will be decoupled from the mechanisms used for acquiring the raw sensor data.

2. MOTIVATING AMI SCENARIOS IN PRECISION FARMING

The olive fly (Bactrocera (Dacus) oleae) (Gmelin) (Diptera: Tephritidae) is the most serious insect pest that affects the olive tree (*Olea europea*) cultivation leading to significant qualitative and quantitative consequences. Olive flies survive best in cooler coastal climate, but are also found in hot, dry regions. The optimum temperature for development is between 20^oC to 30^oC. High temperatures (35^oC or more) are detrimental to adult flies and to maggots in the fruit. However, since the flies are very mobile, they have the ability to seek out cooler areas of the orchard and urban trees. The olive fly has three to perhaps even five generations per year, depending on local conditions. Over-wintered adult populations decline to low levels, however new adults from the over-wintered pupae begin to emerge in the spring.

In order to monitor insect populations adequately, it is important to record both biological and climatic data. This data is collected in real time in order to give information about both adult and larval stages. The use of such information is only worth if it is incorporated into pest management models. In the case of olive fly, monitoring of adults in traps and observations of larval stages in fruit samples are coupled with climatic data (temperature, relative humidity etc.) to make predictions of damage and take preventive measures. Such climatic data is collected from automatic agro-climatic weather stations, which are capable of recording and storing great quantities of such data and then sending it automatically to a central collection point.

Several approaches can treat the above pest problem [7]; however spray applications from the ground seem the most appropriate for several reasons, such as climatic, environmental protection etc.. To avoid failures in the spray treatment, there is a need to ensure that: the experimental plants of olive trees are large enough; the population of olive flies tend to increase significantly; the olives must be in an advanced stage; the female to male insect ratio should be greater than one; the female insects must be in a mature stage; the temperature and the humidity levels should exceed a threshold.

The spray application takes place during a day using, in most of the cases, tractors. Each tractor covers one section of the spraying area. During the spray application, thought, several problems may arise. The first problem is that the air temperature and air speed values are unknown to the spraying attendant. In this case, the spraying could continue even when the meteorological conditions have been violated. The second problem arises because the spraying areas cannot be memorized by the attendant and, therefore, over or under spraying may occur. The third problem is that the spray volume is dependent on the coverage of olive trees and, as a result, the spraying attendant cannot easily determine if the number of the olive trees per area unit is low, medium, or high. The attendant is not aware of the existing areas inside the spraying area which must not be sprayed for some reason (i.e. domestic areas). The fourth problem arises because the olive fly population of the spraying area is not known to the attendant and therefore the spray volume per area cannot be determined at all. The fifth, but not last problem, is the lack of communication between the supervisor and the attendants. Since the exact location of the attendant and the spraying coverage are not always known to the supervisor the whole spraying process can be altered.

3. THE *DACUS_OLEAE* MIDDLEWARE

3.1. The *Dacus_oleae* WSN

The four-tiered architecture of the *Dacus_oleae* Wireless Sensor Network (WSN) shown in Fig. 1 is designed to transmit the data from an array of sensors to a

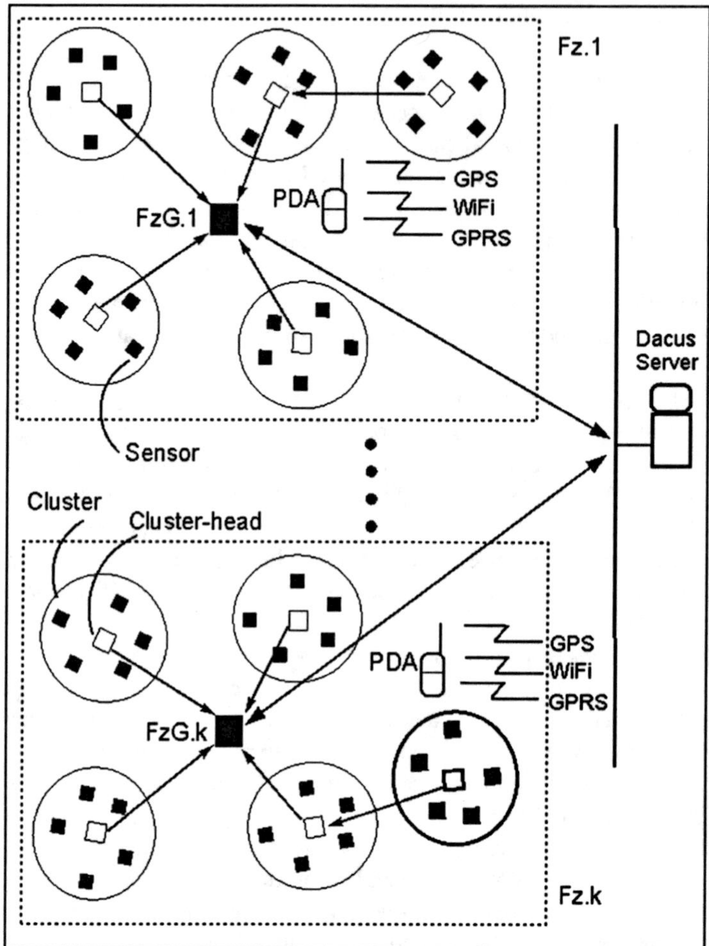

Figure 1. Physical view of the four-tier architecture of *Dacus_oleae* WSN

Field-zone Gateway (FzG), through their cluster-heads(Chs), using a two-way data stream over a wireless link. Depending on the geographical conditions and sensors density the WSN can be divided into several field-zones ($Fz.x$; $x = 1, 2, \ldots k$). To facilitate query dissemination and efficiency in-network processing, a field-zone is further divided into clusters; however a cluster cannot be shared among different field-zones. Thus, field-zones as well as clusters are disjoint. In each cluster, there are many sensors coordinated by exactly one Ch. Sensors are responsible for all sensing-related activities; however sensors do not communicate with other sensors in the same or other clusters and usually are independently operated. Chs have many more responsibilities comparing to sensors. Firstly, a Ch receives raw data from all active sensors in the same cluster. It may also instruct sensors to be in a sleep, idle, or active state, if some sensors are found to always generate irrelevant, inaccurate, or duplicated data, thereby allowing these sensors to be reactivated later when some existing active sensors run out of energy. Secondly, a Ch creates an application-specific local view for the whole cluster by exploring correlations among the data sent from sensors. Excessive redundancy in raw data can be alleviated and the fidelity of captured information should be enhanced. Thirdly, a Ch forwards the composite bit-stream toward a FzG that generates a comprehensive global view for the entire field zone. However, in order to maintain sufficient network connectivity, Chs have no sensing capabilities. They are used as communication relays, aggregating traffic and routing the data to their FzG. The FzG will also play the role of a data repository; thus it will make decisions about what data to pass on, such as local area summaries, and filtering, in order to minimize power use while maximizing information content. FzGs will then forward the data to the *Dacus_server*. To further improve communication, Chs and FzGs are installed at an appropriate height and, optionally, can be involved in inter-relaying, if such activities are applicable and favorable.

The logical four-tiered architecture of the proposed WSN offers a flexible balance among reliability, redundancy, and scalability . Under this architecture, the primary goal of lower - tiers (sensors and Chs) is to gather data as effectively as possible, while the upper-tiers (FzGs and *Dacus_server*) are designed to move information as efficiently as possible. With this functionality partition we can optimize the performance of individual tiers separately.

Finally, additional sensors could be used for environmental monitoring and to take measurements of atmospheric pollution. Thus, each field zone is equipped with a weather station registering the luminosity, air pressure, precipitation, wind strength and direction.

3.2. Design of *Dacus_oleae* Middleware

The proposed architecture focuses on the development and deployment of added value services and combines the capabilities of several technologies. Modules, in a form of agents, communicate with each other to send and receive data, while their collaboration is coordinated via the central receiving point. In this way, communication, coordination and cooperation are ensured. The agents are able to perform some actions on other objects, thus at least partially perceiving their environment and communicating among themselves. Each agent perceives a modification of its environment and receives messages from other agents. It then reacts to these stimuli by acting on the environment and sending messages to other agents according to its own methods and characteristics.

The proposed architecture uses XML configuration files to simply add or remove sensors. It also adds new information recipients or listens to other existing sensors by modifying specified XML files. The different sensors are abstracted by internal software drivers to facilitate the interaction. The architecture provides also an interface to interact with the data base in a simple way. Figure 2 shows a screen

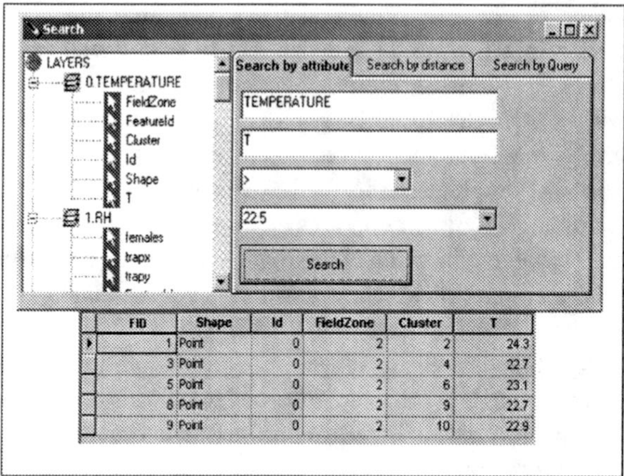

Figure 2. GUI input and output of the SQL query (search by attribute): *select *from* temperature *where* *T* > 22.5

shot of the interface that can be used by an attendant in order to submit an SQL-like query. The input text fields are used to enter the threshold values for the sensor readings. Sensors detect temperature, humidity and air pressure. The values presented here are in a generic form, they do not relate to actual physical units, and only an AND operator is used in the WHERE clause for demonstration purposes. Fig. 2 also shows a sample result for the query made. This result shows the readings received from all the sensors from Fz.2 with temperature greater than 22.5.

The queries can be adjusted according to the request. For example, the users may retrieve the most recent, average, maximum, or minimum value, received from any or all of the sensors. Also, they may interest in periodic sensor readings, namely; "temperatures from sensors of the Fz.3 every twelve hours". In fact queries may consist of SELECT - FROM - WHERE - GROUP BY - HAVING blocks to support selection, joining, projection, aggregation and grouping. For example, in order to count the number of sprayed trees in the Fz.2 we use the query:

```
SELECT
 Count(trees.tree_sprayed) AS CountOftree_sprayed
FROM trees
GROUP BY trees.tree_sprayed, trees.fieldzone,
trees.cluster
HAVING(((trees.tree_sprayed)="yes")AND
((trees.fieldzone)=2));
```

4. COMMUNICATION MODULE

The Communication (Com) module is a layered, hierarchically structured, software classified into different types of mobile agents and in accordance with their functionalities. It is related with the developed WSN, as described in subsection 3.1, the network-based operations, the sensors' hardware, as well as the data acquisition mechanisms [4]. Methods regarding how WSNs can benefit from the use of autonomic techniques of multi-agent systems are discussed in [6] and others. The network operations of the Com module are distributed into four layers, each one controlled by specific task oriented mobile agents. Agents in each layer achieve their assigned task by collaborating with each other in the same or different layers and transmitting the processed data to their destination layer. The architecture consists of the following four layers; the Interface Layer (IL), the Field-zone Layer (FzL), the Cluster Layer (CL), and the Sensor Layer (SL). At the IL, an Interface Agent (IAnt) provides interaction, by means of flexible mediation services, between the users and the *Dacus_oleae* WSN. At the FzL, the Field-zone Agent (FzAnt) acts as a gateway between the clusters of the sensors and the *Dacus_server*, whereas, at the CL, the Query Agent (QAnt) receives the required sensing data captured by the sensor acquisition devices and performs query dissemination and network processing. Finally, the SL is responsible for acquiring the sensor data, namely, it acts as a data provider for the CL.

Fig. 3 presents the architecture, which operates as follows: The IAnt receives and translates the SQL-like user requests, and proceeds them into the appropriate FzL. The corresponding FzAnt accepts the queries and it includes them, through the QAnts, into the CL. The Chs obtain the sensor results from the SL and the QAnts pass the results to the IAnt through the FzAnt.

4.1. Interface Layer

The Interface Layer (IL) is the front end of the WSN, responsible for accepting the user requests, processing them and displaying the results in a predefined format. It is also responsible for the messages sent over to several networks (i.e. SMS, E-Mail, etc.). Since it is possible to send an information structure to a recipient (e.g. an XML file), where the recipient can be a human, a different machine, or simply a software module, the chain can be extended indefinitely. Messages can also be stored in a separate database for further handling. The user interface agents can be implemented by any portable device of the client, such as a laptop, or a PDA; however, for the time being it is running on a laptop. Note that the design of the IAnt does not depend on the size and complexity of the WSN, but it is based on the *Dacus_oleae* application and the user profile.

Form-based queries are submitted by a different type of users to the XML-based IAnt. As soon as the IAnt receives a query, it maintains the connection with the users; for example, if a user uses a web browser, an HTTP connection is assumed (Fig. 3). The IAnt consists of the following, well known, component classes; The Receiving Component (RC), the Request Handler Component (RHC), and the Service Component (SC). The RC is always active and responsible for the processing of all user requests. Since for each user request an RHC instance is created, the total number of RHC instances depends on the number of concurrent

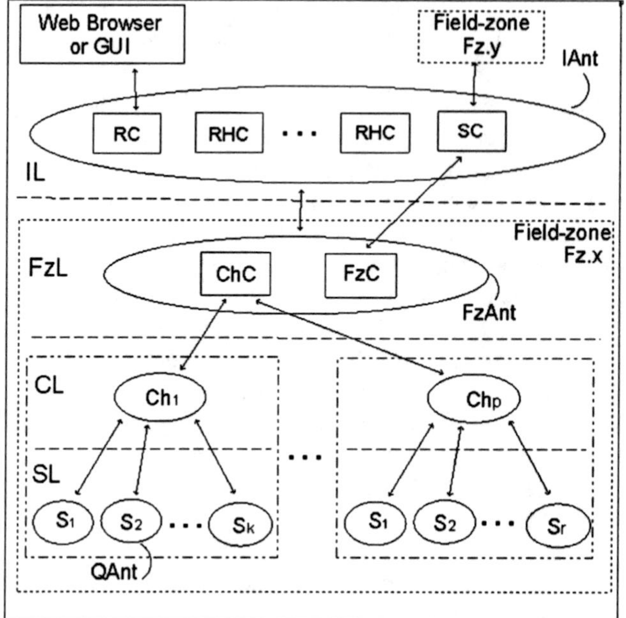

Figure 3. The four layer, agent-based, architecture of the Com Module

user requests. The obligation of the RHC is to process the user request, to format the query result for the user and to maintain the connection in progress (i.e, with the web browser or the GUI). Finally, the IL interacts with the FzL through the SC, which, always active as it is, sends the queries to the desired FzAnts and the results received are combined and forwarded to the appropriate RHC.

4.2. Field-zone Layer

The Field-zone Layer (FzL) is responsible for accepting the user requests from the IL and forwarding them to the CL. It also handles all the information that comes from the CL. The experimental region comprises of several field-zones, each one managed by its own FzAnt in the corresponding FzG. Note that the FzGs are assigned, based on the network topology and the deployment conditions of Chs. They are also responsible for transmitting data either across their field-zone or to pass this data to another field zone, namely to the Field-zone Component (FzC) and Cluster-head Component (ChC) of the FzAnt, respectively. Both component classes are linked (wired or wirelessly) using an appropriate communication protocol (Ethernet based or HTTP). The FzC interacts either with the *Dacus_oleae* server or with the other field-zones, maintaining (in both occasions) the connection with the IAnt. The ChC interacts with the sensors through their Chs, maintaining the connection with the various QAnts and managing the radio communication with the ChC.

4.3. Cluster Layer

The Cluster Layer (CL) is responsible for accepting the user requests from the FzL and forwarding them to the SL. Each field-zone comprises of several clusters, each one managed by its own Ch. For each sensor in a cluster there is a corresponding query agent (QAnt) responsible for the acquisition of sensor data, the filtering of inaccurate and irrelevant data, the aggregation and the processing of useful data, aa well as for the transmission of the desired results. Therefore the QAnts are located on the data collecting sensors and perform data acquisition and local computation in the corresponding cluster (Fig. 3). Their main functionality is to receive queries from the external component of the associated FzAnt and to re-transmit them to the desired sensors through their Ch. Any query data received from other FzAnts is immediately rejected. The sensor data is collected depending on the query. Every QAnt keeps information regarding the types of sensors available, their activity, their current energy level, and other complementary configuration details. Based on this information, as well as on the availability of sensors, query dissemination designs could be developed.

An interesting problem, which is quite common in many agro-environmental applications like the *Dacus_oleae* application considered here, is to periodically check the QAnts to see if any sensor fails to transmit the data. Usually the QAnts take turns in a TDM fashion (other protocols are also applicable), in order to send the aggregation data to the CL and then to the FzL. In case some of the QAnts do not receive any response to the query calls, real-time data can be received only from the functioning sensors. Thus, instead of computing the required parameter values with degradation (due to the non-functioning sensors) using only the available sensor readings, kriking methods may be applied in the available sensor readings to obtain a more accurate result. A similar problem is to predict the required parameter values in different places from those, where the sensors are located. As it will be seen in section 5, kriking can be applied successfully to predict olive fly population.

4.4. Sensor Layer

The Sensor Layer (SL) is responsible for acquiring the data from the sensors, gathering and passing it to the CL. It receives the raw data from the attached sensors and simply transforms it. The sensors themselves can be either software or hardware. Note that in software sensors raw data is taken by a software means e.g. a web service, an aggregation of the data coming form other sensors, or any another software component. The QAnt analyzes the data to maintain consistency and to filter inaccurate and irrelevant data, sending it the to the FzL. The FzAnts perform data aggregation and the resulting data is sent to the IL where the IAnt make the data available to the users, displaying them on a GUI or on a web-based interface. Note that the QAnt has an additional function; as it abstracts the sensors from the other components it has the possibility to create further virtual sensors that rely usually on aggregations. The corresponding FzAnt has no means to find out whether this is a new sensor or a software one. This means that the QAnt multiplexes the data and creates new sensors. The aggregation functions are clearly a defined set of methods (e.g. *sum, avg, max, min*, etc.). The QAnt has no real information about the meaning of the data and simply executes these methods.

5. SIMULATION AND RESULTS

The proposed agent-based architecture is used for the design of a sensor application that monitors temperature, humidity, air pressure, light, and mobility. The clusters, where sensors are deployed, are distributed into three field-zones, Fz1, Fz2, and Fz3, each one comprising of three, four, and four clusters, respectively (Fig. 4). It is useful to remind at this point that the QAnts are deployed on the sensors which provide the data, the FzAnts are deployed on the FzGs, whereas the IAnt is deployed on the client devices. For each incoming user request, a RHC is created in a separate thread and it is managed by the *Dacus* web server. The control of ground spray was performed by taking in account that for each field-zone the system provides the olive fly population per McPail trap, as well as the volume of the required spray. It is important to note that a trap only partially covers the spraying area. In each field-zone a tractor with a portable device (laptop) and wireless Internet access (Wi-Fi and/or GPRS) is assigned to perform the ground spray. The temperature is assumed to be between $14^{o}C$ and $28^{o}C$, whereas the air speed does not not exceed 8m/s. Fig. 5 shows a screen shot interface of the performed simulation which includes the following information layers:

- *T, RH, AIR*: These three layers show the temperature, the relative humidity and the air-speed, respectively.
- *CLICK*: Provides the GIS interactive menu. The user may invoke commands to update the GIS data, to browse though the web pages etc..
- *ROADS*: Provides information about the roads of the spraying area.
- *TRAP*: Provides information on the location of each trap, the number of female and male insects it contains, the total number of insects and the insect population level per trap.
- *TRACTOR*: Provides the spraying areas designated by each tractor.
- *AREA*: Provides the boundaries of the area to be sprayed.
- *SPRAYAREA*: Provides the areas needed to be sprayed. Also, the layer shows the olive-tree density in each field-zone (an index taking values 1,2 and 3).
- *TRAPAREA*: Provides information about the trap area. For example, based on the level of the olive fly population per trap and using kriking methodology, we determine a function, specifying the extreme levels of the spray volume to be applied.

During the spraying performance, the knowledge base becomes aware of the location of the tractors in respect to the designated area of olive trees. As soon as a decision on the volume and the actual spray area is made the tractor attendant proceeds with the operation. If it is not allowed to spray the area, the system reports the attendant for the specific reason, i.e. temperature is too high, air-speed

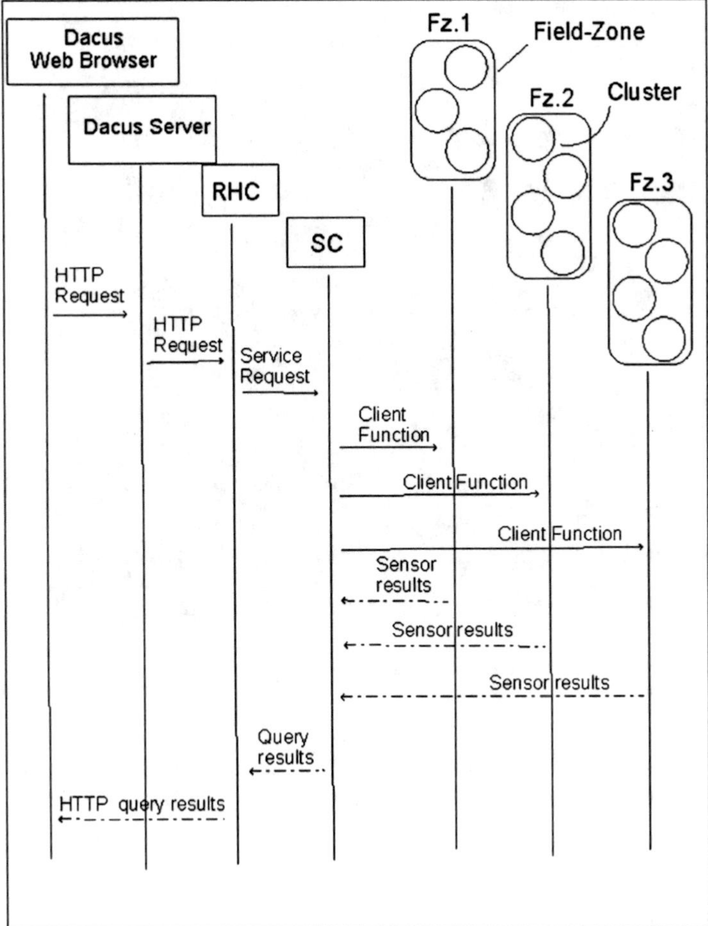

Figure 4. Web-based query process in the pilot deployment of *Dacus_oleae* WSN

is too high, etc.. Note that when a tractor is in the area that is designated to be sprayed by another tractor, a location-based notification is sent to the attendant with the correct number of the tractor. Notifications are also sent by the system informing both the attendant and the supervisor when a tractor is spraying within an area that has been sprayed. In such cases, the supervisor may send a notification to both or to all the attendants. Furthermore, when a trap is within the range of a tractor, the system notifies the attendant with an estimation of the number of insects that have been captured by this trap. An additional capability is that the attendant can use multimedia data, such as photos of the spraying areas, sound files with advises or orders on the spraying operation, textual information, network data concerning the spraying operation etc.. Finally, as it is

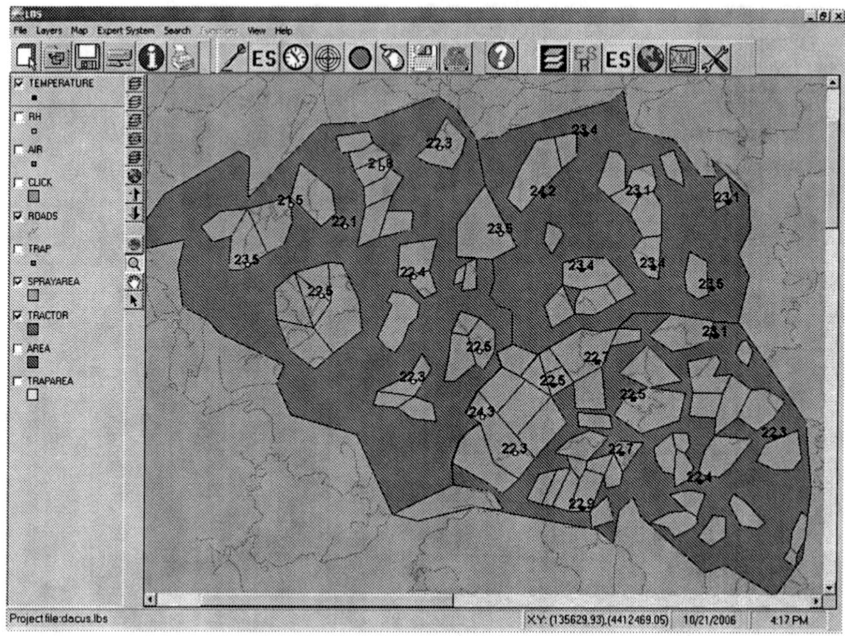

Figure 5. The *Dacus_oleae* personalized options of the attendant

shown in Fig. 5, the attendant has the capability of personalizing the system with his options, namely, to update the database or utilize help on the system's operation.

6. CONCLUSIONS

In this paper we proposed the design of a multi-agent middleware, called *Dacus_oleae*, to support a novel application for spray control and treatment of the olive fly pest problem. It provides a multi-agent design architecture, allowing agents to cooperate and communicate among themselves, disseminating and/or gathering the sensory data on the WSN. The architecture consists of four layers with different types of functionalities. It is open and may adapt future technologies, encapsulating and integrating various modules as agents, by means of communication, with a clearly defined XML schema. The application integrates many interesting characteristics, including the independence of the positioning system, the location model support, the capability of the decision support system, the user friendly environment with multimedia capabilities of the GUI, the flexibility in the development of a new location aware application and services, etc.. To conclude, the system provides efficient data dissemination, in cases where sensors are deployed in large areas with limited power.

REFERENCES

1. Bocur O, Beaune P, Boissier O (2005) Representing Context in an Agent Architecture for Context-Based Decision Making. In: Proceedings of the Workshop on Context Representation and Reasoning (CRR05). Paris France EU.
2. Campo C (2002) Service Discovery in Pervasive Multi-Agent Systems. In: Proceedings of the First International Joint Conference on Autonomous Agents and Multiagents Systems (AAMAS'02). Bologna Italy EU.
3. Chen H, Finin T, Joshi A (2003) An Ontology for Context-Aware Pervasive Computing Environments. The Knowledge Engineering Review 18(3):197–207.
4. Hill J, Szewczyk R, Woo A, Hollar S, Culler D, Pister K (2000) System architecture directions for networked sensors. SIGPLAN Not. 35(11):93–104.
5. Liu J, Jing H, Tang Y (2002) Multi-agent oriented constraint satisfaction. Artificial Intelligence 136:101–144.
6. Marsgh D, Tynan R, O'Kane D, O'Hare G (2004) Autonomic wireless sensor networks. Engineering Applications of Artificial Intelligence 17:741–748.
7. Montiel A, Jones O (2002) Alternative methods for controlling the olive fly, *Bactrocera oleae*, involving semiochemicals. Use of pheromones and other semiochemicals in integrated production. IOBC wprs Bulletin 25.
8. Tweedale J, Ichalkaranje N, Sioutis C, Jarvis B, Consoli A, Phillips-Wren G (2006) Innovations in multi-agent systems. Journal of Network and Computer Applications. In Press, Corrected Proof.

CHAPTER 49

AN ADVANCED METERING INFRASTRUCTURE FOR FUTURE ENERGY NETWORKS

STAMATIS KARNOUSKOS [1], ORESTIS TERZIDIS [1], AND PANAGIOTIS KARNOUSKOS [2]

[1] *SAP Research, Vincenz-Priessnitz-Strasse 1, D-76131 Karlsruhe, Germany. {stamatis.karnouskos , orestis.terzidis}@sap.com*

[2] *Frigoglass S.A.I.C., GR-25200, Kato Achaia, Greece. pkarnouskos@frigoglass.com*

Abstract: We are moving towards a highly distributed service-oriented energy infrastructure where providers and consumers heavily interact with interchangeable roles. Smart meters empower an advanced metering infrastructure which is able to react almost in real time, provide fine-grained energy production or consumption info and adapt its behavior proactively. We focus on the infrastructure itself, the role and architecture of smart meters as well as the security and business implications. Finally we discuss on research directions that need to be followed in order to effectively support the energy networks on the future

Keywords: Service-Oriented Infrastructure, Advanced Metering, Energy Management, Information Services, Business Process

1. INTRODUCTION

In the near future, due to deregulation in the energy sector, a much more decentralized and diversified production and distribution energy infrastructure will emerge. New technologies and increased use of renewables such as biomass, solar energy and wind power will introduce a considerable number of diversified systems into the power grid, in addition to traditional large scale power plants. Consequently, the share of decentralized power generation – by industrial or private producers – will increase and have a dominating effect on existing infrastructure, technologies and business practices.

This paradigm shift will reshape the energy business sector, since new technologies and concepts will emerge as we move towards a more dynamic, service-based, market-driven infrastructure, where energy efficiency and savings can be better

H. Labiod and M. Badra (eds.), New Technologies, Mobility and Security, 597–606.
© 2007 *Springer.*

addressed though interactive distribution networks. A fully liberalized market will advance legacy processes, improve energy sustainability and security, create new business opportunities and have a positive impact on the citizens' everyday life.

New, highly distributed business processes will need to be established to accommodate these market evolutions and fully integrate the distributed electricity sources. The traditionally static customer process will increasingly be superseded by a very dynamic, decentralized and market-oriented process where a growing number of providers and consumers interact. A new generation of fully interactive ICT infrastructure has to be developed to support the optimal exploitation of these changed, complex business processes and to enable the efficient functioning of the deregulated energy market for the benefit of citizens and businesses.

As depicted in Fig. 1, the future energy network is much more dynamic. Households still connect with legacy providers, but also have a number of alternative energy sources and are not simply consumers but are able also to generate electricity. They are able to buy and sell electricity in marketplaces [6], subscribe to services that monitor in real time e.g. the energy consumption of specific devices and can take short-timed decisions based on that info. In this infrastructure energy and its associated products are a commodity that can be traded and managed in at local and global level. A key issue in the process of realizing this goal is the existence of an advanced metering infrastructure (AMI) which heavily depends on sophisticated metering devices referred to as smart meters.

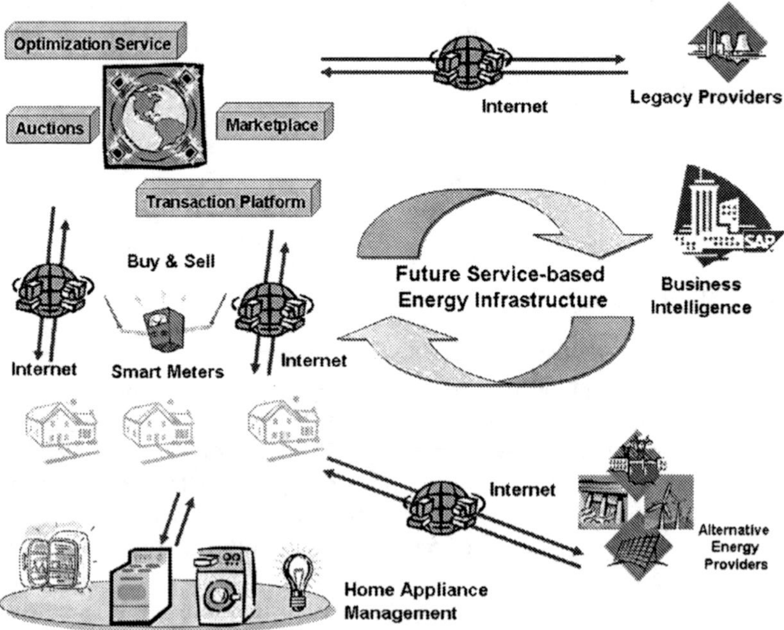

Figure 1. Future service-oriented energy network infrastructure

Several projects [3] have been launched (among them most notably CRISP [5], SESAM [4] and SELMA [1]) that directly or indirectly contribute to aspects that touch this infrastructure. From the European Commission side, a technology platform initiative named Smart Grids [2] was launched in 2006 with the aim to envision the grid infrastructure that needs to be in place for Europe by 2020. In the new European research framework FP7, the identified issues, among of which is the creation of an advanced metering infrastructure, will be further investigated in research projects.

2. THE ROLE OF SMART METERS

We are moving towards the "Internet of things" [9], where almost all devices will be interconnected and able to interact. The same will hold true for energy metering devices. These smart meters will be multi-utility ones, managing not only electricity but also gas, heat etc. New information-dependant intelligent energy management systems will be needed for an infrastructure capable of supporting the deregulated energy market. Smart meters will have to be installed for millions of households and companies and get connected to transaction platforms.

Smart meters provide new opportunities and challenges in networked embedded system design and electronics integration. They will be able not only to provide (near) real-time data but also process them and take decisions based on their capabilities and collaboration with external services. That in turn will have a significant impact on existing and future energy management models. Decision and policy makers will be able to base their actions on real-world, real-time data and not simple predictions. Households and companies will be able to react to market fluctuations by increasing or decreasing consumption or production, thus directly contributing to increased energy efficiency.

In the longer term, smart meters could even be the gateway of communication of household devices with the Internet. It is expected that smart meters will have advanced local communication capabilities (e.g. Bluetooth, IrDA, ZigBee, Wibree etc) and an Internet connection (e.g. via WiFi, DSL, UMTS etc). Therefore they could both participate in local ad-hoc networks with other household devices and in parallel be their communication medium with the outside world. This in turn opens up some interesting issues to be researched as well as the possibility to apply new business models. The replacement of legacy meters will not be linear depending only on the cost or energy provider's intentions, but rather a dynamic one. How fast we will move towards a fully fledged advanced metering infrastructure will depend on the co-evolution of technology and business opportunities.

3. SMART METER ARCHITECTURE

Existing electronic metering devices support in their majority basic electronic characteristics e.g. electronic display of the meter's status and some even have connectivity capabilities e.g. they are able to submit their reading via a wireless

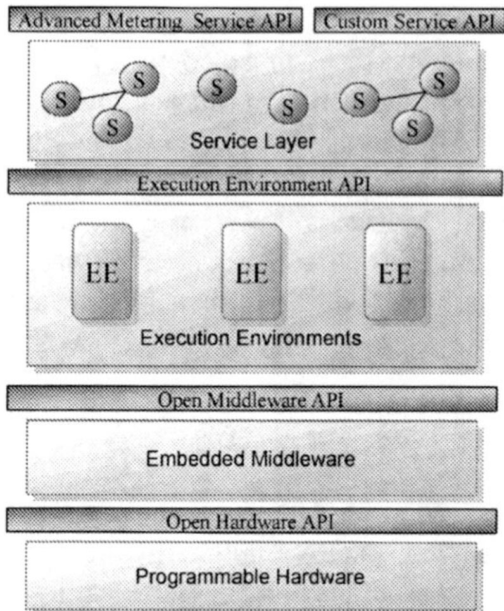

Figure 2. Smart meter architecture

e.g. WiFi/GPRS/IrDA or wired e.g. Ethernet communication channel. However, the existing architectures are closed, allow for very limited interaction with third party services or devices and in order to be integrated in a new application they are usually wrapped in a system-specific way. While this might still work currently, as we move to a service oriented infrastructure, this closed model will not survive. In the near future, meters will transform themselves to embedded devices with CPU, memory, and will have the capability to execute general purpose code that implements third party services. Seeing the meter as a device with computing capabilities, allows us to define a layered open architecture for smart meters as the one depicted in Fig. 2.

As seen, we have several layers that communicate with each other via APIs. These APIs need to be defined and standardized in order to allow for interoperable interaction.

- *Programmable hardware*: This is the lowest layer of the architecture e.g. the electricity meter and the basic software delivered by the manufacturer. In order to ease the integration of the hardware in other systems, the functionality offered has to be captured by the open hardware API. Also via the same API one is able to manage the hardware device i.e. program or configure it according to the capabilities offered.
- *Embedded middleware*: This layer is a general purpose middleware for embedded devices. Its role is to provide the capabilities for creating and support of execution environments (EE). The middleware manages the lifecycle of the EEs and is able to also capture the hardware's capabilities and offer via a multitude of APIs a finer programming environment to the EEs.
- *Execution Environments*: The execution environments (EE) are hosted by the middleware and provide specific capabilities that service providers can use to deploy their services. Each meter is expected to host at least one EE.

- *Service Layer*: Several services run in the different EEs on the metering device and offer a standard API to the applications. One service can be standalone or depend on others to provide its functionality. The API offered by the services is standardized and is a uniform way of accessing, the meter's capabilities and programming it

The main motivation behind this modular approach is that each layer should be agnostic of the other layers and only depend on the specific API below it. In a heterogeneous infrastructure such as that on future energy networks, many programming languages and a plethora of implementations are expected to exist for various reasons e.g. performance, flexibility, advanced capabilities etc. However, as long as the basic standardized APIs are globally implemented, all will have a common basis which will enable their interoperability. This is expected to ease also vertical integration at customer side that may be necessary to create robust and highly distributed deployments. Furthermore the existence of an execution environment implies that the meter can adapt its behavior and be incrementally software-upgraded.

The Business model depicted in Fig. 3 is compliant with the architecture in Fig. 2. We can clearly distinguish:

- *Hardware Manufacturer*: This is the manufacturer of the hardware devices e.g. an electricity meter, which includes also a very generic software capability that fully supports the open hardware API standard.
- *Embedded Middleware Provider*: this role is responsible for delivering the middleware that can be used with the specific hardware architecture.
- *EE provider*: This role specializes delivering EEs in different implementations and for different hardware platforms
- *Advanced Metering Infrastructure Provider*: This role is responsible for deploying and initializing the whole infrastructure. He is also responsible for providing the communication needed.

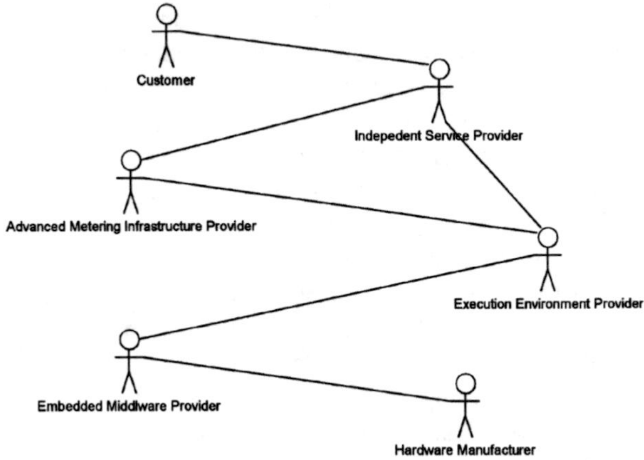

Figure 3. AMI business model

- *Independent Service Provider*: this role is responsible for creating and managing services that depend on an AMI. He is able to deploy services network wide e.g. in EEs, combine them with others e.g. enterprise services and deliver sophisticated services to the customer.
- *Customer*: This is either the end-user of the infrastructure who may be located at the edge of the information system infrastructure or assume other roles e.g. an Internet application, a connection management system etc.

4. DISTRIBUTED BUSINESS INTELLIGENCE

The existence of smart meters that can be also accessed in a seamless and uniform way via standardized methods is a must for the future service oriented infrastructure. Assuming that smart meters will be accessed e.g. via web-services, has far reaching implications, since now business processes can actively integrate them in their execution. Smart meters can provide real-time data which can then be consumed by services, which in their turn now can act based on rapid changing context conditions Furthermore, instead of providing only their data (one-way communication), which limited their usage up to now, they can now be active and host business intelligence (bidirectional communication) which does not have to rely only on the back-end systems.

As depicted on the left side of Fig. 4, a typical business process execution in a backend system that hosts the business intelligence. At some point the meter is interrogated by the business process and its metering status is sent in a time-frame which may vary greatly since the acquisition can be done electronically or even per post. On the left side of Fig. 4 a similar process is depicted, which however assumes an advanced metering infrastructure in place. Since the meters do have computing capabilities, are able to process locally their data and take local decisions, this data does not need to be sent to the backend systems. Therefore we have a part of the business process executed outside the backend system. The business process could be even more distributed since the meter may trigger an external Internet service which will do advance the business process itself. So from the original steps (four are depicted as an example in Fig. 4) in the business process execution only two of them have been done at the backend infrastructure while two others have been executed collaboratively by the meter itself and another Internet based service. The advantages are profound i.e. more lightweight business processes which can outsource or parallelize specific execution steps, we have reduced communication overhead since the data do not have to be transferred to the backend but stay at their original source, and we are able to realize more sophisticated business processes that are highly distributed and may even partially belong to different domains. In an infrastructure where real-time data is constantly generated, needs to be processed and is composed of millions of devices (only in Europe there are more than 225 Million electricity meters), centralized processing (e.g. on backend systems) could be problematic, but such delegation of tasks and distribution of business intelligence may be another step towards more viable and better managed infrastructure.

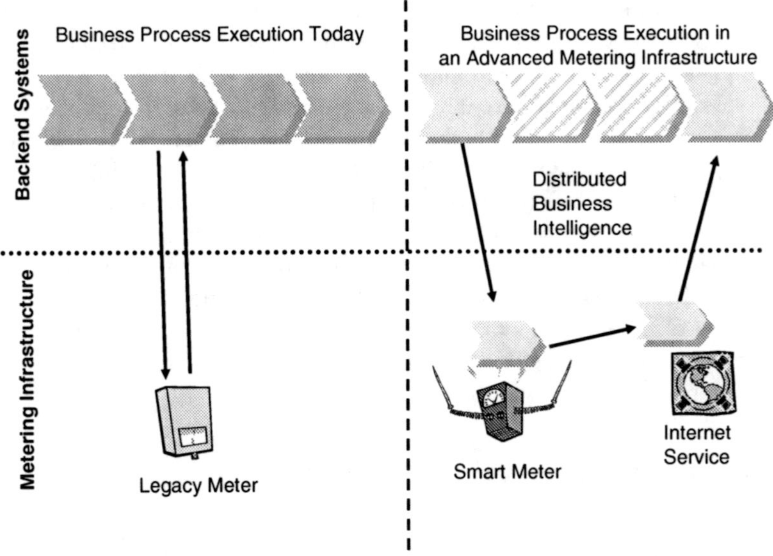

Figure 4. Distributed business process execution

5. SECURITY IMPLICATIONS

Opening up a closed infrastructure as that of energy networks and taking into account the associated business background, can not be done without well-tested security and trust models in place. Furthermore if in the longer term the smart meter evolves to a gateway for household devices, the implications are far reaching, since it is expected that any device will have its own IPv6 address and it will be possible not only to turn it on/off but constantly monitor its behavior. It is clear that several aspects have to be taken care of in order to provide a secure basis for all the implicated actors. In such a heterogeneous infrastructure as the envisioned future energy one, the author of services to be deployed in the smart meters, the entity that deploys a service, the owner of the smart meter, and the owner of the data may be different entities governed by different interest.

A comprehensive threat model needs to be defined. At first it must be secured that the measurement process can not be tampered and the data measured can not be altered (or if this happens there is proof of that). The next step would be to securely transmit the data to the consuming parties. State of the art concepts can be used here e.g. encryption or digital signatures. Projects like SELMA [1] have already tackled parts of this threat as a security architecture that authenticates the measurement data, provides access security and certified software has been developed. However, since the smart meter is able to host execution environments and external entities can deploy services on it, the security model needs to be further elaborated. Issues like repudiation, masquerading, denial of service, unauthorized access need to be successfully tackled. Finally since now via the smart meter private info go beyond simple energy consumption profiling, as their correlation can reveal indication of money flow (e.g. amounts of energy produced/bought/sold), personal habits (e.g. monitoring of energy consumption per device and possibly at very fine level), and other private context info, it has to be assured that there is no misuse or unwanted exploitation of this info. On the other hand, the end-user will be able to enjoy a variety of sophisticated services and with the right tools be in full control of the personal info s/he shares with other parties, something that is not at high degree possible in practice today (but is implied by the legal framework and the contracts between the parties). Furthermore the interactions at global level will have to be investigated and security & trust must be tacked at technology and business model level. Development of appropriate security, safety and risk concepts and architectures for an advanced metering infrastructure for the future energy networks in total is not expected to be trivial.

6. RESEARCH DIRECTIONS

We are still in a very early era of development for the future energy network. The most commercial approaches are slowly moving towards automatic metering readers (AMR) which is the transmission of electronic data, but we are still far away from defining a common information model, standardize APIs for communication among heterogeneous services, hardware devices and applications. However one can clearly identify some issues that need to be resolved in the short and mid term in order to allow for the evolution from AMR to AMI.

First of all AMI will require interoperability at several layers as this was depicted with the proposed architecture. Therefore the need to work and agree upon basic functionality that needs to be provided at hardware level is of high priority. In the short term, developing electronic meters with a web-service interface e.g. DPWS [7] will provide them with support for secure web service messaging, discovery, description, and eventing, which in turn allows easy integration with service-oriented efforts in the business domain. Binding them successfully to enterprise services will allow existing business processes to at least integrate their readings and treat them as smart items along with other similar devices such as RFID and sensor networks. Later, in the long term, one could focus on further developing a more advanced smart meter architecture such as the one presented in this paper. Subsequently as discussed, security and trust issues will have to be investigated as early as possibly and get integrated from the scratch at any solutions to be developed, and not as late add-ons – an approach which has been proven to fain and lead to insecure systems.

Once the smart meters have been integrated in business processes, real-time high volume data will be available. It is doubtful if any backend system will be able to deal with this amount of data, therefore it has to be investigated what are the most promising models for distributing business intelligence at several layers (the last of which is the smart meter itself) and take advantage of local in-device and in-network processing and decision making. One size fits all model is not expected to be found, therefore theoretical models should be tested on real-world task specific scenarios.

Also scenarios developed nowadays for other domains will be possible to migrate to the smart meters world. As the infrastructure envisioned (depicted in Fig. 1) will allow the today's energy consumer to slip into the role of buyer and seller on online marketplaces, this will provide another domain where well-known approaches, such as the software agents, may revive. As an example a lightweight agent platform could be installed on the meter (an agency could be one of the execution environments depicted in Fig. 2), and mobile agents could take over tasks of negotiating, buying and selling electricity in energy marketplaces on behalf of the user [8].

Once the smart meters are seen as part of the business process, it will be easier to link them to production planning and energy management systems. This will allow for a more fine-grained management of the energy network overall and a better control of its multiple generation and consuming entities. It will be also possible to have due to the accurate data better models and move towards a more reliable predictive infrastructure that better manages its requirements, the available resources and their optimal usage.

It is expected that the future energy networks will be a highly dynamic ecosystem of consumer, producers, and services and be highly market-driven. This is a paradigm shift with far reaching technological, social and economic effects, whose interdependencies need to be identified, analyzed and understood.

7. CONCLUSIONS

We slowly move towards AMR, while AMI is still in its majority a research domain. However, the liberalization in the energy domain will speed-up this transition in the mid-term. There are still several research issues to be tackled in order to move towards a dynamic service oriented future energy network infrastructure. AMI based on smart readers could be the direction to follow. We have presented an architecture that was designed with the future requirements in mind, a business model that is compliant to the proposed architecture and have discussed on the security and business aspects that this new approach brings. Finally we have laid out some AMI related research directions that will have to be successfully tackled in order to enable the realization of such a dynamic infrastructure. The energy domain and its combination with ICT present tremendous challenges and opportunities for citizens and businesses in the years to come.

REFERENCES

1. SELMA Project (Sicherer ELektronischer Messdaten-Austausch), http://www.selma-project.de
2. European SmartGrids Technology Platform - Vision and Strategy for Europe's Electricity Networks of the future. European Commission, March 2006, ISBN 92-79-01414-5, http://www.smartgrids.eu

3. Zobel, R., Filos, E. (2006):"The Impact of ICT on Energy Efficiency", eChallenges 2006, 25–27 Oct. 2006, Barcelona, Spain, IOS Press ISBN: 1-58603-682-3.
4. SESAM Project (Selbstorganisation und Spontaneität in liberalisierten und harmonisierten Märkten), http://www.sesam.uni-karlsruhe.de
5. CRISP Project (distributed intelligence in Critical Infrastructures for Sustainable Power), http://www.ecn.nl/crisp
6. Kok, K. Warmer, C., Kamphuis, R., Mellstrand, P., Gustavsson R. (2005): "Distributed Control in the Electricity Infrastructure", Proceedings of IEEE International Conference on Future Power Systems, 16–18 Nov. 2005, Amsterdam, the Netherlands. ISBN: 90-78205-02-4
7. Devices Profile for Web Services (DPWS) specification, Feb 2006, http:// schemas.xmlsoap.org/ ws/2006/02/devprof/
8. Kok, J. K., Warmer, C. J., and Kamphuis, I. G. (2005): PowerMatcher: multiagent control in the electricity infrastructure. In Proceedings of the Fourth international Joint Conference on Autonomous Agents and Multiagent Systems (The Netherlands, July 25 – 29, 2005). AAMAS '05. ACM Press, New York, NY, 75–82.
9. Fleisch, E., Mattern, F. (2005):. Das Internet der Dinge: Ubiquitous Computing Und Rfid in Der Praxis:Visionen, Technologien, Anwendungen, Handlungsanleitungen. June 2005, Springer, Berlin, ISBN: 3540240039.

CHAPTER 50

INTERNET TOPOLOGY BASED IDENTIFIER ASSIGNMENT FOR TREE-BASED DHTS

LECHANG CHENG[1], MABO R. ITO[2], AND NORM HUTCHINSON[3]

[1]*Department of Electrical and Computer Engineering, University of British Columbia*
lechangc@ece.ubc.ca
[2]*Department of Electrical and Computer Engineering, University of British Columbia*
mito@ece.ubc.ca
[3]*Department of Computer Science, University of British Columbia, norm@cs.ubc.ca*

Abstract: The current routing algorithms of DHT-based P2P systems have a large end-to-end delay and inconsistent routing performance because of their random selection of identifiers (IDs). In this paper, an Internet topology based overlay construction method is proposed for tree-based DHTs. The node ID is divided into three parts and assigned according to the autonomous system (AS), IP network prefix, and IP address of the node. This algorithm assigns the AS ID prefix based on the AS-level Internet topology. Proximity Neighbor Selection (PNS) is used with topology based ID assignment so that the overlay routing can match the underlying IP routing path. The assignment of AS ID prefixes also takes into account the node densities of ASes to alleviate the ID space load imbalance. Simulation results show that this method can reduce the routing stretch and the standard deviation of the routing stretch without introducing any single points of failure

Keywords: Peer-to-peer, DHT, Internet Topology

1. INTRODUCTION AND OVERVIEW

In recent years, there has been considerable research effort in the area of structured P2P systems using distributed hash tables (DHTs). Typical structured P2P systems include Chord [9], Pastry [7], Tapestry [11], and CAN [5]. In structured P2P systems, every node and object is assigned a random ID from a numeric space using a consistent hashing algorithm. An object is usually mapped to the node with the closest numeric ID to that of the object. In structured P2P systems, nodes connect to each other based on their IDs to form a routing mesh and every node has a neighbor table of size O(log N). With the neighbor tables of the participating nodes, messages can be efficiently routed to the destination node through O(log N) hops.

H. Labiod and M. Badra (eds.), New Technologies, Mobility and Security, 607–616.
© 2007 *Springer.*

The advantage of structured P2P systems is that their routing and locating methods are scalable. However, they also introduce some problems. Firstly, DHT-based systems assign IDs to nodes randomly. The nodes with close IDs might be widely separated in the Internet. Ignoring the positions of nodes causes high end-to-end delay and inconsistent routing performance. Secondly, the randomness in assigning IDs to nodes and files destroys data locality. Data might be stored far from its users and outside the administrative domain that it belongs to.

P2P systems are usually overlay networks built on top of the Internet. In order to obtain an efficient routing, the overlay routing path must match the underlying Internet routing path as much as possible. Fundamentally, the Internet can be regarded as a decentralized network with a flat structure. However, in order to make routing efficient and scalable, a hierarchical structure has been introduced. The IP address is divided between the network address part and the computer address part. The Autonomous System (AS) was introduced to make the Internet appear as a three-tier structure. This hierarchical structure improves the routing scalability, efficiency and locality of the Internet.

In fact, tree-based DHTs (Pastry and Tapestry) can be considered as having a similar hierarchical routing architecture. To illustrate, if one wants to route a message to the node with ID 356, the message will first be sent to one of the nodes in the group of nodes with prefix 3**, then the group of 35* and finally 356. Therefore, if node IDs can be assigned according to the hierarchical structure of the Internet, the routing of the upper layers can be expected to match the underlying IP routing path. If the node ID bears the information of the network topology, it will become easy to provide data locality.

In this work, an Internet topology based ID assignment algorithm is proposed for tree-based DHTs, such as Pastry and Tapestry. Proximity Neighbor Selection (PNS) is used in combination with the topology based ID method to improve the routing efficiency for DHTs. With topology-based ID assignment, the hierarchical routing structure is introduced into DHTs without using super-nodes. Since the IDs bear information of the location of the nodes, it can provide explicit locality.

The rest of the paper is organized as follows. Section 2 discusses some of the previous work that is closely related to our work. Section 3 presents the details of our algorithm. Section 4 presents some simulation results. Section 5 concludes and presents some future work.

2. RELATED WORK

Recently, there has been a lot of research on improving the routing efficiency and locality of DHTs [2,6,10,4,13,12]. Some work proposed incorporation of topology information in overlay construction to improve the routing efficiency. Ratnasamy et al. [6] proposed three methods: Proximity Neighbor Selection (PNS), Proximity Routing Selection (PRS) and geographic layout. Tapestry and Pastry implement PNS. For nodes with the same prefix, they chose the closest one as the primary neighbor. However, even with PNS, the average routing stretch (overlay end-to-end delay divided by the underlying IP network end-to-end delay) is still high.

Furthermore, as the node ID has no relationship with the position of the node in the Internet, it is difficult to provide data locality. Ratnasamy et al. [4] proposed a binning method for CAN to infer network information and used it construct a topology-aware overlay. As CAN lacks the ability to use PNS, its routing stretch remains significant.

Some other work introduced the hierarchical structure to the DHT-based systems and leveraged the heterogeneity of the nodes to improve routing efficiency [2,10]. In Brocade [10], a set of super-nodes are selected to form a secondary overlay on top of the existing DHT. The second-layer overlay provides a shortcut routing algorithm to quickly route to a remote network. Garces-Erice et al. [2] proposed the idea of a hierarchical DHT system. Instead of building a flat overlay, the participating nodes are divided into a set of groups and each group uses a DHT to build its own intra-group overlay network and look-up service. In each of the groups, the nodes with high bandwidth and capacity are selected as super-nodes to form a top-level overlay network. This two-tier structure can help improve the routing efficiency and locality of structured DHTs, however it also introduces single points of failure as they rely more on super-nodes to forward traffic. With the topology-based ID assignment, our work introduces the hierarchical structure to DHTs without using super-nodes.

The closest work to ours is Zhou et al. [13]. They also propose a hierarchical location-based ID generation method for DHTs. This method divides IDs into pre-defined prefix and random suffix parts. The pre-defined prefix is assigned hierarchically according to the geographical location of the node in the Internet. The length of the prefix varies among different groups and a Huffman like algorithm is used to encode the prefix. The length of the prefix determines the number of keys that the group is responsible for. This method provides coarse grain ID space load balance. Our method differs from [13] in that we try to reduce the routing stretch by matching the overlay routing path with the underlying Internet routing path as much as possible. Instead of assigning ID prefixes with respect to geographical positions, ID prefixes are assigned with respect to the AS-level Internet structure. Furthermore, in order to achieve a low routing stretch, proximity neighbor selection must be used in combination with topology-based ID assignment.

3. TOPOLOGY BASED NODE ID ASSIGNMENT

In this section, the idea of Internet topology based node ID assignment is proposed for tree-based DHTs such as Tapestry and Pastry. Figure 1 shows the structure of node ID. Instead of generating the node IDs randomly using a consistent hashing function, IDs are assigned to the nodes according to their IP address, port number, IP network prefix and AS number. Let the length of node IDs be L and the base of ID be b. The node

← LA →	← LN →	← LE →
AID	NID	EID

Figure 1. Node ID structure

ID is divided into three parts. The first L_A digits are assigned according to the AS that the node resides in. The next L_N digits are assigned according to the node's network prefix. The last L_E digits are assigned according to the IP address and port number. These three parts are called the AID (AS ID), NID (Network ID) and EID (End node ID) respectively. The AID, NID and EID can be of a fixed length or a variable length. The algorithms to generate the AID, NID or EID are illustrated as follows.

AID Assignment Algorithm On the Internet, an AS is the unit of router policy, either for a single network or a group of networks that is controlled by a common network administrator. ASes are usually stable and the connectivities among ASes do not change much with time. An inter-domain routing protocol such as BGP is used to determine the best path among ASes. For IP level routing, a packet is usually first routed to the destination AS, then its destination network prefix and its destination node.

Tree-based DHTs such as Pastry and Tapestry use prefix-based routing. The participating nodes forward a message by gradually resolving the prefix until getting to the node with the closest ID. In order to match the overlay routing path with the underlying Internet routing path, it is desirable to assign AIDs so that the overlay routing mesh will reflect the connetivities among ASes and the first several hops of prefix resolving routing will match the AS-level routing path.

In recent years, a lot of research has been done to infer the AS-level structure of the Internet. Ge et al. [3] proposed a hierarchical structure to model the AS-level Internet topology. In the Internet, the majority of the relationship between ASes can be classified as the provider-customer relationship and ASes in the Internet can be organized into a hierarchical structure with several levels. Each level is called a tier. An AS can be classified as either a tier-k provider or tier-k customer. It is assumed that the provider ASes only contain the core routers, not end nodes.

In this work, a hierarchical AIN assignment algorithm is proposed. The basic idea is to model the ASes in the Internet with a hierarchical structure and assign AIDs according to the position of the AS in the hierarchical structure. In addition, Proximity Neighbor Selection guarantees that the routing path among AS follows the underlying AS-level path.

Firstly, the AID is divided into K sections. Therefore the AID will be in the format of: $D^1 : \cdots : D^K$. D^i is called section-i of the AID and $D^1 : \cdots : D^m$ is called the level-m prefix of the AID. Every section of an AID can be represented by a sequence of digits with base b or the numeric value of the sequence. The AID of a tier-m AS will have the format of $D^1 : \cdots : D^m : 0 : \cdots : 0$. The format is the same for both the provider and customer ASes.

In assigning IDs to the nodes, it is assumed that all the child nodes of a tier-m provider AS share the level-m prefix of their father node. To illustrate, if the AID of a tier-m provider is $D^1 : \cdots : D^m : 0 : \cdots : 0$, all its child ASes will have ID $D^1 : \cdots : D^m : D^{m+1} : 0 : \cdots : 0$. Section D^i of the AIDs is calculated iteratively from the top to the bottom. Figure 2 illustrates an example of the hierarchical structure of a simple network.

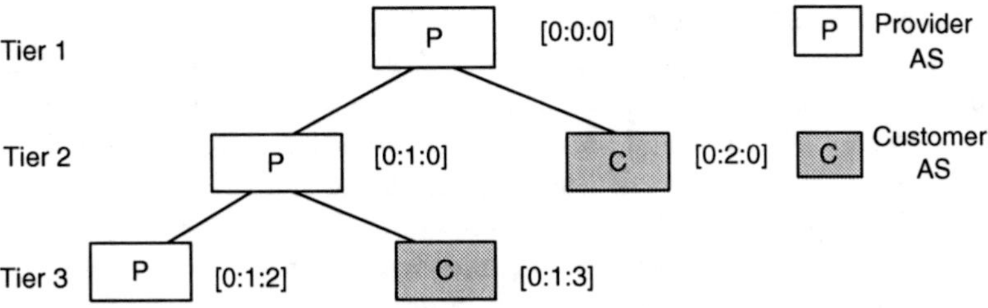

Figure 2. Illusration of hierarchical AID assignment

Address-space Load Balancing DHTs provide a mapping between keys (object IDs) and nodes. Every node is responsible for the set of keys that are numerically closest to itself. For simplicity, it is assumed that closest node ID of a key is the closest node ID that is larger than the key circularly. Random node IDs and object keys will guarantee that the keys are evenly distributed among nodes.

Our AID assignment method could cause imbalance in the key distribution for two reasons. 1) ASes in the Internet have different node densities and the number of keys assigned to each of the ASes is not proportional to its node density. 2) Some high level ID prefixes might be empty (not assigned to any AS). This will cause one of the nodes with the closest prefix to be responsible for the entire group of keys with the empty prefix.

We propose a fixed length method which provides fine grain ID space load balance. Every section of the AID is of fixed length. We try to carefully put the AID prefix in the ID space so that every prefix would be responsible for a portion of the key space proportional to its node density. With the hierarchical structure of ASes, the sections of AIDs are calculated iteratively from the top to the bottom. For the ASes that have the same father node, we sort them with respect to their densities in increasing order. Suppose there are M ASes and their node densities are $C_0 \leqslant C_1 \leqslant \cdots \leqslant C_{M-1}$. The node density of a provider AS is the sum of the node densities of its children ASes.

Let the digit length of the section-m of the AID be p. Then the number of possible p-length digit sequences will be b^p . The portion of sequences that the ith ASes should be responsible for is:

$$B_i = C_i / \sum_{j=0}^{M-1} C_j$$

And the $m+1$ section of the AID of the ith AS will be:

$$D_0^{m+1} = 0, D_i^{m+1} = [b^p * \sum_{j=0}^{i-1} B_j]$$

where [] means the closest integer.

To illustrate, it is assumed that there are three ASes (A1 A2, A3) in a one tier Internet structure. The node densities of the ASes are 100, 600 and 900 respectively. If the base is 16 and the AID length is 1, the AID of A1, A2 and A3 will be 0, 6 and 15 respectively. Therefore, A1 will be responsible for all the keys with prefix 0, and A2 will be responsible for keys with prefix 1–6 and A3 will be responsible for keys with prefix 7–15. This warrants that the keys assigned to A1 and A2 are proportional to their node densities.

In order to eliminate the load imbalance inside each prefix, the mapping function has to be changed so that a key is mapped to the node with the closest AID, closest NID and closest EID. For closest AID, it means that every section of the AID has to be the closest so that the key space of the empty prefix will be evenly distributed among the nodes with the closest prefix. The mapping is deterministic since there is only one node with the closest AID, NID and EID to that of the desired key.

NID and EID Assignment Algorithm In the Internet, different IP network prefixes have different node densities. In order to keep the address-space load balancing, NID is generated using the same method as that of AID in Section 3.2. The only difference is that within an AS, IP network prefixes have 1-tier hierarchical structure. The EID is generated using consistent hashing to preserve randomness.

Neighbor Table and Routing With Internet topology based node ID assignment, the routing table will also reflect the hierarchical structure of the system. In the routing table, the last L_E columns contain nodes whose IDs match both the AID and the NID of the current node. Therefore, they are pointers to nodes in the same network. The entries of column $L_A + 1$ to $L_A + L_N$ contain nodes whose IDs match only the AID of the current node. They point to nodes in the same AS but in different networks. The first L_A columns contain nodes whose IDs do not match either the AID or the NID of the current node. They point to nodes in other ASes.

For simplicity, we use the same digit based incremental routing as that of Pastry. This is equivalent to gradually routing a message to the AS of the destination node, the network of the destination node and the destination node.

IP prefix and AS extraction In order to calculate the AID for a node, a mechanism is needed to determine the AS number of a node given its IP address and to estimate the node density of every AS. We used the method similar to that mentioned in Sen et al. [8]. Router level data is collected using Cisco's NetFlow services [1]. NetFlow caches IP flow information in routers and sends them to a data collector for reporting. For each IP flow, NetFlow maintains a record in the router cache, including the IP addresses, BGP routing prefixes, ASes and port numbers of the source and destination. With this data, all the network prefixes and masks and the corresponding AS can be found, thus building a mapping table. With an IP address, longest prefix matching is used to find the IP prefix, mask and AS number. As the mapping between an IP address and an AS is not changing with time, the process of creating the mapping table can be done offline. A mapping table could be provided to the user when they join the overlay network. With the mapping table, the node density of each of the ASes can also be estimated.

Discussion The main concerns for topology based ID assignment for DHT are its side effects. One is the load imbalance. Some nodes might become the hotspot for both overlay routing and object storage. However, with topology based ID assignment, there is no super-peer. Nodes are regarded as the same during routing construction. Therefore, there is no routing hotspot. The simulation in the following section will also show that distribution of the in-coming link degrees is the same as that with random IDs. Besides, by taking the node density into consideration, the ID assignment method will alleviate the load imbalance with respect to object storage.

The second concern is the failure resilience. In the Internet, correlated failures are not unusual. In this method, the EIDs are assigned randomly. This keeps a certain level of randomness and thus improves the resilience to correlated failures. Another method to improve the failure resilience is to use multiple object IDs. An object can obtain more than one object ID using multiple hash functions. One of the object IDs is the primary ID and the rest are backups. In searching for an object, one can use the primary object ID to search for the object. If the object can not be found, then the backup object ID could be used. With multiple object IDs, one can improve the resilience to correlated failures.

4. SIMULATION RESULTS

In this section, some simulation results are presented to demonstrate that Internet topology based ID assignment can effectively reduce the end-to-end delay of DHTs. The topology-based ID assignment method is applied to Pastry. In this simulation, the routing stretch is measured for 500 randomly selected pairs of nodes. Routing stretch is defined as the overlay routing delay divided by the underlying IP network delay. The average routing stretch is calculated as the metric of the average overlay routing performance. The standard deviation of routing stretch is also computed to find out how the routing stretch varies among different paths in the overlay. The base b of IDs is set as 4.

The simulated network is generated using Georgia Tech's transit-stub network topology model (GT-ITM). As GT-ITM usually generates only routers, not nodes in the stub LANs, L_N (the length of NID) is set as zero. We will compare the routing performance of Pastry with random IDs and topology based ID in network topologies with increasing size.

Routing Performance Figure 3 shows the average routing stretch versus network size. In the following figures, TopID represents Pastry with topology based ID assignment and RandID represents Pastry with randomly selected ID.

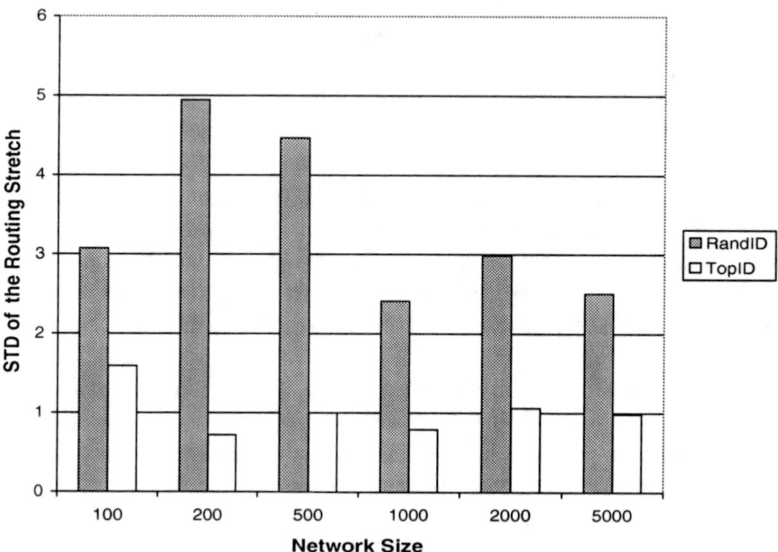

Figure 3. Average routing stretch versus network size

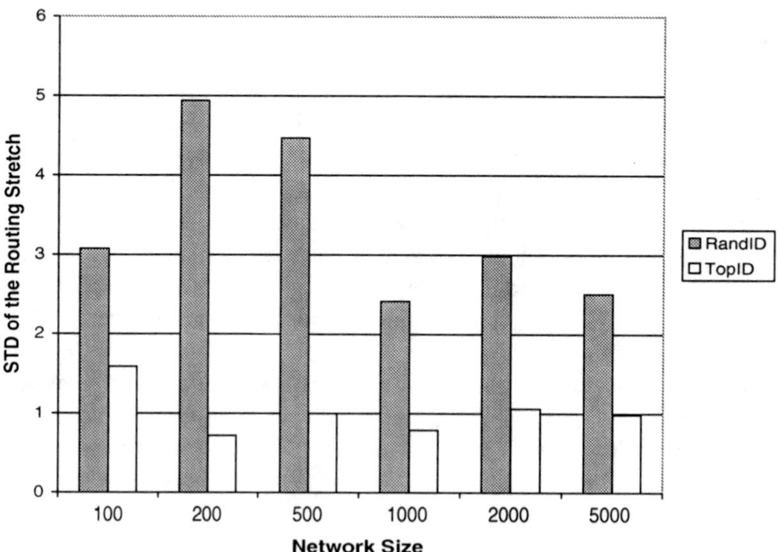

Figure 4. Routing stretch standard deviation versus network size

Table 1. The incoming-link degree versus network size

Size	RandID			TopID		
	Avg	Max	Min	Avg	Max	Min
100	89.7	103	83	89.9	103	82
200	91.1	103	83	90.8	105	83
500	93.1	102	84	93.2	115	82
1000	94.2	109	81	94.3	114	84
2000	95.6	113	86	93.8	116	85
5000	96.8	111	85	95.2	107	87

Figure 3 shows that using TopID decreases average routing stretch by 33%–43%. The standard deviation of routing stretch reflects the consistency in the routing performance. A large standard deviation of routing stretch means that some routing paths will have large routing stretch. Figure 4 shows that the standard deviation of the routing stretch of the original Pastry is 2–4, while the routing stretch standard deviation to topology ID assignment is less than half of that of the original Pastry.

The incoming-link degree of a node D in an overlay network counts the number of nodes that have D in their routing tables. The larger incoming-link degree a node has, the more routing traffic the node will have. Table 1 shows the average, minimum and maximum of in-coming link degrees of all the nodes in Pastry. With topology based ID assignment, the average, the minimum and maximum of the in-coming link degrees are almost the same as that with random IDs. The variance of in-coming link degrees is small compared to the average in-coming link degrees. This shows that topology based ID assignment does not introduce routing hotspot.

5. CONCLUSIONS AND FUTURE WORK

In this work, a topology based ID assignment algorithm is proposed for tree-based DHTs such as Tapestry and Pastry. Simulation results show that the proposed method can reduce the routing stretch and routing stretch standard deviation of Pastry significantly in a variety of network topologies. This work can be applied to delay sensitive P2P applications such as video streaming. As the simulations in this work are only conducted to the scale of 5000 nodes, future work might be larger-scale simulations on networks with millions of nodes.

REFERENCES

1. Netflow services solutions guide. http://www.cisco.com/en/US/products/ps6601/products_ios_protocol_group_home.html, 2002.
2. L. Garces-Erice, E. W. Biersack, K. W. Ross, P. A. Felber, and G.Urvoy-Keller. Hierarchial peer-to-peer systems. *Parallel Processing Letters*, 13(4):643–657, 2003.
3. Z. Ge, D. Figueiredo, S. Jaiwal, and L. Gao. On the hierarchical structure of the logical internet graph. In *Proc. SPIE ITCOM*, August 2001.

4. S. Ratnasamy, M. Handley, R. Karp, and S. Shenker. Topologically-aware overlay construction and server selection. In *Proceedings of IEEE INFOCOM'02*, 6 2002.
5. Sylvia Ratnasamy, Paul Francis, Mark Handley, and Richard Karp andScott Schenker. A scalable content-addressable network. In *Proceedings of the 2001 conference on applications, technologies, architectures,and protocols for computer communications*, pp. 161–172. ACM Press, 2001.
6. Sylvia Ratnasamy, Scott Shenker, and Ion Stoica. Routing algorithms for dhts: Some open questions. In *First International Workshop on Peer-to-Peer Systems (IPTPS)*, 2002.
7. Antony Rowstron and Peter Druschel. Pastry: Scalable, decentralized object location, and routing for large-scalepeer-to-peer systems. In *IFIP/ACM International Conference on Distributed Systems Platforms (Middleware)*, pp. 329–350, 2001.
8. Subhabrata Sen and Jia Wong. Analyzing peer-to-peer traffic across large networks. In *Second Annual ACM Internet Measurement Workshop*, November 2002.
9. Ion Stoica, Robert Morris, David Karger, Frans Kaashoek, and HariBalakrishnan. Chord: A scalable Peer-To-Peer lookup service for internet applications. In *Proceedings of the 2001 ACM SIGCOMM Conference*, pp. 149–160, 2001.
10. Ben Y. Zhao, Yitao Duan, Ling Huang, Anthony Joseph, and John Kubiatowicz. Brocade: Landmark routing on overlay networks. In *Proc. of IPTPS*, pp. 34–44, Mar 2002.
11. Ben Y. Zhao, John D. Kubiatowicz, and Anthony D. Joseph. Tapestry: An infrastructure for fault-tolerant wide-area location and routing. Technical Report CSD-01-1141, U. C. Berkeley, Apr 2001.
12. Shuheng Zhou, Gregory R. Ganger, and Peter Steenkiste. Balancing locality and randomness in dhts. Technical report, School of Computer Science Carnegie Mellon University, November 2003.
13. Shuheng Zhou, Gregory R. Ganger, and Peter Steenkiste. Location-based node ids: Enabling explicit locality in dhts. Technical report, School of Computer Science Carnegie Mellon University, September 2003.

CHAPTER 51

A SECURE DYNAMIC REMOTE USER AUTHENTICATION WITHOUT ANY SECURE CHANNEL

ASHUTOSH SAXENA

Security and Privacy Group, SETLabs, Infosys Technologies Limited, Hyderabad DC, Survey No.210, Lingampally, Hyderabad 500 019. Andhra Pradesh, INDIA. ashutosh_saxena01@infosys.com

Abstract: With the distributed nature of computer and network systems, the achievement of privacy and security has become increasingly important [1]. At the same time Internet applications have become more and more indispensable in today's climate. Web site services, including remote login, are in rapid demand. However the current environment is not secure [2]. Identities could be impersonated, authentication information could be eavesdropped or cracked, and communication contents could be revealed. Therefore, authentication is becoming increasingly essential. Remote user authentication scheme allows the authenticated user to login the remote system for accessing the services offered. Lamport [8] introduced the first well-known hash-based password authentication scheme, but the scheme suffers from high hash computation overhead and password resetting problems. Thereafter, many authentication schemes have been proposed based on hashed password [2,4–13] and on public key cryptography [14–?] some using smart card and some without the use of smart card. It is observed that all the methods require a secure offline channel at least once typically at the time of registration. To the best of the knowledge there is possibly no remote user authentication scheme that avoids the use of secure channel or offline registration. This motivates us to work upon and propose an online registration to avoid secure or offline communication with the remote server, using pre-partial personalized smart card.

The scheme presented here for remote user authentication using smart cards is superior to all the existing remote user authentication schemes as it completely eliminates the requirement of a secure offline communication for registration by the remote entity. In addition, it offers a flexible password change provision and resilient to insider, replay, forgery, stolen-verifier, man-in-the-middle and guessing attacks. The scheme does not require to maintain any password-verifier table to validate the users' login request. The scheme also provides flexible password change option to the users. Security of the scheme relies on the hardness of solving Integer factorisation problem (IFP). The scheme can be easily extended to any other asymmetric key cryptography scheme

Keywords: Smart card; Authentication; Password

H. Labiod and M. Badra (eds.), New Technologies, Mobility and Security, 617–617.
© 2007 *Springer.*

CHAPTER 52

CROSS-LAYER OPTIMIZATION FOR DYNAMIC RATE ALLOCATION IN A MULTI-USER VIDEO STREAMING SYSTEM

C. YAACOUB[1,2], J. FARAH[1], N. RACHKIDY[1], AND B. PESQUET-POPESCU[2]

[1] *Faculty of Sciences & Computer Engineering, Holy-Spirit University of Kaslik, Jounieh, Lebanon*
[2] *Ecole Nationale Supérieure des Télécommunications, Paris, France*

Abstract: In this paper, we present a novel cross-layer optimization system that allocates unequal transmission rates for different users requesting video from a streaming server. The rate allocation technique exploits distortion information provided by the source encoder to take into account the influence of channel impairments, lossy compression, and error concealment. Rate-Compatible Punctured Turbo-Codes are used to vary the transmission rate and achieve unequal error protection of the video streams experiencing different channel conditions. Simulation results show a significant improvement in the overall system performance compared to a traditional system where all users are allocated equal channel resources

Keywords: Cross-layer optimization, dynamic rate allocation, turbo-codes, video streaming

H. Labiod and M. Badra (eds.), New Technologies, Mobility and Security, 619–619.
© 2007 *Springer.*

CHAPTER 53

DESIGN AND IMPLEMENTATION OF A DECENTRALIZED ACCESS CONTROL SYSTEM (DACS) AND AN APPLICATION

TOBIAS HOF[1], ERIC ROBERT[1] ISABELLE BARTHES[2], AND SÉBASTIEN BASTARD[2]

[1] *Thales Architecture Framework Centre, Thales Communications France, SA, 92704 Colombes cedex, France, {surname.name}@fr.thalesgroup.com*

[2] *Avionics and Mission Systems, Thales Avionics, SA, 31036 Toulouse Cedex 1, France, {surname.name}@fr.thalesgroup.com*

Abstract: This document proposes a Decentralized Access Control System (DACS) and presents an implementation of this system. The system aims at enabling any commercial service, e.g. Video on Demand, to verify autonomously and locally users' access rights without relying on a centralized architecture. The DACS was thus designed as simple and efficient decentralized and user-centric authorization infrastructure. In the context of the research project "Mobilizing the Internet", the DACS was integrated and validated in an enhanced version of the In-Flight Entertainment (IFE) system, proposed by Thales Avionics. Nevertheless, this new approach can also be adapted to networks with loose connectivity and to generic infrastructures offering various services from third party providers, e.g. shopping malls, train stations or airports

Keywords: Decentralized access control; mobility; service; authorization, secret sets, privacy protection

H. Labiod and M. Badra (eds.), New Technologies, Mobility and Security, 621–621.
© 2007 *Springer.*

CHAPTER 54

MONITORING AND SECURITY OF CONTAINER TRANSPORTS

JENS OVE LAUF AND DIETER GOLLMANN

Hamburg University of Technology, Security in Distributed Applications, 21079 Hamburg-Harburg, Germany, masc@lauf.cc

Abstract: Equipping transport containers with sensor nodes is today considered in logistics applications both to increase the quality of information about goods in transport and as part of counter-terrorism efforts. This paper briefly surveys how transport chains are assembled, and the security regulations that had been introduced in this sector to counter the threat of terrorism. We then give a high level description of a container monitoring system that uses Recognised Security Organisations as trusted third parties that control access to the data collected by sensor nodes placed in transport containers. We compare the features of this proposal with an alternative system being developed by IBM and Maersk Logistics

Keywords: Monitoring and security of containers, secure transport chains, new application for ubiquitous computing, energy efficiency, sensor network, SPKI/SDSI

H. Labiod and M. Badra (eds.), New Technologies, Mobility and Security, 623–623.
© 2007 *Springer*.

CHAPTER 55

A NEW APPROACH FOR ANOMALIES RESOLUTION WITHIN FILTERING RULES

ANIS YAZIDI AND ADEL BOUHOULA

Ecole Supérieure des Communications de Tunis, Cité Technologique des Communications, Route de Raoued Km 3,5 – 2083 Cité El Ghazala, Tunisia

Email: anis_supcom@yahoo.fr, bouhoula@planet.tn

Abstract: During the last past years, the Internet has been growing at a high pace raising new challenges in the field of network security. Obviously, firewalls are core elements in network security. Firewalls have been regarded as barriers against unauthorized traffic and attacks. However, the effectiveness of a firewall is generally affected by the presence of anomalies within its filtering rules. Anomalies discovery within filtering rules has been a crucial issue. Multiple approaches have been developed aiming at discovering firewalls anomalies. However, no such work, to the best of our knowledge, was invested in studying the correction of these anomalies. From this perspective, our work attempts to fill the void in this field. In this paper, we propose a new scheme to resolve policy anomalies. The correction process is assisted by the network administrator in order to reflect exactly the desired policy. We consider the claim of [1] stating that ordering the rules do not work in all the cases.

Constraints on the rules order are deduced from the process of anomalies discovery. Based on these constraints, we define a model to arrange the rules with the possibility of adding new rules when needed. We have implemented our method and the first results are very promising

Keywords: Firewall, Filtering rules, Firewall policy, anomalies discovery, anomalies correction

H. Labiod and M. Badra (eds.), New Technologies, Mobility and Security, 625–625.
© 2007 *Springer.*

CHAPTER 56

SPMCS: A SCALABLE ARCHITECTURE FOR PEER TO PEER MULTIPARTY CONFERENCING

MOURAD AMAD[1] AND AHMED MEDDAHI[2]

[1] *University of Bejaia, Algeria, mourad amad@yahoo.fr*

[2] *GET/Telecom Lille1, France, ahmed.meddahi@telecom-lille1.eu*

Abstract: IP multiparty conferencing services such as: audio/video conferencing constitute a cost effective and flexible solution compare to TDM based multiparty conferencing. On one hand most of the existing multiparty or N-way conferencing solutions are based on a centralized server (Conference Bridge). On the other hand Peer-to-Peer model is inherently characterized by high scalability, robustness and fault tolerance. With its decentralized and distributed architecture, a P2P network is somehow able to self organized dynamically. Peer-to-peer (P2P) model or architecture are well adapted to N-way conferencing applications, effectively it can greatly benefit from P2P attributes such as: flexibility, scalability and robustness, particularly in critical environments such as: mobile networks. In this paper we propose a novel and original approach for Scalable P2P Multiparty Conferencing. This model combines a call control and signalling protocol (SIP) with a "P2P" protocol (Chord) for maintaining (dynamically) a well stabilized and optimized architecture topology. This model is also based on an "application layer" multicast mechanism. Theoretical analysis shows that our proposed approach benefits from SIP protocol flexibility, with the robustness and scalability of Chord protocol. The use of a multicast mechanism optimizes the overall traffic flow (control and media) and transmission efficiency

Keywords: Multiparty conferencing, Application Layer Multicast, SIP, P2P, Chord, SPMCS

H. Labiod and M. Badra (eds.), New Technologies, Mobility and Security, 627–627.
© 2007 *Springer.*

CHAPTER 57

A NEW AUTHENTICATED KEY AGREEMENT PROTOCOL

PIERRE E. ABI-CHAR, ABDALLAH MHAMED, AND BACHAR EL-HASSAN

UMR CNRS 5157, GET/Institut National des Télécommunications, 9 rue C. Fourier-91011 Evry CEDEX–France, Libanese University, Faculty of Engineering. Tripoli-Lebanon
Email: fpierre.abichar, abdallah.mhamedg@int-evry.fr, elhassan@ul.edu.lb

Abstract: Several protocols have been proposed to provide robust mutual authentication and key establishment for wireless local area network (WLAN). In this paper we present a New Authenticated Key Agreement (NAKA) protocol that provides secure mutual authentication, key establishment and key confirmation over an untrusted network. The main comparative study of our proposed protocol concerns the performance capabilities, including computational and communication load. In addition, the new protocol achieves many of the required security properties. It can resist dictionary attacks mounted by either passive or active networks intruders, allowing even the use of a weak password to be used. It can resist Impersonate attack. It also offers perfect forward secrecy which protects past sessions and passwords against future compromise. Finally, it can resist known-key and resilience to server attack. Our proposed protocol combines techniques of challenge-response protocols with symmetric key agreement protocols and offers significantly improved performance in computational and communication load over comparably many authenticated key agreement protocols such as B-SPEKE, SRP, AMP, PAK-RY, PAK-X, SKA and LR-AKE

H. Labiod and M. Badra (eds.), New Technologies, Mobility and Security, 629–629.
© 2007 *Springer.*

CHAPTER 58

GROWING HIERARCHICAL SELF-ORGANIZING MAP FOR ALARM FILTERING IN NETWORK INTRUSION DETECTION SYSTEMS

AHMAD FAOUR[1], PHILIPPE LERAY[1], AND BASSAM ETER[2]

[1]*Laboratoire LITIS–EA 4051, INSA, Rouen, France, fafaour,plerayg@insa-rouen.fr*

[2]*Lebanese University, Beyrouth, Liban, beter@ul.edu.lb*

Abstract: It is a well-known problem that intrusion detection systems overload their human operators by triggering thousands of alarms per day. This paper presents a new approach for handling intrusion detection alarms more efficiently. Self-Organizing Map (SOM) and Growing Hierarchical Self-Organizing Map (GHSOM) are used to discover interest patterns, signs of potential real attack scenarios aiming each machine in the network. GHSOM addresses two main limits of SOM which are caused, on the one hand, by the static architecture of this model, as well as, on the other hand, by the limited capabilities for the representation of hierarchical relations of the data. The experiments conducted on several logs extracted from the SNORT NIDS, confirm that the GHSOM can form an adaptive architecture, which grows in size and depth during its training process, thus to unfold the hierarchical structure of the analyzed logs of alerts

H. Labiod and M. Badra (eds.), New Technologies, Mobility and Security, 631–631.
© 2007 *Springer.*

CHAPTER 59

AGDH (ASYMMETRIC GROUP DIFFIE HELLMAN) AN EFFICIENT AND DYNAMIC GROUP KEY AGREEMENT PROTOCOL FOR AD HOC NETWORKS

RAGHAV BHASKAR[1], DANIEL AUGOT[2], ĆEDRIC ADJIH[2], PAUL MÜHLETHALER[2], AND SAADI BOUDJIT[3]

[1]*Microsoft Research India*

[2]*INRIA, BP 105, Rocquencourt, 78153 Le Chesnay Cedex, France*

[3]*GET/ENST, 37/39, rue Dareau 75014 PARIS, FRANCE*

Abstract: Confidentiality, integrity and authentication are more relevant issues in Ad hoc networks than in wired fixed networks. One way to address these issues is the use of symmetric key cryptography, relying on a secret key shared by all members of the network. But establishing and maintaining such a key (also called the session key) is a non-trivial problem. We show that Group Key Agreement (GKA) protocols are suitable for establishing and maintaining such a session key in these dynamic networks. We take an existing GKA protocol, which is robust to connectivity losses, and discuss all the issues for the correct functioning of this protocol in Ad hoc networks. We give implementation details and network parameters, which significantly reduce the computational burden of using public key cryptography in such networks

H. Labiod and M. Badra (eds.), New Technologies, Mobility and Security, 633–633.
© 2007 *Springer.*

Printed in the United States
95432LV00001B/44/A

9 781402 062698